Inclusive Smart Cities

Inclusive Smart Cities

Guest Editors

Thomas Bock
Rongbo Hu

MDPI

Basel • Beijing • Wuhan • Barcelona • Belgrade • Novi Sad • Cluj • Manchester

Guest Editors

Thomas Bock
School of Engineering
and Design
Technical University
of Munich
Munich
Germany

Rongbo Hu
Kajima Technical Research
Institute Singapore
Kajima Corporation
Singapore
Singapore

Editorial Office
MDPI AG
Grosspeteranlage 5
4052 Basel, Switzerland

This is a reprint of the Special Issue, published open access by the journal *Smart Cities* (ISSN 2624-6511), freely accessible at: https://www.mdpi.com/journal/smartcities/special_issues/NF3QDR1HRP.

For citation purposes, cite each article independently as indicated on the article page online and as indicated below:

Lastname, A.A.; Lastname, B.B. Article Title. *Journal Name* **Year**, *Volume Number*, Page Range.

ISBN 978-3-7258-5661-9 (Hbk)
ISBN 978-3-7258-5662-6 (PDF)
https://doi.org/10.3390/books978-3-7258-5662-6

Cover image courtesy of Rongbo Hu

Contents

About the Editors

Thomas Bock

Thomas Bock is Professor Emeritus of the Chair of Building Realization and Robotics at the Technical University of Munich (TUM). His research and development work focuses on AI, automation, and robotics for an affordable energy-efficient built environment, its transformation, and demographic change. He obtained his degrees in architecture from the University of Stuttgart in Germany, in engineering from the Illinois Institute of Technology in the USA as Fulbright Scholar, and in construction automation and robotics from the University of Tokyo in Japan. He holds a US FAA commercial pilot license with IFR and MEL ratings. At the French National Centre for Scientific Research (CNRS) in France, he founded the first European commission for building robotics. In 1989, he was appointed to the University of Karlsruhe (now Karlsruhe Institute of Technology), before he was appointed to the Chair of Building Realization and Building Robotics at TUM in 1997. He has worked on over 50 automated and robotized product and production systems in the most innovative construction companies and research institutions in Europe, the USA, and Japan since 1979. He is the co-founder of the International Association for Automation and Robotics in Construction (IAARC) and the Asian Habitat Society and advises governments and academies worldwide. In 2017, the German Research Foundation (DFG) awarded him the Seibold Prize for his 38 years of pioneering research in building robotics. He has co/authored over 550 articles in several languages, has been awarded an honorary doctorate, several honorary professorships, and holds various fellowships, honorary, and visiting professorships. He is co-editor of various specialist journals and co-author of the world's first book series on *Construction Robotics* published by Cambridge University Press since 2015.

Rongbo Hu

Rongbo Hu is Researcher and Owner of the Construction Robotics Lab at the Kajima Technical Research Institute Singapore. He is also a Member of the Board of Directors of the International Association for Automation and Robotics in Construction (IAARC) and Technical Committee Member of European Council on Computing in Construction (EC3). His research interests include construction robotics, lean construction, human factors engineering, smart homes, and smart cities. He obtained his Doctor of Engineering from the Technical University of Munich, Master of Architecture from the University of Illinois Urbana-Champaign, and Bachelor of Architecture from Wuhan University. He previously worked as Senior Consultant at CREDO Robotics GmbH, Germany, Visiting Professor via JSPS Fellowship at Keio University, Japan, Postdoctoral Researcher and Lecturer at Technical University of Munich, Germany, Project Designer for Ko Architects in California, USA, and Design Assistant for artist Ai Weiwei in Beijing, China. He has co/authored more than 60 articles and won the ISARC Best Paper Award twice.

Preface

This is the Reprint edition of the Special Issue "Inclusive Smart Cities" published in *Smart Cities*. This Reprint compiles 14 scholarly articles, including 1 editorial, 4 review articles, 1 project report, and 8 research articles, covering a wide range of relevant topics in inclusive smart cities. The collected articles highlight that inclusivity is both a core value and a practical requirement for sustainable urban transformation, making sure that marginalized groups are heard and digital tools serve everyone. By examining governance, citizen involvement, design, and technology across different scales, they show that inclusive smart cities are feasible when built through collaboration and context-sensitive strategies. As more cities adopt smart solutions, this Reprint offers valuable insights for researchers, planners, policymakers, and innovators to co-create cities that are advanced, fair, green, and, most importantly, accessible to all.

Thomas Bock and Rongbo Hu
Guest Editors

smart cities MDPI

Editorial

Towards Inclusive Smart Cities

Rongbo Hu [1,*] and Thomas Bock [2]

1 Kajima Technical Research Institute Singapore, Kajima Corporation, Singapore 489690, Singapore
2 School of Engineering and Design, Technical University of Munich, 80333 Munich, Germany; thomas.bock@tum.de
* Correspondence: r.hu@kajima.com.sg

1. Introduction to the Applications, Challenges, and Opportunities in Inclusive Smart Cities

Today, due to the widening of the wealth gap, the intensification of climate change, and the acceleration of both population growth and population aging, our cities are being tested by multiple economic, environmental, and social challenges, including, but not limited to, urban sprawl, urban gentrification, marginalization, housing crisis, tent city, urban flooding, urban heat island, environmental migrants, urban slums, tent cities, urban aging, and empty nesters [1–6]. In the past few years, pandemics and regional conflicts have undoubtedly exacerbated these problems [7,8]. To some extent, our cities are increasingly becoming less inclusive.

In recent years, on the other hand, on top of novel concepts and strategies of urban planning and design [9–11], emerging technologies such as big data, cloud computing, Internet of things (IoT), advanced information and communications technology (ICT) like 5G/6G, artificial intelligence (AI), blockchain, robotics, automated vehicles, and mixed reality have undoubtedly provided infinite possibilities for the development of smart cities, and also brought opportunities to solve the problems that plague our cities [12]. Furthermore, in accordance with the political–societal systems, the whole value chain for constructing and maintaining our cities can be optimized for higher return on investment (ROI) by design for manufacturing and assembly (DfMA), robot-oriented design (ROD), and on-site automation in order to achieve better affordability, adaptability, accessibility, resilience, and circularity for more dwellers [13,14]. As a result, many cities in the developed world, including, but not limited to, Barcelona, Copenhagen, Dubai, Hong Kong, London, Singapore, and Tokyo, have started to adopt smart city technologies to improve their residents' quality of life [12], and meanwhile, cities in developing countries are also accelerating the pace of smart city adoption [15,16]. Specifically, these smart city adoptions can be enhanced by a menu of socio-technical categories of applications, including healthcare, governance, environment, mobility, energy, safety/security, infrastructure, and education [17].

However, while the technological advancements of smart cities are emphasized, the inclusivity in smart cities is less discussed [18]. As a newly emerging urban model concept, inclusive smart cities currently lack a clear definition, although they are usually characterized by their core focus on recognizing and addressing the needs of all individuals, particularly older adults and persons with disabilities [19]. Some criticisms point out that current smart city visions may hinder citizens' freedom due to dependence on technologies [20,21], while others worry that vulnerable groups such as older adults and people with disabilities may be left behind due to the digital divide [22,23]. More importantly, it is necessary to remain cautious about the potential misuse of the smart city concept, particularly where inclusivity is overlooked [24]. As some researchers rightly point out, inclusivity is essential to enable an effective and sustainable transition to smart cities [25]. Therefore, advancing research on inclusive smart cities has become an urgent priority.

This Special Issue's purpose is to attract relevant high-quality research articles on the planning, operation, and maintenance processes of inclusive smart cities, or smart cities for all, using technologies that are friendly to all city dwellers, especially those who are vulnerable, regardless of their age, gender, ethnicity, education, wealth, and social status. The scopes of the submissions are on the micro-, meso-, and macro levels of smart cities. As a result, articles from technological, economic, and social perspectives, including critical review articles, are all included.

2. Overview of the Special Issue

The Special Issue, entitled "Inclusive Smart Cities", attracted contributions from many experts around the world associated with the theme, and finally, after rigorous peer review processes, 13 high-quality articles including 8 research articles, 4 review articles, and 1 project report were published in this Special Issue, covering a wide range of relevant topics on the micro-, meso-, and macro levels of inclusive smart cities. This section presents a concise summary of the 13 contributions, aiming to encourage readers to explore them in greater depth. The topics of interest for this Special Issue are reflected in the word cloud in Figure 1, where larger font sizes represent more frequent occurrences of terms.

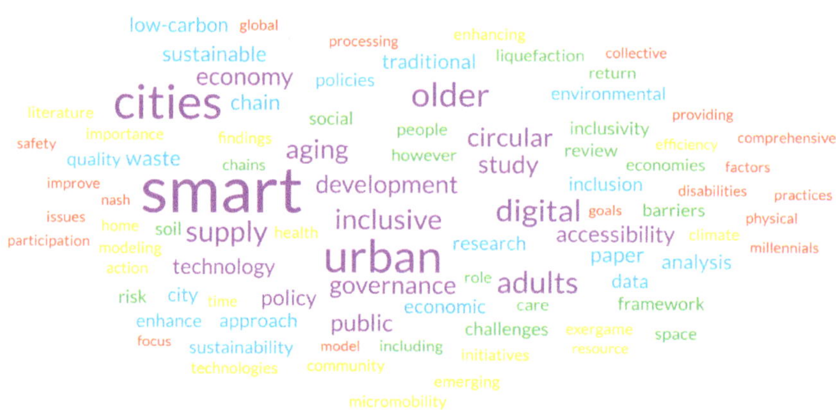

Figure 1. Word cloud based on the metadata of the articles collected in this Special Issue.

The paper titled "The Role of Smart Homes in Providing Care for Older Adults: A Systematic Literature Review from 2010 to 2023" by Vrančić et al. offers a comprehensive analysis of 58 studies to understand how smart home technologies (SHTs) support aging populations. The review explores the benefits, challenges, and ethical implications of SHTs for older adults, focusing on enhancing safety, health monitoring, autonomy, and social connection. It highlights promising innovations such as automated health systems, emergency alerts, and user-friendly interfaces, which can empower older adults to age in place independently. However, significant challenges—such as privacy concerns, system complexity, high costs, and low technological literacy—hinder widespread adoption. The review also addresses ethical considerations like autonomy, informed consent, and human dignity. It calls for user-centered designs, accessible technologies, and educational initiatives to boost older adults' acceptance. Technological integration through IoT, AI, and wearable sensors is emphasized for effective monitoring, while the study underscores the need for multidisciplinary collaboration in designing inclusive and ethical smart home environments. Ultimately, the review offers actionable recommendations for developers, researchers, and policymakers to promote supportive, human-centered smart home solutions for aging societies.

The paper "Inclusive MicroMob: Enhancing Urban Mobility Through Micromobility Solutions" by Rollandi et al. explores the integration of novel micromobility devices into Swiss urban environments, with a primary focus on Lugano. It introduces the Genny Zero, a self-balancing vehicle, and evaluates its performance through urban analysis, real-world experimentation, and participatory processes. The study identifies key challenges such as regulatory gaps, infrastructure limitations, and social acceptance. Data were collected via sensors and depth cameras during urban tests, revealing that the Genny Zero is most effective in areas with designated cycling infrastructure and smaller slopes. The project also highlights the importance of participatory urban governance, showing that public perception shifts positively after hands-on experience. The social study revealed generational differences in attitudes toward micromobility and underscored the need for inclusive, safe, and well-regulated systems. Policy recommendations include adaptive regulations, improved infrastructure, and integrating micromobility with the existing transport networks. Overall, the project demonstrates that micromobility can enhance urban sustainability and inclusivity if technological, spatial, and social factors are addressed collectively.

The paper "Insights from Smart City Initiatives for Urban Sustainability and Contemporary Urbanism" by Veloso et al. provides a comprehensive review of smart city initiatives in four cities around the world—Barcelona, Helsinki, Medellin, and Singapore—evaluating their contributions to sustainable urban development across six domains: mobility, governance, environment, people, living, and economy. Based on 71 peer-reviewed articles, the study finds that most smart city initiatives emphasize technology-driven governance and environmental strategies, but often lack comprehensive, inclusive frameworks addressing social equity and holistic sustainability. Barcelona and Helsinki excel in citizen-centric and participatory governance, while Singapore leads in ICT innovation but relies heavily on top-down strategies. Medellin's focus is on social inclusion and infrastructure renewal in disadvantaged areas. The study highlights an uneven focus across domains, with "smart people," "living," and "economy" receiving less attention despite their relevance to urban equity and resilience. It calls for context-sensitive, integrated approaches that combine digital innovation with inclusive, adaptive planning practices. The authors advocate standardized assessment tools to measure sustainability impacts and stress the need for smart city strategies that prioritize social needs and long-term urban resilience beyond technological deployment.

The paper "Smart Cities, Digital Inequalities, and the Challenge of Inclusion" by Kolotouchkina et al. critically examines how smart city developments often deepen digital inequalities and fail to inclusively engage all city dwellers. While smart cities aim to improve urban governance through digital innovation, they tend to prioritize tech-savvy, affluent, and younger populations, thereby marginalizing vulnerable groups such as older adults, people with disabilities, migrants, and low-income communities. These groups face barriers in digital access, skills, and participation due to factors like affordability, lack of education, limited self-efficacy, and systemic ageism or ableism. To address these challenges, the authors propose a reflective framework focusing on three key questions: Who is included or excluded? What is the value of digital inclusion? How can inclusive governance be ensured? The framework emphasizes digital accessibility, digital citizenship, equal representation, and co-creation in policymaking. It aims to guide the development of inclusive urban policies that foster digital rights, ethical data governance, and participatory practices. Eventually, the paper advocates a citizen-centric smart city model where inclusion is not incidental but foundational to urban digital transformation.

The paper titled "Embedding Circular Economy in the Construction Sector Policy Framework: Experiences from EU, U.S., and Japan for Better Future Cities" by Marzani et al. explores how Circular Economy (CE) principles are embedded in construction sector policies across the EU (specifically Italy and Germany), the US, and Japan, aiming to guide the transition toward sustainable and smart urban development. Through comparative policy analysis, it highlights diverse approaches: the EU integrates CE within comprehensive strategies like the European Green Deal and New Circular Economy Action Plan; Japan anchors CE in national laws focused on resource efficiency and long-term environmental planning; while U.S. initiatives are driven more by city-level actions due to limited federal support. Common policy targets include reducing resource use, increasing recycling rates, promoting the reuse of construction materials, and minimizing construction and demolition waste. Despite growing recognition of CE, most policies still emphasize end-of-life waste management, with limited attention to the entire building lifecycle, particularly the use and maintenance phases. Moreover, the integration of digital technologies—such as Building Information Modeling (BIM), digital twins, and material passports—remains limited, with Japan as an exception. The paper underscores the need for policy harmonization and stronger inclusion of digital tools to enhance material traceability and lifecycle optimization. Furthermore, it advocates research-informed, digitally enabled CE policies as key to achieving more resilient, inclusive, and circular smart cities.

The paper "The Use of Smart Technology for Creating an Inclusive Urban Public Space" by Itair et al. investigates the critical role of urban public spaces in community development and proposes a smart framework to enhance their inclusivity, particularly in developing regions. Using the city of Nablus, Palestine, as a case study, the authors assess public space inclusivity through a framework of eight indicators: spatial distribution, typology, facilities and services, green and humid areas, governance and management, safety, user demographics, and user satisfaction. The findings reveal that public spaces in Nablus are limited and unevenly distributed, particularly disadvantaging low-income populations, women, children, and people with disabilities. Users expressed dissatisfaction with accessibility, safety (especially at night for women), lack of green areas, and inadequate facilities. The study introduces the concept of "smart public spaces," leveraging technologies such as IoT, mobile apps, and AI for real-time monitoring, service optimization, and environmental sustainability. The smart public space framework comprises six layers—governance, physical infrastructure, data collection, communication, server infrastructure, and smart services—designed to promote citizen engagement, safety, and eco-friendliness. This approach aims to support the UN's Sustainable Development Goal (SDG) 11.7 by making urban public spaces more inclusive, accessible, and responsive to community needs.

The paper titled "Fit Islands: Designing a Multifunctional Virtual Urban Community to Promote Healthy Aging for Chinese Older Adults" by Shen et al. addresses the urgent challenge of aging populations in emerging economies, particularly in China. In response to the challenges of rapid population aging, health issues, digital divides, and pandemics in emerging economies, the research team creates a virtual urban community that encourages physical activity, cognitive engagement, and social interaction. Designed to be affordable, inclusive, and adaptable, Fit Islands uses motion capture via standard cameras, avoiding the need for intrusive wearables and costly sensors. It includes age-friendly UI, customizable gameplay, health education, and a reward system. The game is inspired by Chinese seniors' behaviors, such as square dancing, and offers both seated and standing exercises. A standard development framework for Exergames in emerging economies is proposed and applied. Functional tests with Chinese seniors demonstrated the system's feasibility and user-friendliness. The study concludes that Fit Islands can enhance the well-being of older adults and serves as a replicable model for other aging societies.

The paper "Smart Cities for All? Bridging Digital Divides for Socially Sustainable and Inclusive Cities" by Colding et al. explores how smart city development, while offering climate-proofing benefits, risks reinforcing social exclusion due to digital divides. Using Amartya Sen's capability approach, the authors analyze how personal, social, and environmental factors affect individuals' ability to convert digital resources into real opportunities or "functionings." The paper identifies three levels of digital divides: access (first-level), digital skills and participation (second-level), and differential benefits from ICT use (third-level). These divides disproportionately affect marginalized groups such as the elderly, low-income populations, and people with disabilities. The authors argue that digital technologies, often touted as universally beneficial, can serve as "non-choice default technologies," excluding those without digital proficiency or access. To build inclusive and socially sustainable smart cities, the paper recommends creating institutions that enhance capabilities, safeguarding analog options for public services, and integrating citizen feedback. A unified governing body is proposed to monitor inclusivity and promote equity-focused policies. Overall, the paper calls for a shift from a technocratic, one-size-fits-all model to a socially inclusive approach rooted in human capabilities and climate justice.

The article titled "Redesigning Municipal Waste Collection for Aging and Shrinking Communities" by Pandyaswargo et al. addresses Japan's pressing challenge of maintaining municipal waste collection amid rapid aging and depopulation. Researchers surveyed 1000 Japanese residents to understand preferences around contactless, collective waste disposal. Surprisingly, older adults were more willing to walk longer distances (up to 5 min) to waste collection points and were less concerned about disease transmission than younger individuals. Single-person households also preferred disposal sites placed farther from their homes and generated less waste. Based on these findings, two redesign strategies were proposed: (1) centralized large-container collection to reduce labor demand, and (2) automated pickup using robotics and autonomous mobility. While each has trade-offs—such as increased energy needs or potential resistance from users—they support more sustainable, inclusive systems. These insights not only guide Japan's smart city transition under the "Society 5.0" concept but also offer lessons for other aging and shrinking communities worldwide.

The paper titled "AI-Driven Prediction and Mapping of Soil Liquefaction Risks for Enhancing Earthquake Resilience in Smart Cities" by Katsuumi et al. presents an AI-driven model for predicting and mapping soil liquefaction risks to enhance earthquake resilience in smart cities, with Yokohama, Japan, as a case study. Using ensemble machine learning techniques—combining artificial neural networks and gradient boosting decision trees—the model integrates geotechnical and geographical data to assess soil susceptibility during seismic events. The study introduces high-resolution risk maps based on predicted N-values and soil classifications, validated through the Soil Liquefaction Potential Index (LPI). Among three prediction procedures, the bottom-up approach (i.e., predicting from depth upward) proved most accurate. The model demonstrated strong performance despite limited data, highlighting ensemble learning's effectiveness in improving prediction accuracy. This approach offers significant advancement over traditional empirical methods, enabling smarter urban planning, proactive disaster mitigation, and more resilient infrastructure design in earthquake-prone regions. The study underscores the transformative potential of AI in geotechnical engineering and smart city development.

The paper titled "Transitioning to a Low-Carbon Lifestyle? An Exploration of Millennials' Low-Carbon Behavior—A Case Study in China" by Wu et al. explores how Chinese millennials engage in low-carbon behaviors and what drives or hinders their adoption of sustainable lifestyles. Using qualitative methods based on in-depth interviews with 50 well-educated Chinese millennials from diverse sectors and regions, the study identifies key patterns in attitudes and actions. The study finds that millennials generally show a positive attitude toward low-carbon living. Women are more active in domestic low-carbon actions, such as energy conservation and waste reduction, although they are underrepresented in decision-making roles within organizations. Millennials' low-carbon behaviors are evident in clothing choices (e.g., second-hand or sustainable materials), food habits (e.g., reduced meat and takeout), transportation (e.g., use of public transit and electric vehicles), and social interactions (e.g., participating in digital environmental initiatives like Ant Forest). Barriers include limited awareness, perceived inconvenience, quality concerns about sustainable products, and a lack of effective governance or infrastructure. Conversely, enablers include economic incentives, technological advancements, personal values, and governmental policies. The study concludes that systemic support and cross-sector collaboration are essential to foster deeper lifestyle shifts among millennials toward carbon neutrality.

The study titled "Probabilistic Causal Modeling of Barriers to Accessibility for Persons with Disabilities in Canada" by Zakir et al. applies probabilistic causal modeling to analyze accessibility barriers faced by persons with disabilities (PWDs) in Canada using data from the 2022 Canadian Survey on Disability. By integrating exploratory factor analysis, structural equation modeling, and Bayesian networks, the authors identify how demographic factors such as age, gender, and disability type influence the likelihood and frequency of encountering barriers in communication, transportation, and digital services. The study confirms strong causal relationships between disabilities and specific accessibility challenges, with pain, mobility, and mental health conditions being most affected. It also demonstrates the feasibility of generating synthetic data to address limitations in real-world datasets while preserving privacy. The findings aim to inform inclusive smart city planning and policy by offering a data-driven framework to predict and address future accessibility issues, ultimately supporting a more equitable and barrier-free environment for PWDs in urban settings.

The paper "Game-Theoretic Analysis of Policy Impacts in Competition Between Reverse Supply Chains Involving Traditional and E-Channels" by Aghaei et al. investigates the strategic interactions between two types of reverse logistics-based supply chains—one using only traditional channels (T-SCs) and the other using both traditional and e-channels (H-SCs)—within the context of smart cities and circular economy goals. Using game-theoretic models (Nash equilibrium and Nash–Stackelberg scenarios), the study evaluates how five policy levers—incentive pricing, return quality level, advertising investment, transportation investment, and return processing time—affect economic (i.e., profitability) and environmental (i.e., return rates) outcomes. The findings show that the hybrid supply chain (H-SC) consistently outperforms the traditional model in both profitability and return rates due to its dual-channel flexibility. Specifically, traditional supply chain leadership in a Nash–Stackelberg structure yields the highest overall profitability for both systems. The study also provides sensitivity and Pareto analyses, revealing that return quality and processing time are critical for H-SCs, while advertising investment and processing time are most influential for T-SCs. These insights highlight the need for balanced, policy-driven strategies to optimize sustainable supply chain operations in smart cities.

3. Conclusions

In conclusion, this Special Issue highlights that the future of smart cities must be measured not only by their technological advancement but also by their capacity to foster inclusivity, equity, and resilience. The presented contributions demonstrate that inclusivity is both a guiding principle and a practical necessity for sustainable urban transformation, ensuring that marginalized voices are heard and that digital innovation benefits all. By critically examining governance models, citizen participation, design practices, and emerging technologies, the collected papers underline from micro-, meso-, and macro levels that inclusive smart cities are not only promising but also achievable through deliberate, collaborative, context-sensitive, and multi-level approaches. As cities worldwide continue to adopt smart solutions, the insights from this Special Issue provide a foundation for researchers, urban planners, policymakers, and innovators to co-create urban environments that are technologically sophisticated, socially just, environmentally friendly, and most importantly, accessible to everyone.

Author Contributions: Conceptualization, T.B. and R.H.; writing—original draft preparation, R.H. and T.B.; writing—review and editing, R.H. and T.B. All authors have read and agreed to the published version of the manuscript.

Funding: This article received no external funding.

Acknowledgments: The Guest Editors of this Special Issue sincerely thank all the researchers who submitted their research articles, the reviewers who assisted in evaluating these manuscripts, and both the Editor-in-Chief and the Editorial Team for their full support.

Conflicts of Interest: The authors declare no conflicts of interest.

List of Contributions:

1. Vrančić, A.; Zadravec, H.; Orehovački, T. The Role of Smart Homes in Providing Care for Older Adults: A Systematic Literature Review from 2010 to 2023. *Smart Cities* **2024**, *7*, 1502–1550. https://doi.org/10.3390/smartcities7040062.
2. Rollandi, A.; Papandrea, M.; Bignami, F.; Di Maggio, L.; Günther, F.; Quattrini, A.; Minardi, L.; Cocca, M.; Bettini, A. Inclusive MicroMob: Enhancing Urban Mobility Through Micromobility Solutions. *Smart Cities* **2025**, *8*, 69. https://doi.org/10.3390/smartcities8020069.
3. Veloso, Á.; Fonseca, F.; Ramos, R. Insights from Smart City Initiatives for Urban Sustainability and Contemporary Urbanism. *Smart Cities* **2024**, *7*, 3188–3209. https://doi.org/10.3390/smartcities7060124.
4. Kolotouchkina, O.; Ripoll González, L.; Belabas, W. Smart Cities, Digital Inequalities, and the Challenge of Inclusion. *Smart Cities* **2024**, *7*, 3355–3370. https://doi.org/10.3390/smartcities7060130.
5. Marzani, G.; Tondelli, S.; Kuma, Y.; Cruz Rios, F.; Hu, R.; Bock, T.; Linner, T. Embedding Circular Economy in the Construction Sector Policy Framework: Experiences from EU, U.S., and Japan for Better Future Cities. *Smart Cities* **2025**, *8*, 48. https://doi.org/10.3390/smartcities8020048.
6. Itair, M.; Shahrour, I.; Hijazi, I. The Use of the Smart Technology for Creating an Inclusive Urban Public Space. *Smart Cities* **2023**, *6*, 2484–2498. https://doi.org/10.3390/smartcities6050112.
7. Shen, Z.; Hu, R.; Wan, D.; Bock, T. Fit Islands: Designing a Multifunctional Virtual Urban Community to Promote Healthy Aging for Chinese Older Adults. *Smart Cities* **2024**, *7*, 208–232. https://doi.org/10.3390/smartcities7010009.
8. Colding, J.; Nilsson, C.; Sjöberg, S. Smart Cities for All? Bridging Digital Divides for Socially Sustainable and Inclusive Cities. *Smart Cities* **2024**, *7*, 1044–1059. https://doi.org/10.3390/smartcities7030044.
9. Pandyaswargo, A.H.; Shan, C.; Ogawa, A.; Tsubouchi, R.; Onoda, H. Redesigning Municipal Waste Collection for Aging and Shrinking Communities. *Smart Cities* **2024**, *7*, 1149–1168. https://doi.org/10.3390/smartcities7030049.

10. Katsuumi, A.; Cong, Y.; Inazumi, S. AI-Driven Prediction and Mapping of Soil Liquefaction Risks for Enhancing Earthquake Resilience in Smart Cities. *Smart Cities* **2024**, *7*, 1836–1856. https://doi.org/10.3390/smartcities7040071.
11. Wu, Y.; Martens, P.; Krafft, T. Transitioning to a Low-Carbon Lifestyle? An Exploration of Millennials' Low-Carbon Behavior—A Case Study in China. *Smart Cities* **2024**, *7*, 2015–2041. https://doi.org/10.3390/smartcities7040080.
12. Zakir, M.; Wolbring, G.; Yanushkevich, S. Probabilistic Causal Modeling of Barriers to Accessibility for Persons with Disabilities in Canada. *Smart Cities* **2025**, *8*, 4. https://doi.org/10.3390/smartcities8010004.
13. Aghaei, A.; Cai, F.; Wu, T. Game-Theoretic Analysis of Policy Impacts in Competition Between Reverse Supply Chains Involving Traditional and E-Channels. *Smart Cities* **2025**, *8*, 36. https://doi.org/10.3390/smartcities8010036.

References

1. Yin, C.; Xiong, Z.; Chen, H.; Wang, J.; Cooper, D.; David, B. A literature survey on smart cities. *Sci. China Inf. Sci.* **2015**, *58*, 1–18. [CrossRef]
2. Broere, W. Urban underground space: Solving the problems of today's cities. *Tunn. Undergr. Sp. Technol.* **2016**, *55*, 245–248. [CrossRef]
3. Tulier, M.E.; Reid, C.; Mujahid, M.S.; Allen, A.M. "Clear action requires clear thinking": A systematic review of gentrification and health research in the United States. *Health Place* **2019**, *59*, 102173. [CrossRef]
4. Hu, R.; Follini, C.; Pan, W.; Linner, T.; Bock, T. A Case Study on Regenerating Informal Settlements in Cairo using Affordable and Adaptable Building System. *Procedia Eng.* **2017**, *196*, 113–120. [CrossRef]
5. Yang, L.; Qian, F.; Song, D.-X.; Zheng, K.-J. Research on Urban Heat-Island Effect. *Procedia Eng.* **2016**, *169*, 11–18. [CrossRef]
6. Yang, Y.; Miao, Y.; Yang, Y.; Qiong, Y.; Ting, L. Relationships between children-related factors, basic psychological need satisfaction, and multiple happiness among urban empty-nesters in China: A structural equation modeling. *BMC Geriatr.* **2022**, *22*, 925. [CrossRef]
7. Sharifi, A.; Khavarian-Garmsir, A.R. The COVID-19 pandemic: Impacts on cities and major lessons for urban planning, design, and management. *Sci. Total Environ.* **2020**, *749*, 142391. [CrossRef]
8. Elfversson, E.; Höglund, K. Are armed conflicts becoming more urban? *Cities* **2021**, *119*, 103356. [CrossRef]
9. Moreno, C.; Allam, Z.; Chabaud, D.; Gall, C.; Pratlong, F. Introducing the "15-Minute City": Sustainability, Resilience and Place Identity in Future Post-Pandemic Cities. *Smart Cities* **2021**, *4*, 93–111. [CrossRef]
10. Mouratidis, K. Compact city, urban sprawl, and subjective well-being. *Cities* **2019**, *92*, 261–272. [CrossRef]
11. Hu, R.; Pan, W.; Bock, T. Towards dynamic vertical urbanism: A novel conceptual framework to develop vertical city based on construction automation, open building principles, and industrialized prefabrication. *Int. J. Ind. Constr.* **2020**, *1*, 34–47. [CrossRef]
12. Javed, A.R.; Shahzad, F.; Rehman, S.U.; Zikria, Y.B.; Razzak, I.; Jalil, Z.; Xu, G. Future smart cities: Requirements, emerging technologies, applications, challenges, and future aspects. *Cities* **2022**, *129*, 103794. [CrossRef]
13. Bock, T.; Hu, R. Towards Instant Availability and Full Life Cycle Resilience in Vertical Cities: Automated Deployment and Transformation of High-Rise Buildings to Mitigate Social Challenges. *Int. J. High-Rise Build.* **2022**, *11*, 75–86. [CrossRef]
14. Bock, T.; Linner, T. *Robot-Oriented Design: Design and Management Tools for the Deployment of Automation and Robotics in Construction*; Cambridge University Press: Cambridge, UK, 2015. [CrossRef]
15. Zhu, J.; Gianoli, A.; Noori, N.; de Jong, M.; Edelenbos, J. How different can smart cities be? A typology of smart cities in China. *Cities* **2024**, *149*, 104992. [CrossRef]
16. Prasad, D.; Alizadeh, T. What Makes Indian Cities Smart? A Policy Analysis of Smart Cities Mission. *Telemat. Inform.* **2020**, *55*, 101466. [CrossRef]
17. Gracias, J.S.; Parnell, G.S.; Specking, E.; Pohl, E.A.; Buchanan, R. Smart Cities—A Structured Literature Review. *Smart Cities* **2023**, *6*, 1719–1743. [CrossRef]
18. Makkonen, T.; Inkinen, T. Inclusive smart cities? Technology-driven urban development and disabilities. *Cities* **2024**, *154*, 105334. [CrossRef]
19. Tekin, H.; Dikmen, I. Inclusive Smart Cities: An Exploratory Study on the London Smart City Strategy. *Buildings* **2024**, *14*, 485. [CrossRef]
20. Vanolo, A. Is there anybody out there? The place and role of citizens in tomorrow's smart cities. *Futures* **2016**, *82*, 26–36. [CrossRef]
21. Mondschein, J.; Clark-Ginsberg, A.; Kuehn, A. Smart cities as large technological systems: Overcoming organizational challenges in smart cities through collective action. *Sustain. Cities Soc.* **2021**, *67*, 102730. [CrossRef]

22. Nápoles, V.M.P.; Páez, D.G.; Penelas, J.L.E.; Pérez, O.G.; de Pablos, F.M.; Gil, R.M. Social Inclusion in Smart Cities BT. In *Handbook of Smart Cities*; Augusto, J.C., Ed.; Springer International Publishing: Cham, Switzerland, 2021; pp. 469–514. [CrossRef]
23. Wang, C.H.; Steinfeld, E.; Maisel, J.L.; Kang, B. Is your smart city inclusive? Evaluating proposals from the U.S. Department of Transportation's Smart City Challenge. *Sustain. Cities Soc.* **2021**, *74*, 103148. [CrossRef]
24. Vanessa, W. The allure of 'smart city' rhetoric: India and Africa. *Dialogues Hum. Geogr.* **2015**, *5*, 36–39. [CrossRef]
25. Lee, J.; Babcock, J.; Pham, T.S.; Bui, T.H.; Kang, M. Smart city as a social transition towards inclusive development through technology: A tale of four smart cities. *Int. J. Urban Sci.* **2023**, *27* (Suppl. 1), 75–100. [CrossRef]

smart cities

MDPI

Systematic Review

The Role of Smart Homes in Providing Care for Older Adults: A Systematic Literature Review from 2010 to 2023

Arian Vrančić, Hana Zadravec and Tihomir Orehovački *

Faculty of Informatics, Juraj Dobrila University of Pula, Zagrebačka 30, 52100 Pula, Croatia; arian.vrancic@student.unipu.hr (A.V.); hana.zadravec@student.unipu.hr (H.Z.)
* Correspondence: tihomir.orehovacki@unipu.hr

Abstract: This study undertakes a systematic literature review, framed by eight research questions, and an exploration into the state-of-the-art concerning smart home innovations for care of older adults, ethical, security, and privacy considerations in smart home deployment, integration of technology, user interaction and experience, and smart home design and accessibility. The review evaluates the role of smart home technologies (SHTs) in enhancing the lives of older adults, focusing on their cost-effectiveness, ease of use, and overall utility. The inquiry aims to outline both the advantages these technologies offer in supporting care for older adults and the obstacles that impede their widespread adoption. Throughout the investigation, 58 studies were analyzed, selected for their relevance to the discourse on smart home applications in care for older adults. This selection came from a search of literature published between 2010 and 2023, ensuring an up-to-date understanding of the field. The findings highlight the potential of SHTs to improve various aspects of daily living for older adults, including safety, health monitoring, and social interaction. However, the research also identifies several challenges, including the high costs associated with these technologies, their complex nature, and ethical concerns surrounding privacy and autonomy. To address these challenges, the study presents recommendations to increase the accessibility and user-friendliness of SHTs for older adults. Among these, educational initiatives for older adults are emphasized as a strategy to improve technology acceptance, along with suggestions for design optimizations in wearable devices to enhance comfort and adaptability. The implications of this study are significant, offering insights for researchers, practitioners, developers, and policymakers engaged in creating and implementing smart home solutions for care of older adults. By offering an understanding of both the opportunities and barriers associated with SHTs, this research supports future efforts to create more inclusive, practical, and supportive environments for aging populations.

Keywords: smart homes; older adults; aging in place; healthcare; well-being; systematic literature review

1. Introduction

The aging of the population is a global phenomenon that presents challenges for societies, particularly in developed countries. According to UN predictions, by 2050, the proportion of people aged 60 and older will surpass the number of children under 14 years old [1]. This demographic shift will impact various societal sectors, including the economy, social welfare, and healthcare. It is, therefore, crucial to explore innovative solutions to enhance the quality of life for older adults and meet the increasing demand for long-term care. Older adults, typically those close to retirement age, vary in designation from as early as age 60 to as late as age 65, depending on the nation and culture.

One significant aspect of aging is the growing desire among older adults to age in place. This concept refers to the ability of older adults to live independently in their homes and communities, maintaining their lifestyle despite age-related health challenges or the need for assistance with daily activities [2]. Factors influencing this preference include emotional attachment to one's home, preservation of familiar surroundings and social ties, and the belief that institutional care may not provide the desired level of personalized care.

The potential for technology to support older adults in aging in place is increasingly recognized. Smart homes, with their transformative capabilities, emerge as a promising solution. These homes provide functionalities such as automated medication dispensing, medical monitoring, and personal assistance through features such as connectivity, automation, mobile device management, energy and resource efficiency, accessibility, and device integration [3]. By utilizing sensors, actuators, and intelligent systems, smart homes monitor the activities and health of older adults, detecting changes in patterns and providing timely assistance or notifications to caretakers or medical specialists [4]. This proactive approach can prevent accidents, detect early signs of health deterioration, and facilitate timely interventions, reducing the need for hospitalization or institutional care.

Furthermore, smart homes empower older adults by enhancing their independence and autonomy. User-friendly interfaces and voice-activated instructions simplify the control of various aspects of their living environment, such as lighting, temperature, and entertainment systems, thereby maintaining daily routines and enhancing convenience [5]. These features also help tackle social isolation by enabling older adults to maintain relationships through integrated communication technologies, such as video calls, social media, and online activity trackers, fostering a sense of belonging [6].

Additionally, smart homes are pivotal in managing chronic illnesses and promoting healthy aging. They monitor vital signs, medication adherence, and lifestyle trends, sharing these data with healthcare providers for remote monitoring and personalized interventions. This capability allows older adults to better manage their conditions, prevent complications, and improve their overall health [7]. Considering the significant challenges posed by population aging, smart homes offer a comprehensive solution that enhances the quality of life for older adults, enabling them to maintain independence, promote social connections, and manage health-related challenges effectively.

The aim of this study is to provide a comprehensive and interdisciplinary review of smart home technologies (SHTs) for older adults, covering the period from 2010 to 2023. Integrating technological, ethical, and practical perspectives, it captures broader trends and persistent challenges in the development and implementation of SHTs. Previous research has often been limited by narrower focuses or shorter timeframes, failing to encompass the full spectrum of advancements and obstacles in this field [8–24]. Our review not only offers a longitudinal perspective on the evolution of these technologies but also provides a detailed analysis of their impact on the lives of older adults, which is crucial for understanding both progress and enduring challenges.

The originality of our work is highlighted by its dual focus on the benefits and limitations of SHTs, a balance often overlooked in existing literature. While numerous studies document the technological advancements and potential health benefits of smart homes, few simultaneously address the significant ethical and practical challenges accompanying these innovations. Our systematic review goes beyond merely listing the advantages of SHTs, such as enhanced safety, independence, and well-being for older adults. It rigorously examines the high costs, technological complexities, and ethical concerns, particularly those related to privacy and autonomy, that hinder the widespread adoption of these technologies. By addressing both positive and negative aspects, our study provides a more balanced and realistic assessment of the potential of SHTs.

Furthermore, this study fills a critical knowledge gap by presenting targeted recommendations that are both practical and theoretically grounded. Derived from a thorough analysis of existing literature, these recommendations are designed to address the identified challenges directly. For instance, educational initiatives are proposed to improve older adults' technological literacy, thereby enhancing their acceptance and effective use of SHTs. Design optimizations to make these technologies more user-friendly and adaptable to the specific needs of older adults are also suggested, with a focus on comfort and ease of use. Additionally, the development of ethical guidelines to address privacy and consent issues is advocated, ensuring that the deployment of SHTs respects the dignity and autonomy of older adults. These recommendations are not only innovative but also actionable, providing clear pathways for future research, development, and policy formulation.

Our research methodology is noteworthy for its integration of qualitative, quantitative, and mixed-method analyses. This methodological diversity allows for a more thorough understanding of user perceptions and the real-world applicability of SHTs. By synthesizing data from various research approaches, our study offers comprehensive insights into how these technologies are perceived and utilized by older adults. This multifaceted analysis enhances the robustness of our findings and underscores the importance of considering multiple perspectives when evaluating the impact of SHTs. Such a methodological approach distinguishes our study from previous research, which often relies on a single type of analysis, thus providing a less complete picture.

The remainder of the paper is structured as follows. An overview of current relevant and related studies is provided in Section 2. The employed research methodology is described in Section 3. Outcomes of the systematic literature review on the role of smart homes in the lives of older adults are reported in Section 4. Recommendations for enhancing SHTs for older adults are provided in Section 5. Implications for designers, developers, healthcare professionals, researchers, older adults, and the community are discussed in Section 6. Future work directions are proposed in Section 7. Limitations are explained in Section 8. Conclusions are drawn in Section 9.

2. Related Work

Numerous studies have explored the role of smart homes as caregivers. Lattika et al. [8] conducted a review examining technologies to alleviate loneliness and social isolation among older adults, including robots, wearable technology, and smart homes. They concluded that while robotic dogs and social robots can only partially address loneliness, smart homes can effectively identify and anticipate both loneliness and social isolation, thus improving older adults' social well-being and offering guidance for future study and development in this field. Ghafurian et al. [9] evaluated the literature on smart home environments and equipment designed to support older adults, uncovering a range of devices, technical challenges, and their advantages for older adults.

In the context of assistive smart homes, Zulas et al. [10] evaluated the needs of caregivers of older adults, revealing a demand for technology that simplifies monitoring older adults' safety, health, and activities, and facilitates access to and sharing of information with other healthcare professionals. This study underscores the significance of deploying innovative home technology to meet the needs and preferences of older adult caregivers and provide them with adequate support. Additionally, Zulas et al. [11] examined the needs of healthcare workers in smart homes for older adults, identifying essential requirements, such as monitoring vital signs, safety, communication, and medication management. The study emphasizes the importance of customizing each solution to the individual older adult, highlighting challenges, such as user interface difficulties, the need for training, and concerns about privacy and data security.

Ma et al. [12] focused on design strategies for smart homes to support aging in place, identifying numerous benefits and challenges. The benefits included improved safety, enhanced comfort, increased autonomy, and social connectedness, while the challenges encompassed high costs, compatibility issues, privacy and data security concerns, and technological literacy among older adults. Sumner et al. [13] conducted a systematic literature review to explore knowledge about the design of technology used in smart homes, concluding that their active involvement in the technology design process is crucial for ensuring relevance, acceptance, and effectiveness for older adults. Additionally, they emphasized the importance of adaptability of technology to the diverse needs and preferences of older adult users. Zielonka et al. [14] investigated current developments and trends in smart homes, with a focus on how these environments interact with older adults through software and artificial intelligence (AI). Their study concludes that an improvement in older adults' quality of life will not be possible with the current infrastructure of smart homes.

Graybill et al. [15] examined the cost-effectiveness of aging in place, analyzing various studies on the costs associated with home-based care versus other forms of care for older adults. They concluded that aging in smart homes can be more cost-effective compared to options such as institutional care, also identifying factors that influence this cost-effectiveness, including home modification costs, availability of support and services, and social support. The research highlighted that aging in smart homes can positively impact older adults' quality of life by enabling them to remain in a familiar environment and maintain their autonomy. Kachouie et al. [16] explored the role of socially assistive robots in enhancing the well-being of older adults by providing companionship, assistance, and engagement. Ollevier et al. [17] conducted a systematic literature review exploring how technology can support aging in place for older adults, identifying technologies such as sensors, telemedicine, smart home devices, and mobile applications that assist older adults in maintaining their independence, monitoring their health status, ensuring safety, and improving their quality of life. The study emphasizes challenges in using innovative home technology, including technological literacy, acceptance, financial factors, individual differences, and security and privacy concerns.

Solaimani et al. [18] examined the need for achieving a higher level of maturity to enable the mass commercialization of smart homes and increased connectivity, noting that the existing literature often overlooks social and economic aspects. Chadborn et al. [19] examined the advantages and drawbacks of smart homes and innovative health technologies, discovering varying perspectives and concerns among older adults regarding smart homes, including the importance of customizing technology to individual needs and the significance of emotional support and human interaction within technological systems. Privacy and data security emerged as additional significant concerns. The study emphasized the crucial role of involving older adults in the design of innovative home technologies.

Mashal et al. [20] conducted a systematic review of literature spanning ten years to explore the acceptance and adoption of smart homes, finding that psychological factors, such as user attitudes and perceptions, along with personalized applications and services that meet user needs and desires, are crucial for adoption. Wang et al. [21] focused on ethical issues concerning the implementation of smart homes for older adults. This research emphasized the importance of considering ethical aspects while developing innovative home technologies to ensure the safety, privacy, and well-being of aged individuals. It highlighted the significance of involving older adults in technology development to provide transparency and privacy protection, allowing them to participate and ensuring their needs and preferences are respected. Felber et al. [22] conducted a systematic review to map the ethical issues associated with smart home health technologies for the care of older adults. The review aimed to identify and analyze ethical concerns, such as privacy, autonomy, informed consent, data security, and the potential for social isolation, emphasizing the need for an ethical framework to guide the development, implementation, and use of these technologies, considering older adults' values, rights, and well-being.

Zwijsen et al. [23] highlighted the absence of specific guidelines and ethical codes concerning the use of assistive care technologies for older adults, asserting that concepts such as privacy, autonomy, security, consent, dignity, and fairness should form the foundation of ethical deliberations. The research underscored the importance of understanding older adults' perspectives on assistive care technology and their ethical concerns. It advocated for an interdisciplinary approach involving experts from fields such as ethics, law, medicine, and social work to develop appropriate guidelines and practices. Gochoo et al. [24] systematically reviewed the literature to explore privacy preservation in aging in place. The study emphasized the utilization of approaches and technologies, including data encryption, anonymization, identity protection techniques, and access controls, to safeguard the privacy of older adults in smart homes.

Analysis of the aforementioned studies has identified several areas for improvement and directions for future research. First, none of the studies provided detailed instructions on how to effectively communicate the advantages of innovative home technology to older adults. Additionally, while ethical issues related to technology were mentioned, there is a pressing need to develop better solutions for these problems, particularly concerning new legislative frameworks to regulate them. Further studies are needed to address the stigma surrounding older adults using innovative home technology and to develop strategies to encourage their acceptance. More investigation is also needed around socially assistive robots and their potential to alleviate loneliness in older adults, with a focus on developing innovative approaches for emotional support and social interaction. Although the importance of technology customization for each user was emphasized, few studies have explored specific methods to achieve this adaptation. Sensors are crucial in supporting older adults, but more research is needed to address the discomfort or unsuitability of wearing sensors and to develop more comfortable and practical alternatives for monitoring the health status of aging adults.

Building on these identified gaps and challenges, our research provides a detailed examination of the current state of SHTs and offers actionable insights and recommendations for future development and policy formulation. By addressing both the technological and ethical dimensions of SHTs, our study ensures that the findings are not only academically robust but also practically relevant. This dual approach advances the field of smart home technologies for older adults in a meaningful way, contributing to the creation of more inclusive, practical, and supportive environments for aging populations. Our study thus represents a significant step forward in understanding and enhancing the role of SHTs in the care of older adults.

The responsible research and innovation (RRI) [25] framework offers a robust guide for navigating the complex ethical landscape associated with integrating technology into the lives of older adults. Aligning with the principle of anticipation, this research endeavors to proactively identify and contemplate the potential long-term consequences and ethical implications of SHTs. This involves a thorough examination of privacy concerns, the autonomy of older adults, and the risk of social isolation that may arise from an over-reliance on technological solutions. By foreseeing these challenges, the study advocates for the development of technologies that prioritize the dignity and independence of older adults, ensuring that these innovations serve to enhance rather than diminish their quality of life. The principle of reflexivity urges researchers and technologists to critically examine their own assumptions, values, and potential biases in the design and implementation of SHTs. This research recognizes the importance of such introspection and advocates for a design philosophy that is empathetic to the unique needs and preferences of older adults. It calls for ongoing dialogue between technology developers, caregivers, and older adults themselves to ensure that the solutions developed are truly in alignment with the users' desires and best interests. Inclusion is paramount in the responsible development and deployment of SHTs. This research emphasizes the need for a participatory approach, where older adults and their caregivers are actively involved in the co-creation of technological solutions. By fostering an environment of collaboration and open communication, the study aims to ensure that the voices of those most impacted by these technologies are heard and considered, leading to more equitable and acceptable outcomes. Finally, the principle of responsiveness highlights the necessity for flexible and adaptive research and innovation processes that can respond to new insights and societal concerns as they arise. This research underscores the need for continuous monitoring and evaluation of SHTs in real-world settings, allowing for iterative improvements that address unforeseen ethical challenges and user needs. It advocates for the establishment of ethical oversight mechanisms that can guide the responsible evolution of these technologies. By embedding RRI principles into the fabric of this research, the study aims to foster a more ethical, inclusive, and human-centered approach to the integration of SHTs in older adults' care.

3. Research Methodology

This research paper presents a systematic literature review based on the guidelines proposed by Kitchenham et al. [26]. The central focus of our research is shaped by eight pivotal questions, outlined as follows: (1) How can we ensure the accessibility of innovative home technology for all levels of society, regardless of financial stability? (2) What level of support do health ministries provide for innovative home technology to improve people's quality of life? (3) How actively do health ministries promote the use of smart home technology to reduce pressure on nursing homes? (4) To what extent do wearable sensors need to be redesigned to be unobtrusive and not interfere with the daily lives of users? (5) How can we create a smart home application that is equally adaptable to every user? (6) To what extent is it necessary to modify state laws and regulations to align with ethical issues? (7) How can smart home technology address the issue of loneliness among older adults while also facilitating their daily lives? (8) How can we overcome the stigma associated with older adults and introduce them to smart home technology? In addition to addressing these questions, this systematic literature review aims to identify the state-of-the-art in the subject area of our study, specifically pertaining to smart home innovations for older adults' care, ethical, security, and privacy considerations in smart home deployment, integrating technology in smart homes, user interaction and experience in smart homes, and smart home design and accessibility.

To conduct the review, we used the following search string: "smart homes" AND "older adults" AND "quality of life" AND "health" AND "acceptance". The review was performed between November 2022 and May 2023, covering articles from databases including ACM Digital Library, Elsevier ScienceDirect, Emerald Insight, Google Scholar, Hindawi, IEEE Xplore, Inderscience Publishers, MDPI, SAGE Journals, Scopus, SpringerLink, Taylor & Francis Online, Web of Science Core Collection, and Wiley Online Library.

The process of ascertaining the suitability of identified articles for our research topic entailed three steps. Initially, we selected articles based on the relevance of their titles to our study's thematic scope, which encompasses smart home innovations, ethical considerations, technological integration, user experience, and design accessibility in older adults' care. We then obtained all articles that met the criteria of the first phase. If it was not feasible to locate and obtain the complete version of a specific article, it was excluded from further consideration. In the second stage, every article that passed the initial screening underwent a review of its abstract. Articles that met the relevance criteria based on their abstracts, addressing either the specific research questions or the broader thematic focus, progressed to the third stage, where the entire body of each article was scrutinized. The final compilation of relevant studies included articles that fulfilled all the requirements set out by our research questions and thematic exploration. Throughout all three stages of selection, articles written in languages other than English and duplicate articles were excluded.

4. Results

4.1. Identification of Studies

Our search across academic databases yielded a total of 5721 records. However, only a small percentage of these were relevant to our study's subject. Initially, 119 research papers with titles pertinent to our focus advanced to the screening stage. During this phase, we faced challenges with accessibility, leading to the exclusion of 18 papers due to the unavailability of their full texts. Consequently, 101 records proceeded to the next round of evaluation. Out of these, we identified and removed 8 duplicate records, 31 papers that were not relevant to our study, and 4 articles that were not written in English. This refining process left us with 58 studies that were directly related to the topic of our study and were, therefore, included in this research paper. Table 1 presents the breakdown of the initial and final sets of studies across different academic databases, while Figure 1 employs the PRISMA 2020 flow diagram [27] to graphically depict our study selection process, ensuring a standardized report of the methodology used.

Table 1. The initial and final sets of studies found in relevant databases.

Database	Initial Set of Studies	Final Set of Studies
ACM Digital Library	68	1
Elsevier ScienceDirect	227	7
Emerald Insight	40	0
Google Scholar	3479	21
Hindawi	1	1
IEEE Xplore	14	11
Inderscience Publishers	0	0
MDPI	5	4
SAGE Journals	1	1
Scopus	212	0
SpringerLink	1620	7
Taylor & Francis Online	45	1
Web of Science Core Collection	5	0
Wiley Online Library	4	4

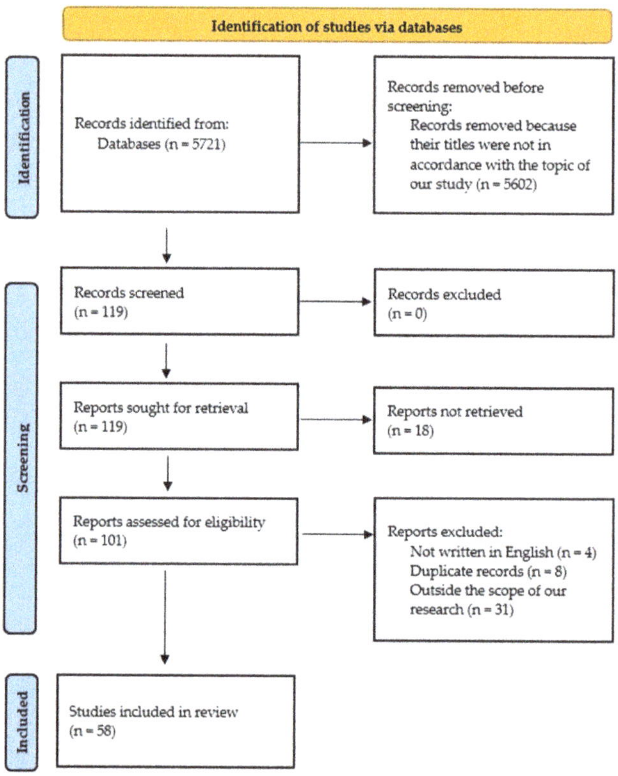

Figure 1. PRISMA flow diagram.

4.2. Bibliometrics of Selected Studies

Of the total 58 studies selected for review, 77.59% were published in peer-reviewed journals, 18.97% appeared in conference proceedings, and 3.45% were master's theses. The largest number of studies (15.52%) relevant to our research was published in 2020, while 10.35% of them were published in each of the years 2015, 2017, and 2019. On the other hand, the smallest number of papers (1.72%) related to the topic of our research was published in

2013. The distribution of papers selected for analysis by year of publication is shown in Figure 2.

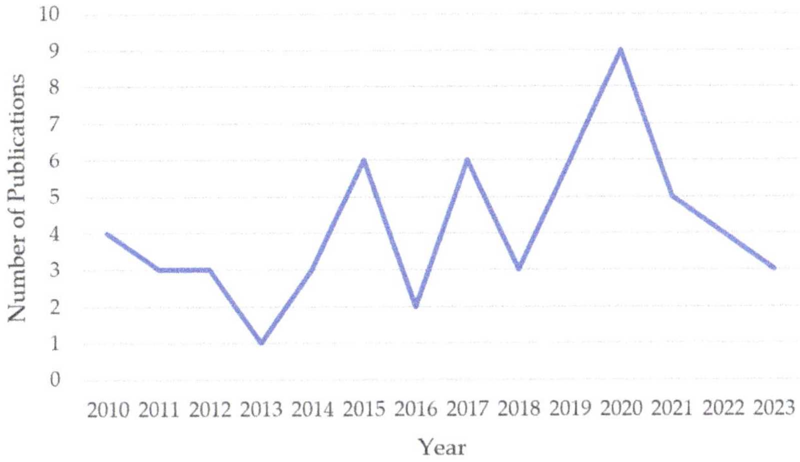

Figure 2. Distribution of studies by year of publication.

Two papers each were published in the International Journal of Advanced Science and Technology, International Journal of Environmental Research and Public Health, and Sensors, while one paper was published in each of the remaining thirty-nine journals. This is summarized in Table 2.

Table 2. The distribution of selected studies across journals.

Name of the Journal	Number of Studies	References
International Journal of Advanced Science and Technology	2	[28,29]
International Journal of Environmental Research and Public Health	2	[30,31]
Sensors	2	[32,33]
Advances in Experimental Medicine and Biology	1	[34]
Annual Review of Nursing Research	1	[35]
Asian Nursing Research	1	[36]
BMC Health Services Research	1	[37]
BMC Medical Informatics and Decision Making	1	[38]
British Journal of Community Nursing	1	[39]
Disability and Rehabilitation: Assistive Technology	1	[40]
Frontiers in Psychology	1	[41]
Future Internet	1	[42]
Geriatric Nursing	1	[43]
Gerontology and Geriatric Medicine	1	[44]
IEEE Access	1	[45]
IEEE Nanotechnology Magazine	1	[46]
IEEE Transactions on Information Technology in Biomedicine	1	[47]
IEEE Transactions on Systems, Man, and Cybernetics, Part C (Applications and Reviews)	1	[48]
Informatics in Primary Care	1	[49]
Innovations in Clinical Neuroscience	1	[50]
International Journal of Computer Networks and Applications	1	[51]
International Journal of Human–Computer Studies	1	[52]
International Journal of Medical Informatics	1	[53]
International Journal of Older People Nursing	1	[54]
International Journal of Technology Assessment in Health Care	1	[55]

Table 2. *Cont.*

Name of the Journal	Number of Studies	References
International Journal of Wireless and Mobile Networks	1	[56]
Internet of Things	1	[57]
JMIR Aging	1	[58]
JMIR mHealth and uHealth	1	[59]
JMIR Research Protocols	1	[60]
Journal of Advanced Nursing	1	[61]
Journal of Electrical and Computer Engineering	1	[62]
Journal of Electrical Engineering and Technology	1	[63]
Journal of Intelligent and Robotic Systems	1	[64]
Journal of Nursing Management	1	[65]
Journal of Physics: Conference Series	1	[66]
Nature Reviews Materials	1	[67]
Personal and Ubiquitous Computing	1	[68]
Primary Health Care Research and Development	1	[69]
Review of Policy Research	1	[70]
Robotics and Autonomous Systems	1	[71]
Technological Forecasting and Social Change	1	[72]

Table 3 illustrates that for each conference, the proceedings included the publication of one paper.

Table 3. The distribution of selected studies across conference proceedings.

Name of the Conference	Number of Studies	References
Ambient Assisted Living	1	[73]
IEEE International Symposium on "A World of Wireless, Mobile, and Multimedia Networks"	1	[74]
IEEE International Symposium on Multimedia	1	[75]
IEEE International Symposium on Technology and Society	1	[76]
International Conference on Computer-Supported Cooperative Work in Design	1	[77]
International Conference on Green Engineering and Technologies	1	[78]
International Conference on Human Aspects of Information Security, Privacy, and Trust	1	[79]
International Conference on Intelligent Environments	1	[80]
International Conference on Pervasive Computing Technologies for Healthcare	1	[81]
International Conference on Software, Knowledge, Information Management, and Applications	1	[82]
International Conference on World Wide Web Companion	1	[83]

As shown in Table 4, both master's theses included in our study were defended in Sweden, with one at Linnaeus University and the other at Lund University.

Table 4. The distribution of selected studies across master's theses.

Name of the University	Number of Studies	References
Linnaeus University	1	[84]
Lund University	1	[85]

4.3. State-of-the-Art in Smart Home Technologies: Insights into Care Innovations and Challenges for Older Adults

This section examines the latest advancements in smart home technologies (SHTs) and their application in caring for older adults. It assesses how these technologies can improve safety, health monitoring, and daily living support, with a focus on their practical implementation. Through a review of current research and technologies, we explore both the opportunities these innovations present for enhancing the lives of older adults and the challenges, such as user acceptance and system integration, that must be overcome to maximize their effectiveness.

4.3.1. Smart Home Innovations for Older Adults' Care: Enhancing Quality of Life and Independence

This section comprises research focused on the application of SHTs to improve the quality of life for older adults. The studies explore various innovations, such as health monitoring systems, emergency response solutions, and daily activity support, all tailored to meet the needs of older adults. The goal is to enhance independence, safety, and overall well-being through technology, making everyday life more manageable and healthcare more accessible in a home setting.

Doyle et al. [81] present findings from their investigation into older adults' attitudes toward self-managing their health and wellness using data from SHTs. The study, conducted within the Great Northern Haven smart homes, involved interviews with seven older adults who had been living with integrated smart home systems for over four years. The research aimed to explore the residents' long-term attitudes toward living with smart home sensors and assistive ambient living (AAL) technologies, their opinions on the utility of the data these technologies provide, and their preferences on data visualization techniques. Key findings revealed that older adults are generally interested in using sensor technology data to aid in self-managing their well-being. They expressed a particular interest in feedback on metrics such as time spent indoors and outdoors, walking times, sleep patterns, activity levels, blood pressure, and weight. However, the study also highlighted the need for data representation to be simple, intuitive, and accompanied by educational support and goal-setting to be genuinely useful. Residents showed a preference for visualizations that clearly and effectively communicate health and wellness data, suggesting that appropriate design and presentation of information can significantly enhance their engagement and the usability of smart home data. The reported findings underscore the potential benefits of SHTs in supporting older adults in managing their health and wellness independently. However, they also point to the necessity for careful consideration of how data are presented and the broader context of use, ensuring that these technologies meet the real needs of their older adult users.

Chernbumroong et al. [82] conducted a study to investigate perceptions, needs, and concerns related to SHTs aimed at assisting older adults. The research surveyed individuals in various settings, including a major hospital and nursing homes, to gather data on the acceptability and perceived usefulness of six specific SHTs. These technologies included systems for cooking safety, sleeping pattern monitoring, emergency alarms, automatic lighting, video monitoring, and activity monitoring. The study found that while there was general support for SHTs, significant concerns were also expressed. Participants appreciated the potential of these technologies to enhance their safety and well-being—particularly through features that prevent accidents (such as automatic stove shut-offs) and monitor health parameters (such as sleeping patterns and activity levels). However, there were notable apprehensions about privacy, the complexity of the technology, and the need for training on new devices. Moreover, the findings indicated a varied willingness to adopt these technologies. While some participants were open to integrating technologies such as automatic lighting systems into their homes, there was reluctance toward devices that might invade privacy or require extensive interaction, such as video monitoring systems. This highlights the importance of addressing user-friendliness and privacy concerns to increase the adoption rates of SHTs among older adults. The study emphasized that the successful implementation of SHTs in older adults' care significantly depends on understanding and addressing the specific perceptions and concerns of the potential users, suggesting that user-centric design and comprehensive stakeholder engagement are crucial.

Ghorayeb et al.'s [52] study on older adults' perspectives of smart home technology (SHT) revealed a complex relationship between user experience and technology acceptance. The findings indicated that familiarity with the technology through direct experience led to increased acceptance among older adults, who initially may harbor concerns about privacy, usability, and the intrusive nature of the technology. Those with experience tended to view these technologies more favorably over time, appreciating their potential to enhance safety and maintain independence. On the other hand, those without direct experience remained skeptical, perceiving these systems as potentially invasive and expressing heightened concerns regarding privacy. Both groups, however, emphasized the importance of customizable and non-intrusive technology solutions that can be adapted to individual needs and preferences. The study underscores the critical role of involving older adults in the design and development processes of SHTs, ensuring that these innovations align effectively with the end-users' expectations and lifestyle requirements, thereby enhancing overall acceptance and utility.

Pal et al.'s [75] systematic review examined the impact of SHTs on the quality of life for older adults. The research assimilated findings from multiple studies, revealing a generally positive reception toward smart homes, particularly valued for health monitoring and assisted independent living. However, significant concerns about privacy, security, and potential social isolation due to increased technological reliance also prominently surfaced. Older adults appreciate smart homes for the autonomy and safety they confer, but simultaneously, they expressed apprehensions about being monitored and the impersonal nature of technology. This dichotomy underscores the need for designing smart home solutions that not only enhance physical safety and health monitoring but also consider the social and emotional well-being of older adults. The study highlighted the crucial balance between technological assistance and maintaining human interactions, pointing toward a future where SHTs are seamlessly integrated into the fabric of older adults' care without compromising personal privacy or social engagement.

Majumder et al. [32] provided a review of recent advancements in SHTs tailored for older adult healthcare, addressing both significant progress and emerging challenges in the field. Their study emphasized the integration of environmental and wearable medical sensors within smart homes to facilitate continuous and remote health monitoring of older adults. This approach not only promises enhanced living standards for the aging population by allowing them to remain in their familiar home environments longer but also aims to alleviate the financial burdens on healthcare systems through cost-effective monitoring solutions. The research highlighted a variety of smart home applications, from basic health monitoring to sophisticated systems that can predict emergency situations, all enabled by the convergence of Internet of Things (IoT) technology, AI, and advanced sensor technology. However, despite these technological strides, the review also pointed out critical hurdles, such as the need for interoperability among different devices, privacy concerns, and the technical challenges in developing user-friendly systems that are both affordable and reliable. The study serves as a backbone for understanding the potential of smart homes in transforming older adults' care, underscoring the importance of multidisciplinary approaches that merge healthcare, technology, and user-centered design to develop solutions that are not only technologically sound but also align with the real-world needs of older adults.

Dermody et al. [61] conducted an exploratory qualitative study to assess the readiness of community-dwelling older adults to adopt SHT. The study identified several key factors that influence this readiness, including knowledge, health and safety, independence, security, and cost. These factors manifested through qualitative themes, such as the level of familiarity with technology, concerns about privacy and data security, and perceived usefulness in enhancing independence and safety. The study underscored the mixed feelings among older adults regarding the adoption of smart homes: while recognizing the potential benefits for health monitoring and independent living, they also expressed apprehensions about privacy invasions and the intrusiveness of constant monitoring. Importantly, the research highlighted the need for increased awareness and information dissemination to enhance acceptance and reassure potential users about the benefits and safeguards of SHTs.

Marikyan et al. [72] conducted a systematic review that explored the burgeoning literature on SHTs from a user perspective, revealing how these innovations are reshaping residential living. Their review encapsulated a multitude of studies, illustrating that while users acknowledge the considerable benefits of smart homes, including enhanced convenience, increased safety, and potential health monitoring, there remain significant reservations, primarily about privacy, technology integration, and cost. The authors detailed that the adoption of SHTs is influenced by several factors, including the level of technology integration possible within existing infrastructures, the perceived ease of use, and the overall reliability of these systems. They argued that for broader adoption, smart home solutions must not only be technologically advanced but also user-friendly, affordable, and respectful of privacy and data security. This review sheds light on the crucial role of user-centric approaches in the design and implementation of SHTs, suggesting that future research should continue to address these user concerns to enhance acceptance and utility.

Bal et al. [77] provided an extensive review of collaborative SHTs aimed at supporting independent living of older adults. Their research synthesized findings from approximately 30 projects, underscoring the critical role these technologies play in enhancing safety, health, and social connectedness for older adults. The review highlighted how smart homes can significantly boost the quality of life by facilitating continuous health monitoring, emergency assistance, and social interaction, which are essential for older adults living independently. However, the authors also pinpointed several challenges that need to be addressed to maximize the benefits of these technologies. These included the integration of systems into existing infrastructures, the usability of the technology by older adults who may have varying levels of comfort and proficiency with digital tools, and the cost associated with implementing such technologies. Moreover, they emphasized the importance of privacy and data security, critical factors that influence the acceptance and success of smart home solutions among older adult users. The review provides a comprehensive perspective on how collaborative intelligence and smart environments can be leveraged to support the autonomy and well-being of older adults, suggesting a pathway toward more sustainable, efficient, and inclusive healthcare solutions.

Lee and Kim [41] critically evaluated the development and effectiveness of smart residential environments for older adults, with a particular focus on enhancing pleasurable experiences. Their review underscored that while smart technologies in residential settings significantly improve the physical and cognitive support for older adults, they often overlook the hedonic and emotional aspects that contribute to a pleasurable living experience. The study identified a prevalent focus on utilitarian benefits, such as safety, monitoring, and energy efficiency, which, while crucial, do not fully address the needs for emotional and social enrichment. The authors argued that smart environments should integrate features that not only ensure physical wellness and safety but also promote fun, social interaction, and emotional engagement. They suggest that a shift toward including hedonic factors would greatly enhance the quality of life for older adults, making the technology more appealing and acceptable. The study presented an evaluation framework consisting of wellness, independence, acceptance, and design, which aims to guide future research and development in creating more holistic smart residential environments that support not just the functional needs but also the psychological and emotional well-being of older adults.

Yusif et al. [53] conducted a systematic review to explore the barriers to the adoption of assistive technologies (ATs) by older individuals, focusing on empirical studies that identify these barriers from the perspective of older adults. Their research highlighted several critical issues influencing older adults' willingness to embrace ATs. A primary concern among older adults was privacy, cited as a major barrier in 34% of the studies reviewed. Other significant barriers included trust in the technology and its added value, both of which impacted the perceived usefulness and efficacy of ATs. The cost of technologies and their ease of use also posed substantial challenges, each noted in 23% of the studies.

Additionally, the perception of not needing assistance, stigma associated with using ATs, and the fear of becoming dependent on technology further hindered adoption rates. These factors were compounded by physical declines related to aging and generational differences in technology acceptance, which are often exacerbated by the design of ATs that may not appeal to older users due to their clinical appearance or perceived complexity. This review clearly demonstrated that while ATs have the potential to significantly improve quality of life for older adults, adoption rates will remain low unless these technologies are designed and marketed in a way that addresses these barriers, emphasizing privacy, trustworthiness, affordability, simplicity, and the preservation of independence and dignity.

Sokullu et al. [57] discussed the design and implementation of an IoT-supported smart home system aimed at improving the living standards of older adults, particularly those with partial memory loss, such as mild cognitive impairment (MCI) or dementia patients. Their study elaborated on the deployment of various sensors throughout a residence to monitor the activities of older adults unobtrusively. The system included an emergency wristband for urgent situations, offering a comprehensive, low-cost solution for enhancing safety and providing early warnings for routine issues faced by older adults. The results highlighted the system's effectiveness in contextual activity detection using ambient sensors, which assess activities without invading privacy. The prototype is praised for its affordability and ease of integration, reflecting a significant advancement in assistive living technologies. By offering real-time monitoring and emergency alerts, the system not only ensures the safety of older adults but also facilitates their independence, aligning with the needs and preferences of its users. This study underscored the potential of IoT in transforming older adults' care through SHTs, fostering a safer and more responsive living environment.

Existing research implies that SHTs significantly contribute to the enhancement of care for older adults by improving safety, health monitoring, and social interaction capabilities. These technologies are tailored to meet the unique needs of older adults, facilitating better management of their healthcare costs and supporting their independence. However, the adoption of smart home solutions faces challenges, including privacy concerns, the complexity of the technologies involved, and the substantial costs associated with implementation. Studies also emphasized the importance of intuitive design and accessible interfaces to ensure that these innovations are user-friendly for an aging population. There is a notable need for educational support to help older adults become more comfortable and proficient with these technologies. Furthermore, the research advocates for a holistic approach that not only focuses on physical wellness but also considers the emotional and psychological health of older adults, promoting a shift from technology-centric solutions to those that are human-centered and enrich the lives of users by integrating elements of fun and social connectivity. Table 5 showcases various smart home innovations, detailing their main advantages, disadvantages, and specific recommendations for enhancing adoption and effectiveness in older adults' care.

Table 5. Enhancing senior care with smart home innovations: benefits, challenges, and recommendations.

Smart Home Innovations	Advantages	Challenges	Recommendations	References
Automatic stove shut-offs, health parameter monitoring systems (sleep patterns, activity levels).	Enhanced safety and well-being through health monitoring and accident prevention.	Privacy concerns and complexity of technology.	Address privacy and user-friendliness to increase adoption rates.	[75,77,81,82]
Customizable health metrics' tracking (time spent indoors/outdoors, walking times, blood pressure).	Increased acceptance and independence with customizable, non-intrusive solutions.	Skepticism and apprehensions about constant monitoring and invasiveness.	Provide clear, simple data representation and educational support.	[32,41,52,53,61]

Table 5. *Cont.*

Smart Home Innovations	Advantages	Challenges	Recommendations	References
IoT-supported monitoring systems, emergency wristbands for urgent situations, wearable sensors.	Continuous health monitoring and real-time emergency alerts.	Cost and interoperability challenges, along with usability issues.	Focus on affordable, user-friendly solutions with strong privacy and data security measures.	[57,72]
Systems facilitating social interaction, entertainment and engagement tools, interactive devices.	Autonomy and safety with potential for enhanced social interaction and emotional engagement.	Potential social isolation and overlooking emotional aspects of living experience.	Integrate features promoting social interaction and emotional engagement, along with physical safety.	[41,75]

4.3.2. Ethical, Security, and Privacy Considerations in Smart Home Deployment: Navigating the Challenges

The works in this section explore the ethical challenges as well as security and privacy concerns associated with the deployment of SHTs. These studies discuss issues such as data protection, user autonomy, informed consent, and the balance between technological benefits and the potential for intrusion. The focus is on ensuring that smart home solutions are developed and used in ways that respect the dignity and rights of users, fostering trust and acceptance.

Zhu et al. [65] conducted a scoping review to address ethical issues in smart-home-based care for older adults, exploring the implications of incorporating advanced technologies, such as AI and information technologies, into care practices. Their study highlighted key ethical concerns revolving around privacy, autonomy, safety, fairness, and the potential reduction in human contact. The findings emphasized significant privacy issues, particularly how smart technologies collect and handle sensitive personal data, raising questions about information privacy and physical privacy. Autonomy is another critical focus, with discussions on the importance of informed consent and user-centered control, ensuring that older adults can make voluntary and well-informed decisions about the technologies that affect their lives. Safety guarantees provided by these technologies are acknowledged as beneficial, enhancing the quality of life by enabling more effective emergency responses and everyday activity monitoring. However, there are concerns about fairness, as access to these technologies is not uniformly available, often limited by economic or social barriers. The review also addressed the reduction in human contact, a significant ethical concern, as SHTs might lead to less human interaction, potentially increasing feelings of isolation among older adult users. The researchers advocate for the careful integration of technology that complements human care rather than replacing it, ensuring that care for older adults remains a deeply human-centric service.

Sanchez et al. [55] conducted a systematic literature review exploring the ethical implications of implementing smart house welfare technology for older adults. Their study highlighted several key ethical challenges, including issues related to privacy, autonomy, informed consent, dignity, safety, and trust. These challenges arise from the potential for smart houses to intrude on personal space and monitor private activities without adequate safeguards, raising concerns about the erosion of autonomy and privacy. Additionally, the need for informed consent is critical, ensuring that users are fully aware of the technology's capabilities and limitations. The authors emphasized that while smart home technology offers substantial benefits, such as enhanced safety and extended independent living, it also presents significant ethical dilemmas that must be addressed through careful design and transparent policies. The review suggested that developers and policymakers need to prioritize these ethical considerations to foster trust and acceptance among users, thereby ensuring that the technology genuinely supports the dignity and quality of life of older adults. This examination of the ethical landscape serves as a crucial guide for advancing smart home technologies in a responsible and user-focused manner.

Chung et al.'s [35] systematic review explored the ethical implications and concerns associated with the use of SHTs to support older adults. Their study identified key ethical issues, including privacy, autonomy, obtrusiveness, equal access, reduction in human contact, and usability, which are crucial in the context of aging, technology, and home environment. The review underscored that while smart homes have the potential to enhance the quality of life, safety, and independence of older adults, they also raise significant ethical concerns. The privacy of users is highlighted as a particularly pressing issue, considering the invasive nature of continuous monitoring technologies that collect and transmit personal health information. The authors argued for the need to balance technological benefits with ethical considerations to ensure that SHTs serve as a supportive, rather than intrusive, solution for aging individuals. The review calls for ongoing research to address these ethical challenges and for the development of guidelines that safeguard the interests and rights of older adults.

Robles and Kim [28] provided a thorough examination of security challenges and technological advancements in smart home development. Their review outlined the significant progress in smart home automation systems, focusing on enhancing security measures to protect homeowners and their properties. They discussed the evolution from basic security systems to advanced smart home systems that offer a myriad of additional benefits, such as remote monitoring, automated alerts, and integration with various household devices through sophisticated network protocols. The study highlighted the critical aspect of security within smart homes, noting that the technological sophistication of these systems also introduces complex security vulnerabilities. These vulnerabilities include network security threats and privacy concerns due to the potential for unauthorized data access. The authors emphasized the necessity for robust security protocols and encryption methods to safeguard data transmitted across smart home networks. They advocated for continuous advancements in SHTs to address these security challenges effectively, ensuring the safety and privacy of users in the increasingly connected and automated home environment.

Current studies indicate that while SHTs offer substantial benefits for improving care for older adults, they also present significant ethical challenges that must be addressed to ensure their responsible implementation. Key concerns include ensuring privacy and data protection, maintaining autonomy and dignity, and providing safety and fairness in technology use. The potential for reduced human contact and increased social isolation also highlights the need for solutions that balance technological benefits with maintaining human connections. Furthermore, the ethical deployment of these technologies requires robust frameworks that include clear guidelines on informed consent, user-centered control, and transparency. It is essential to minimize data collection to what is strictly necessary and ensure that all collected data are securely protected. The role of trust and user perception in the adoption of SHTs is critical, emphasizing the need for developers and policymakers to work toward building and maintaining trust through ethical practices. This entails a multidisciplinary approach that integrates ethical considerations at every stage of technology development and deployment, ensuring that smart home innovations enhance the lives of older adults without compromising their privacy or autonomy. Table 6 outlines various considerations in smart home technology for older adults, along with key points and strategies to address these challenges.

4.3.3. Integrating Technology in Smart Homes: Building Connected and Efficient Environments

This section explores the integration of various technologies within smart homes, highlighting the roles of sensors, data analytics, and networked devices in creating interconnected living environments. The studies examine how these technologies can improve the efficiency, functionality, and adaptability of home systems, from energy management to security and health monitoring, emphasizing the technical underpinnings that make homes smart.

Table 6. Overcoming ethical, security, and privacy challenges in smart home technology for older adults.

Considerations	Key Points	How to Address	References
Privacy Issues	Significant concerns about how smart technologies collect and handle sensitive personal data, raising questions about both information privacy and physical privacy.	Implement robust data privacy measures and transparent policies.	[28,35,55,65]
Autonomy and Informed Consent	Importance of ensuring that older adults can make voluntary and well-informed decisions about the technologies affecting their lives. Emphasizes the need for user-centered control and informed consent.	Ensure clear, comprehensive informed consent processes and user-centered control.	[35,55,65]
Safety Guarantees	Benefits of enhanced safety and quality of life through effective emergency responses and everyday activity monitoring.	Develop reliable safety protocols and integrate comprehensive emergency response systems.	[35,55,65]
Fairness and Equal Access	Concerns about the uneven availability of technologies, often limited by economic or social barriers. Highlights the need for equitable access to ensure fairness in the deployment of smart home technologies.	Promote equitable access through subsidies or community programs and ensure cost-effective solutions.	[35,65]
Reduction in Human Contact	Ethical concern that smart home technologies might lead to less human interaction, potentially increasing feelings of isolation among older adults. Advocates for technology that complements rather than replaces human care.	Design technology to complement human care, not replace it.	[35,65]
Dignity and Trust	Challenges related to the potential for smart homes to intrude on personal space and monitor private activities without adequate safeguards, raising concerns about the erosion of autonomy and dignity. Emphasizes building trust.	Prioritize transparency in data usage and implement strong data protection.	[55]
Obtrusiveness	Ethical issues related to the intrusive nature of continuous monitoring technologies, emphasizing the need for a balance between technological benefits and ethical considerations.	Ensure that technologies are non-intrusive and maintain a balance between monitoring and privacy.	[35]
Security Measures and Encryption	Critical need for robust security protocols and encryption methods to safeguard data transmitted across smart home networks, addressing complex security vulnerabilities and network security threats.	Implement advanced security protocols and continuous updates to mitigate vulnerabilities.	[28]

Mohammed et al. [66] presented a review of the integration of Internet of Things (IoT) technologies in smart homes and healthcare systems. Their study detailed the transformative impact IoT has on enhancing healthcare management by facilitating remote monitoring and management of health data with minimal human intervention. The review elaborated on various IoT applications, such as real-time health monitoring, patient information management, and emergency management systems, that improve the efficacy of healthcare services and patient care. The authors highlighted the potential of IoT to alleviate the strain on healthcare systems by reducing costs, increasing the efficiency of medical services, and enhancing patient outcomes through continuous monitoring. However, they also acknowledged significant challenges, including data security, privacy concerns, and the need for robust network infrastructure to support the seamless integration of IoT devices. The review underscored the necessity of addressing these challenges to fully harness the benefits of IoT in smart healthcare environments, suggesting future research directions to overcome these barriers and enhance system implementations.

Robles and Kim [29] provided an analysis of SHTs, focusing on their applications, systems, and methods. Their review underscored the integration of technology and services through home networking as a means to enhance the quality of life. The study revealed that smart home systems facilitate improved convenience and safety for residents, particularly older adults and disabled individuals, through automated systems that can control various aspects of the home environment, such as lighting, security, and temperature. The authors discussed several technologies, such as X10, Z-Wave, and ZigBee, which are crucial for the operation of these smart homes. These technologies enable devices within the home to communicate over both electrical wiring and radio frequencies, creating a mesh network that ensures commands are received and executed, even in the presence of signal interference. The use of such advanced protocols enhances the reliability and efficiency of smart home systems. Furthermore, Robles and Kim highlighted the potential for these technologies to provide significant energy savings and enhanced security measures. They illustrated how smart homes can adapt to the residents' habits and needs, providing not only operational efficiency but also a personalized living experience. The review pointed to a future where SHT plays a pivotal role in everyday living, offering greater comfort, security, and energy efficiency.

Maswadi et al. [45] conducted a systematic literature review to evaluate smart home monitoring technologies for older adults, emphasizing IoT integration. The review underscored the transformative potential of these technologies in enhancing older adults' care by facilitating health monitoring, emergency responsiveness, and daily activity management, which are essential for maintaining independence and safety. However, the authors highlighted a critical need for standardized research methodologies to reliably assess and implement these technologies. This work not only maps the current landscape but also calls for future research to adopt rigorous approaches for validating smart home solutions, thereby ensuring their effectiveness and reliability in real-world applications.

Kim et al. [74] presented an examination of the U-Health Smart Home project, focused on enhancing older adults' care through technology. Their work elaborated on the implementation of an autonomic system within smart homes, using semantic web technologies and wireless sensor networks (WSNs) to monitor and assist older adults in real time. The study highlighted the integration of a knowledge-based framework that uses semantic reasoning to understand and adapt to the needs of older adults based on continuous data collection from environmental and body sensors. This approach enables the smart home to autonomously manage various home devices and systems in response to the identified needs of its residents, promoting safety and independence. The autonomic system is designed to operate with minimal human intervention, providing a proactive health management solution that significantly reduces the need for traditional healthcare interventions. The project's findings demonstrated the potential of integrating advanced technologies in smart homes to significantly enhance the quality of life for older adults, offering a promising direction for future innovations in older adult care technology.

Agoulmine et al. [46] presented a study on innovative solutions for managing older adults' care and chronic diseases through the integration of nanotechnology, biotechnology, and information and communication technologies. Their research discussed the development of a U-Health Smart Home system designed to support older adults and people with chronic diseases within their own homes, thus reducing the necessity for hospitalization and specialized care facilities. This U-Health Smart Home system is engineered to allow autonomous monitoring and assistance using advanced sensor technologies, including wearable devices and environmental sensors that continuously track health metrics. The system is connected to a back-end medical institution, which enables real-time updates on the patient's status to healthcare providers, potentially allowing for an immediate medical response in emergency situations. The research illustrated how such technological integrations could alleviate the financial burden on healthcare systems by enabling older adults' independence and proactive health management. It emphasized the role of a layered architecture in the smart home system, including sensors and actuators, a home communication network, and an autonomic decision-making system. This approach aims to provide a sustainable, cost-effective solution for enhancing older adults' care through technology-driven innovations.

Paolini et al. [33] developed an IoT-based digital platform for smart homes under the HABITAT project, aiming to integrate advanced technologies, such as Radio Frequency Identification (RFID), wearable electronics, wireless sensor networks, and AI, to support older adults' independence both at home and in retirement settings. The platform's novelty lies in its system-level flexibility, enabling the interoperability of various smart devices, which allows for the customization of home environments to better serve the needs of older adult users, thereby reducing healthcare costs by minimizing the need for personal assistance. The study emphasized the potential of such integrated systems to enhance the quality of life for older adults by embedding technology in everyday objects, making them part of an unobtrusive assistive environment. This approach not only promotes aging in place by enabling older adults to live independently for longer but also integrates health monitoring and emergency management seamlessly into the living space. The HABITAT project showcases a forward-thinking application of IoT in gerontechnology, focusing on user-centered design to ensure the technology is adaptable, practical, and acceptable to its end-users.

Obaid et al. [56] developed a ZigBee-based voice-controlled smart home system intended to enhance the autonomy of older adults and disabled individuals. Their system utilizes voice commands to operate home appliances, offering a user-friendly and adaptable home automation solution. The study showcased how voice recognition integrated with ZigBee technology allows for efficient command execution with low power consumption, significantly improving accessibility for users with limited mobility or disabilities. The proposed system demonstrated a high degree of reliability in recognizing and executing commands, overcoming common challenges, such as noise interference and speaker accent variation. This advancement in SHT not only improves the quality of life for its users by providing them with more control over their living environment but also paves the way for further innovations in assistive home technologies. The study emphasized the potential for scalable smart home solutions that prioritize user interface simplicity and efficient communication protocols to address the specific needs of vulnerable populations.

Brulin et al. [47] developed a computer vision-based posture recognition method for home monitoring of older adults, integrating a fuzzy logic approach to categorize four static postures: lying, sitting, squatting, and standing. Their system begins with human detection, focusing on human silhouettes to ensure robustness against environmental variations, such as lighting changes and object movements. The posture recognition process utilizes fuzzy logic to handle the ambiguity in human posture, enhancing the system's reliability in real-world applications. The system achieved a satisfactory accuracy of 74.29% in recognizing the correct postures, proving its efficacy in identifying emergency situations, such as falls, within a smart home environment. This capability indicates significant potential for enhancing older adults' care by promptly detecting and responding to falls, thereby improving safety and potentially reducing the need for constant human supervision. The approach underscores the importance of incorporating adaptive, context-aware technologies in developing assistive tools for the aging population.

Kim et al. [63] developed a vision-based Human Activity Recognition (HAR) system utilizing depth silhouettes to monitor the daily activities of older adult residents within smart homes. The system employs depth video cameras to capture human silhouettes, which are then processed to extract skeleton joint features. These features include centroid points, joint distances, and magnitudes, forming a robust dataset for training Hidden Markov Models (HMMs). The experimental results demonstrated a high mean recognition rate of 84.33% across nine routine activities, showcasing the system's effectiveness in identifying a wide range of daily actions accurately. This innovative approach highlighted the potential of depth-based video surveillance in enhancing older adults' care by providing continuous, real-time monitoring that supports independent living while ensuring safety and well-being in smart home environments.

Neßelrath et al. [73] introduced a novel gesture-based interaction system for smart homes that significantly enhances the interface for older adult users by utilizing contextual sensitivity. Their system employs the WiiMote to recognize gestures, which then translate into commands for home appliances through a middleware based on the ISO 24752 standard [86]. The study tested two interaction concepts: one that adapts the gesture's function based on its contextual use and another that triggers predefined scenarios, such as "go to bed", simplifying complex sequences of actions into single gestures. The research demonstrated that contextual interactions allow a single gesture to perform various functions depending on the active appliance, reducing the cognitive load and number of gestures users need to remember. For example, a lifting gesture might increase TV volume or a fan's speed depending on the device being controlled at the time. The scenario-based approach further simplifies interaction by allowing a single gesture to execute multiple related commands across different devices. User testing shows that the context-sensitive model was preferred, with 85% of participants responding favorably, highlighting its effectiveness in simplifying the control of various home functionalities without overwhelming the user. This approach not only enhances user experience but also supports independent living for older adults by making technology more accessible and intuitive.

Existing research demonstrates that smart home technologies (SHTs) have significantly progressed in functionality, offering substantial benefits in terms of convenience, safety, and health monitoring, particularly tailored to meet the needs of older adults. These technologies each play a distinct role in creating an effective and supportive living environment for older adults, enabling them to maintain their independence while ensuring their safety and well-being. The foundation of modern smart homes is the Internet of Things (IoT), which connects various devices and appliances within the home to the internet. This integration allows for seamless interaction and automated control over the home environment, adapting to the occupants' habits and preferences. For example, smart thermostats can automatically adjust the temperature based on the weather and the residents' schedule, while smart locks can provide secure, keyless entry based on digital permissions.

One of the earliest home automation systems, X10, utilizes the home's existing electrical wiring to transmit control signals to various connected devices. This technology enables older adults to easily control lighting, security cameras, and other appliances without the need for complex installations or setups, making it a cost-effective option for enhancing home automation. Both Z-Wave and ZigBee employ low-energy radio waves to establish a network of connected smart home devices. Z-Wave is particularly valued for its robust security features and the creation of a large mesh network, enhancing the system's reliability across a larger area. Conversely, ZigBee operates at higher frequencies and can handle more data, making it ideal for tasks that require significant data throughput, such as advanced security systems or streaming high-definition audio and video content.

The U-Health Smart Home Project specifically targets health monitoring integration into living spaces for older adults. It employs sensors and smart medical devices to continuously track vital signs and other health metrics. For instance, bed sensors can monitor sleep quality and duration, while wearable devices can track heart rate and physical activity, providing alerts to caregivers or medical professionals if abnormalities are detected. In smart homes, RFID technology is often used for security and personal identification. Embedding RFID tags in wearable devices, such as bracelets or key cards, enables the system to recognize individuals and tailor environmental settings to their preferences. Additionally, RFID can enhance home security by restricting access to certain areas or making it easier for residents to interact with their environment without physical contact.

Wearable electronics, such as smartwatches and fitness trackers, are central to personal health monitoring in smart homes. These devices continuously collect data on vital signs and physical activity, which can be analyzed to monitor health trends and potentially detect early signs of medical issues. For example, a smartwatch might detect irregular heart rhythms and alert both the user and their healthcare provider. Wireless sensor networks (WSNs) involve multiple sensors distributed throughout the home to monitor various environmental and structural parameters, such as temperature, humidity, and motion. These sensors facilitate automated responses, such as activating the heating, ventilation, and air conditioning (HVAC) system when a room reaches a certain temperature or triggering security alarms if unexpected movement is detected at night.

Artificial intelligence (AI) technologies analyze the data collected from various sensors and devices within the smart home to learn about the residents' behaviors and needs. This insight allows AI to automate routine tasks, such as adjusting lighting and temperature, suggesting meal plans based on dietary restrictions and available ingredients, or even managing medication schedules. Computer vision allows smart home systems to interpret visual information from cameras and sensors. It is particularly useful for security applications, enabling the system to recognize faces, detect unusual movements, and provide alerts about potential threats or emergencies, such as detecting a fall or unauthorized entry. Semantic web technologies improve the interoperability of smart devices by using common data formats and protocols, ensuring that devices from different manufacturers can communicate effectively. This integration facilitates a more user-friendly and cohesive operation of the smart home environment.

Nanotechnology and biotechnology are pushing the boundaries of health monitoring in smart homes by developing devices that can interact directly with biological systems. For instance, microscopic sensors embedded in wearable devices can continuously analyze blood samples to monitor glucose levels or detect pathogens, offering real-time health insights and preemptive care options. Depth video cameras add another dimension to home security and functionality by capturing not only images but also the spatial relationships within a scene. They can be particularly valuable for applications that require precise motion detection, such as advanced security surveillance or assisting older adults with mobility challenges to navigate their homes safely. Gesture-based interaction systems allow older adults to control various aspects of their home environment through simple hand gestures. This technology reduces reliance on physical controls or complex interfaces, making it easier for users with limited mobility to interact with their home environment effectively.

Integrating various SHTs can significantly enhance the quality of life for older adults by optimizing their living environment for comfort, security, and health. Beginning with IoT devices, AI, and wearables, a typical morning routine for an older adult might include waking up, taking medications, and having breakfast. IoT-enabled appliances, such as coffee makers, can automatically start brewing as the resident wakes up. Simultaneously, wearable devices monitor sleep patterns and vital signs, and AI interprets these data to adjust lighting, temperature, and even play soothing music if it detects any abnormalities, such as a significant change in heart rate. This integration ensures a smooth start to the day by automating routine tasks and monitoring health metrics. Next, the integration of RFID and computer vision enhances security. RFID-enabled key cards or bracelets can be assigned to family members and caregivers, allowing them access to the home. As someone approaches the door, RFID confirms the card's validity, while computer vision analyzes the visitor's face to verify their identity. If there is a mismatch or unauthorized attempt, the system alerts the homeowner and local security services. This dual-layer security ensures that only recognized and authorized individuals can enter, providing peace of mind to the resident. Environmental control is crucial for comfort and health, achieved through the integration of Z-Wave/ZigBee with WSNs. Sensors throughout the home monitor temperature, humidity, and air quality, communicating with the HVAC system via Z-Wave or ZigBee to adjust settings automatically. For example, if a room becomes too cold or overly humid, the system can modify the heating or air conditioning settings accordingly, maintaining optimal comfort without any manual intervention. For older adults with mobility limitations, depth video cameras and gesture-based interaction systems provide an invaluable service. These cameras detect specific gestures, allowing residents to control lighting, televisions, window blinds, and other home devices with simple hand movements. This system eliminates the need for physical reach and manual operation, enabling easy and accessible control of the home environment. Semantic web technologies integrated with IoT ensure that devices from different manufacturers can operate together seamlessly. This integration allows residents to control all connected devices through a single interface, enhancing the usability of the smart home system. For example, a resident could adjust a thermostat, check security cameras, and manage lighting through a unified app, regardless of the brand or model of the devices. Health monitoring and emergency response are addressed by integrating the U-Health Smart Home Project with wearables. Continuous monitoring of health metrics through wearable devices and embedded sensors in the home allows for an immediate response if a fall is detected or abnormal vitals are recorded. The system can automatically contact emergency services and inform family members, ensuring that timely medical help is provided. By weaving all these systems together, a smart home can provide a supportive, adaptive, and secure environment that caters specifically to the needs of older adults, enhancing independence and ensuring safety and health in daily living. This cohesive approach demonstrates the powerful potential of SHTs to transform living spaces into safe, comfortable, and health-conscious environments for older adults. Table 7 presents an overview of various smart home technologies, their integration methods, benefits, and associated challenges, highlighting how these technologies can enhance healthcare management, improve convenience and safety, and support independent living for older adults.

Table 7. Integration of technologies in smart homes: advantages and challenges.

Technology	Integration Methods	Advantages	Challenges	References
IoT for Healthcare Management and Older Adults' Care	Remote monitoring and management of health data, health monitoring, emergency responsiveness, and daily activity management.	Enhances remote health monitoring, reduces strain on healthcare systems, improves patient outcomes, promotes independence and safety for older adults.	Data security, privacy concerns, robust network infrastructure needed, standardized research methodologies needed to ensure reliability and effectiveness.	[45,66]
Advanced Home Networking, Voice Control, and Gesture-Based Interaction (X10, Z-Wave, ZigBee)	Automated control of lighting, security, temperature, and home appliances using home networking protocols, voice commands, and context-sensitive gestures.	Improves convenience and safety, facilitates energy savings, ensures reliable communication even with interference, increases autonomy for older adults and disabled individuals, user-friendly automation, simplifies user interactions, and reduces cognitive load.	Signal interference, network reliability issues, noise interference, speaker accent variation, user learning curve, ensuring intuitive and accessible gestures.	[29,56,73]

Table 7. *Cont.*

Technology	Integration Methods	Advantages	Challenges	References
Semantic Web Technologies, WSNs, Nanotechnology, RFID, Wearable Electronics, AI	Real-time assistance and health management through autonomic systems, continuous data collection, and customizable home environment integration.	Autonomic management of home devices, proactive health solutions with minimal human intervention, supports older adults and chronic disease management, and enhances quality of life through personalized assistive environments.	Continuous adaptation required, ensuring seamless integration of technologies, financial burden, layered architecture complexity, ensuring interoperability, and customizing technology to individual needs.	[33,46,74]
Computer Vision for Posture and Activity Recognition	Monitoring of static postures and daily activities using computer vision, depth video cameras, and Hidden Markov Models.	Enhances safety by identifying emergency situations, improves fall detection, high recognition rate of daily activities, and supports independent living through continuous monitoring.	Variations in environmental conditions, accuracy of posture recognition, ensuring accuracy across various activities.	[47,63]

4.3.4. User Interaction and Experience in Smart Homes: Understanding and Improving Engagement

Studies in this section focus on user interaction with SHTs and the overall user experience. These works assess how users perceive, interact with, and benefit from smart home systems, with an emphasis on usability, acceptance, and satisfaction. The research aims to identify factors that influence the adoption of smart technologies and how these systems can be designed to better meet the needs and preferences of users, particularly older adults.

Turjamaa et al. [54] conducted an integrative review analyzing how SHTs support older adults in their daily lives. They collated data from sixteen studies, emphasizing the benefits of smart homes in enhancing daily living and safety for older adults. The review revealed that smart homes facilitate everyday activities, promote healthier lifestyles, and provide a sense of security, which is crucial for older adults who wish to live independently. The technologies discussed include automated systems for routine tasks and emergency alerts, as well as devices for health monitoring and social communication. Despite these advancements, the review identified a critical lack of older adult involvement in the development of these technologies. The authors argued that the perspectives and experiences of older adults are essential for tailoring SHTs to better meet their needs. Furthermore, while these systems improve physical safety and support daily activities, they also need to be developed in a way that considers maintaining and enhancing the social relationships of older adults, thereby avoiding the risk of social isolation. The study underscored the potential of SHTs to transform living environments for older adults, but it also calls for more inclusive and participatory approaches in technology development to fully realize this potential. This includes continuous collaboration between technical developers, healthcare professionals, and older adults to ensure that smart home solutions are both effective and aligned with the users' actual needs.

Liu et al.'s [43] systematic review and meta-analysis examined the impact of SHTs on older adults with chronic conditions. This study, which included fifteen randomized controlled trials (RCTs), assessed the effects of smart homes on hospital admissions, emergency department visits, days spent in hospitals, quality of life, and satisfaction with the technology. The meta-analysis revealed no significant reductions in hospital or emergency department admissions, nor in the days spent in hospital due to tele-monitoring. There was also no significant improvement in the quality of life associated with the use of these technologies. However, the study noted a positive effect of tele-monitoring on physical functioning and a reduction in depression among older adults, suggesting that while

SHTs may not significantly alter healthcare usage or perceived quality of life, they could contribute positively to specific health outcomes. The majority of participants expressed satisfaction with the smart home setup, highlighting its acceptance among the older adult population. This suggests that despite the lack of significant findings in some health-related metrics, the perceived benefits and satisfaction could encourage continued use and further refinement of SHTs in older adults' care.

Pietrezak et al. [49] conducted a literature review to assess the efficacy of SHT in preventing falls among community-dwelling older adults. Their systematic review included studies that explored the use of various SHTs, such as fall detectors and motion sensors, aimed at preventing falls or reducing the fear of falling. The results from these studies, however, provided limited evidence that SHTs are effective in reducing fall rates or the fear of falling among older adults. Despite the lack of significant reductions in falls, the review found that older adults generally had positive attitudes toward the technology, valuing the increased sense of security and independence it provided, which could contribute to prolonging independent living. Privacy concerns and the intrusiveness of technology were noted but were deemed less significant compared to the perceived benefits. The review also highlighted the importance of user-friendly designs and adequate technical support to facilitate the adoption and effectiveness of these technologies.

Suresh and Sruthi [78] provided a review of SHT, focusing on its applications, technological challenges, and future research directions. Their study examined the integration of various SHTs aimed at enhancing the quality of life for older adults and disabled individuals through automation and energy conservation. They explored the increasing use of home automation systems that facilitate independent living by managing everyday tasks and monitoring health and safety. The authors highlighted the critical role of these technologies in reducing energy consumption through intelligent management systems and discussed the broader implications for e-health and telemedicine within smart homes. However, they also pointed out significant barriers to adoption, such as high costs, technological complexities, and concerns about privacy and data security. The authors concluded that while SHTs hold significant promise for improving living conditions, particularly for vulnerable populations, there is a pressing need for continued research and development to address these challenges and ensure the widespread acceptance and effectiveness of these systems.

Peetoom et al. [40] conducted a literature review focusing on the use of monitoring technologies to assist older adults in living independently. Their review identified various technologies designed to monitor daily activities and significant events, such as falls, highlighting the promising role these technologies play in supporting older adults. Five main types of monitoring technologies were described: passive infrared motion sensors, body-worn sensors, pressure sensors, video monitoring, and sound recognition, with additional mentions of multicomponent and SHTs. While the potential of these technologies to prolong independent living and reduce hospital visits and healthcare costs is acknowledged, the review also pointed out that research in this area is still in its early stages, predominantly featuring small-scale and short-term studies. The effectiveness, user acceptance, and cost-efficiency of these technologies remain under-examined in long-term scenarios. The authors suggested that further research is required to validate these technologies' effectiveness in real-world settings and over longer periods, which is crucial for their broader adoption and to realize their full potential in aiding older adults.

Elers et al. [59] conducted a qualitative study exploring user requirements for technology that supports aging in place for older adults and their informal support networks. The research revealed three main themes influencing technology adoption: social enablers, technology concerns, and information desired. The "social enablers" theme highlighted the positive role of informal support networks in encouraging technology use, which aids in maintaining older adults' independence and safety. Timing was identified as crucial, with technology adoption seen as beneficial either currently or in the near future, depending on the individual's needs. The "technology concerns" theme addressed significant issues, such as cost, usability, security, and privacy, noting that these factors can act as barriers to technology adoption. Concerns were also raised about technology potentially replacing face-to-face interactions, which are valued for their depth and emotional connection. Lastly, the "information desired" theme examined the types of information that should be collected and shared via technology, emphasizing the importance of respecting older adults' privacy and autonomy in deciding what information is shared and with whom. This study not only underscored the potential of technology to enhance the well-being of older adults by supporting their ability to live independently but also highlighted the need for designs that address both practical and interpersonal aspects of technology use in aging populations.

Gudmundsson [84] carried out a qualitative analysis of older adult citizens' perceptions and acceptance toward SHT aimed at supporting their independence and welfare. Conducted in a care housing setting, this study integrated the Unified Theory of Acceptance and Use of Technology (UTAUT2) [87] to frame the investigation, focusing on how the technology aligns with older adults' desires to age in place. The findings revealed a complex relationship between older adults and the technology. While some participants saw potential benefits, especially in terms of safety and maintaining independence, there were notable hesitations concerning the usability and intrusion of technology into daily life. The study underscored that despite the promise of SHT in enhancing older adult living, the actual acceptance hinges significantly on individual attitudes toward technology, perceived complexity, and the tangible benefits it offers. Moreover, Gudmundsson [84] highlighted the influence of social enablers, such as health professionals and peers, in shaping older adults' attitudes toward technology. Older adults expressed a conditional acceptance of smart technology—if it is easy to use and clearly beneficial without being intrusive. This nuanced view suggests that while technology can support aging in place, its design and implementation need careful consideration to meet the actual needs and preferences of older adult users. The study thus calls for a more user-centered approach in the development of SHTs for older adults, ensuring they enhance rather than complicate the lives of the intended users.

Tural et al. [44] examined adults' attitudes and intentions toward adopting SHTs, which could contribute to safe and active aging in place. The study employed a sequential mixed-methods approach, combining online and in-person surveys with a focus group discussion to gather comprehensive insights from community-dwelling older adults aged 50 and above. The research revealed that perceived usefulness is a consistent predictor of positive attitudes and the willingness to use smart home products. While smart fire prevention devices received favorable views due to their potential safety benefits, the affordability of these devices significantly influenced the older adults' intentions to incorporate them into their homes. The focus group discussions highlighted several key determinants impacting attitudes toward smart home products, including skepticism about technology, privacy concerns, and concerns about the return on investment. These factors contribute to a nuanced understanding of the barriers and facilitators to SHT adoption among older adults. This study provided vital insights into designing and marketing SHTs that are aligned with the needs and preferences of older adults, emphasizing the necessity of user-friendly designs that address privacy and affordability to enhance adoption rates.

Cao et al. [58] explored the factors influencing the decision-making process for the adoption of smart home sensors (SHS) by older adults in Singapore. The research identified both intrinsic and extrinsic factors that impact older adults' decisions to adopt such technologies. Intrinsic factors included the users' understanding of the technology, its perceived usefulness, and the benefits, while extrinsic factors included affordability and community care support. The findings revealed that 74% of the participants adopted SHS after an intervention period, which involved installing SHS in their homes. The main reasons for adoption were the perceived benefits of increased safety and the convenience of monitoring health and activity without intrusion. However, 26% of participants chose not to adopt SHS due to privacy concerns, lack of family support, and perceived complexity or irrelevance of the technology. The study highlighted the necessity of involving older adults in the design process of SHS to ensure these systems are user-friendly and meet the actual needs of the end-users. Furthermore, it suggested that providing clear and comprehensive training on how to use these technologies and demonstrating their direct benefits can enhance adoption rates. This research emphasized the complex interplay of personal understanding, perceived benefits, and socioeconomic factors in the adoption of SHTs by older adults, pointing toward the need for strategies that address these diverse aspects to improve technology uptake in this growing population segment.

In their qualitative study, Zhao et al. [36] explored the expectations and acceptability of smart nursing home models among older Chinese individuals and their families. The study identified key expectations, including the desire for quality care supported by both government and social systems, the integration of smart technology applications, the presence of skilled healthcare professionals, access to basic medical services, and comprehensive integration of these services. The participants saw smart nursing homes as potentially enhancing the quality of care and improving their quality of life through innovative technology applications, such as health monitoring and emergency management systems. They also expected these homes to provide a link between traditional healthcare facilities and at-home care through smart technologies, thereby enhancing the continuity and accessibility of care. Acceptability of the smart nursing home model was largely influenced by its perceived efficacy, usability, and the tangible benefits it could offer, such as improving care quality and personal independence. Factors such as age, economic status, health condition, and openness to technology among older adults also significantly influenced their acceptance of smart nursing homes. The study underscored the importance of considering these factors when planning and implementing smart nursing home models to ensure they meet the needs and expectations of older adults and their families effectively.

Current research emphasizes the critical role SHTs play in enhancing the daily lives of older adults, primarily through health monitoring, safety features, and enabling independent living. Studies consistently point to the need for smart home solutions to be user-centric, incorporating the direct input and preferences of older adults to ensure the technologies align with their unique needs. Furthermore, the emphasis on intuitive design, ease of use, and robust technical support highlights an essential requirement for smart homes to be widely accepted and effectively used by the older adult population. However, the outcomes from the integration of these technologies are mixed, with significant benefits in physical performance and safety contrasted against less clear impacts on overall quality of life and hospitalization rates. This suggests a nuanced interaction between user expectations and technology performance, underscoring the importance of personalization in smart home applications. The research advocates for continuous adaptation and refinement of these technologies, focusing on enhancing their accessibility, reliability, and relevance to older adults' lifestyles. Moreover, the studies reveal that successful adoption of SHTs among older adults not only hinges on the functionality of the devices but also on the broader support systems that include family and community engagement. This integrated approach ensures that SHTs are not merely tools but part of a strategy to improve the quality of life for older adults, addressing both their physical and social needs. The exploration of these technologies, therefore, continues to evolve, pointing toward a future where smart homes could become integral to older adults' care, provided they are developed with a deep understanding of the end-users' preferences and everyday challenges. Table 8 provides an overview of different smart home technologies, their advantages and disadvantages, and recommendations for improving their adoption and effectiveness among older adults, with a focus on user interaction and experience.

4.3.5. Smart Home Design and Accessibility: Creating Inclusive and User-Friendly Solutions

This section addresses the design and accessibility aspects of smart homes, focusing on how architectural, ergonomic, and technological design can enhance the livability and functionality of smart environments for all users, with a particular focus on aging populations. The works discuss the need for homes that not only incorporate advanced technology but also prioritize user-friendly, inclusive, and adaptable designs that support the diverse needs of individuals, fostering independence and improving quality of life through thoughtful innovation.

Table 8. User interaction and experience in smart homes: key insights and recommendations for improvement.

Interaction Type	Advantages	Challenges	Recommendations	References
Smart Home Technologies (SHTs)	Enhances daily living, promotes healthier lifestyles, provides a sense of security.	Lack of older adult involvement, risk of social isolation.	Inclusive, participatory approaches, collaboration with stakeholders.	[54,78]
Tele-monitoring and Health Monitoring Technologies	Positive impact on physical functioning, reduction in depression, high user satisfaction.	No significant reduction in hospital admissions or quality of life.	Further refinement and assessment of specific health outcomes.	[40,43]
Fall Prevention Technologies	Increased sense of security and independence among older adults.	Limited evidence of effectiveness in reducing falls or fear of falling.	User-friendly designs, adequate technical support.	[40,49]
Home Automation and Energy Conservation	Facilitates independent living, reduces energy consumption, supports health and safety monitoring.	High costs, technological complexities, privacy and data security concerns.	Continued research and development to address barriers.	[44,78]
Aging in Place Support	Maintains independence and safety, informal support networks encourage technology use.	Cost, usability, security, privacy concerns, potential replacement of face-to-face interactions.	Design addressing both practical and interpersonal aspects.	[36,59,84]
Adoption of SHTs	Positive attitudes when perceived useful, importance of affordability.	Skepticism, privacy concerns, return on investment.	User-friendly designs addressing privacy and affordability.	[44,58]
Smart Nursing Homes	Enhances quality of care, improves quality of life, supports continuity of care.	Acceptance influenced by age, economic status, health condition, openness to technology.	Consideration of individual needs and expectations, government and social support, integration of skilled healthcare professionals, access to basic medical services.	[36]

Yu et al. [30] evaluated the diverse needs of older adults for SHTs in South Korea, categorizing them into five distinct lifestyle-based groups. Their study, based on a questionnaire that assessed aspects such as home activities, social events, and overall quality of life, revealed significant variations in the preferences and priorities for smart home functionalities across these groups. One group comprised older adults who are 'Tired of Housework' and seek technologies to alleviate the burden of daily chores and house maintenance. Another set, referred to as 'Just Rest', included individuals who favor technologies that enhance comfort and relaxation to aid leisure and rest. There was also an 'Active at Home' segment of older adults who appreciate technologies that support physical activity and engagement within their living space. Conversely, the group labeled 'Poor Quality of Life' benefit from functions that focus on health monitoring and emergency management to improve their living conditions. Lastly, older adults with a 'High Quality of Life and Good Health' still value technologies that help maintain their well-being and support efficient health management. The reported findings indicated that the need for 16 out of 26 smart home functions varied notably across these groups, with some prioritizing assistance for daily tasks and health monitoring while others valued enhancements in social connectivity or environmental management. This underscored the critical importance of adopting a user-centered design approach in the development of SHTs to cater effectively to the varied lifestyles and health needs of the aging population, ensuring that the solutions provided genuinely support their independence and enhance their quality of life.

Liu and Tamura's research [80] investigated the integration of smart home technologies (SHTs) to support the aging population in maintaining independence and improving connectivity. Their study focused on the "New elders" in China, defined as older adults who remain active in society post-retirement. The central finding was that the essential need of these older adults is 'connectivity', which includes both physical and psychological aspects. The researchers argued that the primary goal of SHTs for older adults should be to enhance this connectivity, thereby improving their quality of life and enabling independence, safety, and happiness. They highlighted the potential of smart homes to meet the diverse needs of older adults, including basic needs of connectivity, health management, and social interaction, through technology. By integrating intelligent systems within the home, these technologies can streamline daily routines, offer health monitoring, and support social interactions that might otherwise decrease with age. The study emphasized the importance of designing smart home systems that are technologically advanced yet attuned to the specific needs and preferences of the older adult population, aiming to create a comprehensive living experience that promotes well-being and satisfaction.

Kon et al. [83] explored the effectiveness of SHTs in assisting older adults in six key areas: safety, health and nutrition, physical activity, personal hygiene and care, social engagement, and leisure. Their findings indicated that while smart homes are adept at enhancing safety and health management, including fall detection and vital signs' monitoring, there are notable deficiencies in personal hygiene, care, and leisure activities. The researchers stressed that smart homes should not only address essential functions but also enhance the overall quality of life by fostering social engagement and leisure activities, which are often neglected. Despite technological advancements, the adoption of these systems by older adults largely depends on their ease of use, perceived benefits, and how well they address privacy concerns.

By juxtaposing Liu and Tamura's [80] findings with those of Kon et al. [83], we gained a more comprehensive understanding of smart home technologies' (SHTs) potential to enhance the lives of older adults. Both studies underscored the pivotal role of connectivity and social engagement in improving the elderly's quality of life. Liu and Tamura [80] underscored a broad concept of connectivity that encompasses physical, psychological, and social dimensions. In contrast, Kon et al. [83] identified specific gaps in social engagement and leisure within current SHT implementations. Both sets of researchers converged on the necessity for SHTs to provide a holistic and inclusive approach, addressing all facets of elder care beyond mere safety and health monitoring. Integrating Liu and Tamura's emphasis on holistic connectivity [80] with Kon et al.'s detailed examination of daily life assistance [83] offers a clear direction for the evolution of SHTs. Future designs should not only advance technological capabilities but also cater to the distinct needs and preferences of older adults, ensuring that these systems promote comprehensive connectivity. This approach will result in smart homes that support overall well-being, foster social interactions, and enhance contentment for the aging population.

Humphreys [39] critically examined the potential of SHT to support older adults living independently. The study evaluated various smart home components, such as health monitoring, fall detectors, and daily living activity monitors, that are integral to aiding older adults in their own homes. Despite the advanced technological integrations aimed at facilitating independent living and preventing hospitalizations, Humphreys highlighted a significant gap in the evidence concerning the effectiveness of these technologies on actual health outcomes, such as reducing falls or improving the overall quality of life. The review underscored the challenges in assessing the real-world impact of these technologies, noting a scarcity of rigorous implementation studies that go beyond the feasibility and functionality to actually measure the impact on health outcomes. Furthermore, it pointed out the psychological aspects, indicating that while some users find the technology reassuring, increasing their confidence and sense of security, others may feel heightened awareness and anxiety regarding their risks in daily activities. Thus, while SHTs hold potential benefits in supporting aging in place by monitoring health and ensuring safety, Humphreys calls for more high-quality research and large-scale demonstrative projects to thoroughly ascertain their effectiveness and impact. This approach will help in overcoming barriers to adoption and utilization, ensuring that SHTs can truly meet the needs of older adults in a practical, reassuring, and health-promoting manner.

In their systematic literature review, Facchinetti et al. [31] explored the impact of SHTs on managing chronic diseases in older adults. The study categorized these technologies into three groups: smart home systems with environmental sensors, external memory aids involving mobile apps, and hybrid technologies combining multiple devices with telemedicine capabilities. The review evaluated the efficacy of these technologies in maintaining or improving various health outcomes, such as vital signs, medication management, daily living activities, mobility, falls, and overall quality of life. The findings indicated that smart homes can significantly aid in monitoring and managing health, potentially reducing the exacerbation of chronic conditions and improving safety by enabling independence. However, the study highlighted a crucial need for more comprehensive and long-term research to fully understand and quantify these benefits. Despite the promising potential, gaps in effectiveness, user acceptance, and cost-efficiency remain, suggesting that further technological advancements and rigorous evaluations are necessary to ensure these systems meet the complex needs of older adults effectively.

Miller et al. [76] examined the intersection of aging and technology, specifically through the lens of aging in place within smart environments. Their study highlighted the crucial role of SHTs in supporting older adults' desire to live independently in their familiar environments. They addressed how different living arrangements—such as personal homes, assisted living facilities, nursing homes, and family housing—present unique and overlapping challenges that can be mitigated through targeted technological solutions. The research discussed the development and application of various technologies aimed at enhancing the quality of life for older adults. These technologies include sensors to prevent accidents, robotic vacuums for automated cleaning, and digital assistants to support routine tasks. The study critically analyzed these technologies, pointing out that while they offer significant benefits, they must be designed to meet the specific needs and preferences of older adults across different living situations. This work emphasized the importance of integrating advanced technology with a deep understanding of the social and physical aspects of aging. This approach aims to ensure that aging in place not only becomes a feasible option but also supports a high quality of life. The findings suggested that while technology can provide critical support, it must be developed in close consultation with its end-users to truly address the diverse challenges faced by older adults.

Lê et al.'s [42] study provided a detailed examination of the potential of SHTs to facilitate positive aging for older individuals. Their research assessed how these advanced systems can enhance the quality of life for older adults by supporting their independence and health, integrating various functionalities, such as health monitoring, emergency management, and daily living assistance, within a home environment. The study highlighted the benefits of smart homes, such as increased safety, improved communication, and greater convenience, which are crucial for older adults who may face physical and cognitive challenges. However, the authors also discussed several significant challenges related to the adoption and use of SHTs by older adults, including issues with accessibility, technology acceptance, and ethical concerns, such as privacy and autonomy. Overall, the authors advocated for a user-centered approach in the development of SHTs. They stressed the importance of considering the needs and preferences of older adults in the design process to ensure these systems are both beneficial and acceptable to the end-users. The research underscored the need for ongoing dialogue between developers, users, and policymakers to address these challenges and fully realize the potential of smart homes in supporting the aging population.

In their study, Alam et al. [48] provided a review of SHTs, outlining their evolution from basic automated systems to advanced networks capable of enhancing healthcare, security, and energy management in residential settings. They explored the multifaceted applications of smart homes, particularly emphasizing their role in supporting older adults through enhanced monitoring and health management systems. The review discussed the integration of various technological components, such as sensors, communication protocols, and specialized algorithms, which collectively improve the functionality and efficiency of smart homes. The authors highlighted the significant benefits of smart homes in facilitating independent living for older adults, improving comfort through automation, and ensuring safety through advanced monitoring systems. However, the authors also addressed the challenges facing the widespread adoption of SHTs, including issues related to interoperability, privacy, and user acceptance. They suggested that future research should focus on developing user-centric designs, improving the robustness of communication systems, and ensuring the security of data to fully realize the potential of smart homes in enhancing the quality of life for their users.

Mallinson and Shafi [70] explored the interplay between SHT and collaborative governance in the context of public policy, underscoring both the potential benefits and challenges. They highlighted how SHT can improve the quality of life for aging populations by promoting independence and reducing long-term healthcare costs. However, the authors identified significant governance and policy challenges, particularly concerning the protection and use of vast amounts of sensitive data collected by smart home devices. The study proposed the use of a collaborative governance framework to better integrate public and private efforts in regulating and advancing SHT. This approach aims to address the multifaceted issues of data privacy, security, and ethical concerns while also promoting innovation and user-friendly technology development. By fostering greater engagement from policy scholars and stakeholders, Mallinson and Shafi argued that more effective strategies can be developed for governing this rapidly expanding sector, thus ensuring that SHTs are implemented in ways that are both beneficial and ethical.

Moraitou et al. [34] provided a systematic review of smart home care technologies designed for older adults and those with chronic diseases, emphasizing the technological, psychosocial, ethical, and economic challenges associated with their implementation. The authors explored various applications of these technologies, including physiological and functional monitoring, safety and security assistance, and social interaction enhancement. The study found that while SHTs have significant potential to improve the quality of life for these populations by enhancing health monitoring and daily living support, they also raise considerable challenges. Technologically, the integration of various devices and systems within a home environment requires advanced networking capabilities and standardization to ensure seamless functionality and user-friendly interaction. Psychosocially and ethically, the concerns about privacy, autonomy, and the potential reduction in human interaction highlight the need for careful consideration in technology design and policy formulation. Economically, the authors discussed the cost implications of SHTs, arguing that while they may reduce healthcare expenses by decreasing the need for in-person care services, the initial setup and maintenance costs can be significant. They advocated for a balance between cost and benefit, emphasizing the importance of designing systems that are both affordable and effective in meeting the health and social needs of older adults and chronic disease patients. The research underscored the complexity of implementing SHTs and the need for multidisciplinary approaches to address the diverse challenges they present.

Das et al. [51] presented a study on the design and implementation of a smart home system specifically tailored for older adults and disabled individuals, leveraging a network of sensors and IoT technology to enhance living conditions. Their research outlined how SHT not only supports daily activities through automation but also provides critical monitoring for health and safety. The system includes features such as environmental controls, security measures, and emergency responses, all controlled via a centralized software platform that ensures real-time monitoring and intervention when necessary. The study highlighted the integration of a variety of sensors, such as motion detectors and appliance usage monitors, which collectively help to maintain an environment that is both safe and responsive to the needs of its residents. The results showed potential for reducing the need for constant human supervision, thereby increasing independence for users and reducing overall care costs. However, the authors also discussed the challenges faced in implementing such systems, including technical barriers, the need for continuous updates, and the adaptation of the technology to individual homes. They emphasized the importance of user-friendly designs and the need for systems to be scalable and adaptable to different home environments. The research underscored the critical role of SHTs in improving the quality of life for older adults and disabled individuals, suggesting ongoing development and refinement are key to their broader adoption.

In the study by Deen [68], a comprehensive exploration of information and communications technologies (ICT) for ubiquitous healthcare within smart home environments for older adults was presented. The research highlighted the growing need for innovative healthcare solutions due to increasing life expectancy and the rising costs associated with traditional healthcare models, particularly in the context of an aging global population. Deen introduced several low-cost, non-invasive, and user-friendly sensor and actuator systems that are integrated into smart homes to address the pressing healthcare challenges faced by older populations. These technologies are designed to monitor health conditions in a non-intrusive manner, allowing older adults to continue living independently at home. Key functionalities included environmental sensing, health monitoring through wearable devices, and emergency management systems. The author discussed the potential of SHTs to significantly enhance the quality of life for older adults by improving safety, increasing independence, and reducing healthcare costs. These systems enable remote monitoring of vital signs and physical activity, potentially decreasing the need for frequent hospital visits. Additionally, the integration of intelligent computing and sensors facilitates the early detection of health deteriorations, allowing for prompt medical interventions. This research provided valuable insights into the integration of advanced ICT in smart homes, proposing a future where older individuals can receive comprehensive care within the comfort of their own homes, thereby addressing both economic and social challenges associated with aging populations.

Do et al. [71] presented a study on a Robot-Integrated Smart Home (RiSH) system, specifically designed to enhance care for older adults through advanced assistive technologies. This system integrates a home service robot with a home sensor network, body sensors, mobile devices, cloud servers, and remote caregivers, providing a multi-layered approach to support for older adults. Their research demonstrated the capability of RiSH to effectively recognize and track human body activity using inertial measurement units (IMU) and to perceive the environment through audio signal processing. The development of specific applications, such as human localization and tracking, as well as human activity recognition, extends the robot's perception beyond its onboard sensors, utilizing particle filter-based and Dynamic Bayesian Network-based methods. These applications achieved notable accuracies in tracking and activity recognition in their testbed experiments. The functionalities of the RiSH system include human activity recognition, which helps in monitoring daily activities and ensuring the well-being of older residents. It is also equipped for emergency detection and response, particularly geared toward detecting falls, thereby ensuring timely medical attention or intervention. Sound event detection utilizes sound-recognition technologies to understand various household sounds, playing a crucial role in activity recognition and environmental understanding. The integration with a home service robot allows it to act as a mobile sensor and actuator, able to interact with residents, provide companionship, and perform simple tasks. Additionally, environmental sensing monitors conditions, such as temperature and light, ensure the environment is safe and comfortable for older adults, while cloud connectivity facilitates remote monitoring and data-driven insights into the health and activities of the residents. These functionalities combine to create a comprehensive support system within the home, focusing on safety, monitoring, and interactive assistance. The high-level application of the system is designed to detect and respond to human falls, showing promising results in managing care for older adults more effectively. The experiments conducted with twelve human subjects validated the system's auditory perception services, human body activity recognition, position tracking, sound-based activity monitoring, and fall detection and rescue capabilities. The authors reported that the robot successfully recognized 37 human activities through sound events, with an average accuracy of 88% and detected falling sounds with an accuracy of 80% at the frame level, indicating robust operation of various components within the RiSH system.

In the study conducted by Torta et al. [64], the effectiveness of a small, socially assistive humanoid robot in smart home environments for care of older adults was evaluated. This research explored both short-term and long-term impacts of the robot on older adults' acceptance and emotional engagement, offering insights into its utility as a proactive interface in a smart home setting. The robot was tested in real-world scenarios within smart homes, aiming to provide multimodal communication channels and generate positive emotional responses among older adults. The study demonstrated that the robot was trusted by participants, who might engage in an emotional relationship with the technology over time, although perceived enjoyment decreased. Notably, the robot effectively recognized and tracked human activities using inertial measurement units and environmental perception through audio signal processing, which included capabilities such as human localization, activity recognition, and fall detection. The findings suggested that while the robot was generally well-received and trusted, the novelty and enjoyment factors might diminish over time, indicating the need for continuous adaptation and enhancement in functionalities to sustain user interest and emotional engagement.

Norell Pejner et al. [60] presented a mixed-methods study protocol for developing a smart home system designed to support information sharing, health assessments, and medication self-management for older adults. This protocol outlines a participatory design approach involving various stakeholders, including older adults, caregivers, and industrial partners to ensure the system meets the actual needs of its users. The smart home system, referred to as the Intelligent Age-Friendly Home (IAFH), is intended to evolve and adapt to the changing needs, preferences, and behaviors of its users. It integrates various components, such as health behavior assessments, communication of health data, and personalized medication reminders. These functionalities are supported by advanced home sensors and digitalized systems that facilitate accurate health monitoring and effective communication between care receivers and providers. The study's phased approach, from conceptualization to full-scale intervention, aims to iteratively develop and refine the system, with a strong focus on user participation and real-world applicability. This methodical progression ensures that the system not only supports medication management but also enhances overall healthcare decision-making through improved information accuracy and accessibility. The ultimate goal is to promote independence among older adults by allowing them to manage their medications more effectively and safely within their home environments.

Existing research underscores a significant focus on enhancing the quality of life for older adults through adaptive and user-friendly environments. Studies suggest a strong need to develop smart homes that are not only technologically advanced but also tailored to meet the diverse needs and preferences of older adults. These environments often integrate a variety of functionalities, from health monitoring and fall detection to medication management and emergency response, highlighting the versatility and potential of smart homes to support independent living. The research highlights several key areas of focus: the importance of designing interfaces that are easy to use and understand, the need for systems that can be customized to individual preferences, and the critical role of security and privacy in the adoption of SHTs. Additionally, the engagement of older adults in the design process is crucial for ensuring that these technologies are practical and relevant to their daily lives. Moreover, the studies reveal that while smart homes have the potential to significantly improve safety and autonomy for older adults, there are still challenges related to technology acceptance and integration. These challenges include overcoming technical complexities, ensuring adequate support and training for users, and addressing potential issues related to privacy and data security. Table 9 provides an overview of key focus areas and insights on smart home design and accessibility, highlighting inclusive and user-friendly solutions for the aging population.

Table 9. Key focus areas and insights on smart home design and accessibility.

Key Focus	Findings	Challenges	Recommendations	References
Categorization of older adults' needs	Identified five distinct groups with varying smart home functionality needs.	Variations in preferences and priorities among these groups.	Adopt a user-centered design approach to cater to diverse needs.	[30]
Enhancing connectivity and social engagement	Connectivity is crucial for quality of life, supporting social interaction and engagement.	Gaps in social engagement and leisure activities.	Design comprehensive living experiences that promote holistic connectivity.	[80,83]
Health monitoring, safety, and chronic disease management	SHTs improve safety and health monitoring, potentially reducing chronic condition exacerbations.	Lack of evidence on actual health outcomes, privacy, and security concerns.	Conduct large-scale studies, improve health monitoring systems, address ethical concerns.	[31,34,39,68]
User-friendly designs for aging in place	Enhances quality of life and independence, addresses specific needs and preferences.	Accessibility issues, technology acceptance, ethical concerns.	Engage users in the design process, ensure ongoing dialogue between stakeholders, create user-friendly designs.	[42,60,64,76]
Integration and evolution of smart home technologies	Advanced technologies support health management, safety, automation.	Interoperability, need for continuous updates, user adaptation.	Develop scalable, adaptable systems, improve communication and data security.	[48,51,71]
Policy and ethical considerations	Promotes independence, reduces healthcare costs, but raises data privacy and security issues.	Ethical concerns, lack of rigorous implementation studies, data governance challenges.	Implement collaborative governance framework, conduct high-quality research, balance costs and benefits.	[70]

4.4. Answers to Research Questions

4.4.1. How Can We Ensure the Accessibility of Innovative Home Technology for All Levels of Society, Regardless of Financial Stability?

Innovative home technology should be accessible to people of all socioeconomic classes, regardless of their financial stability. Most older adults expressed concerns about financial hurdles in the study by Elers et al. [59]. However, to achieve this goal, concrete steps must be taken. First, innovative home technology manufacturers should prioritize the development of a wide range of products at different price points to provide consumers of all budgets with access to affordable options [62]. In addition, manufacturers should explore partnerships with local governments or nonprofit organizations to offer subsidized or sponsored programs to financially disadvantaged individuals who cannot afford this technology. Governments and regulators should also actively promote the accessibility of innovative technologies by taking initiatives that make technology more accessible to all citizens. In most countries, many older adults receive very low or even below-average pensions. Older adults expressed concern about paying additional rent for SHT due to their already meager pension and high rental costs [84]. Those efforts could include subsidies or tax incentives for smart device buyers and setting up financial assistance programs for low-income and vulnerable people. Another critical action is education and awareness of the potential benefits of innovative home technology, as education plays a crucial role in ensuring affordability. Information programs and resources should be developed to educate people about the benefits and possibilities of innovative home systems. One action could be to encourage the recycling and refurbishment of smart devices to reduce costs and make them accessible to people with limited resources. Refurbished devices can also be a sustainable solution for reducing electronic waste. Providing internet access to rural areas and less developed communities is vital for intelligent technologies to be available across the country. The expansion of infrastructure and the provision of low-cost or free internet access contribute to the introduction of new technologies.

4.4.2. What Level of Support do Health Ministries Provide for Innovative Home Technology to Improve People's Quality of Life?

The level of support from health ministries for intelligent technologies, aimed at improving people's quality of life, varies by country and the specific policies implemented. According to Cao et al. [58], some countries recognize the potential of innovative technologies to enhance healthcare, which is reflected in strong support from health ministries. These countries often provide financial support, resources, and guidance for implementing smart homes within the healthcare system to achieve measurable user benefits. In such cases, health ministries can initiate pilot projects, engage in research, and collaborate with smart device manufacturers to develop programs and solutions focused on health and well-being. Increasingly, countries worldwide are realizing the benefits of smart homes and are actively working to implement them. Support from health ministries may include research funding, project implementation, training of health professionals, and raising awareness about the benefits of smart technologies in healthcare [37].

It should be noted that the availability of Ministry of Health support for smart homes is country-specific and may depend on factors such as healthcare sector priorities, available financial resources, and political will to implement technological innovations. Obaid et al. [55] illustrated that the level of support can vary within a single country, influenced by regional or local initiatives and policies. Despite these variations, health ministries, in general, are increasingly supporting smart homes, raising hopes that this technology will be increasingly used to improve people's quality of life, particularly in the context of healthcare.

4.4.3. How Actively do Health Ministries Promote the Use of Smart Home Technology to Reduce Pressure on Nursing Homes?

The degree to which ministries of health encourage the use of intelligent technologies to alleviate the burden on nursing homes and caregivers varies between countries, based on their specific circumstances. In some countries, health ministries actively support the use of intelligent technologies to enhance the quality of care for older adults and reduce the reliance on institutional care. The Netherlands, for example, has developed plans that include financial support, training on technology use, and fostering collaboration among stakeholders [38]. Health ministries promote the adoption of intelligent technologies through campaigns aimed at informing and educating the public, healthcare professionals, and older adults about the benefits and opportunities of smart homes. These campaigns utilize media channels, public events, and educational materials to raise awareness about how smart technologies can enhance the quality of life for older adults and facilitate operations in care homes.

In addition to promotional campaigns, health ministries may issue guidelines and recommendations for implementing smart technologies in care homes. These guidelines typically cover device selection, onboarding instructions, privacy, and system security [59]. They also provide information on training providers in the use of smart technologies and practices for successful healthcare implementation. Furthermore, health ministries can promote collaboration among healthcare sectors, technology companies, and suppliers, including financial support for purchasing smart devices.

It is important to note, however, that the extent to which ministries of health encourage the use of intelligent technologies can vary depending on each country's political will, financial resources, and priorities. Some countries may lack the resources or political support to actively promote and implement smart technology in nursing homes. Therefore, it is essential to monitor and compare approaches from different countries to identify best practices and opportunities for improving older adults' care through smart technologies.

4.4.4. To What Extent Do Wearable Sensors Need to Be Redesigned to Be Unobtrusive and Not Interfere with the Daily Lives of Users?

Integrating wearable sensors unobtrusively and seamlessly into users' daily lives requires a major overhaul focused on comfort, aesthetics, and practicality. Wearable sensors are essential to intelligent medical technology, providing valuable data on vital signs, activity, and other health parameters. According to a study by Ates et al. [67], the design of wearable sensors is crucial for achieving high user adoption and sustained usage. The design strategy includes several approaches: reducing size and weight to make the sensors less visible and more comfortable, using flexible and soft materials that adapt to the body to ensure comfort and minimize skin irritation, and integrating sensors into existing clothing or accessories, such as bracelets or watches, to increase their discretion.

One of the critical aspects in the redesign of wearable sensors is to consider the user's daily activities [77]. Sensors should be designed to not restrict the user's mobility and allow for everyday activities to be carried out without interference. It is also important that the sensors are moisture and dust resistant to withstand a wide range of conditions and environments. According to findings from Kachouie et al. [16], an improved design of wearable sensors that emphasizes discretion and convenience could be the key to increasing the acceptance and integration of this technology into users' daily lives. This approach allows users to receive helpful information about their health without experiencing any inconvenience or limitations. It also ensures that wearable sensors are acceptable to a wide range of users, including older adults and those with special needs.

One notable example that encapsulates the outlined design principles and addresses the user's daily activities is the development of the 'SmartFabric Sensor Shirt' [88]. Designed for older adults in smart homes, this shirt incorporates lightweight, flexible sensors seamlessly woven into the fabric. These sensors monitor vital signs such as heart rate and body temperature and detect falls, sending alerts to caregivers when anomalies occur. The shirt is moisture and dust resistant, suitable for various conditions, but its reliance on technology might reduce personal interaction with healthcare providers and present maintenance challenges due to regular washing. Similarly, the 'Flexi-Watch Health Monitor' [89] combines traditional watch aesthetics with health monitoring functionalities. It tracks heart rate, activity levels, and sleep patterns, encouraging routine wear due to its familiar design and ease of use. However, it has limitations in the range of health metrics monitored and requires frequent charging, which can be inconvenient. The 'VitaBand' [90] is a discreet wristband that monitors vital signs and physical activity while serving as a digital carrier for emergency contact information and medical IDs. It promotes continuous wear due to its unobtrusive design but raises data security concerns and relies heavily on smartphone connectivity, which can be problematic in areas with poor connectivity. 'Smart Socks' [91] are designed for diabetic patients, monitoring foot temperature to detect early signs of ulcers. These socks offer targeted preventative care and integrate seamlessly into daily life. However, they require meticulous maintenance to prevent sensor damage and may be cost-prohibitive for some users. The 'E-Textile Sleep Shirt' [92] monitors respiratory and heart rates during sleep using sensors embedded in the fabric. It offers non-invasive monitoring with comfortable materials, but the sensors require special care during washing and might cause discomfort for sensitive individuals.

Comparing these wearables, the common thread is their integration of health monitoring into everyday items, promoting seamless use. The SmartFabric Sensor Shirt and E-Textile Sleep Shirt provide comprehensive health monitoring without compromising daily activities or sleep comfort. The Flexi-Watch and VitaBand blend functionality with familiarity, reducing stigma and encouraging routine use. Smart Socks offer targeted care for diabetic patients, transforming a daily necessity into a health tool. However, each device faces challenges: the need for durability and maintenance in the SmartFabric Sensor Shirt and E-Textile Sleep Shirt, data security concerns with VitaBand, and the high maintenance and cost of Smart Socks.

These examples illustrate how the careful consideration of wearable sensor design, focusing on user comfort, discretion, and functionality, can lead to high user adoption and long-term usage, particularly among older adults who value both the practicality and unobtrusiveness of health monitoring solutions. The challenges presented, however, highlight the need for continued innovation and consideration in the development of such technologies to ensure they meet the needs of all potential users without adding undue complexity or dependency.

4.4.5. How Can We Create a Smart Home Application That Is Equally Adaptable to Every User?

To create an innovative home application that works equally well for every user, a significant overhaul is necessary to ensure the unobtrusive and seamless integration of wearable sensors into users' everyday lives, emphasizing convenience, aesthetics, and functionality [67]. As an essential component of intelligent medical technology, wearable sensors provide valuable data on vital signs, activity, and other health parameters. When developing such an application, the most crucial factor is an intuitive and user-friendly interface that allows easy management of smart home devices, regardless of the user's technical level. Studies have highlighted that a well-organized layout of functions, icons, and intuitive controls is associated with a better user experience [45].

Flexibility within the application is essential so it can adapt to each user's preferences. This adaptability should allow users to customize the layout, prioritize features, and adjust settings to meet their specific needs [29]. It can include personalizing controls, scenarios, and automation based on users' daily activities and habits. Moreover, the application must include features that cater to various user needs, including accessibility options for people with disabilities. This could involve support for alternative controls, such as voice commands or custom interfaces for users with special needs.

Regular collection of user feedback is critical. This feedback helps identify gaps, areas for improvement, and the potential for new features that users would like to see in the app. The application can be developed according to users' needs and preferences through the systematic analysis of these opinions. Engaging users in the smart home application development process is crucial for ensuring adaptability and user satisfaction [85].

By fostering continuous cooperation with users and actively involving them in the development process, a flexible, smart home application that suits every user can be created. This approach makes the app intuitive, personalized, and usable for a broad audience, thus enhancing the overall experience and acceptance of the smart home.

4.4.6. To What Extent Is It Necessary to Modify State Laws and Regulations to Align with Ethical Issues?

The ethical alignment of government laws and regulations is essential to ensure justice, protect individual rights, and promote social well-being. Depending on the complexity of the ethical issues involved, changes in laws and regulations may be necessary, especially in fields such as AI, biotechnology, and autonomous systems. Zhu et al. [65] highlighted the need to amend laws and regulations to ensure patient privacy, transparency in data collection and use, and fair distribution of benefits. As technology advances, it is crucial to adapt laws and regulations to meet new challenges.

According to Sánchez et al. [55], data protection reforms are needed to address ethical issues arising from the collection and processing of large amounts of data. This may include creating stricter rules on how data are collected, used, and shared, making data processing transparent, and strengthening individuals' rights over their data. It is also vital to ensure that technological advances are applied with respect for human rights, avoiding discrimination, and considering the safety and well-being of users. Moreover, openness and public participation in the legislative and regulatory processes are critical to ensuring compliance with ethical principles.

Laws and regulations must evolve to reflect the ethical implications of new technologies before their widespread adoption [79]. Citizens, experts, and organizations should have opportunities to express their opinions, participate in discussions, and contribute to policymaking. Through dialogue and collaboration, balanced laws and regulations that consider different perspectives and values can be enacted. It is crucial to continuously monitor and review existing laws to ensure they adequately regulate technological advances and remain up-to-date with innovations, research, and societal developments. Furthermore, cooperation between state institutions, experts in various fields, and the public is essential to create a compelling and fair legal framework.

Balancing government laws and regulations with ethical considerations is crucial to protect the interests of individuals and society amid rapid technological change. Legislative changes must ensure fairness, transparency, and respect for human rights while maintaining a balance between technological advances and ethical principles. This requires constant dialogue, the adjustment of regulations, and the promotion of openness and social participation.

4.4.7. How Can Smart Home Technology Address the Issue of Loneliness among Older Adults While Also Facilitating Their Daily Lives?

Innovative technology can provide significant support in nursing homes through sensors and activity monitoring systems. For example, motion detectors can track the activities of older adults and alert caregivers or healthcare professionals to unusual or emergency situations. Additionally, smart devices, such as smart thermostats, lighting, and security systems, can enhance the comfort and security of older adults in their homes. Studies have demonstrated that older adults utilize technology to connect with others, maintain social relationships, and overcome feelings of loneliness [72]. It is important to note, however, that smart technology cannot replace human interaction and social support.

Properly designed and adapted smart home apps can be a valuable tool that provides additional support to older people and simplifies everyday life. The research by Cotten et al. [50] highlighted the benefits of innovative technology in addressing the issue of loneliness among older people. By integrating technology that promotes social contact and facilitates everyday activities, a smart home can help reduce loneliness and improve the quality of life for older individuals.

To achieve maximum adaptability of smart home applications, it is essential to develop such technology in cooperation with users, taking into consideration their needs and opinions. Continuous improvement and updating of the application based on user experiences and feedback are essential to ensure adaptability and satisfaction.

Smart technology in the home can be a crucial support in addressing the loneliness of older people and facilitating their daily lives. Assistive technology enables them to connect with others and access information and services that would otherwise be unavailable [75]. Adequately designed and customized smart home applications, developed in collaboration with users and in compliance with ethical guidelines, can provide effective tools to maintain social connections, improve safety and comfort, and facilitate the daily activities of older adults.

4.4.8. How Can We Overcome the Stigma Associated with Older Adults and Introduce Them to Smart Home Technology?

Overcoming the stigma associated with older adults adopting smart home technology (SHT) requires careful customization and a sensitive approach to counteract negative perceptions and ensure acceptance. It is essential to educate older adults about the benefits and potential of innovative technologies, providing concrete examples of how these can enhance their quality of life and simplify daily activities. Suresh et al. [78] emphasized the importance of education and support in familiarizing older adults with the advantages of SHT. Through specialized approaches, it is crucial that older adults understand how the system can be personalized and tailored to their needs, with a focus on intuitive user interfaces.

Older adults should be aware that smart technology complements, rather than replaces, human interaction and serves as a supportive tool for facilitating everyday activities. Emphasizing the importance of privacy and data security is crucial for building trust when using smart technology. Organizing informational events can be beneficial for enhancing the acceptance of smart technologies among older adults. At these events, the details of SHT can be discussed more thoroughly, and testimonials from people who already use this technology can be shared.

Addressing why older people might feel stigmatized for owning smart homes requires understanding societal perceptions and age-related biases. Often, there is a misconception that older adults may not be technologically savvy or that they are too dependent on traditional ways of living. This can lead to feelings of inadequacy or embarrassment among older individuals when they adopt new technologies. To combat these feelings, it is important to showcase how smart homes are not just about technology but about enhancing personal autonomy and safety. Public discussions and media representation that positively reflect the successful integration of older adults with smart technology can also help to change these negative perceptions.

Furthermore, it is essential to explore how the physical and emotional comfort provided by smart homes can significantly reduce the social isolation often experienced by older adults. Smart homes enable older adults to maintain their independence while staying connected with their community and loved ones through advanced communication tools. This aspect should be highlighted to mitigate any feelings of shame associated with their living environment.

Working with health professionals can also play a crucial role in promoting smart technologies to enhance the well-being of older adults and reduce the burden on the social protection system. Peek et al. [69] highlighted the importance of accessibility and the integration of smart technologies for older adults. Adapting user interfaces and application designs to meet the specific needs of older adult users can make it easier for them to interact with technology and help break down stigma barriers. By fostering a community that understands and supports the use of smart technologies by older individuals, we can help ensure that these tools are seen as beneficial and dignifying enhancements to their lives, rather than sources of embarrassment or shame.

Table 10 summarizes the key points and actions suggested to address each research question.

Table 10. Summary of research questions and responses.

Research Question	Response Summary	References
How Can We Ensure the Accessibility of Innovative Home Technology for All Levels of Society, Regardless of Financial Stability?	To ensure accessibility, manufacturers should develop a range of affordable products, explore partnerships for subsidies, and governments should promote technology through subsidies, tax incentives, and financial assistance. Education on benefits, recycling, and providing internet access are also crucial steps.	[59,62,84]
What Level of Support do Health Ministries Provide for Innovative Home Technology to Improve People's Quality of Life?	Support varies by country. Some health ministries provide financial support, resources, and initiate pilot projects, while others offer research funding and training. The level of support depends on healthcare priorities, financial resources, and political will.	[37,55,58]

Table 10. *Cont.*

Research Question	Response Summary	References
How Actively do Health Ministries Promote the Use of Smart Home Technology to Reduce Pressure on Nursing Homes?	Promotion levels vary by country. Some health ministries actively support smart technology use through financial aid, training, public campaigns, and guidelines. The extent of promotion depends on political will, financial resources, and country-specific priorities.	[38,59]
To What Extent do Wearable Sensors Need to be Redesigned to be Unobtrusive and Not Interfere with the Daily Lives of Users?	Wearable sensors need redesigning for comfort, aesthetics, and practicality. They should be small, lightweight, flexible, moisture and dust resistant, and integrated into daily wearables, such as clothing or accessories, to ensure high adoption and usability.	[16,67,77]
How Can We Create a Smart Home Application that is Equally Adaptable to Every User?	An adaptable smart home application requires an intuitive, user-friendly interface, customizable features, accessibility options for disabilities, and continuous user feedback for improvements. Engagement with users in the development process is crucial for adaptability.	[29,45,67,85]
To What Extent is it Necessary to Modify State Laws and Regulations to Align with Ethical Issues?	Modifications are essential to address ethical issues in AI, biotechnology, and autonomous systems. This includes patient privacy, data transparency, and fair benefit distribution. Laws should evolve with technology, ensuring public participation and constant review to reflect ethical implications.	[55,65,79]
How Can Smart Home Technology Address the Issue of Loneliness Among Older Adults while also Facilitating Their Daily Lives?	Smart technology can support social connections and enhance daily life through sensors and smart devices. It should be designed in collaboration with users, regularly updated based on feedback, and comply with ethical guidelines to effectively reduce loneliness and improve quality of life.	[50,72,75]
How Can We Overcome the Stigma Associated with Older Adults and Introduce Them to Smart Home Technology?	Overcoming stigma requires education on benefits, focusing on personalization, emphasizing privacy and security, and involving health professionals. Public events, positive media representation, and community support are also essential to change negative perceptions and promote acceptance.	[69,78]

5. Recommendations for Enhancing Smart Home Technologies for Older Adults

This section presents a carefully curated series of strategic recommendations aimed at enhancing smart home technologies (SHTs) to precisely cater to the distinct needs of the aging population. As societies grapple with the increasing demands of supporting a growing number of older adults, SHTs emerge as vital tools to aid in their everyday living. However, achieving their full potential requires substantial advancements in design, accessibility, security, and seamless integration with current systems and policies. The objective of these recommendations is to cultivate an environment in which SHTs not only bolster the daily lives of older adults but also enrich their capacity to live independently and with dignity.

5.1. User-Centric Design and Customization

User-centric design in smart homes involves developing technologies that are accessible and intuitive for older adults. This requires understanding the physical and cognitive changes associated with aging, such as reduced mobility, vision impairment, and diminished hearing. Design features might include voice-activated controls, high-contrast visual interfaces, and tactile buttons for easier manipulation. Additionally, customization should allow users to tailor settings to their individual needs, such as adjusting the thermostat via voice commands or customizing alert sounds for better audibility. Furthermore, engaging older adults in the design process can significantly improve the end product. This can be achieved through participatory design workshops, where older adults can try out prototypes and provide feedback. Such involvement not only ensures the functionality of the technology aligns with their needs but also increases their comfort level and familiarity with the technology, thereby promoting greater acceptance and satisfaction.

5.2. *Cost-Effectiveness and Accessibility*

The high cost of SHTs can be a significant barrier to their adoption among older adults. Strategies to address this include developing subsidized pricing models through partnerships with government agencies or private organizations focused on senior welfare. Additionally, insurance companies could be encouraged to include SHTs in their coverage, especially those that contribute to preventive healthcare, which could ultimately reduce overall medical costs by enabling older adults to manage their health at home. Moreover, offering scalable solutions can provide basic access to SHTs with optional upgrades. This ensures that all older adults, regardless of income level, can access essential features, with the possibility to expand functionalities as their comfort with the technology grows or as their needs change. This approach not only makes SHT more accessible but also allows for greater flexibility in adapting to individual requirements over time.

5.3. *Enhanced Privacy and Security Measures*

Implementing robust privacy and security measures is crucial in smart home environments to protect sensitive personal data of older adults. This includes the use of advanced encryption methods to secure data transmission and storage, as well as regular updates to security protocols to guard against potential cyber threats. Privacy policies should be transparent and easily understandable, providing clear information on how data are collected, used, and shared. In addition to technical measures, it is important to foster a culture of privacy within organizations that handle personal data. This involves training employees on privacy practices and developing policies that prioritize user confidentiality. Users should have control over their data, including who can access them and how they are used, empowering them to manage their privacy settings according to their personal preferences and security needs.

5.4. *Interoperability and Standards*

Ensuring interoperability among smart home devices from different manufacturers can greatly enhance the user experience. This requires the adoption of universal standards and protocols that allow devices to communicate seamlessly, regardless of their brand or model. Standardization not only simplifies the setup and maintenance of smart home systems but also enhances their reliability and functionality. The development of open platforms where developers can create compatible applications and tools can further enrich the smart home ecosystem. Such platforms encourage innovation and competition, leading to the development of more advanced, user-friendly solutions. Promoting interoperability also helps in future-proofing investments in SHT, as newer devices can integrate smoothly with existing systems.

5.5. *Education and Training*

To mitigate fears and build confidence among older adults in using SHTs, comprehensive education and training programs are essential. These programs should include hands-on workshops that allow older adults to interact with the technologies in a guided setting, which can help demystify the technology and illustrate its practical benefits. Online tutorials and printed manuals designed for cognitive and visual clarity can also support learning at home. Furthermore, community centers and senior organizations can play a pivotal role in providing ongoing support and training. By offering regular courses and updates on new features and devices, these organizations can help older adults stay current with technology developments. Peer-led training sessions can also be particularly effective, as older adults may feel more comfortable learning from someone who understands their experiences and challenges.

5.6. Monitoring and Evaluation

Regular monitoring and evaluation are necessary to ensure that SHTs continue to meet the needs of older adults effectively. This involves not only technical assessments to ensure devices are functioning as intended but also user feedback mechanisms to gather insights on user satisfaction and areas for improvement. Continuous user feedback can be facilitated through digital surveys, direct interviews, and usability testing sessions. Evaluating the impact of these technologies on the quality of life for older adults is also crucial. This can be performed through longitudinal studies that assess changes in independence, health outcomes, and social engagement as a result of using SHTs. The findings from these studies can help in refining existing products and shaping future innovations in the field.

5.7. Ethical Considerations

Ethical considerations in the deployment of SHTs involve ensuring that these innovations enhance rather than undermine the autonomy of older adults. Technologies should be designed with the consent and input of the users, providing them with clear information about what the technology does and any associated risks. It is also vital to ensure that these technologies do not replace human care but rather supplement it, maintaining essential human contact. Ethical deployment also involves considering the inclusivity of technology. This means designing solutions that are accessible to all older adults, including those with disabilities or those who are not technologically savvy. Efforts should be made to ensure that no segment of the older adult population is left behind as SHTs become more prevalent.

5.8. Community and Social Integration

Integrating community and social functions into SHTs can help address issues of loneliness and social isolation among older adults. Technologies that facilitate easy communication with family and friends, community engagement platforms, and social media can help maintain strong social connections. Features such as simple video calling interfaces and social networking tools specifically designed for older users can enhance their social participation and engagement. Moreover, integrating technologies that enable participation in virtual community events or online classes can help older adults feel more connected to their communities. These technologies can open up new avenues for engagement and interaction, ensuring that older adults remain active participants in social life despite any physical limitations.

5.9. Legislative and Policy Advocacy

To effectively integrate SHTs into the lives of older adults, comprehensive legislative and policy frameworks are needed. These frameworks should support the development and deployment of these technologies by addressing issues such as funding, standards, and privacy regulations. Policymakers can influence the scalability and sustainability of smart home solutions through incentives for innovation and strict guidelines that protect end-users. It is essential for government bodies to play an active role in promoting SHTs as part of aging-in-place initiatives. This can include providing grants for research and development, offering tax incentives for companies and consumers, and creating pilot programs that test the efficacy and safety of these technologies in real-world settings. Developing clear policies that guide the collection, use, and management of data from smart home devices is crucial. Policies should also ensure that these technologies are accessible to all, regardless of socioeconomic status. This might involve setting price controls or providing these technologies through government-funded healthcare services. Establishing and enforcing standards can help ensure that SHTs are safe, reliable, and effective. Regulatory bodies can set requirements not just for the physical and software components of these technologies, but also for their ethical implications, such as ensuring they do not infringe on the rights or dignity of older adults. Raising awareness among legislators and the general public about the benefits and challenges of SHTs can foster a more supportive environment for their adoption. Advocacy groups can play a vital role in ensuring that the voices of older adults are heard in the policymaking process.

6. Implications

For technology developers and designers, this research emphasizes the imperative of integrating human-centered design principles specifically tailored to the capabilities and preferences of older adults. The nuanced understanding of older adults' interaction with technology necessitates designs that prioritize ease of use, intuitive interfaces, and minimalistic aesthetics to mitigate cognitive overload. Developers are encouraged to engage in participatory design processes, involving older adults and their caregivers in the developmental stages to ensure that the end products resonate with their actual needs and usage patterns. Furthermore, the findings advocate for the incorporation of adaptive technologies that can be customized to suit varying levels of physical ability and technical proficiency, ensuring broader accessibility. This approach not only enhances the user experience but also fosters a sense of ownership and comfort with the technology, ultimately leading to higher adoption rates.

Healthcare professionals and caregivers stand to significantly benefit from the integration of SHTs into older adult care routines. The implications for this stakeholder group involve a paradigm shift toward more proactive and preventive care models. By leveraging real-time data from smart homes, healthcare providers can monitor patients' health more continuously, allowing for timely interventions before acute episodes occur. However, this transition necessitates upskilling and continuous education to effectively interpret data from smart home devices and integrate this information into care plans. Moreover, caregivers must develop a nuanced understanding of the ethical considerations associated with surveillance and autonomy to navigate the delicate balance between oversight and intrusion, ensuring that the dignity and independence of older adults are upheld.

The research outlines a clear mandate for policymakers and regulatory bodies to create an enabling environment that fosters the widespread adoption of SHTs among older adults. This includes crafting policies that incentivize the development of affordable and accessible smart home solutions, ensuring that the benefits of such technologies are not confined to the economically affluent. Equally important is the establishment of stringent data protection and privacy regulations to safeguard the sensitive information collected by smart home devices. Policymakers must also consider the broader socioeconomic implications of this technological shift, including potential impacts on employment within the caregiving sector and the need for social programs that facilitate digital literacy among older adults.

For the academic and research community, this study underscores the rich interdisciplinary potential that the field of smart homes and older adult care presents. Researchers are encouraged to further investigate the socio-technical intersections of this domain, exploring the psychological, ethical, and social dimensions of technology adoption among older adults. There is a pressing need for longitudinal studies that assess the long-term impacts of smart home living on the well-being and autonomy of older adults, providing empirical evidence to guide future developments. Moreover, the academic community plays a crucial role in training the next generation of technologists, designers, and healthcare professionals, imbuing them with a holistic understanding of the aging process and the role technology can play in enhancing the quality of life in later years.

For older adult users and their families, the integration of SHTs promises a transformative impact on daily living and care. This research advocates for systems that empower older adults with greater independence and self-efficacy, reducing reliance on caregivers for routine tasks. Families can find solace in enhanced safety features and emergency response systems, ensuring that their loved ones are protected and supported, even in their absence. However, the adoption of such technologies also necessitates open dialogues within families about privacy, consent, and the boundaries of technological intervention, ensuring that the dignity and wishes of older adults are at the forefront of any technological integration.

Community and social services organizations play a pivotal role in bridging the gap between technological advancements and the everyday lives of older adults. The implications for these organizations include the development of educational programs and support services that facilitate older adults' understanding and use of SHTs. Collaborative efforts with technology developers can lead to the creation of pilot programs that demonstrate the tangible benefits of smart homes, serving as models for broader adoption. Furthermore, these organizations can advocate for policy changes that address the socioeconomic barriers to technology access, ensuring that the most vulnerable among older adults are not left behind in the digital divide.

7. Future Work Directions

Smart homes as caregivers for older adults is still a relatively new topic and, as such, requires further development. To address the gaps in the current literature, this section outlines key directions for future exploration and development, focusing on broadening the capabilities and effectiveness of SHTs in older adults' care.

7.1. Ethical Challenges

The current ethical challenges include cost-effectiveness, privacy, autonomy, informed consent, dignity, security, and trust [93]. Since these issues are not unique to SHT but prevalent in many other technologies available today, it is necessary to establish laws that appropriately address these ethical challenges. Future research could explore developing guidelines and regulations to protect older adults' rights and the moral acceptability of using smart homes. Additionally, future studies should focus on finding innovative technological solutions that provide security, privacy, and autonomy to older adults in smart homes. It is also essential to further investigate older adults' perceptions of ethical issues related to smart homes to better understand their perspectives and needs.

7.2. Knowledge and Acceptance

Several factors influence the readiness of older adults to embrace SHT, one of which is their general knowledge of the technology itself [94]. Future research should focus more on addressing the knowledge gap among older adults regarding SHT, as their lack of understanding hinders their everyday use of smart homes. It is necessary to devise various approaches to transfer this knowledge, such as personalized training, supporting materials, community engagement, and peer-to-peer learning. A training program evaluation should also be created to determine the extent of knowledge achieved by older adults through education.

7.3. Cost Barriers

Due to the high costs involved, only a small number of people have adopted automation that transforms their residences into smart homes [95]. Future research should focus on developing cost-effective solutions to make innovative home technology more accessible to older adults and gain wider acceptance. Considering how to obtain government support for launching financial schemes or subsidies is necessary. Additionally, exploring partnerships between technology companies and organizations for older adults should be investigated to offer more affordable packages and options, reducing user costs. Future research should also analyze the needs and preferences of older adults to develop targeted strategies for cost reduction in smart homes, including the development of adaptable systems that allow for the gradual adoption of technology according to users' financial capabilities.

7.4. Cultural Considerations

Many technologies in the design process do not account for cultural limitations [96]. Therefore, in future research, it is essential to focus on understanding the influence of cultural and geographical factors on the adoption of smart homes among older adults. To achieve this goal, in-depth cross-cultural research must analyze how different cultures impact the perception, acceptance, and use of SHT. What needs to be achieved in the future is the collaboration between cultural organizations and technology companies to tailor the development of smart homes to different cultures in different countries. Furthermore, there is a need to adapt the design of user interfaces to align with the specific needs, values, and preferences of older adults in different countries and cultures.

7.5. Wearable Sensors

Wearable sensors are directly or indirectly attached to the human body and provide continuous information [97]. However, there are challenges related to the comfort and practicality of wearing these sensors. In future research, it is crucial to focus on developing user-friendly wearable sensors that do not cause discomfort or limitations to the users. This includes improving the sensor design to reduce size, weight, and pain during wear. Furthermore, careful consideration should be given to privacy issues regarding data collected through wearable sensors, ensuring secure and protected processing and storage. Additionally, future research should explore integrating wearable sensors with other SHTs to provide comprehensive and personalized support for older adults. It is essential to create wearable sensors that are useful, easy to use, comfortable to wear, and socially and ethically aligned.

7.6. Healthcare Professional and Caregiver Perspectives

While our analysis has thoroughly examined the benefits and challenges smart homes present to older adults, it is evident that the perspectives of healthcare professionals and caregivers have not been as prominently featured. This oversight is significant, as these stakeholders play a crucial role in the adoption and successful integration of SHTs into the lives of older adults. Their viewpoints, informed by direct caregiving experience and medical expertise, are essential for understanding the practicality, ethical considerations, and potential impacts of these technologies on older adults' care. For instance, healthcare professionals can offer invaluable insights into how smart home features could be optimized for medical monitoring and emergency responses, potentially transforming the way older adults' care is administered. Similarly, caregivers' daily interactions with older adults place them in a unique position to assess the usability and effectiveness of smart homes from a practical standpoint. Their feedback can highlight areas for improvement, ensuring that smart homes are not only technologically advanced but also user-friendly and supportive of older adults' independence. To address the identified gap in our understanding of smart homes for older adults, future research should aim to investigate the perceptions of healthcare professionals and caregivers. It is recommended that subsequent studies undertake qualitative methodologies, such as interviews and focus groups, to capture the nuanced opinions of these key stakeholders regarding SHTs. The focus should be on elucidating the perceived benefits and concerns, as well as garnering suggestions for making SHTs more compatible with professional caregiving practices. The insights obtained from such studies are anticipated to be instrumental in refining smart home designs to better meet the needs of both older adults and their caregivers, ultimately fostering a more holistic approach to aging in place.

7.7. Theoretical Frameworks

To enrich the analysis within this systematic review, it is pertinent to incorporate an analytical lens provided by key theoretical frameworks, such as the Technology Acceptance Model (TAM), Innovation Diffusion Theory (IDT), and the Socio-Technical Systems (STS) perspective. These frameworks offer a multifaceted understanding of the adoption and impact of SHTs in older adults' care, thus enhancing the depth and comprehensiveness of the review. The TAM, as proposed by Davis [98], posits that the perceived usefulness and perceived ease of use are significant determinants of an individual's intention to use technology. In the context of smart homes for older adults, perceived usefulness might relate to the technologies' potential to enhance safety, health monitoring, and overall quality of life, while perceived ease of use underscores the importance of user-friendly interfaces and systems designed to accommodate the unique needs of older adults. This framework suggests future research directions aimed at understanding how these perceptions influence older adults' willingness to adopt SHTs and identifying design modifications that could improve usability and acceptance. Rogers' IDT [99] provides further insights into how new technologies are adopted within a community, focusing on attributes such as relative advantage, compatibility, complexity, trialability, and observability. Applying IDT to SHTs for older adults could elucidate factors such as the perceived benefits versus traditional older adult care methods, the compatibility of these technologies with older adults' lifestyles and values, and the role of observable benefits in promoting wider adoption among the older adult population. Lastly, the STS [100] perspective emphasizes the interplay between technology, individuals, and the societal context, highlighting the need to consider not only the technical features of smart homes but also the social environment in which they are implemented. This includes the support systems available to older adults, such as family, caregivers, and healthcare providers, and the importance of designing technologies that facilitate not only physical well-being but also social connectedness and psychological health. Incorporating these theoretical perspectives into the discussion on smart homes for older adults not only addresses the multifaceted nature of technology adoption but also aligns with the goal of creating inclusive, practical, and supportive environments for aging populations. Further exploration of these frameworks in future research could lead to more effective design, implementation, and utilization of SHTs, ultimately enhancing the lives of older adults.

7.8. Collaboration with Healthcare Agencies and Technology Companies

Healthcare organizations enable the exchange of knowledge and experiences and will be able to identify the critical needs of older adult patients living in smart homes. It is imperative to align protocols between healthcare systems and technology companies to enable rapid data exchange and system integration, ultimately establishing clear communication channels between healthcare workers and smart home users for timely decision-making when needed. Additionally, it is crucial to ensure secure data sharing among various systems and prioritize the needs of older adults.

8. Limitations

This study is subject to several limitations that merit consideration. First, the systematic literature review was inherently limited by the articles available for analysis. Studies not published in the searched databases or published outside the timeframe of 2010 to 2023 were not included, potentially omitting relevant innovations or findings. Secondly, the review predominantly relied on published academic literature, which may reflect publication bias favoring studies with positive results. This could skew the overall depiction of SHTs' effectiveness. Thirdly, while this research spans a significant temporal range, the rapid evolution of technology means that newer innovations may not be fully represented in the literature reviewed. Additionally, the scope of this review was restricted to articles written in English, which excludes potentially valuable research published in other languages. Finally, the study's focus on older adults may not capture the nuances of technology use among diverse subgroups within this demographic, such as varying levels of technology literacy or socioeconomic differences that could influence technology adoption and utilization.

9. Conclusions

This systematic literature review, covering a wide array of sources from 2010 to 2023, identified crucial advancements as well as substantial barriers that these technologies face in the context of older adults' care.

The potential of smart home technologies to improve safety, health monitoring, and the overall independence of older adults is evident. Innovations in IoT and integrated smart systems have demonstrated significant benefits, including enhanced ability to age in place, reduced need for institutional care, and better management of health conditions through continuous monitoring. These technologies not only support the physical health of older adults but also address psychological and social well-being by enabling connections with caregivers and loved ones, thus reducing feelings of isolation.

However, the review also highlighted several persistent challenges that need to be addressed to realize the full potential of these technologies. High costs and technical complexities make these solutions less accessible to a broader segment of the aging population. Privacy and ethical concerns, particularly related to data security and user autonomy, continue to pose significant hurdles. Moreover, the rapid pace of technological change means that findings related to the effectiveness and user acceptance of older systems may quickly become outdated, necessitating ongoing research and adaptation of recommendations.

The findings from this review also indicated a need for a more inclusive approach to technology development. Current smart home solutions often fail to fully consider the varied capabilities and preferences of older adult users, suggesting a gap in design practices that could be addressed by involving end-users more directly in the development process. Additionally, policy and regulation have not kept pace with technological advancements, indicating an urgent need for frameworks that better protect users while promoting innovation.

Educational initiatives aimed at both users and caregivers could enhance technology acceptance and utilization, ensuring that more older adults can benefit from these innovations. Future research should focus on longitudinal studies to assess the long-term impacts of smart home technologies on aging in place, and on developing scalable, cost-effective solutions that can be widely implemented.

While the adoption of smart home technologies in older adults' care presents notable advantages, the full integration of these systems into everyday use will require concerted efforts from researchers, developers, and policymakers. Only through sustained collaboration and commitment to addressing the outlined challenges can we ensure that smart home technologies effectively contribute to the care and well-being of the aging global population. This study sets a foundation for further exploration and development in the field, aiming to bridge the gap between technological potential and practical, everyday application for older adults.

Author Contributions: Conceptualization, A.V., H.Z. and T.O.; methodology, H.Z. and T.O.; validation, A.V., H.Z. and T.O.; investigation, A.V.; writing—original draft preparation, H.Z.; writing—review and editing, A.V. and T.O.; visualization, H.Z. and T.O.; supervision, T.O. All authors have read and agreed to the published version of the manuscript.

Funding: This research received no external funding.

Data Availability Statement: Not applicable.

Conflicts of Interest: The authors declare no conflicts of interest.

References

1. UN. UN Global Issue: Ageing. Available online: http://www.un.org/en/globalissues/ageing/ (accessed on 4 February 2023).
2. Eurostat. Eurostat Regional Yearbook 2017. Available online: https://ec.europa.eu/eurostat/web/products-flagship-publications/-/ks-ha-17-001 (accessed on 12 April 2023).
3. United States Census Bureau. Senior People. Available online: https://www.census.gov/topics/population/senior-aging.html (accessed on 23 March 2023).

4. Arif, M.J.; El Emary, I.M.M.; Koutsouris, D.-D. A Review on the Technologies and Services Used in the Self-Management of Health and Independent Living of Elderly. *Technol. Health Care* **2014**, *22*, 677–687. [CrossRef] [PubMed]
5. Liu, L.; Stroulia, E.; Nikolaidis, I.; Miguel-Cruz, A.; Rios Rincon, A. Smart Homes and Home Health Monitoring Technologies for Older Adults: A Systematic Review. *Int. J. Med. Inform.* **2016**, *91*, 44–59. [CrossRef]
6. Aldrich, F.K. Smart Homes: Past, Present and Future. In *Inside the Smart Home*; Harper, R., Ed.; Springer: London, UK, 2003; pp. 17–39.
7. Ding, D.; Cooper, R.A.; Pasquina, P.F.; Fici-Pasquina, L. Sensor Technology for Smart Homes. *Maturitas* **2011**, *69*, 131–136. [CrossRef] [PubMed]
8. Latikka, R.; Rubio-Hernández, R.; Lohan, E.S.; Rantala, J.; Nieto Fernández, F.; Laitinen, A.; Oksanen, A. Older Adults' Loneliness, Social Isolation, and Physical Information and Communication Technology in the Era of Ambient Assisted Living: A Systematic Literature Review. *J. Med. Internet Res.* **2021**, *23*, e28022. [CrossRef] [PubMed]
9. Ghafurian, M.; Wang, K.; Dhode, I.; Kapoor, M.; Morita, P.P.; Dautenhahn, K. Smart Home Devices for Supporting Older Adults: A Systematic Review. *IEEE Access* **2023**, *11*, 47137–47158. [CrossRef]
10. Zulas, A.L.; Crandall, A.S.; Schmitter-Edgecombe, M. Caregiver Needs from Elder Care Assistive Smart Homes: Children of Elder Adults Assessment. In Proceedings of the Human Factors and Ergonomics Society Annual Meeting, Chicago, IL, USA, 27–31 October 2014; Volume 58, pp. 634–638.
11. Zulas, A.L.; Crandall, A.S.; Schmitter-Edgecombe, M.; Cook, D.J. Caregiver Needs from Elder Care Assistive Smart Homes: Nursing Assessment. In Proceedings of the Human Factors and Ergonomics Society Annual Meeting, Boston, MA, USA, 22–26 October 2012; Volume 56, pp. 125–129.
12. Ma, C.; Guerra-Santin, O.; Mohammadi, M. Smart Home Modification Design Strategies for Ageing in Place: A Systematic Review. *J. Hous. Built Environ.* **2022**, *37*, 625–651. [CrossRef]
13. Sumner, J.; Chong, L.S.; Bundele, A.; Wei Lim, Y. Co-Designing Technology for Aging in Place: A Systematic Review. *Gerontologist* **2021**, *61*, e395–e409. [CrossRef]
14. Zielonka, A.; Woźniak, M.; Garg, S.; Kaddoum, G.; Piran, M.J.; Muhammad, G. Smart Homes: How Much Will They Support Us? A Research on Recent Trends and Advances. *IEEE Access* **2021**, *9*, 26388–26419. [CrossRef]
15. Graybill, E.M.; McMeekin, P.; Wildman, J. Can Aging in Place Be Cost Effective? A Systematic Review. *PLoS ONE* **2014**, *9*, e102705. [CrossRef]
16. Kachouie, R.; Sedighadeli, S.; Abkenar, A.B. The Role of Socially Assistive Robots in Elderly Wellbeing: A Systematic Review. In *Cross-Cultural Design, Proceedings of the 9th International Conference, CCD 2017, Held as Part of HCI International 2017, Vancouver, BC, Canada, 9–14 July 2017*; Rau, P.-L.P., Ed.; Springer International Publishing: Cham, Switzerland, 2017; pp. 669–682.
17. Ollevier, A.; Aguiar, G.; Palomino, M.; Simpelaere, I.S. How Can Technology Support Ageing in Place in Healthy Older Adults? A Systematic Review. *Public Health Rev.* **2020**, *41*, 26. [CrossRef]
18. Solaimani, S.; Keijzer-Broers, W.; Bouwman, H. What We Do—And Don't—Know about the Smart Home: An Analysis of the Smart Home Literature. *Indoor Built Environ.* **2015**, *24*, 370–383. [CrossRef]
19. Chadborn, N.H.; Blair, K.; Creswick, H.; Hughes, N.; Dowthwaite, L.; Adenekan, O.; Pérez Vallejos, E. Citizens' Juries: When Older Adults Deliberate on the Benefits and Risks of Smart Health and Smart Homes. *Healthcare* **2019**, *7*, 54. [CrossRef]
20. Mashal, I.; Shuhaiber, A.; AL-Khatib, A.W. User Acceptance and Adoption of Smart Homes: A Decade Long Systematic Literature Review. *Int. J. Data Netw. Sci.* **2023**, *7*, 533–552. [CrossRef]
21. Wang, R.H.; Tannou, T.; Bier, N.; Couture, M.; Aubry, R. Proactive and Ongoing Analysis and Management of Ethical Concerns in the Development, Evaluation, and Implementation of Smart Homes for Older Adults With Frailty. *JMIR Aging* **2023**, *6*, e41322. [CrossRef]
22. Felber, N.A.; Tian, Y.J.A.; Pageau, F.; Elger, B.S.; Wangmo, T. Mapping Ethical Issues in the Use of Smart Home Health Technologies to Care for Older Persons: A Systematic Review. *BMC Med. Ethics* **2023**, *24*, 24. [CrossRef]
23. Zwijsen, S.A.; Niemeijer, A.R.; Hertogh, C.M.P.M. Ethics of Using Assistive Technology in the Care for Community-Dwelling Elderly People: An Overview of the Literature. *Aging Ment. Health* **2011**, *15*, 419–427. [CrossRef]
24. Gochoo, M.; Alnajjar, F.; Tan, T.-H.; Khalid, S. Towards Privacy-Preserved Aging in Place: A Systematic Review. *Sensors* **2021**, *21*, 3082. [CrossRef]
25. Von Schomberg, R. A Vision of Responsible Research and Innovation. In *Responsible Innovation*; Owen, R., Bessant, J., Heintz, M., Eds.; Wiley: Hoboken, NJ, USA, 2013; pp. 51–74.
26. Kitchenham, B.; Charters, S.M. *Guidelines for Performing Systematic Literature Reviews in Software Engineering. EBSE Technical Report*; Software Engineering Group, School of Computer Science and Mathematics, Keele University and Department of Computer Science, University of Durham: Durham, UK, 2007.
27. Page, M.J.; McKenzie, J.E.; Bossuyt, P.M.; Boutron, I.; Hoffmann, T.C.; Mulrow, C.D.; Shamseer, L.; Tetzlaff, J.M.; Akl, E.A.; Brennan, S.E.; et al. The PRISMA 2020 statement: An updated guideline for reporting systematic reviews. *BMJ* **2021**, *372*, n71. [CrossRef]
28. Robles, R.J.; Kim, T.-H. A Review on Security in Smart Home Development. *Int. J. Adv. Sci. Technol.* **2010**, *15*, 13–22.
29. Robles, R.J.; Kim, T.-H. Applications, Systems and Methods in Smart Home Technology: A Review. *Int. J. Adv. Sci. Technol.* **2010**, *15*, 37–48.

30. Yu, J.; de Antonio, A.; Villalba-Mora, E. Older Adult Segmentation According to Residentially-Based Lifestyles and Analysis of Their Needs for Smart Home Functions. *Int. J. Environ. Res. Public Health* **2020**, *17*, 8492. [CrossRef] [PubMed]
31. Facchinetti, G.; Petrucci, G.; Albanesi, B.; De Marinis, M.G.; Piredda, M. Can Smart Home Technologies Help Older Adults Manage Their Chronic Condition? A Systematic Literature Review. *Int. J. Environ. Res. Public Health* **2023**, *20*, 1205. [CrossRef] [PubMed]
32. Majumder, S.; Aghayi, E.; Noferesti, M.; Memarzadeh-Tehran, H.; Mondal, T.; Pang, Z.; Deen, M.J. Smart Homes for Elderly Healthcare-Recent Advances and Research Challenges. *Sensors* **2017**, *17*, 2496. [CrossRef] [PubMed]
33. Borelli, E.; Paolini, G.; Antoniazzi, F.; Barbiroli, M.; Benassi, F.; Chesani, F.; Chiari, L.; Fantini, M.; Fuschini, F.; Galassi, A.; et al. HABITAT: An IoT Solution for Independent Elderly. *Sensors* **2019**, *19*, 1258. [CrossRef] [PubMed]
34. Moraitou, M.; Pateli, A.; Fotiou, S. Smart Health Caring Home: A Systematic Review of Smart Home Care for Elders and Chronic Disease Patients. *Adv. Exp. Med. Biol.* **2017**, *989*, 255–264. [PubMed]
35. Chung, J.; Demiris, G.; Thompson, H.J. Ethical Considerations Regarding the Use of Smart Home Technologies for Older Adults: An Integrative Review. *Annu. Rev. Nurs. Res.* **2016**, *34*, 155–181. [CrossRef] [PubMed]
36. Zhao, Y.; Sazlina, S.-G.; Rokhani, F.Z.; Su, J.; Chew, B.-H. The Expectations and Acceptability of a Smart Nursing Home Model Among Chinese Older Adults and Family Members: A Qualitative Study. *Asian Nurs. Res.* **2023**, *17*, 208–218. [CrossRef] [PubMed]
37. Raja, M.; Bjerkan, J.; Kymre, I.G.; Galvin, K.T.; Uhrenfeldt, L. Telehealth and Digital Developments in Society That Persons 75 Years and Older in European Countries Have Been Part of: A Scoping Review. *BMC Health Serv. Res.* **2021**, *21*, 1157. [CrossRef]
38. Aloulou, H.; Mokhtari, M.; Tiberghien, T.; Biswas, J.; Phua, C.; Kenneth Lin, J.H.; Yap, P. Deployment of Assistive Living Technology in a Nursing Home Environment: Methods and Lessons Learned. *BMC Med. Inform. Decis. Mak.* **2013**, *13*, 42. [CrossRef]
39. Humphreys, C. Can Smart Home Technology Support Older People? *Br. J. Community Nurs.* **2018**, *23*, 435–438. [CrossRef]
40. Peetoom, K.K.B.; Lexis, M.A.S.; Joore, M.; Dirksen, C.D.; De Witte, L.P. Literature Review on Monitoring Technologies and Their Outcomes in Independently Living Elderly People. *Disabil. Rehabil. Assist. Technol.* **2015**, *10*, 271–294. [CrossRef] [PubMed]
41. Lee, L.N.; Kim, M.J. A Critical Review of Smart Residential Environments for Older Adults With a Focus on Pleasurable Experience. *Front. Psychol.* **2020**, *10*, 504641. [CrossRef] [PubMed]
42. Lê, Q.; Nguyen, H.B.; Barnett, T. Smart Homes for Older People: Positive Aging in a Digital World. *Future Internet* **2012**, *4*, 607–617. [CrossRef]
43. Liu, P.; Li, G.; Jiang, S.; Liu, Y.; Leng, M.; Zhao, J.; Wang, S.; Meng, X.; Shang, B.; Chen, L.; et al. The Effect of Smart Homes on Older Adults with Chronic Conditions: A Systematic Review and Meta-Analysis. *Geriatr. Nurs.* **2019**, *40*, 522–530. [CrossRef]
44. Tural, E.; Lu, D.; Austin Cole, D. Safely and Actively Aging in Place: Older Adults' Attitudes and Intentions Toward Smart Home Technologies. *Gerontol. Geriatr. Med.* **2021**, *7*, 233372142110173. [CrossRef] [PubMed]
45. Maswadi, K.; Ghani, N.B.A.; Hamid, S.B. Systematic Literature Review of Smart Home Monitoring Technologies Based on IoT for the Elderly. *IEEE Access* **2020**, *8*, 92244–92261. [CrossRef]
46. Agoulmine, N.; Deen, M.J.; Lee, J.-S.; Meyyappan, M. U-Health Smart Home. *IEEE Nanotechnol. Mag.* **2011**, *5*, 6–11. [CrossRef]
47. Brulin, D.; Benezeth, Y.; Courtial, E. Posture Recognition Based on Fuzzy Logic for Home Monitoring of the Elderly. *IEEE Trans. Inf. Technol. Biomed.* **2012**, *16*, 974–982. [CrossRef]
48. Alam, M.R.; Reaz, M.B.I.; Ali, M.A.M. A Review of Smart Homes—Past, Present, and Future. *IEEE Trans. Syst. Man Cybern. Part C (Appl. Rev.)* **2012**, *42*, 1190–1203. [CrossRef]
49. Pietrzak, E.; Cotea, C.; Pullman, S. Does Smart Home Technology Prevent Falls in Community-Dwelling Older Adults: A Literature Review. *Inform. Prim. Care* **2014**, *21*, 105–112. [CrossRef]
50. Howe Iii, E.G.; Elenberg, F. Ethical Challenges Posed by Big Data. *Innov. Clin. Neurosci.* **2020**, *17*, 24–30.
51. Das, R.; Tuna, G.; Tuna, A. Design and Implementation of a Smart Home for the Elderly and Disabled. *Int. J. Comput. Netw. Appl.* **2015**, *2*, 242–246.
52. Ghorayeb, A.; Comber, R.; Gooberman-Hill, R. Older Adults' Perspectives of Smart Home Technology: Are We Developing the Technology That Older People Want? *Int. J. Hum.-Comput. Stud.* **2021**, *147*, 102571. [CrossRef]
53. Yusif, S.; Soar, J.; Hafeez-Baig, A. Older People, Assistive Technologies, and the Barriers to Adoption: A Systematic Review. *Int. J. Med. Inform.* **2016**, *94*, 112–116. [CrossRef]
54. Turjamaa, R.; Pehkonen, A.; Kangasniemi, M. How Smart Homes Are Used to Support Older People: An Integrative Review. *Int. J. Older People Nurs.* **2019**, *14*, e12260. [CrossRef]
55. Sánchez, V.G.; Taylor, I.; Bing-Jonsson, P.C. Ethics of Smart House Welfare Technology for Senior Adults: A Systematic Literature Review. *Int. J. Technol. Assess. Health Care* **2017**, *33*, 691–699. [CrossRef]
56. Obaid, T.; Rashed, H.; El Nour, A.A.; Rehan, M.; Saleh, M.M.; Tarique, M. Zigbee Based Voice Controlled Wireless Smart Home System. *Int. J. Wirel. Mob. Netw.* **2014**, *6*, 47–59. [CrossRef]
57. Sokullu, R.; Akkaş, M.A.; Demir, E. IoT Supported Smart Home for the Elderly. *Internet Things* **2020**, *11*, 100239. [CrossRef]
58. Cao, Y.; Erdt, M.; Robert, C.; Naharudin, N.B.; Lee, S.Q.; Theng, Y.-L. Decision-Making Factors Toward the Adoption of Smart Home Sensors by Older Adults in Singapore: Mixed Methods Study. *JMIR Aging* **2022**, *5*, e34239. [CrossRef]
59. Elers, P.; Hunter, I.; Whiddett, D.; Lockhart, C.; Guesgen, H.; Singh, A. User Requirements for Technology to Assist Aging in Place: Qualitative Study of Older People and Their Informal Support Networks. *JMIR Mhealth Uhealth* **2018**, *6*, e10741. [CrossRef]

60. Norell Pejner, M.; Ourique De Morais, W.; Lundström, J.; Laurell, H.; Skärsäter, I. A Smart Home System for Information Sharing, Health Assessments, and Medication Self-Management for Older People: Protocol for a Mixed-Methods Study. *JMIR Res. Protoc.* **2019**, *8*, e12447. [CrossRef]
61. Dermody, G.; Fritz, R.; Glass, C.; Dunham, M.; Whitehead, L. Factors Influencing Community-Dwelling Older Adults' Readiness to Adopt Smart Home Technology: A Qualitative Exploratory Study. *J. Adv. Nurs.* **2021**, *77*, 4847–4861. [CrossRef]
62. Chakraborty, A.; Islam, M.; Shahriyar, F.; Islam, S.; Zaman, H.U.; Hasan, M. Smart Home System: A Comprehensive Review. *J. Electr. Comput. Eng.* **2023**, *2023*, e7616683. [CrossRef]
63. Kim, K.; Jalal, A.; Mahmood, M. Vision-Based Human Activity Recognition System Using Depth Silhouettes: A Smart Home System for Monitoring the Residents. *J. Electr. Eng. Technol.* **2019**, *14*, 2567–2573. [CrossRef]
64. Torta, E.; Werner, F.; Johnson, D.O.; Juola, J.F.; Cuijpers, R.H.; Bazzani, M.; Oberzaucher, J.; Lemberger, J.; Lewy, H.; Bregman, J. Evaluation of a Small Socially-Assistive Humanoid Robot in Intelligent Homes for the Care of the Elderly. *J. Intell. Robot. Syst.* **2014**, *76*, 57–71. [CrossRef]
65. Zhu, J.; Shi, K.; Yang, C.; Niu, Y.; Zeng, Y.; Zhang, N.; Liu, T.; Chu, C.H. Ethical Issues of Smart Home-Based Elderly Care: A Scoping Review. *J. Nurs. Manag.* **2022**, *30*, 3686–3699. [CrossRef]
66. Mohammed, M.N.; Desyansah, S.F.; Al-Zubaidi, S.; Yusuf, E. An Internet of Things-Based Smart Homes and Healthcare Monitoring and Management System: Review. *J. Phys. Conf. Ser.* **2020**, *1450*, 012079. [CrossRef]
67. Ates, H.C.; Nguyen, P.Q.; Gonzalez-Macia, L.; Morales-Narváez, E.; Güder, F.; Collins, J.J.; Dincer, C. End-to-End Design of Wearable Sensors. *Nat. Rev. Mater.* **2022**, *7*, 887–907. [CrossRef]
68. Deen, M.J. Information and Communications Technologies for Elderly Ubiquitous Healthcare in a Smart Home. *Pers. Ubiquit. Comput.* **2015**, *19*, 573–599. [CrossRef]
69. Aggarwal, B.; Xiong, Q.; Schroeder-Butterfill, E. Impact of the Use of the Internet on Quality of Life in Older Adults: Review of Literature. *Prim. Health Care Res. Dev.* **2020**, *21*, e55. [CrossRef]
70. Mallinson, D.J.; Shafi, S. Smart Home Technology: Challenges and Opportunities for Collaborative Governance and Policy Research. *Rev. Policy Res.* **2022**, *39*, 330–352. [CrossRef]
71. Do, H.M.; Pham, M.; Sheng, W.; Yang, D.; Liu, M. RiSH: A Robot-Integrated Smart Home for Elderly Care. *Robot. Auton. Syst.* **2018**, *101*, 74–92. [CrossRef]
72. Marikyan, D.; Papagiannidis, S.; Alamanos, E. A Systematic Review of the Smart Home Literature: A User Perspective. *Technol. Forecast. Soc. Chang.* **2019**, *138*, 139–154. [CrossRef]
73. Neßelrath, R.; Lu, C.; Schulz, C.H.; Frey, J.; Alexandersson, J. A Gesture Based System for Context—Sensitive Interaction with Smart Homes. In *Ambient Assisted Living: 4. AAL-Kongress 2011, Berlin, Germany, January 25–26, 2011*; Wichert, R., Eberhardt, B., Eds.; Springer: Berlin/Heidelberg, Germany, 2011; pp. 209–219.
74. Kim, J.; Choi, H.; Wang, H.; Agoulmine, N.; Deerv, M.J.; Hong, J.W.-K. POSTECH's U-Health Smart Home for Elderly Monitoring and Support. In Proceedings of the 2010 IEEE International Symposium on "A World of Wireless, Mobile and Multimedia Networks" (WoWMoM), Montreal, QC, Canada, 14–17 June 2010; pp. 1–6.
75. Pal, D.; Triyason, T.; Funikul, S. Smart Homes and Quality of Life for the Elderly: A Systematic Review. In Proceedings of the 2017 IEEE International Symposium on Multimedia (ISM), Taichung, Taiwan, 11–13 December 2017; pp. 413–419.
76. Miller, J.; McDaniel, T.; Bernstein, M.J. Aging in Smart Environments for Independence. In Proceedings of the 2020 IEEE International Symposium on Technology and Society (ISTAS), Tempe, AZ, USA, 12–15 November 2020; pp. 115–123.
77. Bal, M.; Shen, W.; Hao, Q.; Xue, H. Collaborative Smart Home Technologies for Senior Independent Living: A Review. In Proceedings of the 15th International Conference on Computer Supported Cooperative Work in Design (CSCWD), Laussane, Switzerland, 8–10 June 2011; pp. 481–488.
78. Suresh, S.; Sruthi, P.V. A Review on Smart Home Technology. In Proceedings of the 2015 Online International Conference on Green Engineering and Technologies (IC-GET), Coimbatore, India, 27 November 2015; pp. 1–3.
79. Kolkowska, E. Privacy Principles in Design of Smart Homes Systems in Elderly Care. In *Human Aspects of Information Security, Privacy, and Trust, Proceedings of the Third International Conference, HAS 2015, Held as Part of HCI International 2015, Los Angeles, CA, USA, 2–7 August 2015*; Tryfonas, T., Askoxylakis, I., Eds.; Springer International Publishing: Cham, Switzerland, 2015; pp. 526–537.
80. Liu, Y.; Tamura, R. How Can Smart Home Help "New Elders" Aging in Place and Building Connectivity. In Proceedings of the 16th International Conference on Intelligent Environments (IE), Madrid, Spain, 20–23 July 2020; pp. 100–107.
81. Doyle, J.; Caprani, N.; Bond, R. Older Adults' Attitudes to Self-Management of Health and Wellness through Smart Home Data. In Proceedings of the 2015 9th International Conference on Pervasive Computing Technologies for Healthcare (PervasiveHealth), Istanbul, Turkey, 20–23 May 2015; pp. 129–136.
82. Chernbumroong, S.; Atkins, A.; Yu, H. Perception of Smart Home Technologies to Assist Elderly People. In Proceedings of the 4th International Conference on Software, Knowledge, Information Management and Applications, Paro, Bhutan, 25–27 August 2010; pp. 90–97.
83. Kon, B.; Lam, A.; Chan, J. Evolution of Smart Homes for the Elderly. In Proceedings of the 26th International Conference on World Wide Web Companion, Perth, Australia, 3–7 April 2017; International World Wide Web Conferences Steering Committee: Geneva, Switzerland, 2017; pp. 1095–1101.

84. Gudmundsson, M. Aging-in-Place with ICT: A Qualitative Study of Senior Citizen Users' Perception and Acceptance towards Smart Home Technology. Master's Thesis, Linnaeus University, Växjö, Sweden, 2017.
85. Boyner, E.; Brodin, J. Smart Homes: Design & Development of an Application from a User Centered Perspective. Master's Thesis, Department Design Sciences, Faculty of Engineering LTH, Lund University, Lund, Switzerland, 2021.
86. *ISO/IEC 24752:2008*; Information Technology—User Interfaces—Universal Remote Console. ISO: Geneva, Switzerland, 2008.
87. Venkatesh, V.; Thong, J.Y.; Xu, X. Consumer Acceptance and Use of Information Technology: Extending the Unified Theory of Acceptance and Use of Technology. *MIS Q.* **2012**, *36*, 157. [CrossRef]
88. Yang, K.; Isaia, B.; Brown, L.J.E.; Beeby, S. E-Textiles for Healthy Ageing. *Sensors* **2019**, *19*, 4463. [CrossRef]
89. Dai, B.; Gao, C.; Xie, Y. Flexible Wearable Devices for Intelligent Health Monitoring. *View* **2022**, *3*, 20220027. [CrossRef]
90. McHugh, S.; Yarmey, K. Health Care. In *Near Field Communication: Recent Developments and Library Implications*; McHugh, S., Yarmey, K., Eds.; Springer International Publishing: Cham, Switzerland, 2014; pp. 39–44.
91. Najafi, B.; Mohseni, H.; Grewal, G.S.; Talal, T.K.; Menzies, R.A.; Armstrong, D.G. An Optical-Fiber-Based Smart Textile (Smart Socks) to Manage Biomechanical Risk Factors Associated With Diabetic Foot Amputation. *J. Diabetes Sci. Technol.* **2017**, *11*, 668–677. [CrossRef]
92. Shuvo, I.I.; Hoque, M.S.; Shadhin, M.; Ador, A.I. Intelligent Textiles for Monitoring Physiological Parameters. In *Smart Textiles from Natural Resources*; Mondal, M.I.H., Ed.; The Textile Institute Book Series; Woodhead Publishing: Sawston, UK, 2024; pp. 641–686.
93. Peek, S.T.M.; Luijkx, K.G.; Rijnaard, M.D.; Nieboer, M.E.; van der Voort, C.S.; Aarts, S.; van Hoof, J.; Vrijhoef, H.J.M.; Wouters, E.J.M. Older Adults' Reasons for Using Technology While Aging in Place. *Gerontology* **2016**, *62*, 226–237. [CrossRef]
94. Chang, V.; Wang, Z.; Xu, Q.; Golightly, L.; Liu, B.; Arami, M. Smart Home Based on Internet of Things and Ethical Issues. In Proceedings of the 3rd International Conference on Finance, Economics, Management and IT Business, Online Streaming, 25–26 April 2021; SCITEPRESS—Science and Technology Publications: Setúbal, Portugal, 2021; pp. 57–64.
95. Pirzada, P.; Wilde, A.; Doherty, G.H.; Harris-Birtill, D. Ethics and Acceptance of Smart Homes for Older Adults. *Inform. Health Soc. Care* **2022**, *47*, 10–37. [CrossRef]
96. Hanson, V.L. Age and Web Access: The Next Generation. In Proceedings of the 2009 International Cross-Disciplinary Conference on Web Accessibililty (W4A), Madrid, Spain, 20–21 April 2009; ACM: New York, NY, USA, 2009; pp. 7–15.
97. Ni, Q.; García Hernando, A.B.; De la Cruz, I.P. The Elderly's Independent Living in Smart Homes: A Characterization of Activities and Sensing Infrastructure Survey to Facilitate Services Development. *Sensors* **2015**, *15*, 11312–11362. [CrossRef]
98. Davis, F.D. Perceived Usefulness, Perceived Ease of Use, and User Acceptance of Information Technology. *MIS Q.* **1989**, *13*, 319–340. [CrossRef]
99. Rogers, E.M. *Diffusion of Innovations*; The Free Press: New York, NY, USA, 1962.
100. Trist, E. *The Evolution of Socio-Technical Systems: A Conceptual Framework and an Action Research Program*; Ontario Quality of Working Life Centre: Toronto, ON, Canada, 1981.

smart cities

MDPI

Project Report

Inclusive MicroMob: Enhancing Urban Mobility Through Micromobility Solutions

Annalisa Rollandi [1,*,†], Michela Papandrea [2,†], Filippo Bignami [3], Laura Di Maggio [2], Felix Günther [1], Andrea Quattrini [2], Luca Minardi [2], Michele Cocca [2] and Albedo Bettini [4]

1. Institute of Earth Sciences, Department of Environment Constructions and Design, University of Applies Sciences and Arts of Southern Switzerland, CH-6850 Mendrisio, Switzerland; felix.guenther@supsi.ch
2. Institute of Information Systems and Networking, Department of Innovative Technologies, University of Applies Sciences and Arts of Southern Switzerland, CH-6962 Lugano-Viganello, Switzerland; michela.papandrea@supsi.ch (M.P.); laura.dimaggio@supsi.ch (L.D.M.); andrea.quattrini@supsi.ch (A.Q.); luca.minardi@supsi.ch (L.M.); michele.cocca@supsi.ch (M.C.)
3. Labour, Urbanscape and Citizenship Research Area, Competence Centre for Labour, Welfare and Social Research, Department of Business Economics, Health and Social Care, University of Applies Sciences and Arts of Southern Switzerland, CH-6928 Manno, Switzerland; filippo.bignami@supsi.ch
4. Institute of Applied Sustainability to the Built Environment, Department of Environment Constructions and Design, University of Applies Sciences and Arts of Southern Switzerland, CH-6850 Mendrisio, Switzerland; albedo.bettini@supsi.ch

* Correspondence: annalisa.rollandi@supsi.ch

† These authors contributed equally to this work.

Highlights:

This project explored the micromobility role in four Swiss urban contexts, focusing on Lugano as a key testing ground.

What are the main findings?

- Interdisciplinary approach: this project combined urban analysis, experiments in the urban context, and participatory processes to evaluate the impact of micromobility.
- Integration of an innovative micromobility device: this study introduced Genny Zero, a self-balancing micromobility vehicle, to assess its feasibility in real urban environments.

What is the implication of the main finding?

- Identification of conflicts and synergies: this research investigated potential conflicts with pedestrians and vehicles, while also demonstrating synergies with existing cycling infrastructure and public transport.
- Policy and infrastructure recommendations: this study underscores the need for adaptive regulations, multimodal integration to ensure safe and efficient micromobility adoption, and guidelines for participatory processes.

Abstract: This manuscript investigates the integration of micromobility solutions in four Swiss contexts, with a primary focus on Lugano as the testing area. Through urban analysis, urban experiment, and social testing, the Inclusive MicroMob project introduced Genny Zero, an innovative micromobility device, to assess its impact on urban mobility. The findings highlight key factors for successful micromobility integration, including the need for an interdisciplinary approach that includes a holistic urban overview, an improved road safety analysis, and clear regulatory frameworks. This study also emphasizes the importance of participatory urban governance in addressing community needs and ensuring accessibility. Policy recommendations are provided to support the development of inclusive and sustainable micromobility systems in medium-sized cities.

Keywords: inclusion; micromobility; urban mobility; experimentation; participation; Lugano

1. Introduction

The rapid growth of urban populations and the effects of climate change have deep implications for how people use public spaces in contemporary cities. These intertwined challenges underscore the need for holistic approaches to urban planning and mobility systems that prioritize the sustainability, inclusion, and well-being of individuals [1,2]. Addressing such complex issues requires not only technical innovations, but also collaborative solutions that involve diverse stakeholders and aim to create cities that are resilient and fair for all.

An inclusive city ensures that all inhabitants -regardless of age, gender, ability, or socioeconomic status- have equitable access to public spaces, resources, and opportunities. Key criteria for inclusive cities include the promotion of social integration, accessibility to infrastructure and services, participatory urban governance, and the ability to foster a sense of belonging among diverse populations [3,4]. The concept of inclusivity in cities extends beyond physical accessibility, emphasizing equity in housing, education, healthcare, employment, and public spaces [5]. Inclusive urban planning aims to reduce social disparities by fostering participatory decision-making processes and creating environments that promote diversity, social cohesion, and resilience [6].

In this context, inclusion is intrinsically linked to sustainability, as cities that accommodate diverse needs and perspectives are better positioned to thrive in the face of environmental, economic, and social challenges [7].

In response to these challenges, the Inclusive MicroMob project has explored the role and impact of emerging mobility solutions within urban centers. Specifically, the project examined the introduction of a new micromobility device in a real context like Lugano, Switzerland. Micromobility—which includes modes of transport such as e-scooters, bicycles, and other small lightweight vehicles—has gained significant attention in recent years for its potential to address challenges of urban mobility, reduce environmental impacts, and improve accessibility [8,9]. In addition, the role of micromobility in urban transportation appears to have a significant impact on the implementation of the Sustainable Development Goals (SDGs) of the United Nations and supports the management of the transition from energy to decarbonization [10]. This is particularly relevant within the framework of sustainable urban development outlined in SDG 11, where micromobility serves not only as a tool to promote accessible and low-emission transport, but also as a catalyst for participatory governance, fostering inclusive decision-making processes that improve urban resilience and livability. Therefore, the integration of these new technologies into existing urban systems presents unique challenges and opportunities that require careful analysis.

This study began by referencing existing research on micromobility, with particular attention to the integration of these solutions into urban planning and potential conflicts with other transport modes. In particular, studies examining the role of micromobility in improving connectivity between neighborhoods were analyzed, such as that of Vizmpa et al. [11], which highlights the importance of safe urban pathways to ensure equitable access to services and reduce car dependency. At the same time, the need for data-driven solutions to optimize the use of micromobility in complex urban contexts was considered, as demonstrated by Tamagusko et al. [12], whose approach leverages computer vision techniques to identify risk factors in cycling infrastructure. While data are crucial, so is citizens' perception, which can emerge through participatory processes.

The combination of these two factors -quantitative and qualitative- constitutes the method adopted by this research to investigate the opportunities and needs of micromobility in a real urban context. This approach aligns with recent studies highlighting the importance of integrating objective data with a participatory perspective to better understand urban mobility dynamics [8,13]. In particular, the quantitative dimension allows for the analysis of data on the demand and supply of micromobility services, infrastructure usage, and impacts on traffic and emissions [14]. However, several studies emphasize that quantitative data alone can be limiting, as they do not always capture user perceptions, adoption barriers, and preferences. For this reason, integrating a qualitative perspective -based on interviews, observations, and participatory processes- is essential for identifying social, cultural, and economic factors influencing the adoption of micromobility. Building on these references, this study aims to bridge a gap in existing research by providing a more comprehensive framework on the opportunities and challenges of micromobility integration in urban spaces. The objective is to combine data analysis with an inclusive and participatory perspective, contributing to a better understanding of the dynamics of accessibility, sustainability, and social acceptance of micromobility solutions in contemporary urban contexts.

The central hypothesis of the Inclusive MicroMob project is that the introduction of new micromobility devices can address both emerging and unmet mobility needs within urban settings. By allowing more flexible and adaptive forms of transportation, these devices have the potential to bridge accessibility gaps, reduce dependency on private vehicles, and foster greater inclusivity in how people move through and interact with public spaces [15].

This study seeks to address the following key questions: How can new forms of micromobility be effectively integrated into the current urban mobility scenario? What types of conflict arise from their implementation and how can these conflicts be mitigated? Furthermore, what lessons can be learned from the Lugano case study to inform the broader development of inclusive and sustainable urban mobility systems?

Through an interdisciplinary approach, this paper aims to contribute to the ongoing discussion on urban mobility and inclusivity, offering insights and strategies for the implementation of micromobility solutions that align with the evolving needs of cities and their inhabitants.

State of the Art: Micromobility

Mobility represents a crucial aspect of modern life. Although the movement of populations across the Earth's surface is one of humanity's oldest characteristics, urban environments have become the primary context where mobility systems have developed and manifested in all their physical and cultural complexity [16].

Following the recent pandemic, new lifestyles and technological advancements have led urban mobility to undergo increasingly profound changes, shaping movement and behavior within cities. The rise of micromobility offers a flexible transportation option aimed at reducing the individual use of vehicles while enabling intermodal approaches [17]. This shift holds the potential to decrease congestion, the space required for parking, and both atmospheric and noise pollution.

According to the definition provided by the Italian Encyclopedia Institute [18], micromobility refers to mobility over short distances, primarily within cities, characterized by the use of lighter, less cumbersome, and potentially less polluting transportation modes compared to traditional ones. Although there is no universally recognized definition of micromobility, the International Transport Forum (ITF) uses several parameters to classify lightweight vehicles: speed (a maximum of 45 km/h), weight (up to 350 kg), and power output (a maximum of 2000 W). Subsequently, individual countries define national parameters that may differ but cannot exceed the international limits [19].

Micromobility devices can be categorized into human-powered and electric-assisted vehicles. Human-powered options include traditional bicycles, kick scooters, roller skates, and similar devices, while electric-powered options encompass electric scooters, e-bikes (including foldable ones), and self-balancing devices such as segways, monowheels, and hoverboards, as well as electric skateboards [20].

Another key distinction lies in the mode of use: private or shared. Shared vehicles are made available to the public through mobile applications, enabling users to locate, activate, and deactivate them as needed. There are two main types of trips: docked and dockless. The former involves starting and ending trips at predefined docking stations specifically designed for micromobility vehicles, while the latter allows for free-floating trips, where vehicles can be picked up and dropped off (activated and deactivated) anywhere.

According to Oeschger et al. [19], it is essential to consider the development of micromobility devices as complementary to existing public transport networks. The preferred use case for these services is the "first and last mile" of a journey, with the primary goal of reducing motorized transportation modes. While these services allow users to organize their movements flexibly according to their needs, Rossato [20] highlights a potential challenge: micromobility devices may be used as an alternative to walking, which could have significant implications for urban space planning, management, and allocation.

Switzerland has seen a rapid expansion of micromobility services, particularly in the main cities (Zurich, Geneva, and Basel), where shared e-scooters and bicycles complement public transportation [21]. The country's strong commitment to sustainability and well-developed cycling infrastructure support this growth. However, challenges persist, including regulations, safety, and public space management [14].

After a rapid rise in the use of private and shared micromobility devices, some cities (e.g., Paris) have experienced growing dissatisfaction and criticism. One major concern relates to parking behavior and the inappropriate use of public spaces, often leading to sidewalk clutter and accessibility issues for pedestrians, especially those with reduced mobility. Additionally, road safety remains a significant issue, with reports of accidents involving e-scooters and bicycles raising questions about infrastructure adequacy and the enforcement of traffic regulations [22]. The competition for limited space in urban areas further makes tensions between different categories of road users, including pedestrians, cyclists, micromobility users, and drivers. In recent times, adjustments and experimentation have been implemented in regulatory frameworks to address these challenges. To develop inclusive solutions, it is essential to broaden the discussion to include the design and management of public spaces, as well as the perceptions and needs of inhabitants. Effective regulation and thoughtful design of public spaces are crucial to mitigating conflicts and improving user experience. Furthermore, understanding the perceptions and expectations of inhabitants can support the development of policies that enhance the social acceptance and long-term viability of micromobility solutions.

2. Method and Materials

2.1. The Interdisciplinary Approach

The Inclusive MicroMob project was based on an interdisciplinary approach, integrating urban analysis, urban experimentation, and social testing. This study reflects on micromobility in Switzerland, with particular attention to the city of Lugano, which served as a privileged observatory for real-world testing. These tests were made possible through the collaboration of the Lugano Living Lab and Genny Factory, the company behind a new micromobility device, Genny Zero. This approach enabled a comprehensive evaluation of the integration of the device in the urban context, considering both technical performance and social acceptance, ensuring that the findings could inform inclusive and sustainable urban mobility strategies.

Building on a dual methodology, namely merging data collection focused on urban micromobility with a qualitative social analysis of urban spaces, this article helps bridge a knowledge gap by focusing on three specific conceptual areas.

The first conceptual area concerns urban analysis and regulatory adaptation. This study examines the regulatory challenges, conditions, and policy adjustments required to integrate new forms of micromobility into existing urban transport networks. By comparing Swiss and Ticino regulations, it identifies gaps in planning and policies that must be addressed to ensure safety, multimodal integration, and efficient infrastructure planning. It provides several policy recommendations, including adaptive regulations, improved cycling infrastructure, and the importance of participatory initiatives to accommodate emerging transport modes.

The second conceptual area focuses on urban analysis and inclusivity. The method highlights how micromobility devices can enhance accessibility for all citizens, extending beyond those with reduced mobility, and ensuring that urban transport systems are more inclusive. It explores the role of micromobility in urban spaces and its interaction with pedestrians and other transport modes (e-bikes, electric scooters, etc.). Moreover, this study examines the contribution of micromobility to fostering a more participatory and equitable urban environment, where all individuals -regardless of physical ability or economic status- can engage in city life and access public spaces. The research aligns with the concept of urban citizenship by emphasizing how mobility rights and a participatory approach in policy-making are crucial components of social inclusion and belonging in urban settings.

The third conceptual area addresses public participation. This research underscores the need for participatory urban governance, in which residents actively engage in shaping micromobility policies and infrastructure. Through initiatives like the Ensemble-IMM Citizenship Laboratory, this project demonstrates how combining digital platforms with in-person actions can foster community involvement in decision-making. This study innovatively highlights generational and social differences in attitudes toward micromobility, reinforcing the importance of inclusive and transparent dialogue between urban policymakers and residents.

Figure 1 illustrates the strategy and the key elements developed throughout the project. It provides a comprehensive overview of the process, highlighting how each component contributes to the overall methodology and analysis.

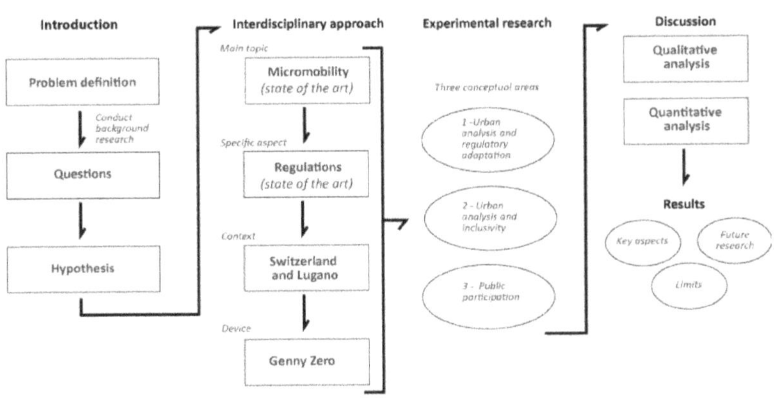

Figure 1. The strategy and the key elements developed throughout the project.

2.2. Comparison of Swiss and Ticino Regulations on Micromobility

The regulation of micromobility is a pressing challenge that stems from its growth in recent years. The primary objectives are to improve road safety, adapt urban development and infrastructure, manage device placement and rental agreements, and reduce the environmental impact [23].

At the federal level in Switzerland, micromobility devices are classified as motorized vehicles. Specifically, fast e-bikes are considered single-track vehicles, while slow e-bikes, electric scooters, and similar devices fall under the category of light motorized bicycles. A third category includes motorized wheelchairs, while self-balancing devices are classified as self-balancing scooters. Roller skates, skateboards, and similar devices are considered "vehicle-like means of transport".

This project focuses on self-balancing devices. These vehicles are subject to specific limitations, such as a maximum speed of 20 km/h, a weight limit of 200 kg, and a power output of up to 2000 W [24]. Like other single-track vehicles, self-balancing devices must have a control plate. For driving licenses, the requirements are the same as those for light motorized bicycles: individuals aged 14 to 16 years must have a license of the M class, while no license is required for users aged 16 and older, as specified in the Ordinance on Admission to Road Traffic [25]. Helmets are not mandatory but are strongly recommended.

The traffic rules for motorized bicycles align with those for conventional bicycles. This means that users should use bike lanes or bike paths where available, as outlined in the Ordinance on Traffic Rules [26]. The use of micromobility devices in pedestrian areas is allowed only for individuals with reduced mobility; otherwise, it is strictly prohibited (ONC, 1962). Devices like unicycles and hoverboards are excluded from the listed categories because they do not have all the technical requirements. Consequently, they are not authorized for use on public roads as they lack official approval [27].

In Ticino, the regulation [28] on micromobility vehicles subjects them to the following conditions:

- They must adhere to the traffic regulations for bicycles;
- They are required to use bike lanes when available;
- They are prohibited from operating in pedestrian areas;
- Their use is permitted from the age of 14 with an M-class (or G-class) license and without a license from the age of 16, except for fast e-bikes;
- Helmet use is recommended and mandatory for fast e-bikes;
- Segways, fast e-bikes, and motorized wheelchairs (except those with speeds under 10 km/h) must have a control plate.

Unlike the federal regulations, the cantonal provisions in Ticino allow micromobility devices to be operated by individuals with a G-class license (for agricultural vehicles) in addition to an M-class license. This difference reflects a unique adaptation of federal guidelines to local needs, highlighting a degree of regulatory flexibility at the cantonal level.

In general, while both Swiss and Ticino regulations emphasize safety, accessibility, and integration into existing traffic rules, the specific distinctions in licensing and access to certain devices underscore the importance of tailoring regulations to regional contexts and priorities.

2.3. Genny Zero

The micromobility device at the center of the Inclusive MicroMob project is called Genny Zero (Figure 2). This is the evolution of Genny 1.0, a medical tool for people with reduced mobility that is now also offered as a micomobility tool, which in Switzerland may be used with a license plate or without, depending on legislative development. Its technical specifications are as follows:

- Minimum weight: 50 kg, maximum weight: 110 kg
- Maximum power: 2000 W
- Maximum speed: 15 km/h
- Maximum incline: 20%

Figure 2. Genny Zero. © Genny Factory SA, Switzerland.

Considering the characteristics of micromobility vehicles, it can be stated that this device is classified as a segway. This means it must have a control plate, and an M-class license is required for operation between the ages of 14 and 16.

2.4. Data Collection: Urban Analysis

The urban analysis was based on several key datasets.

- Existing supply and demand: evaluated through an analysis of the resident population and total job opportunities.
- Street gradients: the Genny Zero device is suitable for slopes with a maximum gradient of 20%, equivalent to approximately 11.3 degrees. Roads were categorized as either accessible (gradient below 11.3°) or requiring exchange points or alternative solutions (gradient above 11.3°).
- Public facilities: local public transport systems, parking areas, railway stations, bike-sharing systems, and charging stations were mapped to evaluate potential integration with micromobility devices like Genny Zero.

The urban analysis was conducted on two distinct levels. First, four Swiss cities were selected as case studies to represent diverse geographic contexts: Chur, a valley-floor city; Fribourg, a flat city intersected by a meandering river; Andermatt, a rapidly growing mountain resort; and Lugano, a lakeside city. Subsequently, Lugano was analyzed in depth, supported by specific tests (see Section 2.5).

2.5. Data Collection: Urban Experiment

In addition to the urban analysis, tests were conducted in a real urban context. The urban experiments took place on Friday, 8 March 2024, after obtaining authorization from the Traffic Section of the Canton of Ticino. The test started from the East Campus of USI-SUPSI in Viganello and traversed the city of Lugano and the municipality of Paradiso before returning to the starting point. Specifically, the route included reaching the Cassarate river mouth, passing through Ciani Park, and continuing to the bus shelter on Via Pretorio. From there, the Genny Zero vehicles boarded a bus to the Lugano FFS train station and then took a train to Paradiso FFS station. The return journey followed the lakeside promenade. One Genny Zero vehicle was equipped with Realsense D455 cameras for data collection.

The goal of this test was to obtain real-world data on Genny Zero's behavior in an urban environment, verifying its qualities, its multimodal integration potential, and possible conflicts with other road users (pedestrians, cyclists, cars, etc.). To ensure a dense mobility environment, the test was conducted around midday to coincide with peak lunch-hour traffic.

Methods and Mathematical Reference Models

The analysis methodology and models used to collect, process, and analyze the data measured from the Genny device are detailed below, and related results are reported in Section 3.

- Analysis of most impactful road characteristics
  An analytical study on machine collected data was performed to identify the road characteristics which have the highest impact on the vehicle's speed and acceleration, based on the result of a supervised machine learning model prediction.As a first step, a preliminary statistical and correlation analysis was performed over the collected machine data (univariate correlation). In particular, we used the following statistical metrics as relevant for the present study:
  - Mean, standard deviation, minimum, maximum, and the 25th, 50th and 75th percentiles for the speed and forward acceleration distributions (Figure 3);
  - Pearson coefficients for linear correlation [29] computed for all the merged datasets, whose result is shown in Figure 4 for the most relevant features, which were also used to focus on some use-cases, such as for the speed and forward acceleration and the tilt back and forward acceleration).

  After that, a model-based analysis on machine data was performed to identify the most impactful road characteristics on vehicle speed and acceleration, as a result of a supervised machine learning model prediction (Figure 5). More specifically, to build such a model, input data, distributed as shown in Figure 6, were considered. These road characteristics were fed into an eXtreme Gradient Boosting (https://xgboost.readthedocs.io/en/stable/, accesed on 1 April 2024) model [30], selected to reliably assign feature importance scores to the input data. One of the advantages behind the choice of such an ensemble method is that, unlike simple correlation-based methods, XGBoost captures non-linear relationships and feature interactions, making it more effective for selecting informative features. To optimize the performance of the model, we performed hyperparameter tuning, which led to the selection of the following values: 1000 estimators, ensuring a sufficient number of boosting rounds to capture complex patterns in the data; a learning rate of 0.01, which balances convergence speed and performance to prevent overfitting; and a maximum depth of 100.

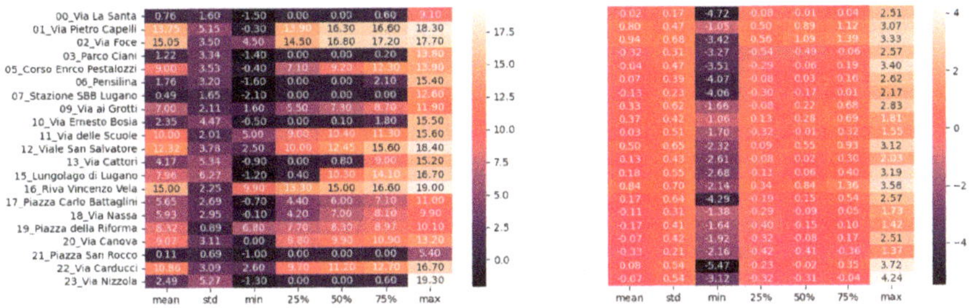

Figure 3. Speed (**left**) and forward acceleration (**right**) distribution per area: numerical report.

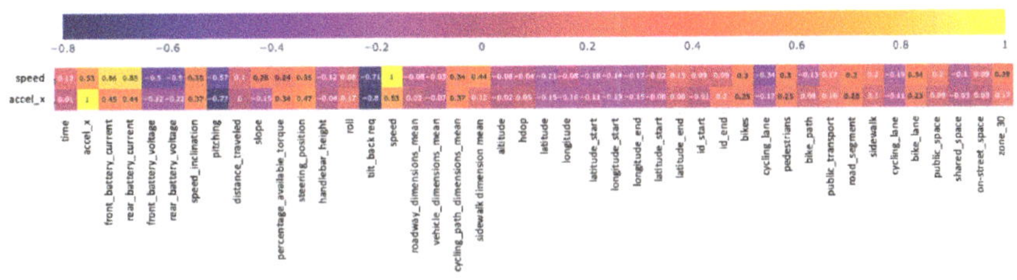

Figure 4. Feature correlation for the most relevant features.

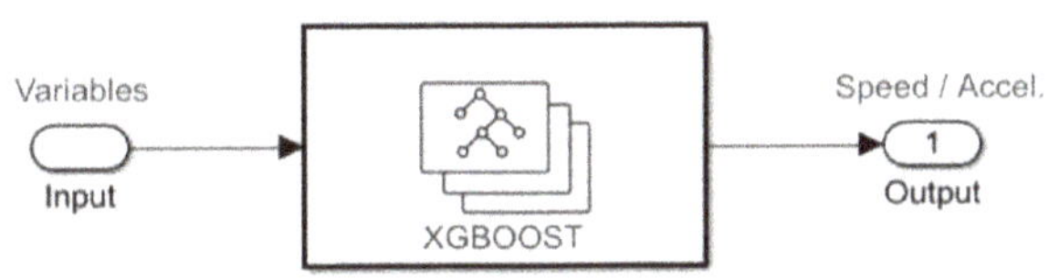

Figure 5. Prediction of speed and acceleration modeled via non-linear regression.

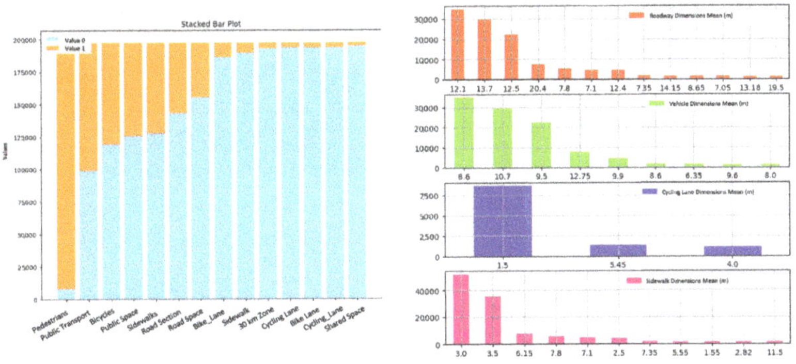

Figure 6. Variable distribution bar plots.

- Analysis of RealSense camera data for the inference of encounters and conflicts
  The RealSense D455 cameras are RGB-D cameras, meaning they capture both color (RGB) and depth (D) information to generate a three-dimensional representation of the surrounding environment. Depth measurement was achieved using stereoscopic vision, employing two grayscale sensors and an infrared projector. Additionally, an RGB sensor was used to capture a standard color image. Although the cameras were equipped with an active projector operating at 850 nm, they were used in passive mode for this study. This choice was made because the active light source

would be significantly weaker compared to sunlight, reducing its effectiveness in outdoor environments. Using the cameras in passive mode also minimized power consumption, thereby extending battery life. The depth image, captured with the stereoscopic depth sensor, was inherently misaligned with the RGB image due to the spatial separation of the sensors. However, a factory calibration process enabled a post-processed alignment between the Field of View (FoV) of the depth map and the RGB image. This alignment ensured that any model applied to the RGB image could be accurately transferred at a pixel level to the corresponding depth map, facilitating precise spatial analysis and scene understanding. An example of aligned depth-image pictures is shown in Figure 7. To perform object detection and tracking, we employed the latest version of the You Only Look Once (YOLO) framework, specifically https://docs.ultralytics.com/it/models/yolov8/, accessed on 1 April 2024. Renowned for its speed and accuracy, YOLO is particularly well suited for real-time applications such as video-based road data analysis. In this study, it was applied to the RGB images and then transferred to the aligned depth images, allowing the estimation of the distance between the detected person and the camera. This estimation was performed using the median of the depth values within the bounding box, mitigating the influence of any background pixels that might be present within the detection area. Among the pre-trained classes available in YOLO, we selected those relevant to urban mobility, including pedestrians, cars, bicycles, and motorcycles, to ensure effective detection and tracking in road environments. Based on the objects tracked by this analysis, a heuristic was applied to differentiate between encounter and conflict events. These events were correlated to the characteristics of the roads visited.

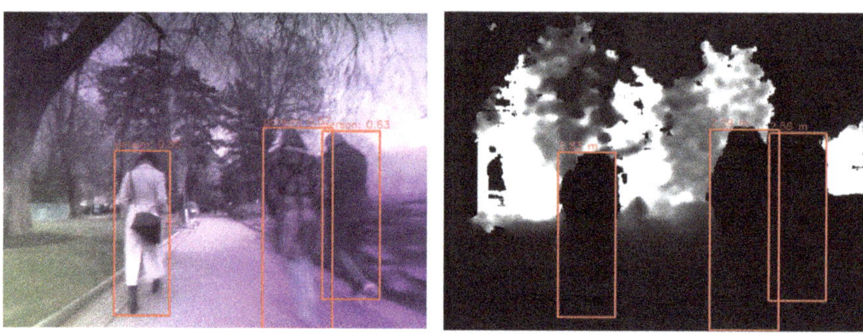

Figure 7. On the (**left**), the RGB image with YOLO-detected bounding boxes; on the (**right**), the depth map where the same bounding boxes were applied, leveraging the alignment between the two modalities.

2.6. Data Collection: Social Testing

Regarding the social and policy analysis of urban micromobility, an event was held on 14 October 2023, at Ciani Park in Lugano. The event aimed to conduct an in-depth analysis of the usability of the Genny Zero device, the social perception of its use, and a qualitative understanding of the subjective reception of micromobility in the urban context. During the project, there was an additional research opportunity through the synergy with the Ensemble project. Ensemble, an internal interdepartmental SUPSI initiative, seeks to identify and test a methodology that fosters a citizenship participatory approach through a digital platform (Decidim-Ensemble). This platform encourages involvement, discussion, and collaborative decision-making. Ensemble is designed for both the university community and the broader public, supporting social activation, community aggregation, political engagement, and social innovation. The project includes the establishment of the ENSEMBLE4IMM Citizenship Laboratory, which integrates three in-person meetings with online activities on the Decidim-Ensemble platform. Through this hybrid approach, guidelines for improving urban mobility and urban policy-making can be defined. The fruitful synergy between the projects emerged because the October 2023 IMM test of the Genny Zero device at Ciani Park had potential applicability within the Ensemble-IMM Citizenship Laboratory as a case study. On 15 April 2024, at the SUPSI East Campus, a second additional test of the Genny Zero device was conducted as part of one of the three in-person Citizenship Laboratory events. This test aimed to evaluate both the usability of the device and the soft mobility approach it promotes while also reflecting on the layout of urban spaces in relation to the broader spectrum of micromobility solutions.

3. Results

The data collected during the urban, experimental, and social tests aimed to explore the state of the art in mobility within specific Swiss contexts, with a particular focus on Lugano. As a privileged research hub, Lugano was central to the investigation thanks to the collaboration with the city of Lugano and the Lugano Living Lab. The analysis below illustrated (set out in urban, experimental, and social layers) is project report-oriented. Therefore, this section exposes key findings and the description of their achievement. In addition, this section does not provide comparison with other micromobility means, since its goal is to assess the feasibility of introducing new micromobility devices, identify potential conflicts with other transportation modes, and evaluate user perceptions, farming urban micromibility policies linked with urban spaces, and limitations.

3.1. Data Analysis: Urban

As previously mentioned, the urban analysis was carried out on two distinct levels. First, four Swiss cities were chosen as case studies to represent diverse geographic contexts: Chur, a city located on a valley floor; Fribourg, a flat city with a meandering river; Andermatt, a rapidly developing mountain resort; and Lugano, a lakeside city. Following this, a more detailed analysis was conducted specifically on Lugano. The data analysis, as outlined in Section 2.4, was performed using QGIS software, https://qgis.org/ accessed on 1 April 2024. To facilitate the understanding of the analyses presented below, please refer to the legend in Figure 8.

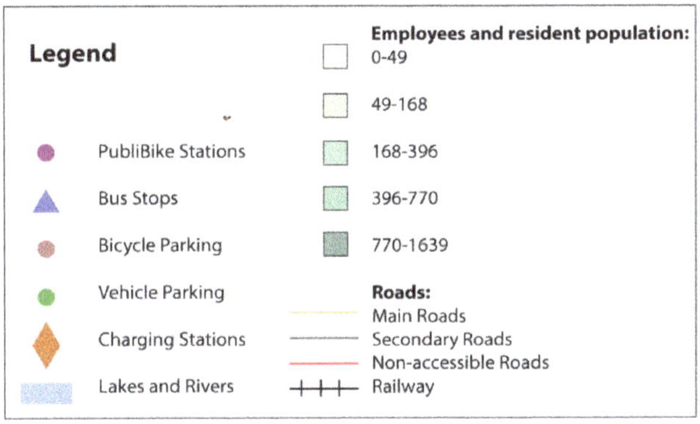

Figure 8. Legend of the symbols and layers used in the maps.

The city of Chur is located in a valley at the confluence of two rivers: the Plessur and the Rhine. The analysis did not identify any gradient-related issues within the urban center. Restrictions on the use of the Genny Zero device are primarily associated with the surrounding slopes (Figure 9).

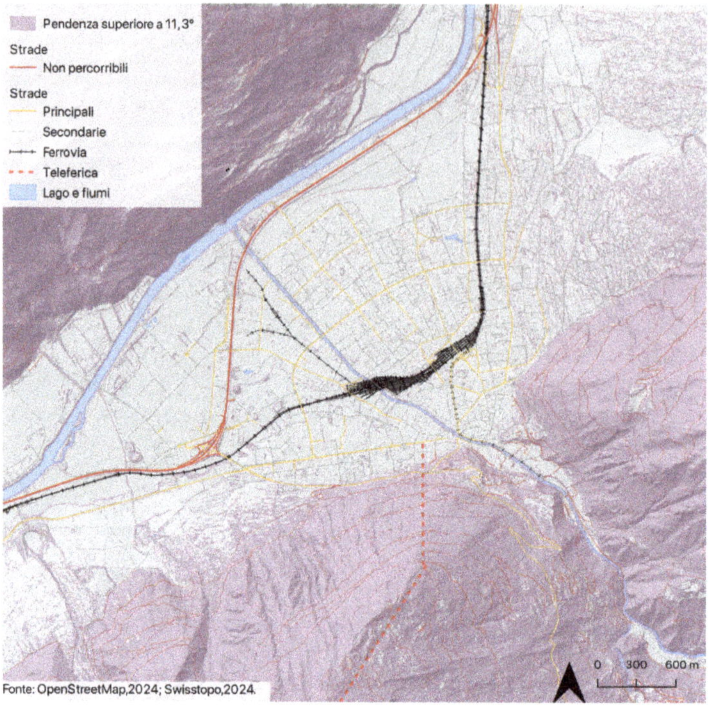

Figure 9. Slopes greater than 11.3° and roads not passable in the city of Chur.

Chur is well equipped with parking facilities (mapped in green) and local public transport stops (mapped in blue) (Figure 10). However, the number of bicycle parking spaces (105) is significantly lower compared to car parking spaces (867).

The city recently introduced a bike-sharing service. Following the success of a pilot project launched in September 2023—managed by Pro Velo Graubünden and the Canton of Graubünden—with 10 initial stations, the service was expanded to 20 stations, each including five e-cargo bikes. Lastly, several charging stations are available, primarily concentrated in the city center.

The city of Fribourg is characterized by a multi-level topography (Figure 11). The Sarine River has eroded the molasse of the plateau, creating distinct levels: a lower level along the riverbanks, a higher level to the west above the erosion, and an even higher level to the east (the Schoenberg district). These elevation differences are easily navigable thanks to the city's intermodal transport network.

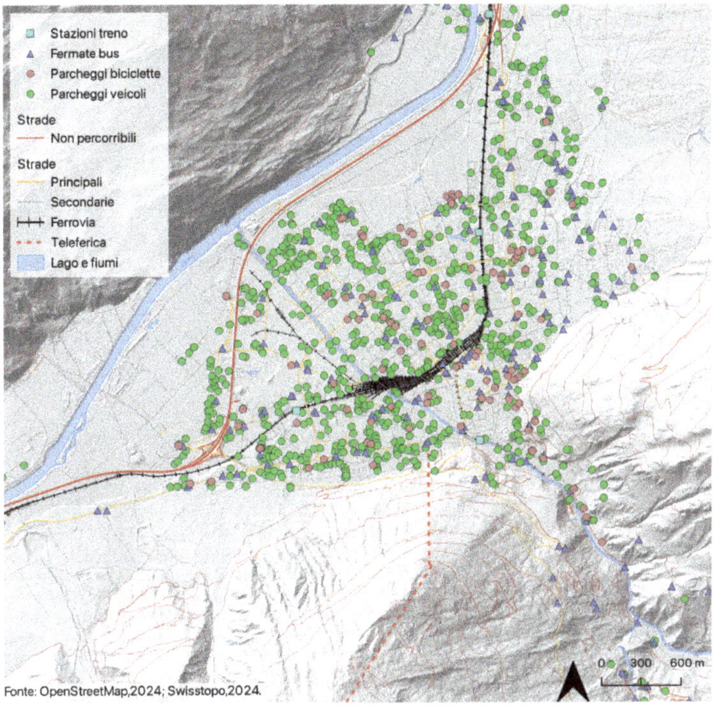

Figure 10. Parkings, public transports, and train station in the city of Chur.

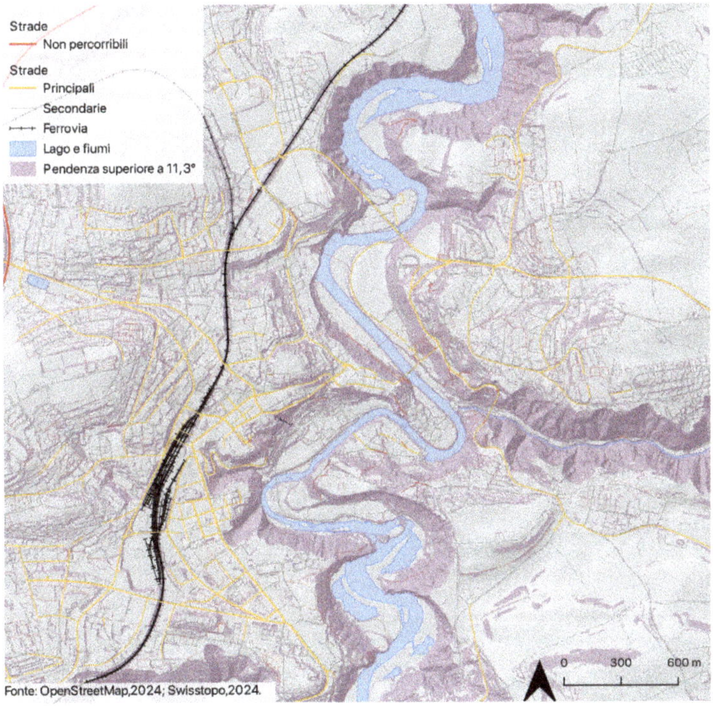

Figure 11. Slopes greater than 11.3° and roads not passable in the city of Freiburg.

Fribourg boasts an extensive public transport system (mapped in blue), including a funicular, numerous parking facilities (green), bicycle parking areas (pink), and two railway stations (light blue) (Figure 12). The upper levels of the city are connected by bridges, ensuring accessibility without significant elevation changes.

In terms of sharing services, Fribourg has a well-developed offering through the Publibike service.

Finally, several charging stations are available, though they are primarily concentrated in the western and northern parts of the city. These stations support the use of electric vehicles and micromobility devices in the region, although their distribution is limited to specific areas within Fribourg.

For the mountain case study, the municipality of Andermatt was selected, a rapidly developing alpine destination with a growing ski resort.

The main roads in Andermatt are accessible for devices like Genny Zero, as they do not exceed a gradient of 11.3°. Inaccessible routes in this context include primarily mountain roads, hiking trails, and via ferrata. A unique feature of this municipality is the presence of a car shuttle train that connects Andermatt to Sedrun.

Recently, Andermatt launched the first bike-sharing service in the Alps, featuring seven stations across the valley, five of which are located in the center (Figure 13). This initiative aims to introduce shared sustainable mobility to alpine tourist regions (Andermatt Swiss Alps, 2024; PubliBike Velospot, 2024). However, gradient-related challenges become significant outside the urban center and valley. As a result, the use of the Genny Zero device is not feasible across the entire territory and requires integration with other modes of transportation, such as buses or trains, particularly for the northeastern slopes.

Vehicle charging stations are scarce in the area (Figure 14).

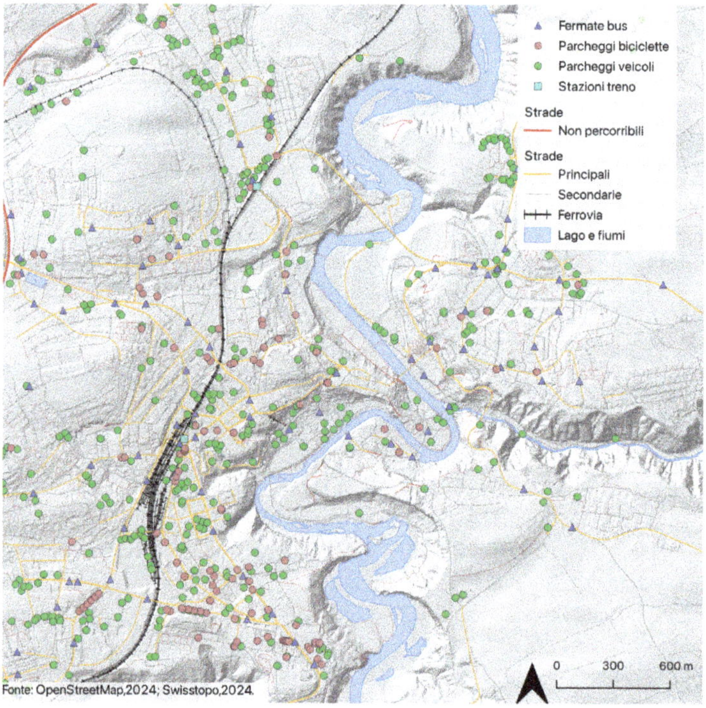

Figure 12. Parkings, public transports, and train station in the city of Freiburg.

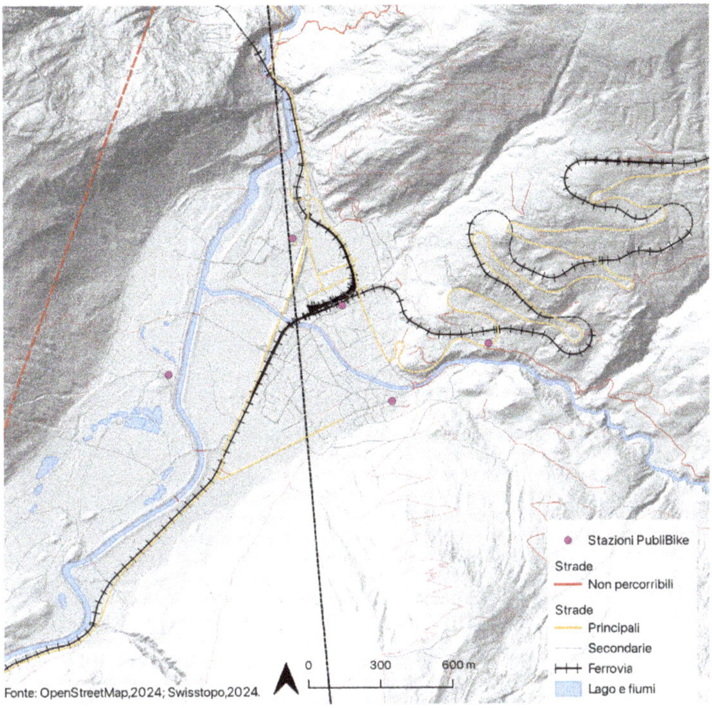

Figure 13. Bike-sharing service in the city of Andermatt.

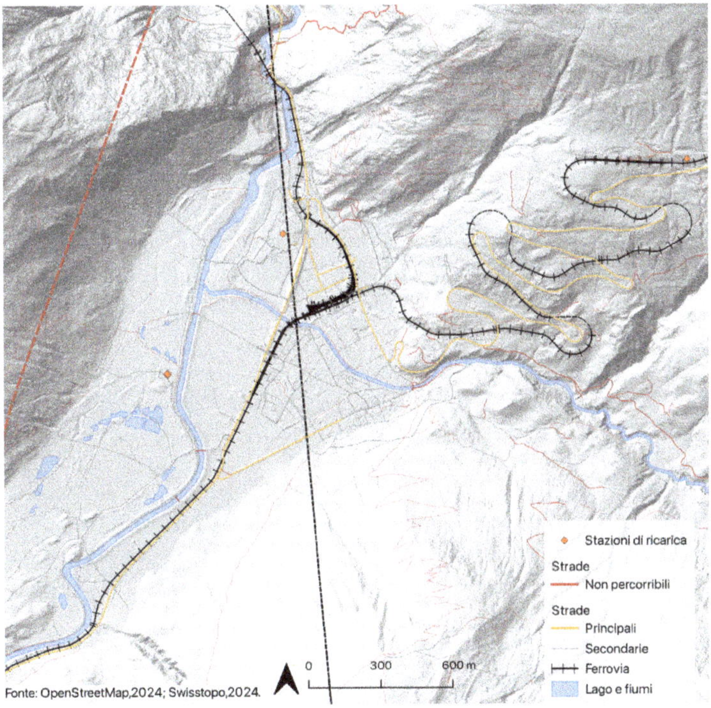

Figure 14. Charging stations in the city of Andermatt.

Lugano is located on the shores of the lake that shares its name. Today, it is the main economic center of the Canton of Ticino. Over the past 50 years, Lugano has undergone significant transformation. In the 1970s, it was a town with 27,000 residents covering an area of 11.6 km^2. Between 1972 and 2013, it grew into a city with more than 67,000 inhabitants, a territorial reach of 75 km^2, around 146,000 residents in the region, and over 100,000 jobs. Within the area analyzed, 16.2% of the roads are classified as non-navigable due to gradients exceeding 11.3° (Figure 15). To address this limitation, the analysis of local public transport, parking areas (both for bicycles and cars), and intermodal hubs identified key locations where the Genny Zero device could not be used due to steep gradients (Figure 16).

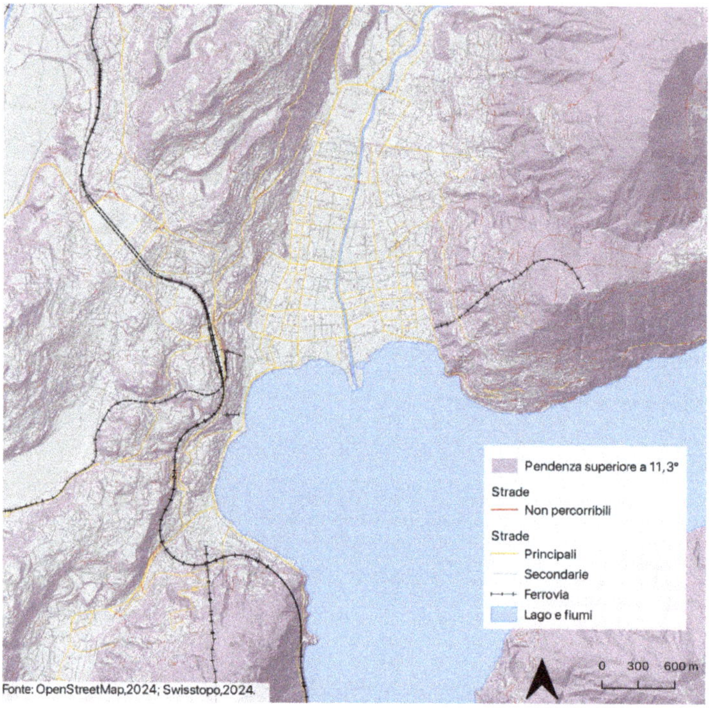

Figure 15. Slopes greater than 11.3° and roads not passable in the city of Lugano.

An area with a significant number of roads that exceed the 11.3° gradient threshold is the route connecting the city center to Lugano's railway station. Fortunately, this issue is mitigated by the presence of a funicular, which provides an alternative means of transport. Charging stations and PubliBike stations (Figures 17 and 18) were also mapped to explore potential integrated solutions that could include micromobility devices in the future.

Lugano served as the primary case study for in-depth analysis. One aspect focused on assessing potential demand, represented by the combined population of residents and total jobs, identifying inhabitants who might be interested in using the device. The map highlights these areas with a gradient scale of green, helping to pinpoint zones of greater interest for device deployment (Figure 19).

By overlaying all elements analyzed, specific areas can be identified as optimal locations to deploy devices. The area around Lugano's railway station (Figure 20) emerges as a highly suitable site for planning device deployment. This sector features high demand, numerous PubliBike stations, and parking facilities for bicycles and motor vehicles, as well as two electric charging stations that could support device recharging (Figure 21). In addition, the availability of public transport options helps overcome the gradient-related challenges of reaching the historic center.

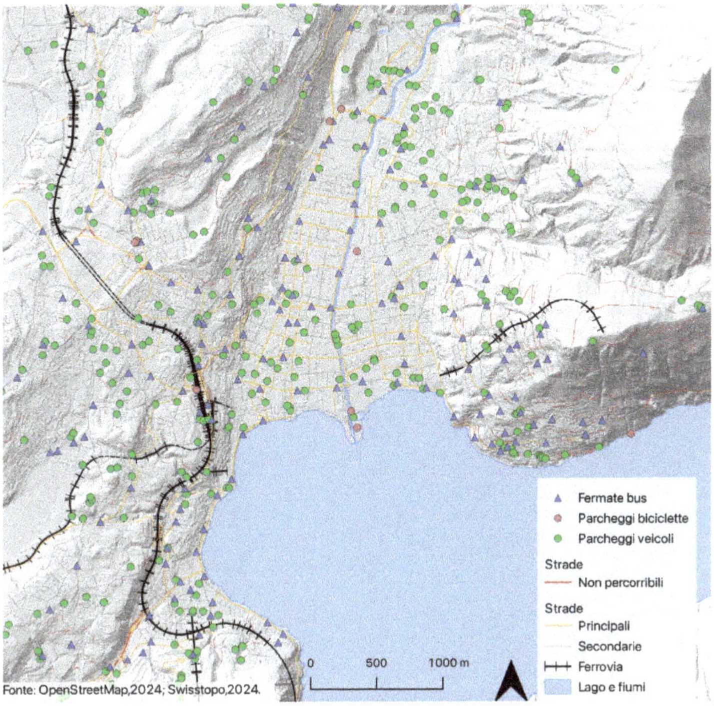

Figure 16. Parkings, public transports, and train station in the city of Lugano.

The analysis of the four Swiss cities as case studies highlighted five key aspects common to all of them:

- Presence of slopes limiting the use of the Genny Zero device. In all the cities analyzed, gradients exceeding 11.3° present a barrier to the use of micromobility devices such as Genny Zero.
- Integration with public transport networks. Each urban area is equipped with well-developed public transportation systems (including buses, trains, and funiculars), facilitating intermodal mobility and helping to overcome challenges related to steep gradients.
- Active or expanding bike-sharing services. All four cities have implemented or are expanding bike-sharing services, including e-bikes and, in some cases, e-cargo bikes.
- Availability of charging stations for electric vehicles. Charging infrastructure is present in all cases, supporting the use of electric vehicles and micromobility devices, although coverage may vary between city centers and peripheral areas.
- Analysis and provision of parking facilities. Each city offers parking areas for both cars and bicycles, although the number of bicycle parking spaces is generally lower compared to car parking spaces.

The analyses were useful in providing a comprehensive overview of the state of the art in mobility across the four reference cities and the potential integrations with micromobility devices. In the case of Lugano, the analyses enabled the selection of the characteristics for the experiments described in Section 3.2.

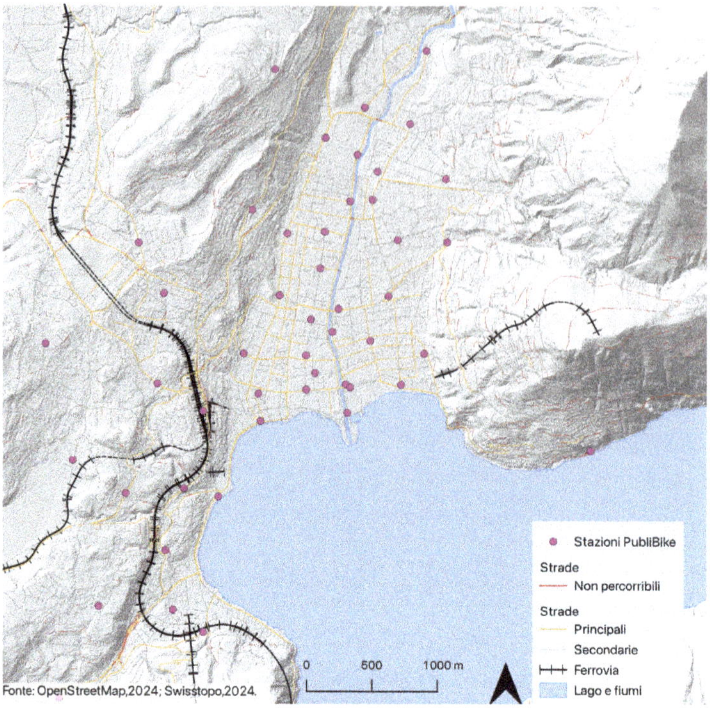

Figure 17. Bike-sharing service in the city of Lugano.

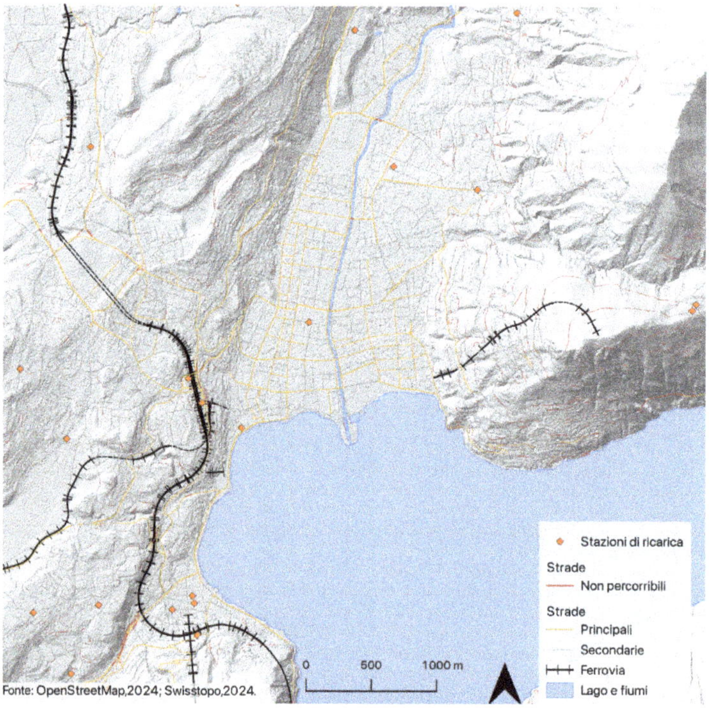

Figure 18. Charging stations in the city of Lugano.

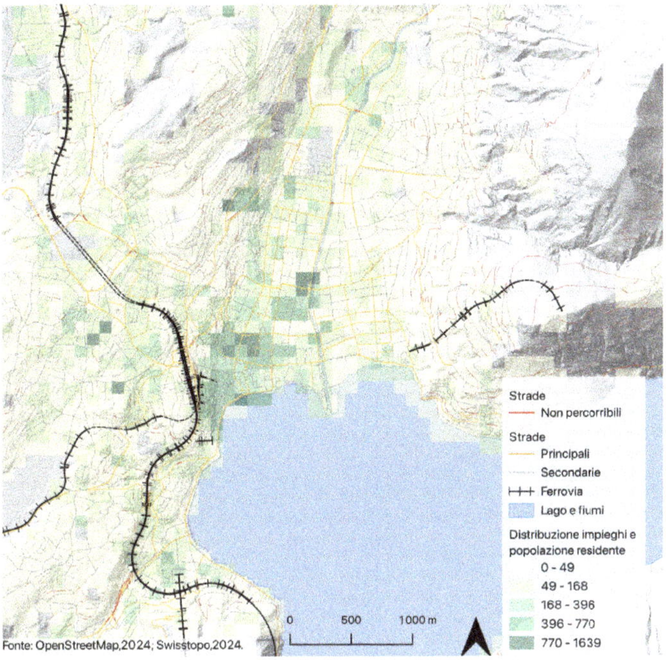

Figure 19. Potential demand in Lugano.

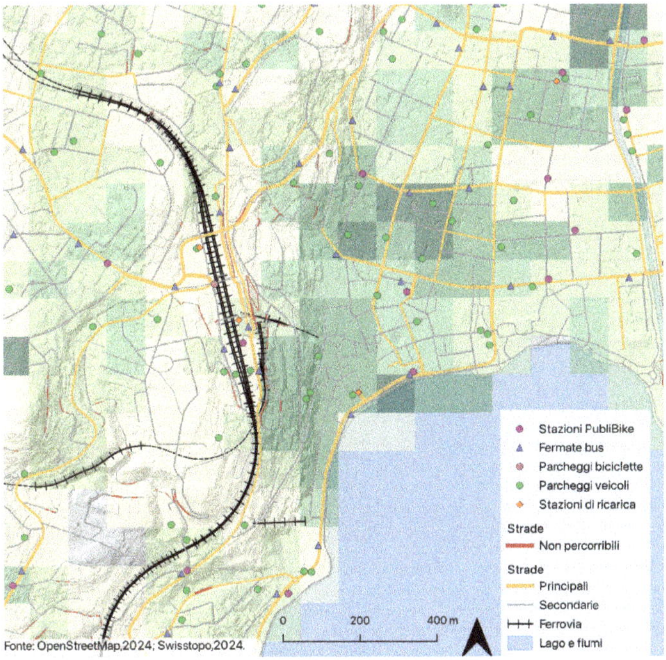

Figure 20. Potential demand, parkings, public transports, and bike sharing around Lugano's railway station.

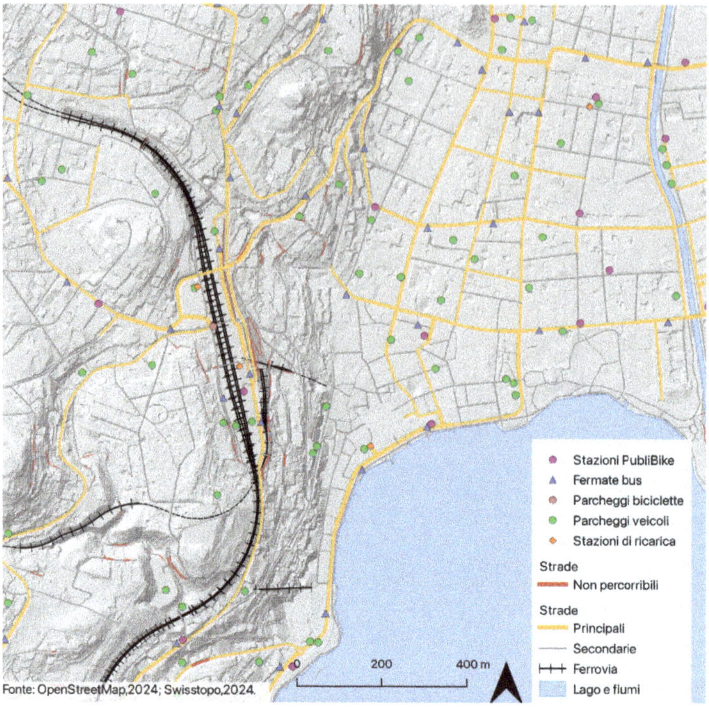

Figure 21. Parkings, public transports, and bike sharing around Lugano's railway station.

3.2. Data Analysis: Experiment

During the execution of the experiment, the Genny Zero device collected machine data (directly through the device), as well as inertial and GPS data (via sensors installed on the device) (Figure 22). Specifically, below is the detailed dataset collected during the urban test, whose traveled route is represented in Figure 23.

- Inertial sensor data, sampling frequency 50 Hz (3D acceleration, 3D gyroscope, front and rear battery voltage, tilt speed, and front and rear battery current).
- Vehicle inertial data, sampling frequency 10 Hz (pitch, distance traveled, slope, torque percentage, steering position, roll, and handlebar height speed).
- GPS data, sampling frequency 1 Hz (latitude, longitude, and altitude).

The data were processed, where a merging operation of the various sources generated a unified dataset, using a timestamp as the synchronization key for the join operation. Specifically, a function was implemented to associate the closest sampled data points (nearest neighbors) in time (in both directions).

During the pre-processing phase of the collected data, operations were performed for data normalization, removal of anomalous values (outliers), feature selection, and evaluation of the correlation between the collected machine data. In particular, the analysis of data quality led to the exclusion of certain sources (e.g., gyroscope). Figure 24 shows the distribution of the data collected by the Genny Zero device.

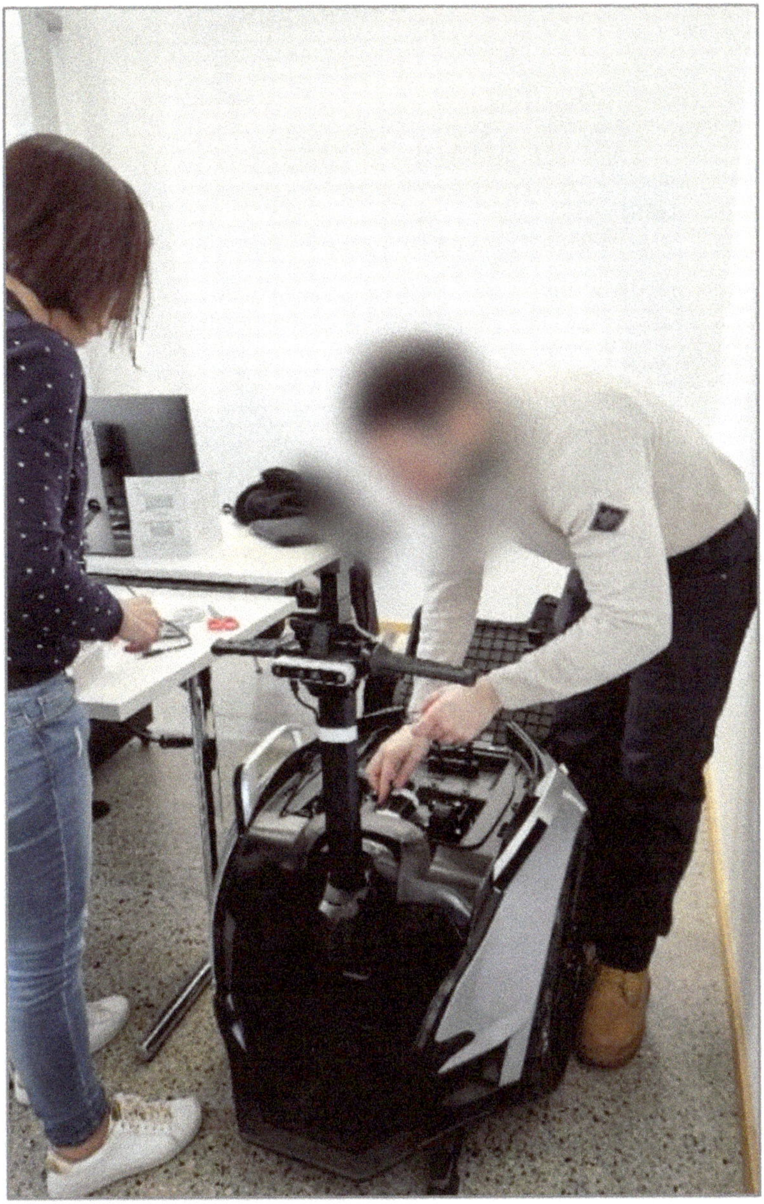

Figure 22. Sensor installation on Genny Zero.

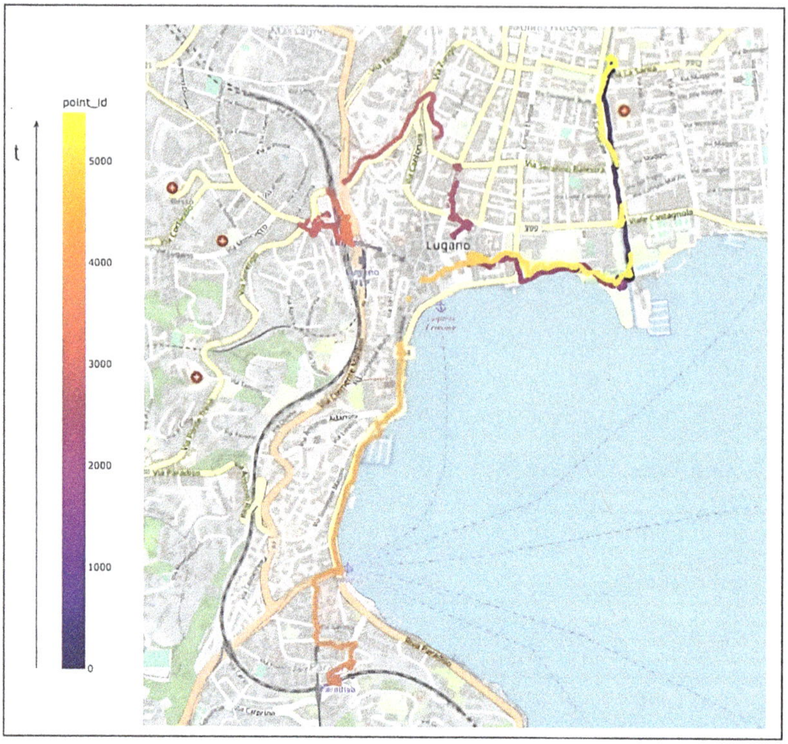

Figure 23. Experiment in Lugano: path overview.

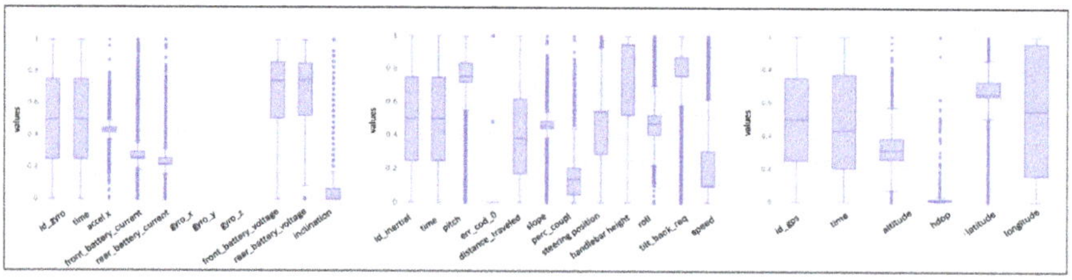

Figure 24. Value distribution for machine, inertial, and GPS data.

A preliminary analytical study was conducted on the collected data to derive insights and real evidence from the dataset. Figure 25 represents the distribution of speed for each traveled segment during the experiments (a unit of measurement specifically defined by the device but useful for comparative analysis). From the figure, it can be observed that the route included six low-speed segments, four medium-speed segments, and six high-speed segments: the lowest-speed segment corresponds to the SUPSI Campus, while the highest-speed segment corresponds to the route connecting LAC to Piazza Carlo Battaglini.

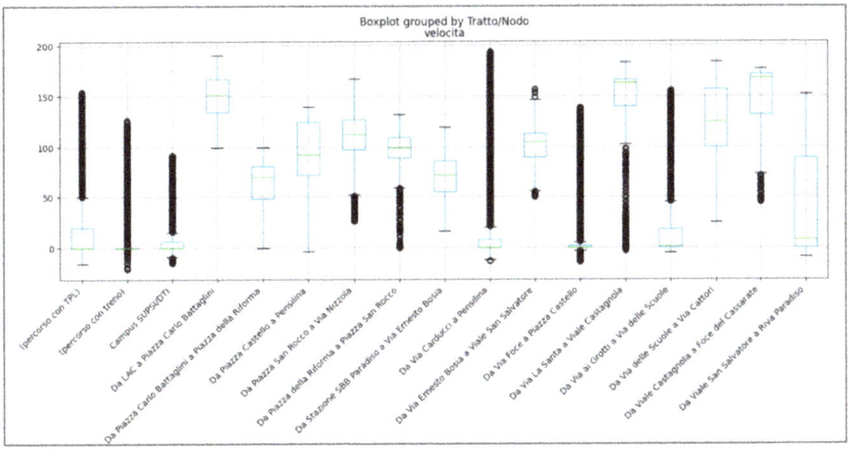

Figure 25. Speed distribution per traveled segment.

Figure 26 underlines the details of speed (top graph) and acceleration (bottom graph) for each street traveled during the route. The speed graph presents higher variance, allowing segmentation of the streets based on speed spread: streets where the device traveled at a quasi-constant speed (boxes with smaller areas) and streets where the driving pattern was more inconsistent (boxes with larger areas). This variability in movement is mainly due to the presence of pedestrians and crossing points along the route.

The key observations which emerge from our analysis are the following:

- The Genny device demonstrated sustained and quasi-constant speed on the 'Bike Path' and 'Bike Lane' segments. This suggests that cycling infrastructure does not pose a potential conflict for Genny Zero (Figure 26);
- In areas where vehicular traffic is restricted and in 30 km/h zones, Genny Zero moves at high speeds without encountering obstacles, allowing free and unobstructed movement (Figure 26).

The project also analyzed the collected data to identify which road characteristics and urban environmental factors significantly impact vehicle speed and acceleration. To achieve this, we leveraged the XGBoost algorithm ([30], a supervised machine learning model), exploiting its ability to model complex relationships between the target output (speed and acceleration) and characteristics of the traveled road, which were used as input features: potential conflict with bicycles, pedestrians, and public transport; the presence of public space; bike lane and sidewalk availability; road width. The importance of each input feature was determined based on its contribution to reducing the model's predictive error. The distribution of the target outputs values measured is represented in Figure 27, where we visualize the distribution of speed and acceleration values segmented per location category. In particular, it is evident from this picture that the Genny vehicle was able to travel at higher speeds when moving through shared-use zones (shared pedestrian and bicycles) or cycle paths, while it adapted by decreasing its speed in pedestrian areas (e.g., TPL bus stops).

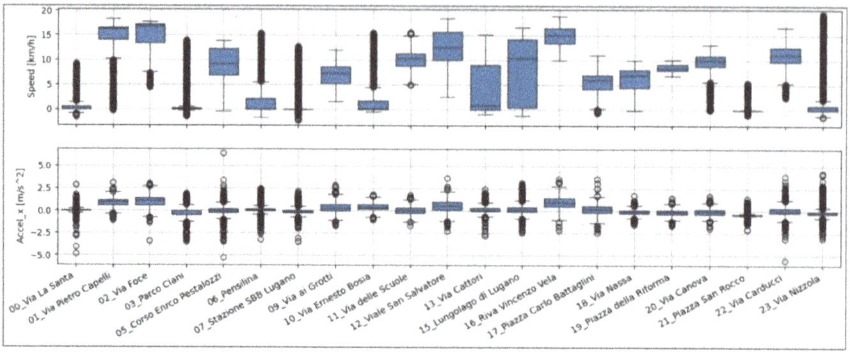

Figure 26. Speed and acceleration of the device as a function of the traveled street.

The objective function in this analysis was designed to optimize the XGBoost model's predictive performance for vehicle speed and acceleration based on input features related to the urban environment and road characteristics. Specifically, the function aims to minimize the prediction error by evaluating how well the model captures the relationship between these inputs and the target variables. Since the targets of our analysis were both continuous, a regression loss function was used (Mean Squared Error) during the training procedure. The objective function employed incorporated feature importance scoring (gain-based importance) to quantify how much each input variable contributes to the prediction of speed and acceleration. We optimized the hyperparameters of the model in order to prevent overfitting, and to ensure that only the most impactful features significantly influence predictions.

As shown in Figure 28, the model is capable of predicting the target variables with a very low average error, ensuring the reliability of the insights derived from our analysis. Specifically, our trained model predicts the vehicle speed with a median error of -0.87 km/h and the acceleration with a median error of -0.09 m/s^2. In order to ensure the XGBoost model was reliable, minimized overfitting, and achieved strong generalization performance on unseen data, a grid search of the best model hyperparameters with a cross-validation strategy was implemented.

Figure 28 shows the distribution of the prediction residuals calculated over the best-performing cross-validation iteration. As visible from the figure, the median of this error is near zero for both cases (-0.87 km/h for the speed and -0.09 m/s^2 for the acceleration), and the distribution of the error values is restricted to a limited range of size 7.6 km/h for the speed and 1.47 m/s^2 for acceleration. This small calculated error confirmed the reliability of our analysis, hence confirming that the selected input features significantly influenced vehicle speed and acceleration in a meaningful and interpretable manner.

To predict the most impactful road characteristics, through the model-based approach described in Section 2, we considered as input data a set of road characteristics whose distribution is shown in Figure 6 and obtained the results reported in Figure 29.

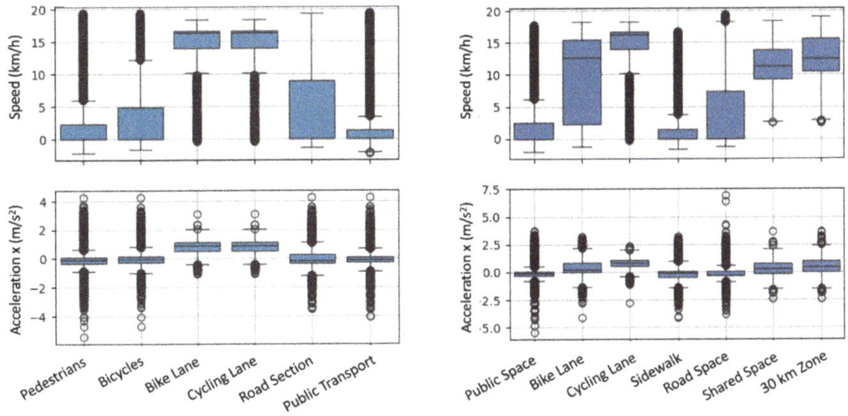

Figure 27. Speed and acceleration distribution of measured values per location category.

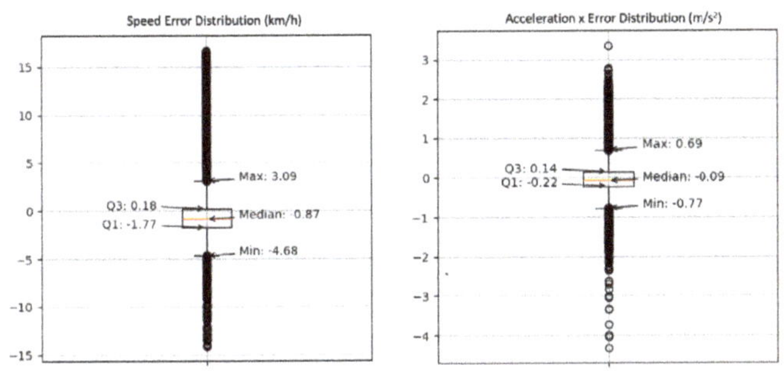

Figure 28. Prediction error distribution for vehicle speed and acceleration.

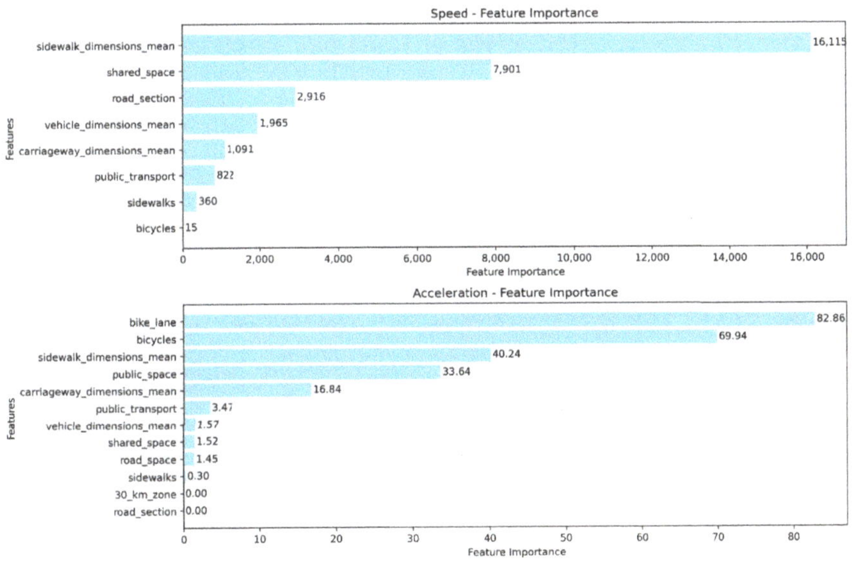

Figure 29. Impact of features on speed and acceleration.

More specifically, Figure 29 illustrates the analysis of the importance of input features for the two different model outputs, speed (top graph) and acceleration (bottom graph). The importance values indicate how much each feature contributes to the predictions of the model. For speed, this kind of assessment showed the following:

(i) The most influential feature is sidewalk width, which has a significantly higher importance score than any other feature.
(ii) The second-most important feature, also with a significant impact, is the presence of shared space.
(iii) All other features, such as, for instance, the vehicle's dimensions, have a much smaller impact on the target variable.
(iv) In this case, the presence of a bicycle has minimal impact.

In contrast, for vehicle acceleration, we observed the following:

(i) The most relevant feature for acceleration is the type of route, e.g., including the presence of a bike lane, followed by the presence of a bicycle on the route.
(ii) Sidewalk width, the presence of public space, and roadway width also play a role but with lower importance.
(iii) Many other features, including vehicle dimensions and shared space, have much smaller contributions.

In summary, the width of the sidewalk appears to be an important factor in predicting both speed and acceleration, suggesting that the dimensions of the sidewalk influence the dynamics of movement. Some other features, such as roadway width, significantly affect both predictions and, in addition, the presence of bike lanes significantly affects acceleration, possibly due to changes in road-sharing conditions (bike lanes generally create smoother transitions for cyclists and vehicles, and also, indirect effects, such as traffic-calming measures in bike-friendly areas, may be considered). Finally, some features have minimal contributions, indicating that they may not be strong predictors in the model.

During the experiments, data were collected from cameras mounted on Genny Zero to enable an automated analysis of potential conflicts with other users in urban spaces. Intel RealSense depth cameras were selected to achieve conflict detection parameters, including the following: the identification and counting of people and vehicles, and the measurement of their distance from the device. A total of four cameras were mounted on Genny Zero, one on each side, providing a 360-degree field of view around the device. The cameras, synchronized during sampling, recorded at a frequency of 5 FPS and a resolution of 480 × 640 pixels.

Figure 30 illustrates the pipeline implemented for processing the sampled video data. The video processing consisted of the following steps:

1. Frames were processed and rotated if necessary.
2. Frames were processed on a backend server, where analyses such as object (human) detection, tracking, and distance estimation were performed.
3. Raw and processed data were stored in a database infrastructure for further analysis.
4. Data were exported in Parquet format to facilitate high-volume structured data analysis.

The processing of video data enabled various types of analysis. In particular, during this process, objects of different types were identified, and their distances were calculated as the median distance of points within the bounding box defining the object in the image (Figure 31).

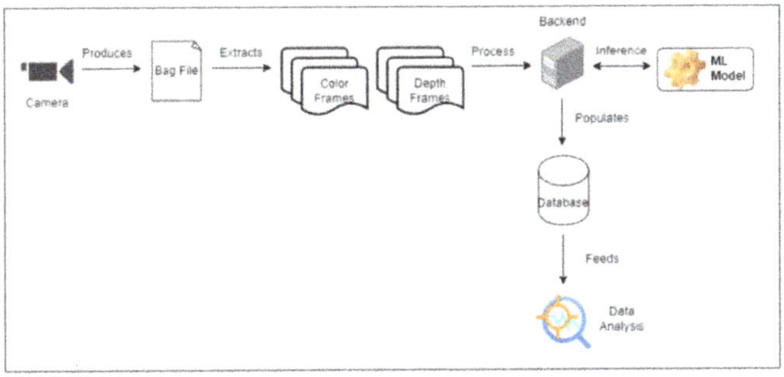

Figure 30. Processing pipeline of video data sampled by Intel RealSense cameras.

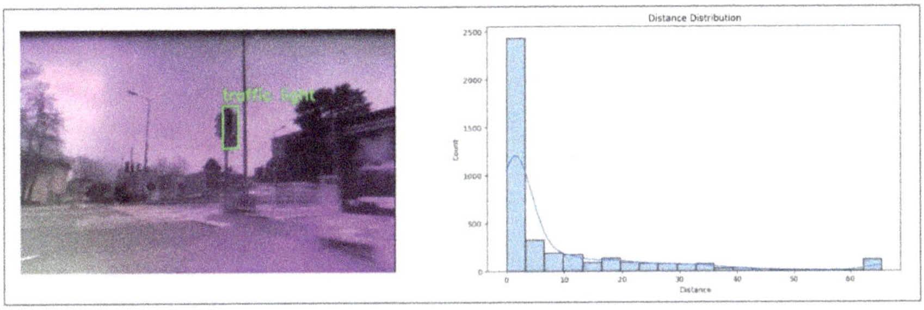

Figure 31. On the (**left**), an example of an Object Identification Bounding Box is shown, while on the (**right**), the distribution of object distances in meters is represented.

The tracking of moving obstacles allowed us to generate time series from the video data, representing the distance between the Genny Zero device and the tracked object. This type of analysis enabled us to identify atypical behaviors of pedestrians or slow mobility devices in situations of potential conflict with Genny Zero, distinguishing between two fundamental concepts.

- **Conflict**: when the vehicle approaches an obstacle (distance below a certain threshold) and brakes. This condition was tested with distance thresholds of 0.5, 1.0, 2.0, and 3.0 m.
- **Encounter**: when the vehicle approaches an obstacle (distance below a certain threshold) but does not brake. This condition was also tested with distance thresholds of 0.5, 1.0, 2.0, and 3.0 m.

Before analyzing episodes of conflict or encounter, the project focused on an exploratory analysis of the sampled data throughout the experiments. Figure 32 represents the number of encounters (count per unique individual/vehicle using tracking analysis), segmented by street and type of road. As shown in the graphs, the highest number of encounters occurred with pedestrians, and the most crowded areas were Lugano Train Station and Via Nassa. It is observed that in areas where the vehicle encounters fewer obstacles, the movement speed is higher, and the time spent traveling in those areas is reduced. However, the measured correlation between the number of people encountered and the average speed is very low and negative (correlation = 0.21). This numerical result is influenced by situations where the device had to stop due to the presence of a red pedestrian traffic light.

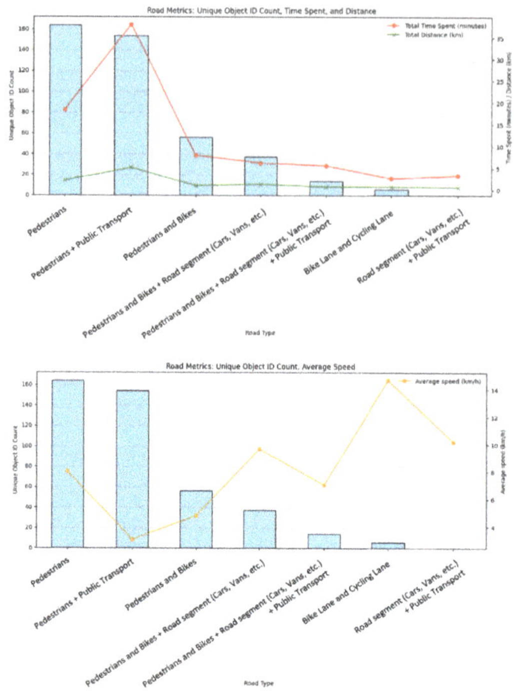

Figure 32. Number of encounters by road type. y-axis: number of people encounters (**left**, bars), time spent in the visited area, and distance traveled in km (**right**, lines).

A similar analysis was conducted by segmenting by street, as shown in Figure 33. The figure accurately represents the measured city environment. For example, as illustrated in the graphs, the highest number of encounters occurred with pedestrians, with the most crowded areas being Lugano Train Station (Stazione SBB Lugano) and Via Nassa. At the Lugano Train Station, the device remained in the area for an extended period while covering only a short distance at a very low average speed. This was due to its movement on the crowded train platform, where it had to navigate within a confined space while waiting for the train's arrival. In contrast, at Via Nassa, a pedestrian street in the city center, the device encountered a high number of pedestrians but was able to cover a greater distance in a shorter time compared to the train station. While still requiring a reduced speed due to pedestrian presence, the street was less congested than the train platform, allowing for more fluid movement. In this case, the linear correlation between the number of people encountered in a specific area (street) and the time spent is positive and significant (correlation = 0.61). Meanwhile, the correlation between the number of people encountered and the average speed is again negative but not significant (correlation = −0.21).

The analysis conducted and presented above allows us to classify the traveled areas as shown in Figure 34.

In Figure 35, the blue paths represent routes traveled in non-crowded areas (with an average speed of 4.13 km/h), while the red paths indicate crowded areas (with an average speed of 1.47 km/h). The dashed line represents the section traveled by train.

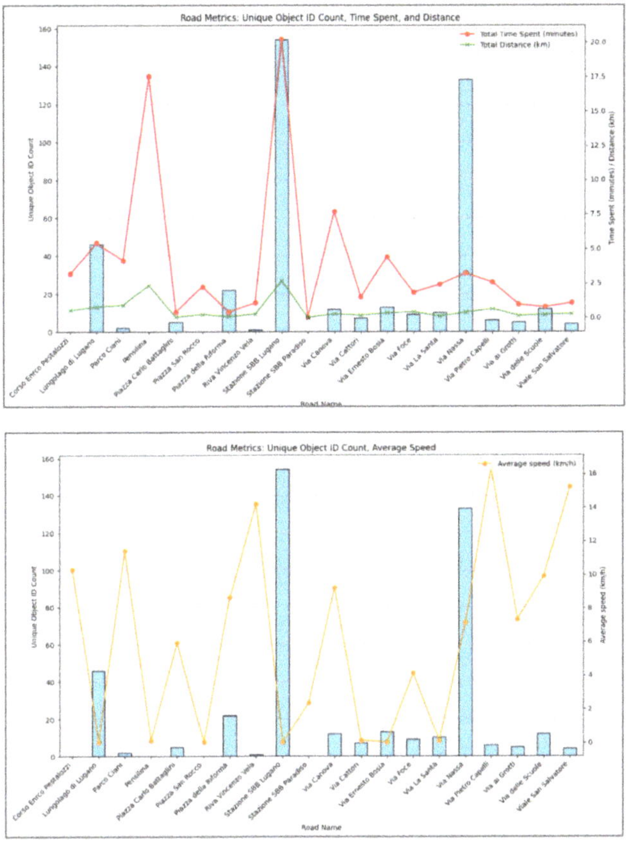

Figure 33. Number of encounters by street and road type.

		Crowdedness	
		Low	**High**
Speed	**Low**	*Possible traffic lights which lowered the median speed ("Via la Santa")*	Many people, hard to move faster in those sections (SBB station)
	High	The Genny can move freely because there are no people around ("Ciani")	Many people, but there is plenty of space for the Genny to move (Via Nassa)

Figure 34. Classification of traveled areas based on proximity analysis.

The analyses presented above refer to the concept of encounter, as defined in the introductory section of this paragraph. However, the analysis was further extended by also focusing on the concept of conflict. The number of conflicts that occurred during the execution of the urban test was calculated. Figure 36 represents the occurrences of conflicts and encounters, varying according to the predefined distance threshold. In general, the vehicle never approached any obstacle within 1 m in situations that required a change in driving mode (braking). Even when moving through crowded areas, the device came very close to obstacles (even below 0.5 m) without generating hazardous situations or discomfort in the traveled area. The number of detected conflict events is limited and does not allow a clustering procedure or a causal analysis.

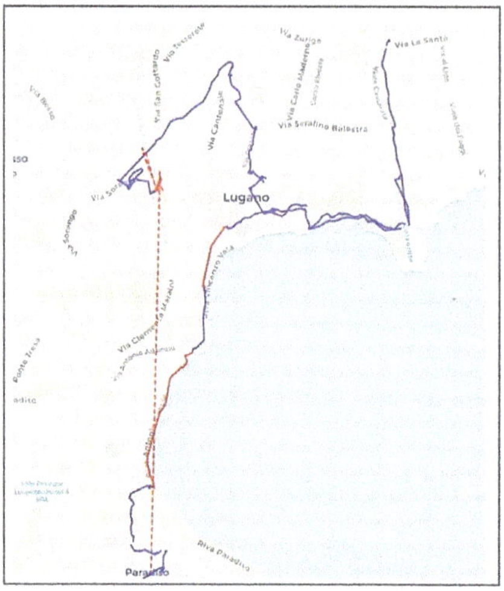

Figure 35. Path in crowded area (red) and non-crowded area (blue).

Most of the conflicts occur in public spaces such as parks, stations, and university entrances. On the other hand, encounters mainly take place along roads and pedestrian crossings. The map shown in Figure 37 visualizes the locations of these conflicts and encounters, with specific markers indicating where they are concentrated. Conflicts are represented with red circles, and each number represents how many conflicts there were in that area, while encounters are marked with blue circles.

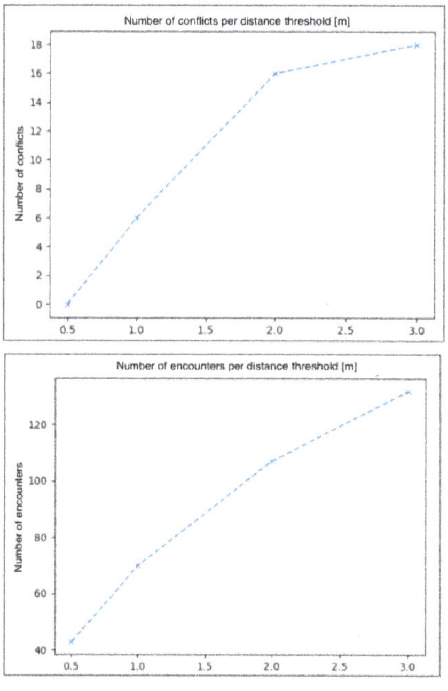

Figure 36. Number of conflicts (**top**) and encounters (**bottom**) based on varying distance thresholds (measured in meters).

Figure 37. Map of conflicts where the distance of the object is less than one meter, requiring braking.

3.3. Data Analysis: Social

The sample size of participants (N = 21) is small, which limits the generalizability of the findings. Despite the quantitative shortage, this study's qualitative validity remains strong due to the diverse range of participants and the in-depth, mixed-method approach combining surveys, focus groups, and real testing. The sample was recruited on a voluntary basis in two ways: (a) 5 interviews with participants in the Ensemble-IMM Citizenship Laboratory recruited among SUPSI students, post-docs, and researchers who also participated in one of the Genny Zero device tests; (b) 16 interviews collected from participants in the two device tests. The interviewed group is not representative of a specific population of the territory or city (Lugano) but rather qualitatively of various age groups, education levels, and origins. The questionnaires were administered in person, in two stages: pre- and a post-test questionnaires. The former questionnaire aimed to assess a general attitude towards mobility, micromobility, and awareness of devices. So the questionnaire contained questions such as the range and frequency of distances, the means used, justification for that device, awareness of micromobility and devices, etc. The latter questionnaire and focus group aimed to more broadly evaluate urban policy-making related to mobility plans, the usability of means and infrastructures, views on the use of means and micromobility, potential for inclusion and reducing pollution and traffic, the need for framing urban mobility policies, etc.

The results of two pre- and post-test surveys, aggregating responses from 21 participants across the social analysis, were encouraging. More than half of the respondents were in the 19–39 age group. The whole sample generally preferred walking for urban mobility (14/21), had a reasonable understanding of micromobility concepts (13/21), but made little use of micromobility solutions (only 5/21). Reasons for non-use were either unspecified (11/21) or attributed to a lack of available services (9/21). Before the test, despite having received information and seen the device, only 7 out of 21 respondents indicated a high or very high probability of using Genny Zero. However, post-test responses showed a significant shift, with 14 out of 21 respondents expressing a high or very high likelihood of using Genny Zero if available. Feedback, while not detailed here, was overwhelmingly positive, particularly regarding braking, comfort, and safety, often favorably compared to scooters. Subsequent online activities, restricted to participants who completed the full Laboratory cycle, revealed strong support for an innovative device like Genny Zero for short daily trips (e.g., last-mile commuting, errands). However, participants emphasized the importance of affordability, widespread availability across the city, safety, and the existence of well-regulated routes (e.g., obstacle-free cycle lanes). During the abovementioned Ensemble-IMM Citizenship Laboratory and the device tests, the 21 sample pre- and post-test surveys did not include questions about gender, since the approach was mainly devoted to capturing opinions on policies and urban mobility strategies. One of the test participants had problems of reduced mobility, using a wheelchair for mobility. This person was strongly positive towards the adoption of modern devices such as Genny Zero.

The interviews and focus group discussions, corroborated by the Ensemble-IMM Citizenship Laboratory's outputs, highlight the multifaceted nature of transportation choices. Various factors influence mobility decisions, ranging from family habits and spatial constraints (such as the necessity to travel long distances for work) to economic availability, family needs, convenience, safety, environmental impact, and individual or social identity. These diverse factors shape mobility choices, yet symbolic, social, political, and identity-driven elements often play a predominant role. For example, the choice of bicycles is based on rational factors or personal histories, but sometimes also reflects an identity decision, linked to environmental consciousness. Conversely, in two interviews, SUVs were identified as symbolic and identity-driven mobility devices, perceived as a contrast to bicycles.

These reflections set the stage for analyzing and envisioning the future of micromobility. While it is generally considered a viable alternative to current mobility systems, concerns about urban infrastructure and road safety persist. Participants expressed apprehension about navigating traffic: one person noted how being surrounded by cars creates unease, while another questioned where micromobility users should travel: on sidewalks, bike lanes, or the road? Safety concerns and the lack of appropriate infrastructure frequently emerged, alongside social considerations such as the need to accommodate family transport. Some respondents acknowledged that, for practical reasons, cars remain the most convenient option, particularly for transporting elderly relatives or children.

Despite these concerns, there is significant interest in expanding micromobility systems. Many participants associated micromobility with improved physical well-being and urban quality of life. Some emphasized that adopting less polluting transport methods and redefining public space is essential, particularly in light of climate change and evolving, modern participatory perceptions of urban life. However, generational differences emerged regarding the urgency of transformation. Younger participants, citing the drastic changes experienced during the COVID-19 lockdown, stressed the need for immediate measures to reduce ecological impact and redesign public spaces responding to future visions, rather than catering to current political lobbies or needs that in the next few years will not be relevant. Older generations, in contrast, expressed concerns about economic feasibility and compatibility with existing commercial activities and daily routines. Three participants advocated for gradual change that balances diverse interests, emphasizing the role of technology in accelerating the transition. Generational perspectives were thus distinct on this issue.

Despite these differences, consensus emerged on policies that could encourage micromobility. These include lowering speed limits, implementing public awareness campaigns, expanding restricted traffic zones, introducing financial incentives, and expanding participation and transparency in policy-making. Some interviewees stated that micromobility is indispensable for making cities more livable. One respondent emphasized their preference for public transportation and walking to minimize their individual carbon footprint. However, another participant pointed out that Lugano still has a long way to go in this regard, noting the scarcity of bike lanes, as well as green and public spaces, which are often reserved for luxury tourism or aesthetic considerations, limiting their broader usability.

Some participants critiqued local strategies, arguing that merely eliminating parking spaces without providing viable alternatives discourages car use inefficiently. Some respondents highlighted how public space allocation is dictated by lobbying interests, making it difficult to advocate for alternative approaches. Others emphasized the importance of infrastructure for micromobility, asserting that having designated spaces is crucial. The need for functional bike lanes, not just for leisure but for daily commuting, was stressed. The inadequacy of Lugano's cycling infrastructure was a recurring concern. Participants pointed out that a lack of sufficient bike lanes impedes micromobility adoption. Some suggested that micromobility should be encouraged for last-mile transport solutions but noted that the necessary infrastructure, such as safe cycling routes, as well as green and semi-pedestrian areas, is still lacking. Specific proposals included creating a dedicated bike route connecting the train station to the city center. While there is recognition that long-term planning is required to improve infrastructure, many participants expressed skepticism about the political willingness to make these changes due to vested interests resistant to altering the status quo. One person cited recent urban projects as examples of this reluctance, pointing out the absence of public spaces, green areas, and bike lanes in new developments.

Many interviewees argued that Lugano should actively promote micromobility by increasing peripheral park-and-ride facilities and adopting a more ambitious transportation masterplan capable of challenging entrenched lobbying interests. Suggestions included permanently closing specific streets to car traffic and prioritizing pedestrian-friendly urban design. One participant noted that recent initiatives have primarily eliminated free public parking while expanding paid parking options, ultimately maintaining car dominance.

Interviewees expressed a widespread awareness of the urban implications of micromobility. Genny Zero was mentioned as a potential catalyst for public space innovation. A semi-pedestrianized city center, similar to those in Bern, Basel, or Zurich, was envisioned as a desirable goal. Some respondents suggested that micromobility solutions could be integrated into multimodal/exchange transport systems alongside bicycles and electric scooters. However, concerns arose regarding competition with public transportation and potential conflicts with pedestrian and cycling traffic. The discussion emphasized the need for comprehensive investment in infrastructure and accessibility measures to support diverse transport options.

A central theme was the necessity of public participation in urban policy-making. One interviewee underscored the need for inclusive decision-making processes concerning public spaces, criticizing their privatization and reduction. Some proposed mechanisms to ensure public spaces serve diverse needs rather than being subjected to commercial interests.

Sustainability was another focal point. Participants widely supported discouraging car use in cities and suggested various strategies, including economic incentives, rescheduling urban activities, and adapting micromobility options to individual needs. Public awareness campaigns and safety training programs were proposed to facilitate micromobility adoption. Some participants even suggested implementing a licensing system for micromobility users, though opinions diverged on whether stringent regulations might hinder widespread adoption.

There was consensus on the need for park-and-ride facilities where users could leave their cars out of city center and access multiple transport options, such as bicycles, scooters, or Genny Zero vehicles. Some suggested that certain areas should be entirely restricted to car access, while others proposed residential zones with strict speed limits. One participant noted that contemporary society, along with the consolidated vision of the city, is conditioned to view cars as essential, but widespread adoption of micromobility solutions could fundamentally alter future perceptions of urban transport and the entire conception of urban spaces.

Divergent views emerged on the pace of transition. While most agreed on reducing motorized traffic, some emphasized protecting economic activities, while others argued that the COVID-19 lockdown demonstrated society's ability to adapt rapidly to new mobility patterns. Participants stressed the importance of long-term urban planning (5–10 years) to support behavioral shifts through infrastructure investments.

The survey captured nuanced perspectives on micromobility, reflecting generational, social, and practical considerations. This rich, contextual insight supports a deeper understanding of user perceptions and urban mobility challenges beyond statistical representation. Four key themes remain central to the debate: citizen participation in public policy and urban space management; sustainability; the relationship between business/industry and public administration; and the interplay between human and goods mobility.

Addressing these social and political issues will be crucial in shaping a turning point toward a vision of the city as more a inclusive, sustainable, and livable urban future. Rethinking urban spaces as multifunctional environments that foster social inclusion and exchange seems essential. Encouraging community engagement in the planning process can reinforce collective responsibility for urban development [31]. Digital platforms could be leveraged to facilitate participatory governance and real-time feedback from citizens. Education and public discourse on sustainable urbanism must be promoted to foster a shared vision of mobility and livability. Policies could be crafted to ensure economic and social equity in access to new transportation systems, with micromobility being a component. Finally, a long-term urban identity that embraces adaptability, resilience, and innovation should be cultivated to shape the future of modern urban citizenship [32].

4. Discussion

The findings of the Inclusive MicroMob project provide a multifaceted perspective on the integration of new micromobility solutions in urban environments. The Swiss case studies, particularly that of Lugano, highlight both the opportunities and challenges associated with the adoption of inclusive micromobility solutions such as the Genny Zero device. By analyzing current regulatory frameworks, the state of urban contexts, and user perceptions, this study contributes to a broader understanding of how these technologies can enhance urban accessibility and sustainability.

From an urban planning perspective, this study confirms that the integration of micromobility solutions within existing mobility networks requires careful consideration of spatial, functional, and regulatory aspects. Regarding public space, the introduction of new micromobility solutions has implications for urban planning, management, and maintenance. If the adoption of devices like Genny Zero aligns with recent rapid shifts in urban mobility trends, it will be necessary to reassess the allocation of public space dedicated to slowing mobility. This shift would also have significant consequences for the management and maintenance of urban centers. A well-developed cycling infrastructure, expanded pedestrian areas, and appropriate communication strategies could stimulate demand in urban centers such as Lugano, as analyzed in the project. Functionally, micromobility serves as an alternative to public transport or last-mile travel, particularly in areas with moderate topographical constraints. While steep road gradients may pose accessibility challenges, the compact size and technical characteristics of the Genny Zero device allow for its integration into the existing transport system. However, to facilitate effective integration, an expansion of charging infrastructure is necessary. The analysis of four Swiss case studies (Chur, Freiburg, Andermatt, and Lugano) revealed a limited and often sparse charging network. Additionally, if Genny Zero were to be implemented as a shared mobility service, the operating model (docked or dockless) must be defined. Assuming integration with the existing bike-sharing system, which primarily operates in a docked mode, appropriate space planning is required to ensure accessibility, inclusivity, and efficient management and maintenance of the service.

Another crucial consideration in micromobility integration is the sustainability of its use throughout the entire life cycle. The comparative life cycle assessment (LCA) of the Genny Zero electric vehicle indicates a low environmental impact, estimated at 45 gCO2eq per passenger-kilometer, assuming a lifespan of 40,000 km [33]. However, this impact is highly dependent on the electricity source, with the Swiss energy mix (predominantly hydro and nuclear) resulting in lower emissions. The sustainability of electric vehicles like Genny Zero also depends on the transport modes they replace; while substituting internal combustion vehicles is beneficial, replacing public transport or walking may lead to higher emissions. Additionally, vehicle longevity is a key factor, as longer lifespans contribute to a reduced overall environmental footprint. The widespread adoption of micromobility could significantly contribute to the reduction in CO_2 emissions from road transport, supporting climate goals while alleviating urban congestion and pollution [34].

Nevertheless, the regulatory framework remains a critical determinant of feasibility and implementation. A comparison of Swiss and Ticino regulations suggests that while national policies provide general guidelines, regional adaptations play a decisive role in shaping the practical deployment of micromobility solutions.

From a technological perspective, urban tests provided valuable insights into the performance and adaptability of the Genny Zero device. Data collected through inertial sensors, GPS tracking, and RealSense cameras indicate that the device successfully navigates diverse urban scenarios while ensuring safety and maneuverability. A key concern in the current discourse on micromobility is safety and potential conflicts with other road users. The Inclusive MicroMob project analyzed speed distribution and acceleration patterns, revealing that designated cycling infrastructure enhances the operational efficiency of micromobility devices. Notably, interaction analysis indicated minimal conflicts with other road users, suggesting that with appropriate urban planning, these devices can be effectively integrated into shared mobility environments.

From a social perspective, user feedback underscores the importance of addressing public perceptions and fostering participatory urban governance. Results from the IMM event and the Ensemble-IMM Citizenship Laboratory reveal a combination of enthusiasm and skepticism regarding micromobility adoption. While participants acknowledge the potential benefits of these solutions for sustainable and inclusive urban transport, concerns persist regarding infrastructure readiness, safety, and the role of micromobility within existing mobility hierarchies. Furthermore, generational differences in attitudes suggest that younger users are more receptive to new mobility technologies, whereas older participants emphasize the need for gradual policy shifts that accommodate established commuting behaviors. The Lugano case study highlights the necessity of engaging the public in a collective learning process to explore emerging mobility needs in relation to future micromobility offerings. Participatory processes, such as the Citizenship Laboratory, serve as critical platforms for structuring discussions and investigating user preferences.

Despite these positive findings, several limitations must be acknowledged. First, the tests and the Citizenship Laboratory were conducted in a specific urban context, and while Lugano provides an informative case study, further research in different geographic and regulatory environments is necessary to generalize the results. Additionally, this study primarily focuses on short-term usability and user acceptance; long-term behavioral shifts and the economic feasibility of micromobility integration remain areas for further exploration. Finally, this study highlights the necessity of continued stakeholder engagement, particularly in policy formulation, to ensure that micromobility solutions align with broader urban sustainability objectives.

5. Conclusions

The findings of this study reinforce the notion that achieving inclusive urban mobility requires both technological innovation and an adaptive regulatory framework that considers infrastructural, social, and policy dimensions. The Inclusive MicroMob project demonstrates that micromobility solutions, such as the Genny Zero device, can enhance urban accessibility, particularly when integrated with existing public transport systems and supported by inclusive urban planning strategies.

A key insight from this research is that the successful implementation of innovative mobility solutions depends on collaborative learning processes involving multiple stakeholders. The findings suggest that even the most advanced mobility solutions must be introduced within a structured framework of public awareness, participatory governance, and regulatory support to prevent political and psychological barriers to adoption. Public perception remains a crucial determinant of success; therefore, future micromobility policies should prioritize community engagement and co-design methodologies.

Future research should explore the scalability of micromobility models in diverse urban contexts, assessing long-term behavioral impacts and economic sustainability. Additionally, further work could concentrate in the collection of data not only on different urban contexts but also from different users driving the Genny device: this would allow a deeper behavioral analysis related to Pedestrian Safety. Furthermore, comparative analyses between cities with varying infrastructural readiness and regulatory flexibility could provide additional information on the optimal conditions for the integration of micromobility. Investigating the intersection of micromobility and emerging digital governance tools, such as participatory platforms for urban decision-making, could further enhance the inclusivity of urban transport planning.

In conclusion, creating inclusive cities requires a holistic approach that extends beyond infrastructure and technology; it necessitates active citizen engagement, responsive regulatory frameworks, and a commitment to sustainable and equitable mobility solutions. Micromobility has the potential to redefine urban movement, but its full benefits will only be realized through a collective commitment to fostering adaptable, accessible, and people-centered urban spaces.

Author Contributions: Conceptualization, A.R., M.P., and F.G.; data curation: A.Q. and L.M.; formal analysis, A.R., M.P., F.B., A.Q., L.M., and M.C.; funding acquisition, A.B.; investigation: A.R., M.P., F.B., A.Q., and A.B.; methodology, A.R., M.P., F.B., F.G., and A.B.; project administration, A.B.; software, A.Q., L.M., and M.C.; supervision, A.R., M.P., and L.D.M.; validation, A.R., F.B., and A.B.; visualization, L.D.M., L.M., and M.C.; writing—original draft: A.R., M.P., F.B., and L.D.M.; writing—review and editing, A.R., M.P., F.B., L.D.M., and A.Q. All authors have read and agreed to the published version of the manuscript.

Funding: This research was funded by Federal Office for Spatial Development; the Federal Department of the Environment, Transport, Energy and Communications; and the Section for Mobility and International Affairs.

Informed Consent Statement: Informed consent was obtained from all subjects involved in the study.

Data Availability Statement: Dataset available on request from the authors.

Acknowledgments: We would like to express our heartfelt gratitude to Aurelio Vigani, at the Federal Office for Spatial Development, for his invaluable support and contributions; Elena Marchiori, at Lugano Living Lab, for her invaluable support and expertise; Giovanni Fulgoni, at Genny Factory SA, for his insightful contributions and guidance; Alex Bordini, at Genny Factory SA, for his technical expertise and valuable assistance; Niccolò Cuppini, at the University of Applied Sciences and Arts of Southern Switzerland (SUPSI), for his thoughtful perspectives and valuable input. Their collective efforts and encouragement have been instrumental in shaping the direction and outcomes of this research.

Conflicts of Interest: The authors declare no conflicts of interest. The funders had no role in the design of the study; in the collection, analyses, or interpretation of data; in the writing of the manuscript; or in the decision to publish the results.

References

1. Gehl, J. *Cities for People*; Island Press: Washington, DC, USA, 2010.
2. Newman, P.; Kenworthy, J. *Sustainability and Cities: Overcoming Automobile Dependence*; Island Press: Washington, DC, USA, 1999.
3. United Nations. *World Cities Report 2020: The Value of Sustainable Urbanization*; World Cities Report; UN Habitat: Nairobi, Kenya, 2020.
4. Lefebvre, H.; Kofman, E.; Lebas, E. *Writings on Cities*; Wiley: Hoboken, NJ, USA, 1996.
5. Turok, I.; McGranahan, G. Urbanization and economic growth: The arguments and evidence for Africa and Asia. *Environ. Urban.* **2013**, *25*, 465–482. [CrossRef]
6. Fainstein, S.S. *The Just City*; Cornell Paperbacks, Cornell University Press: Ithaca, NY, USA, 2010.
7. Agyeman, J. *Sustainable Communities and the Challenge of Environmental Justice*; NYU Press: New York, NY, USA, 2005.
8. Shaheen, S.; Cohen, A. *Shared Micromobility Policy Toolkit: Docked and Dockless Bike and Scooter Sharing*; Technical Report; Institute of Transportation Studies, UC Berkeley: Berkeley, CA, USA, 2019.
9. Banister, D. The sustainable mobility paradigm. *Transp. Policy* **2008**, *15*, 73–80. [CrossRef]
10. Olabi, A.; Wilberforce, T.; Obaideen, K.; Sayed, E.T.; Shehata, N.; Alami, A.H.; Abdelkareem, M.A. Micromobility: Progress, benefits, challenges, policy and regulations, energy sources and storage, and its role in achieving sustainable development goals. *Int. J. Thermofluids* **2023**, *17*, 100292. [CrossRef]
11. Vizmpa, C.; Botzoris, G.; Lemonakis, P.; Galanis, A. Micromobility in Urban Trail Paths: Expanding and Strengthening the Planning of 15-min Cities. *Land* **2023**, *12*, 2181. [CrossRef]
12. Tamagusko, T.; Gomes Correia, M.; Rita, L.; Bostan, T.C.; Peliteiro, M.; Martins, R.; Santos, L.; Ferreira, A. Data-Driven Approach for Urban Micromobility Enhancement through Safety Mapping and Intelligent Route Planning. *Smart Cities* **2023**, *6*, 2035–2056. [CrossRef]
13. Nikolaeva, A.; Adey, P.; Cresswell, T.; Lee, J.Y.; Nóvoa, A.; Temenos, C. Commoning mobility: Towards a new politics of mobility transitions. *Trans. Inst. Br. Geogr.* **2020**, *45*, 1–18. [CrossRef]
14. Gössling, S. Integrating e-scooters in urban transportation: Problems, policies, and the prospect of system change. *Transp. Res. Part D Transp. Environ.* **2020**, *79*, 102230. [CrossRef]
15. Fishman, E. Bikeshare: A review of recent literature. *Transp. Rev.* **2015**, *36*, 132–147. [CrossRef]
16. Nuvolati, G. *Mobilità Urbana: Complessità Fisiche e Culturali*; Firenze University Press: Florence, Italy, 2007.
17. Di Fazio, C.; Paradiso, M. Mobilità urbane emergenti: Valutazione critica della micromobilità di bici e monopattini tra individualità, pubblico e privato. *Doc. Geogr.* **2023**, *2*, 515–532.
18. Treccani Istituto della Enciclopedia Italiana. *Definition of Micromobility*; Treccani Istituto della Enciclopedia Italiana: Rome, Italy, 2020.
19. Oeschger, G.; Caruso, G.; Gauvin, C. *Parameters and Policies for Micromobility Devices*; Forum Internazionale dei Trasporti: Rome, Italy, 2020.
20. Rossato, S. Human-Powered and Electric-Assisted Micromobility: Challenges and Opportunities. Master's Thesis, University of Padua, Padova, Italy, 2021.
21. Becker, H.; Ciari, F.; Axhausen, K.W. Comparing car-sharing and ride-sourcing services: A case study in Switzerland. *Transp. Res. Part A Policy Pract.* **2020**, *131*, 54–70.
22. Meng, S.; Brown, A.; Klein, N.; Thigpen, C.; Haydu, B.; Stout, N. *Shared Scooter Parking: The Role of Parking Density and Land Use in Compliance and Demand*; University of Oregon: Eugene, OR, USA, 2024.

23. United Nations Development Programme. *Urban Mobility and Regulatory Frameworks for Micromobility*; United Nations Development Programme: New York, NY, USA, 2021.
24. OETV. *Ordinanza Concernente le Esigenze Tecniche per i Veicoli*; Swiss Federal Council: Bern, Switzerland, 1995.
25. OAC. *Ordinanza Sull'ammissione Alla Circolazione di Persone e Veicoli*; Swiss Federal Council: Bern, Switzerland, 1976.
26. ONC. *Ordinanza Sulle Norme Della Circolazione Stradale*; Swiss Federal Council: Bern, Switzerland, 1962.
27. Touring Club Switzerland. Technical and Regulatory Requirements for Micromobility Devices in Switzerland. Available online: https://www.tcs.ch/it/ (accessed on 20 March 2025).
28. Ticino Cantonal Regulations. (n.d.). Normative Cantonali per la Regolamentazione Della Micromobilità. Canton of Ticino. Available online: https://www4.ti.ch/di/strade-sicure/mobilita-lenta/mobilita-pedonale (accessed on 20 March 2025).
29. Pearson, K. Note on regression and inheritance in the case of two parents. *Proc. R. Soc. Lond.* **1895**, *58*, 240–242.
30. Chen, T.; Guestrin, C. XGBoost: A scalable tree boosting system. In Proceedings of the 22nd ACM SIGKDD International Conference on Knowledge Discovery and Data Mining, KDD'16, New York, NY, USA, 13–17 August 2016; pp. 785–794.
31. Lemanski, C. *Citizenship and Infrastructure: Practices and Identities of Citizens and the State*; Routledge Studies in Urbanism and the City; Routledge: Oxfordshire, UK, 2019.
32. Bignami, F.; Hanakata, N. Platform urbanisation, infrastructures and techno-politics: The turn towards urban citizenship. *J. Eur. Stud.* **2024**, *54*, 42–58. [CrossRef]
33. Reck, D.J.; Martin, H.; Axhausen, K.W. Mode choice, substitution patterns and environmental impacts of shared and personal micro-mobility. *Transp. Res. Part D Transp. Environ.* **2022**, *102*, 103134. [CrossRef]
34. de Bortoli, A. Environmental performance of shared micromobility and personal alternatives using integrated modal LCA. *Transp. Res. Part D Transp. Environ.* **2021**, *93*, 102743. [CrossRef]

smart cities

MDPI

Review

Insights from Smart City Initiatives for Urban Sustainability and Contemporary Urbanism

Águeda Veloso, Fernando Fonseca and Rui Ramos *

Centre for Territory, Environment and Construction (CTAC), University of Minho, 4800-058 Guimarães, Portugal; aguedafilipa.25@gmail.com (Á.V.); ffonseca@civil.uminho.pt (F.F.)
* Correspondence: rui.ramos@civil.uminho.pt; Tel.: +351-253-510-200

Highlights:

- Smart city initiatives are varied but primarily focus on governance and environment, with less attention to other critical dimensions needed for sustainable urban practices.
- Smart city initiatives heavily rely on ICT for managing urban operations but lack comprehensive and inclusive assessment studies on their sustainable and social benefits.

Implications of the Main Findings:

- The findings highlight the need for more holistic frameworks in smart city approaches to efficiently address urban sustainability challenges and improve urban practices.
- Standardized assessment methods are needed for evaluating the benefits of smart technologies on sustainable development and urban practices.

Abstract: Urbanization growth poses various challenges, such as congestion, pollution, and resource consumption, prompting city planners and governments to adopt smart systems to manage these issues more efficiently. Despite widespread adoption, there is no consensus on the defining attributes of smart cities, particularly regarding their role in urban sustainability and contemporary urbanism. This paper provides a literature review to understand the implications of smart city initiatives for sustainable urban planning, focusing on practices in Singapore, Helsinki, Barcelona, and Medellin. Based on 71 publications surveyed from Scopus and Web of Science, this paper evaluates smart, sustainable initiatives undertaken in these four cities across six smart domains: mobility, governance, environment, people, living, and economy. This review shows that most studies focus on Barcelona and Singapore, particularly in the domains of smart environment and governance. Despite differing urban contexts, the notion of "smart" is closely tied to using information and communication technologies to drive urban operations. This analysis identifies a lack of assessment studies on the benefits of smart cities in terms of urban sustainability and a lack of holistic approaches to address the complex challenges cities face in achieving sustainable development.

Keywords: smart cities; sustainability; urban planning; planning practices

1. Introduction

Today, 56% of the world's population lives in urban areas [1]. Projections indicate that urbanization and global population growth could add another 2.2 billion people to urban areas over the next few decades, potentially raising the world's urbanization rate to 68% by 2050 [1]. This rapid urbanization presents a myriad of global challenges, including congestion, pollution, greenhouse gas (GHG) emissions, depletion of natural resources, environmental degradation, inadequate infrastructure, urban poverty, and high land prices [2]. These challenges have prompted urban planners, decision-makers, and researchers to explore ways to make cities more livable and sustainable.

Urban sustainability is defined as a city's ability to successfully operate across environmental, social, and economic domains simultaneously [3]. Urbanism plays a crucial role in shaping the development of urban areas to address these challenges. Generally, urbanism concerns itself with the form and function of cities, providing a framework for analyzing and understanding urban environments through integrated and design-oriented approaches [4,5].

A smart city (SC) is a contemporary urban approach that emerged in the late 1990s [6,7], leveraging technology and data-driven solutions to enhance urban efficiency, sustainability, and livability [8,9]. SCs have been characterized through tools such as information and communication technologies (ICTs), Internet of Things (IoT), open data, public–private collaboration, and competition. They often comprise six components, which include smart mobility, smart environment, smart governance, smart living, smart people, and smart economy [10–13]. Although SCs are widely recognized for their potential to bring multiple urban benefits to all these dimensions, there is no clear consensus on their precise definition or defining characteristics [7,14]. The concept of SCs has evolved alongside technological advancements and changing urban needs, resulting in a wide range of interpretations and initiatives [6,7]. It encompasses a variety of attributes, depending on the context in which it is applied [14–16]. This diversity often leads to varied priorities and objectives within SC initiatives, ranging from environmental sustainability and social equity to economic development and administrative efficiency [17,18]. As a result, SC initiatives encompass not only technological innovations but also broader dimensions of urban development and sustainability [7,19,20].

The emergence of the term "smart sustainable" seems to indicate a strong association between the concepts of smartness and urban sustainability. According to the International Telecommunication Union (ITU), a smart sustainable city is "an innovative city that uses ICT and other means to improve quality of life, the efficiency of urban operation and services, and competitiveness, while ensuring that it meets the needs of present and future generations with respect to economic, social, environmental as well as cultural aspects" [21]. Within this context, SCs have been viewed as a way of achieving more sustainable urban development [22], while other authors argued that to be smart, cities need to be sustainable first [20]. However, the extent to which a SC contributes to urban sustainability is unclear [23] and there is a substantial lack of research exploring the benefits of SC initiatives in urban sustainability [24]. Much of the existing SC literature is concentrated in the fields of computer and engineering sciences, leaving urban sustainability issues underexplored [10]. It has been argued that the SC literature frequently prioritizes technological and economic concerns over environmental considerations, creating tensions between the concepts of smart and sustainable cities [10,25]. Thus, there is a need to review and examine the contribution of SC approaches on urban sustainability [22] and, more specifically, whether smart urbanism, which serves as a valuable framework for evaluating contemporary urbanism [24], truly promotes sustainable development. There are various reviews on the definition and practice of SCs [26–28], but there have not been any systematic review studies to discuss the implications of SC initiatives on urban sustainability and smart urbanism, particularly through a comparative lens across diverse cities and SC domains. Further, there is a need for a deeper understanding of how SC policies affect different urban groups and how open data platforms can foster innovative applications and practices for smart urban development [29]. Furthermore, while the key focus of SC agendas has been on the replicability and scalability of urban solutions [18], this emphasis has exposed significant limitations as universal solutions often fail to account for the specific needs and contextual conditions of individual cities. Müller et al. [18] suggest that rather than prioritizing scalability, it would be more effective to focus on the learning potential of diverse urban contexts. Context-specific insights and adaptive strategies are better suited to addressing the unique challenges each city faces.

In response to the research gaps identified, this paper seeks to explore the implications of SC initiatives for sustainable urban planning, emphasizing the learning potential from different urban contexts. By analyzing SC initiatives in four cities—Singapore, Helsinki, Barcelona, and Medellin—across different continents, this study seeks to uncover how these strategies can be tailored to varied urban environments, providing valuable insights into their broader impact on urban sustainability and contemporary urban practices. The selection of these cities provides an opportunity to explore the distinct challenges each urban context presents and evaluate how local conditions shape the implementation and outcomes of SC initiatives. This approach aligns with Müller et al.'s [18] argument, which advocates for focusing on context-specific insights and learning opportunities rather than replicating one-size-fits-all solutions. By exploring how different cities adapt SC strategies to their unique socioeconomic, cultural, and environmental conditions, this study aims to deepen the understanding of adaptive smart urbanism. This approach can shed light on how SC practices intersect with sustainable and smart urbanism, offering insights into strategies that promote more adaptive and inclusive urbanism. The findings have the potential to promote SC strategies that are not only technologically innovative but also contextually relevant, advancing sustainable planning and urbanism worldwide. By contributing to the discourse on integrating smart technologies with sustainability and highlighting the learning potential from diverse urban contexts, this research supports the development of SC initiatives that better address the unique needs and challenges of cities in varying global settings.

2. Materials and Methods

2.1. Rationale for Selecting the Case Studies

This paper aims to review and analyze SC initiatives in Barcelona, Helsinki, Medellin, and Singapore, and to assess their contributions to urban sustainability. The selection of these cities as case studies is based on four main reasons. First, they are located on three different continents, ensuring a broad geographic representation. Second, the four selected cities represent both developed and developing countries, providing a diverse range of urban contexts and experiences. The goal was to capture the varied challenges and opportunities that different economic contexts present in SC development. Third, Singapore and Helsinki are among the top-ranking SCs globally, placing within the top ten in the recent Smart City Index of 2024, while Barcelona occupies a lower, though still notable, position in that ranking [30]. Although Medellin does not rank as highly, it is a significant case study due to its innovative approach to SC development in the Global South. Fourth, each of these cities has pioneered unique approaches to sustainable urban planning: Singapore is renowned for its efficient use of technology in managing urban infrastructure; Helsinki, a leading example from Northern Europe, is recognized as the continent's most innovative city and a top destination for foreign investments, boasting the best digital twin in the Kalasatama neighborhood [26]; Barcelona provides a Southern European perspective, aiming to become the most connected city globally through heavy investments in IoT applications [31]; and finally, Medellin has gained prominence amongst cities in the Global South due to urban management initiatives, such as social urbanism, which have transformed the city [32]. Thus, this selection was intended to include cities at different stages of SC maturity that can potentially offer a comprehensive understanding of how SC initiatives and sustainable urban planning practices have been implemented across diverse urban contexts.

2.2. Literature Review Process

The review process followed the PRISMA (Preferred Reporting Items for Systematic Review and Meta-Analysis) guidelines. Figure 1 illustrates the key steps: (i) identification of the relevant literature; (ii) screening based on predefined criteria; (iii) data extraction from selected studies; and (iv) categorization of articles by city and SC domain.

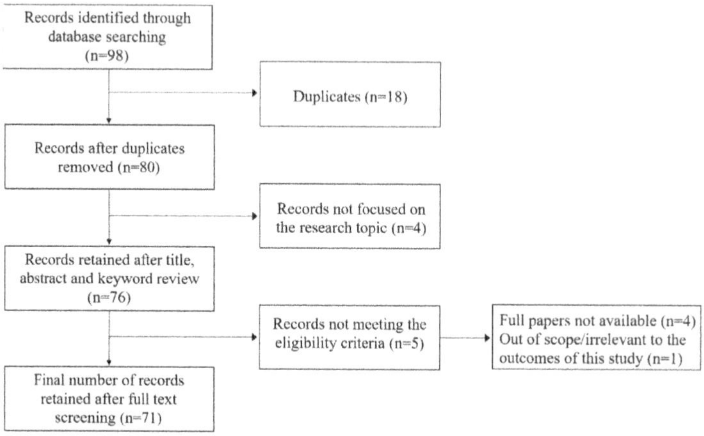

Figure 1. Flow diagram of literature review.

To initiate the search process, Scopus and Web of Science databases were selected and queried due to their established reputation and comprehensive coverage [24]. The search employed the following keywords in the title, abstract, and keywords fields: (smart cities) and (Singapore or Helsinki or Barcelona or Medellin) AND (urban sustainability or smart sustainable or smart sustainability), across all publication fields. The search scope was limited to peer-reviewed English-language journal articles, spanning from the inception of electronic bibliographic databases to 31 May 2024.

In the second step, the reference lists of the selected articles were reviewed to identify any additional studies that may have been missed out. The screening process involved evaluating the eligibility of the retrieved articles. Firstly, duplicate publications from the two databases were removed. Then, the titles, abstracts, and keywords were manually checked to determine if the publications predominantly addressed the topic of SCs in the four selected case studies. Papers not focused on the research topic or where the four analyzed cities appeared incidentally were excluded. Subsequently, the papers underwent independent double-screening by two members of the research team. Documents without full texts were also removed from the analysis, as well as studies that were out of scope or irrelevant to the objectives of this study. Any disagreements were resolved through discussion and consensus among the authors.

After selecting the articles, relevant data were extracted. This included the article title, authors, affiliations, keywords, year of publication, publication title, number of citations, and full texts. The extracted information was then used to conduct a concise bibliometric analysis to assess research trends in this area and identify primary sources, authors, and affiliated research institutions.

Lastly, the extracted articles underwent classification based on their respective SC domains and initiatives adopted in each city. Initially, the literature was categorized by city, and then the findings were organized into the following six major SC domains described in the Introduction, which have been widely used in the literature to evaluate SC initiatives: smart mobility, smart environment, smart governance, smart people, smart living, and smart economy [10–13]. This categorization was performed to extract key urban sustainability initiatives and practices implemented in each of the four SCs.

Following the described steps, a total of 98 references were identified from the Scopus and Web of Science databases. After removing 18 duplicates, 80 documents underwent individual screening based on their titles, abstracts, and keywords. Four papers were excluded because they focused on cities in Abu Dhabi, Iraq, South Korea, and Venezuela, which were not within the scope of this study. Thus, 76 records underwent eligibility assessment through full-text screening. Of these, four papers could not be retrieved due to lack of full access, while another paper was excluded after full-text screening because its content was deemed irrelevant or out of scope for the objectives of this review. As a result, a total of 71 records were eligible for inclusion in this review.

3. Results

3.1. Overview of the Selected Papers

The 71 eligible papers were published between 2013 and 2023. As depicted in Figure 2, the topic has garnered increasing attention in recent years, with 59% of the documents published in the last four years. All retrieved documents were published as articles in a total of 49 journals spanning various subject areas. The journals with the most publications included *Sustainability* (10 publications), *Sustainable Cities and Society* (4), *Cities* (3), *IEEE Access* (3), *Journal of Urban Technology* (3), and *Sensors* (3).

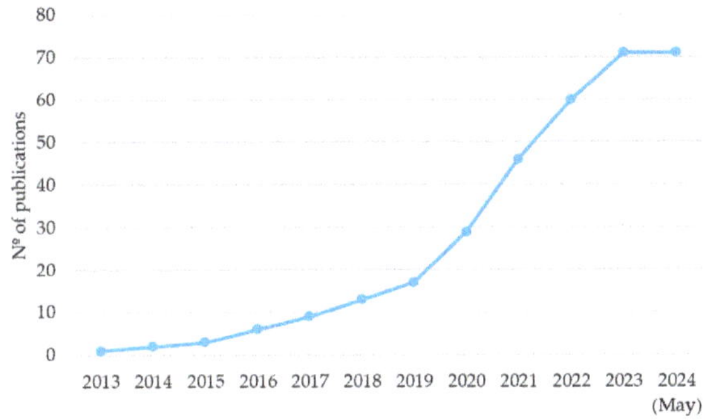

Figure 2. Number of papers (accumulated) published by year.

The 71 documents contained about 380 keywords. The most frequently occurring keywords were *smart cities* (appearing in 62 papers), *sustainability* (12), *sustainable cities* (10), *urban governance* (8), *environment* (6), *Internet of Things* (6), *Singapore* (6), and *smart sustainable cities* (6).

The universities in the four selected cities are among the institutions with the most publications on the researched topic. From Spain, these include the University of Barcelona, the Open University of Catalonia, the Autonomous University of Barcelona, and Ramon Llull University; in Finland, they are the University of Helsinki, Aalto University, the University of Vaasa, and the Lappeenranta-Lahti University of Technology; in Colombia, they are the Colombian American University Corporation and the National University of Colombia; and from Singapore, this includes the National University of Singapore.

According to the Scopus and WoS databases, the most cited documents in this field until May 2024 are the article by Zygiaris [33] with 465 citations, followed by the studies by Angelidou [34] with 249 citations, and Mora et al. [35] with 206 citations.

Focusing on the analysis of the four selected cities, Table 1 presents the distribution of the retrieved papers by city.

Table 1. Returned studies that include the four cities.

Analyzed Cities	References
Barcelona	[10,27,33–63]
Helsinki	[13,26,35,36,64–73]
Medellin	[32,59,74–76]
Singapore	[3,11,12,25,26,28,34,43,52,56,62,74,75,77–93]

Barcelona was featured in 40% of the retrieved documents, Singapore in 37%, Helsinki in 17%, and Medellin in 6%. This suggests that Barcelona and Singapore have significant involvement and leadership in SC initiatives. The search also revealed that 52% of the studies focused on more than one city (either one of the three or other cities altogether), indicating that comparative studies across various SCs are common within this research topic.

3.2. Findings

3.2.1. Smart City Initiatives and Urban Sustainability Practices

This subsection analyzes the SC initiatives and corresponding urban sustainability practices adopted in the four cities across the six described domains: smart mobility, smart environment, smart governance, smart people, smart living, and smart economy. Table 2 provides a summary of the analyzed papers categorized by these domains (papers that cover more than two domains are classified as "multidimensional").

Table 2. Smart city domains explored in the searched documents.

References	Mob.	Env.	Gov.	Peo.	Liv.	Eco.	Multi.
Al-Saffar [37]							X
Ang-Tan and Ang [77]			X				
Angelidou [34]							X
Anttiroiko [64]			X				
Sánchez-Vaquerizo [38]	X						
Bibri and Krogstie [39]		X					
Bibri and Krogstie [40]							X
Bolici and Mora [41]							X
Calzada [42]			X	X			
Chib et al. [78]				X		X	
Crumpton et al. [12]		X	X				
Csukás and Szabó [36]							X
Della Corte et al. [36]			X			X	
Demirel [43]			X				
Demirel and Mülazımoğlu [44]			X				
Drapalova and Wegrich [45]			X				
Echendu and Okafor [79]							X
Fernández and Peek [10]		X					
Fonseca et al. [46]	X			X			
Garau et al. [65]		X					
Górna and Górny [80]							X
Härkönen et al. [66]					X		
Hartmann and Garcia [81]		X				X	
Hernández et al. [67]							X
Icasiano and Taeihagh [82]			X				
Iryna et al. [83]							X
Javed et al. [68]							X
Joo [84]			X				
Kenger et al. [85]							X
Kutty et al. [69]					X		
Lapinskaitė et al. [70]							X

Table 2. *Cont.*

References	Mob.	Env.	Gov.	Peo.	Liv.	Eco.	Multi.
Lim and Taeihagh [86]	X						
Lopes [74]			X				
Mancebo [47]				X			
Mann et al. [48]							X
March and Ribera-Fumaz [49]		X					
Martinez and Mahajan [75]		X					
Mora et al. [35]							X
Nicolas et al. [13]							X
Noori et al. [50]			X	X			
Papyshev and Yarime [87]	X						
Peyman et al. [51]	X						
Phumpiu and Rivera-Kuri [52]							X
Pira [27]					X		
Quijano et al. [71]							X
Ricci and Mariano [53]							X
Ringenson and Höjer [54]		X					
Sanyal et al. [55]							X
Sethi and Mittal [88]		X					
Shamsuzzoha et al. [26]							X
Shmelev and Shmeleva [3]		X				X	
Shmelev and Shmeleva [56]							X
Smith and Martín [57]			X	X			
Smith et al. [32]							X
Soe et al. [72]							X
Soriano-Gonzalez et al. [58]	X						
Sotirelis et al. [89]							X
Steenmans et al. [59]			X				
Šulyová and Vodák [90]				X			
Tan et al. [91]			X				
Tomàs [60]			X				
Valencia-Arias et al. [76]							X
Vardopoulos et al. [61]							X
Wang et al. [28]							X
Wey and Peng [92]		X				X	
Wirtz et al. [93]							X
Woods [25]				X			
Yoo [62]			X				
Zaman and Hertweck [73]							X
Zambrano-Prado et al. [63]		X					
Zygiaris [33]							X

Mob.: Mobility; Env.: Environment; Gov.: Governance; Peo.: People; Liv.: Living; Eco.: Economy; Multi.; Multidimensional.

Smart Mobility

Transportation plays a crucial role in urban sustainability due to its significant contribution to GHG emissions, air pollution, and noise [94]. Smart mobility encompasses the integration of ICT infrastructure and the development of efficient transportation systems that are sustainable, safe, and interconnected [11,13,54]. This review indicates that the four cities are dedicated to promoting smarter and more sustainable mobility solutions.

In Barcelona, the city has implemented data-driven technologies for mobility, such as a GPS monitoring system to track public transportation and a smart traffic light system that prioritizes public transportation [40,61]. The city's public transportation system was designed on an orthogonal grid, strategically placing bus stops to facilitate connections between buses, trams, metro, trains, and bicycles. In addition, the bus fleet utilizes hybrid vehicles to minimize emissions, and bus stops provide real-time waiting information and USB ports. The city has implemented a sensor-based and app-enabled free car parking system, reducing the time spent searching for parking spots. The city has invested in expanding and enhancing its cycling path network, and promoted various initiatives to promote cycling, including a bicycle-sharing system with 6000 bicycles, strategically located near transportation hubs and parking areas [40,58,61]. Approximately 33% of the city's technological solutions are dedicated to mobility [35], including popular apps like Trànsit for real-time traffic updates and Bicing for bike sharing [40]. Some of the reviewed documents are specifically focused on enhancing urban mobility in the city [38,46,51].

In Helsinki, Nicolas et al. [13] showed that the SC strategy prioritizes smart mobility to mitigate the adverse environmental effects of motorized transportation. This is further supported by Mora et al. [35], who reported that 43% of Helsinki's technological solutions are dedicated to mobility. These initiatives primarily aim to combat climate change and enhance energy efficiency in transportation. The living lab Forum Virium has played a proactive role in Helsinki's smart mobility initiatives, particularly in achieving low/neutral carbon energy and developing advanced vehicle technologies and services such as electric-assisted vehicles (EVs), autonomous vehicles, robot buses, and drone transportation, among others [26]. One selected study [68] focuses on this domain, establishing an ecosystem for electric vehicle charging stations in Helsinki to facilitate the seamless integration of this infrastructure.

In the case of Medellin, this review highlights the city's commitment to leveraging ICTs to enhance urban mobility and public transportation [32]. This effort includes initiatives to connect hillside and socially marginalized neighborhoods to the city center through escalators and a Metrocable system, aimed at reducing travel time, alleviating traffic congestion, and lowering emissions [74]. Medellin is also enhancing transportation integration by combining metro, buses, trams, and bike-sharing (EnCicla) to make public transportation more sustainable and accessible. The city has installed intelligent traffic light sensors and monitoring systems to optimize traffic flow and reduce congestion [32].

In Singapore, the most developed smart services are in intelligent transportation systems, positioning the city as a leader in smart mobility [28,79]. The Singaporean transportation system is based on three pillars: a compact urban environment with integrated transit, and restrictions on private vehicles [79]. Over the years, policies have focused on integrating land use and transportation planning to minimize travel demand and optimize the use of limited space [26,52,83]. The government has developed various incentives to discourage personal vehicle use, such as an electronic road pricing scheme and high tax rates for new cars [11,52]. The goal is to ensure that 70% of all journeys in the city are made by public transportation [56]. A smart card is available for various modes, collecting travel data that can be used to improve public transportation services [56]. Enhanced technology is in place to ensure the safety of vulnerable road users, including electronic devices that extend pedestrian crossing times and alert drivers to the presence of pedestrians at night [11,79]. Singapore has also adopted a hands-off approach to regulating disruptive transportation technologies and ridesharing [82], which has enabled the city to benefit from these innovations. The city has conducted various trials with autonomous vehicles, including self-driving cars, autonomous shuttle buses, and driverless trucks [26,86].

Smart Environment

Smart environment is a domain closely linked with urban sustainability [43,44], encompassing issues such as natural conditions, urban green spaces, environmental protection, and biodiversity [26,43]. It also includes actions to monitor and reduce GHG emissions [39], air and noise pollution [89,93], and promote sustainable resource management through efficient use of energy and water [10,11,13,39,43,44,89], along with effective waste management practices [10,44,89,93]. Additionally, policies aimed at mitigating climate change impacts are also integral to this domain [10]. This review demonstrates that all four cities have implemented various measures to advance smart environmental practices.

In Barcelona, various smart environment initiatives related to energy efficiency, air quality, water, and waste management have been adopted [34,35]. The city has made significant investments in implementing smart grids and advanced metering infrastructure, such as smart meters, sensor networks, automated control systems, and cyber–physical systems. These technologies optimize energy distribution and consumption in real time [39]. One of the most innovative energy-efficient projects is the 22@ district, which utilizes centralized energy sources from municipal waste incineration to provide heating and cooling to buildings [10]. The city aims to reduce GHG emissions by 40% by 2030 (compared with 2005) and achieve energy self-sufficiency. To meet these goals, Barcelona launched a program to promote solar energy generation [39]. In waste management, the city has installed bin sensors to monitor fill levels and optimize collection routes, reducing both collection frequency and related emissions [43]. Barcelona also maintains a network of approximately 1800 sensors (Sentilo) that measure real-time environmental indicators such as air quality, humidity, and temperature [10]. In terms of climate change, the city faces challenges such as urban flooding, heat waves, water scarcity, and coastal erosion. The city has increased urban green infrastructure, including parks and green corridors, to mitigate these threats [10]. Additionally, infrastructure development is underway to manage stormwater and reduce flood risks, involving the establishment of a network of stormwater tanks [10].

The city of Helsinki is pursuing sustainable urban development through smart environment initiatives aligned with the UN Sustainable Development Goals [73]. Helsinki has set ambitious goals to reduce its climate emissions with the City Strategy 2017–2021, aiming for carbon neutrality by 2035 through an 80% reduction in GHG emissions [66]. The city launched Climate Watch, an online tool enabling residents to monitor Helsinki's progress toward carbon neutrality [26]. Additionally, the Climate Street project [64], involving businesses, citizens, and public agencies, aims to collectively reduce GHG emissions and energy consumption to enhance climate change adaptation policies. One of the most intriguing smart environment initiatives is linked to the Kalasatama district project. According to Härkönen et al. [66], the goal is to transform Kalasatama into an energy-efficient neighborhood by implementing solutions that enhance energy savings and increase renewable energy utilization through home automation in energy management. The plan includes replicating this energy model across other parts of Helsinki.

Regarding Medellin, the reviewed documents indicate that the city has implemented several initiatives in the environmental sector, which are primarily in the stages of development and systematization [32]. These initiatives encompass projects and tools aimed at monitoring various environmental factors such as air quality and climate change. For example, the SIATA platform is an early warning system designed to monitor and respond to environmental and climatic conditions, continuously monitoring various environmental parameters. The real-time data provided by SIATA can support decision-making processes for urban planning and emergency response [32]. According to Valencia-Arias et al. [76], critical challenges for sustainable development in Medellin include sustainability education and raising environmental awareness, as environmental sustainability depends not only on political decisions but also on individual behavior.

Regarding Singapore, the city is committed to branding itself as the leading livable and sustainable city [11]. Singapore's sustainable development strategy is bolstered by initiatives like the Singapore Green Plan, considered a pioneering document in biophilic urbanism. This plan focuses on integrating nature into the built environment [56,80], promoting features such as green façades, roofs, corridors, and urban green spaces like Gardens by the Bay [11,43]. The city also has a climate plan aimed at reducing GHG emissions by 36% from 2005 levels by 2030, focusing on energy efficiency and the adoption of low-carbon technologies [88]. Energy consumption has been reduced by 20% through the implementation of solar energy, low-power lighting, and efficient home management systems [43]. During this century, the proportion of electricity generated from natural gas has increased from 19% to 79%, resulting in reduced CO_2 emissions [56]. Singapore aims to have 50% of its electricity demand from renewable sources by 2065 [56]. In terms of water efficiency, Singapore has developed a desalination plant and a water recycling scheme to fulfill up to 30% of the city's water needs [56]. Additionally, Singapore aims to achieve a 70% recycling rate by 2030 [56]. The city has also implemented a smart waste management program that provides information on the content and location of bins, enabling the optimization of waste collection routes [28]. Singapore is also committed to greening its built infrastructure, and around 1650 buildings have been made environmentally friendly, with plans for 80% of buildings to meet sustainable standards by 2030 [11]. In terms of climate resilience, Singapore has implemented flood-control projects [88]. All these initiatives have earned Singapore broad recognition as one of the world's leading SCs in environmental performance [56,92].

Smart Governance

Smart governance is also defined as a critical SC dimension [74,86]. Smart governance involves the use of digital technologies and data-driven approaches to enhance the efficiency, transparency, and responsiveness of government services, as well as to increase citizen participation in decision-making [11,50,74]. This literature review reveals that the four cities have implemented various initiatives to make their governance smarter.

In Barcelona, the SC strategy has been structured around collaboration among government, industry, academia, and citizens [34,35]. In 2016, the smart-citizen policy framework was significantly enhanced with the publication of the document "Barcelona Ciutat Digital: A Roadmap Toward Technological Sovereignty" [42]. This new smart-citizen policy was implemented using various ICT tools, such as the platform Decidim and the app Ethics and Good Governance Mailbox, which enable citizens to participate in decision-making and report mismanagement practices to the municipality, respectively [42,44]. There are also non-digital initiatives focused on fostering smart governance, such as (i) participatory budgeting experiences [45]; (ii) the Cibernarium, a training program to increase professionals' and citizens' knowledge and skills in using digital devices; (iii) the Smart City Expo World Congress, a leading global platform for experts working in the field of SC development; (iv) smart education programs to introduce ICT-driven urban innovation and smartness to children through practical activities; and (v) the organization of hackathons and competitions to actively involve civil society in the SC domain [35]. Barcelona's SC development strategy includes more than 100 collaborative projects spanning solutions and applications in domains such as mobility, air quality, energy, water, healthcare, education, and e-government, among others [34,41]. Comparative studies show that Barcelona has more participatory public policies than cities like Singapore [43].

Helsinki is also described in the literature as a model city in terms of open governance and citizen empowerment [13]. The city is actively committed to promoting openness and citizen participation in developing solutions to shape the SC vision through a bottom-up approach [26,64]. Living labs have played a pivotal role in such a strategy. The living lab Forum Virium has played a key role in developing needs-based and internationally competitive digital services in collaboration with private businesses, public organizations, and citizens [64]. The ultimate goal is to make Helsinki the most functional SC in the world [26]. Forum Virium has supported more than 80 projects for co-created SC solutions, involving 750 companies and 60 partner cities [26]. Smart governance has been promoted through various platforms and initiatives such as Helsinki Region Infoshare (HRI), an open-data service available to all; Open Ahjo, an app providing access to the city's decision-making material; Open Helsinki Hack-at-Home, a platform encouraging the use of open data; and Helsinki Loves Developers, a regular meeting of app developers [35,64]. Helsinki was also involved in the Six City Strategy (6Aika), a joint sustainable urban initiative by the six largest Finnish cities [31].

Medellin has been promoting community labs to encourage citizens to use and benefit from ICTs [59]. The Medellin Innovation District and Ruta N are among the entities most actively engaged in promoting collaboration between businesses, academia, citizens, and government to drive innovation, economic development, and entrepreneurship [76]. To improve transparency and encourage citizen participation in decision-making, Medellin has been adopting various initiatives such as participatory budgeting, digital government services, participation in national and international city networks, and the publication of data about specific urban indicators [59]. One example is the Open Data Council, a platform to collect and make open governmental data [32]. However, Smith et al. [32] argued that the adoption of smart governance in Medellin has been constrained by issues such as the lack of a culture of data collection and management, challenges in data disaggregation, concerns about losing power, and participatory processes that often focus narrowly on specific urban issues like flooding and landslides.

According to this review, Singapore is reported as an open-source society renowned for its levels of transparency, openness, and trust, with a particular focus on smart governance [79]. The Smart Nation living lab itself acts as a comprehensive database accessed through an extensive service network, offering a variety of services [79,93]. Around 98% of electronic services, primarily in healthcare, mobility, and public administration, are available online, including 385 e-services and e-citizen portals [43,80], which position Singapore as an "information city" [79]. Various public platforms disseminate data, such as OneMap and the Urban Redevelopment Authority (URA) Map portal [93], while others like REACH (Reaching Everyone for Active Citizenry@Home) and Singapore Together are designed for collecting citizen opinions and shaping public policies [43]. In terms of collaboration, the Singapore Government and the private sector are working together to enhance and position Singapore as a more attractive and competitive destination [11]. In Singapore, living labs like Smart Nation have played a crucial role in promoting smart governance and fostering collaboration among global companies, high-tech startups, and public institutions [84]. Lastly, the city is engaged in several transnational development projects, including the ASEAN Smart Cities Network (ASCN), which Singapore promoted to enhance environmental sustainability and strengthen regional collaboration [12,91].

Smart People

This domain primarily focuses on the skillset and education level of the city's human capital [11,89], and on social inclusion and equality [13,36] in aspects such as ethnic diversity, cosmopolitanism, openness, flexibility, creativity, and public participation [11,13,34,89]. Smart people are considered crucial because SC initiatives often demand innovative, creative, and collaborative solutions [89]. In general, highly educated individuals drive the development of SCs [34,91] but it has been shown that smart solutions can create new forms of social division and exclusion [25,36]. For these reasons, inclusiveness has become a central element in SC policies, focusing on reducing disparities between groups, such as in education and wealth, and addressing social barriers, particularly in learning and participation processes [36].

Barcelona is known for its emphasis on citizen participation, digital inclusion, and education within its SC initiatives. These include, for example, smart education programs for primary and secondary schools to introduce the concept of SCs and innovation to young students and help them develop smart solutions [35,40], and courses, seminars, training sessions, competitions, and hackathons to enhance digital literacy among professionals and citizens, with the Cibernàrium serving as the city's main technological training facility [33,35]. In terms of public participation, the Decidim platform has engaged 120,000 participants, generated 27,000 citizen proposals, and 3500 physical meetings as part of over 50 participatory processes [95]. As noted by Calzada in 2018 [42], many of these citizen proposals (70%) were successfully implemented as public policies across a wide range of initiatives. Barcelona has been particularly active in fostering inclusiveness and citizen empowerment, providing ICT tools to assist elderly people in their everyday lives [35,49,53]. These efforts helped Barcelona in performing better than other cities in this domain including London [40], Singapore, New York, and Istanbul [43].

Helsinki emphasizes smart people through its focus on citizen engagement, education, innovation, and inclusiveness [64]. Helsinki is regarded as a model for citizen empowerment and open governance, especially with its bottom-up approach that involves living labs in designing citizen-centric services [13]. Through companies like Forum Virium, Helsinki has fostered a collaborative and experimental organizational culture aimed at democratizing innovation and enhancing public participation [64]. Consequently, Helsinki consistently ranks better than other SCs, such as Singapore and London, in promoting bottom-up approaches [26] and needs fewer communication initiatives than other cities due to the already high level of public participation achieved in the city [13]. Helsinki is placing a strong emphasis on equitable access to digital services and infrastructure, ensuring that technological progress does not exclude any segment of the population. Within the Kalasatama project, the city is also implementing a pilot age-friendly housing community, which is planned to be rolled out to the rest of the district.

Medellin has been addressing historical inequalities through the integration of social innovation into urban planning, making sure that SC benefits reach low-income and marginalized communities. These include social urbanism initiatives to connect peripheral neighborhoods to the city center and to provide citizens with educational and cultural facilities, like public libraries, to enhance their social inclusion and digital skills [32]. However, the city still has considerable socioeconomic barriers that hinder citizens from less privileged groups from participating in the development of SC initiatives [32]. For these social groups, the most significant barriers include less accessibility and usability of technologies, technological illiteracy, and digital solutions that often overlook their needs [76]. Within this domain, Smith et al. [32] argued that a redefinition of SC approaches to effectively respond to the needs of Medellin's population and foster social inclusion is urgently needed. Reinforcing the investment in ICT infrastructure and reducing technological illiteracy is critical to achieving that goal [76].

Singapore's focus has been in creating a tech-empowered society and the city has invested significantly in integrating ICT into its education system [80]. However, several challenges restrict the effectiveness of smart people initiatives, including a governmental top-down approach, an aging society, and the complexities of a mixed-race population. First, the Smart Nation strategies have been based on a top-down approach controlled by the government [26,43,84]. The strategy encourages citizens to participate in envisioned SC initiatives, but does not allow them to truly shape these initiatives [26]. The goal has been to prioritize the interests of the country over those of its citizens [84]. Second, the Smart Nation plan does not adequately consider the aging process and the diverse needs of the elderly population in the city [25,78]. Some seniors in Singapore are less fluent in English, have lower educational backgrounds, and possess less experience with technology, which diminishes their motivation to participate in SC initiatives. Third, Singapore has a mixed-race population with significant cultural diversity, which poses challenges in including their vision into SC initiatives and in mobilizing their public participation [11]. Thus, despite extensive ICT coverage and the high digitalization of public services that are accessible to all citizens, Singapore performs worse than other cities in issues related with public participation and social inclusion [43,84].

Smart Living

Smart living is an SC domain focused on enhancing citizens' quality of life by providing more efficient, comfortable, and satisfying living conditions [11,13,27]. This domain encompasses various components such as housing quality, healthcare, education, security, social cohesion, culture, and recreation [11,13,36,89]. Enhancing citizens' quality of life is a main goal of SCs, making smart living a fundamental aspect of this concept [13,70]. As noted by Bibri and Krogstie [40], advancements in ICT across other SC domains, such as mobility and governance, also contribute to improving overall quality of life.

Based on the concept of being a city for its people, Barcelona's SC goal aims to enhance well-being and quality of life through a wide array of applied technological solutions [39,43]. To achieve this, the city has made significant investments in its ICT infrastructure, which includes an extensive IoT sensor network that collects data across various sectors impacting quality of life, such as energy, environment, transportation, health, and security [40]. A SC model involving active public participation was developed to analyze the extensive data collected and efficiently manage both public and private services and their access. The Municipal Institute of Information Technology and the municipal company 22@ Barcelona have spearheaded various projects, such as smart parking and bin sensors, aimed at creating a sustainable city with an enhanced quality of life [39,43,61]. Barcelona also adopted non-digital smart living initiatives including incentive programs for urban housing renewal [52,60] and efforts to reduce social inequalities and exclusion [10,60]. For these reasons, Barcelona ranks well among the top European cities for quality of life and smart living [43,52].

The literature also confirms that Helsinki is a leading city in smart living, with numerous urban initiatives aimed at improving residents' quality of life through technology and sustainable practices. For example, preserving the natural environment is a key aspect of enhancing the well-being of its inhabitants [13]. The Smart Kalasatama project, in particular, has tested and implemented various sustainable smart solutions. This includes renewable and energy efficiency solutions to make the neighborhood self-sufficient, as well as the adoption of building automation in commercial buildings and apartments for energy-related services and electric demand management [66]. Additionally, mobile health platforms like Kalasatama Wellbeing allow citizens to record their diets and receive immediate dietary advice [31]. Another example is the provision of QR codes, which enable visitors to access free guided tours of Helsinki's sights via their smartphones [35].

In Medellin, the reviewed documents focused on initiatives taken to mitigate spatial segregation, social inequalities, and urban violence. As described by Smith et al. [32], the first initiatives aimed to address the spatial segregation and urban violence that characterized the city in the early 1990s. This was achieved through a range of social urbanism measures designed to improve poor neighborhoods, such as the PRIMED program, and connect the peripheral hillside neighborhoods to the city center through the Metrocable system. Community labs and projects, such as Makaia, have been used to build partnerships and relationships oriented toward social and economic development [32,59]. However, these improvements have been limited, and socio-spatial segregation between the center and the periphery remains, partly due to accelerated and unplanned urban growth. According to Valencia-Arias et al. [76], the lack of security and access to essential services such as healthcare and education are among the most critical urban issues faced by the city in this domain.

Singapore's SC strategy has been developed with the ultimate goal of enhancing the quality of life for all citizens and developing as a sustainable city [11,56]. As mentioned before, Singapore provides a wide variety of services to its citizens and 95% of homes and businesses have ultra-fast Internet connectivity [11]. More than 100 mobile services are available to citizens, including those from government agencies, public entities, and private sectors. As shown by Wirtz et al. [93], smart social services encompass a diversity of healthcare, education, and cultural services, with healthcare being the most representative and analyzed in the literature. According to Iryna et al. [83], the Singaporean healthcare system is highly digitized, for example, practitioners routinely engage with patients remotely and the platform MyDoc provides various health functions such as virtual consultations. Assistance for lonely patients is possible with AI-based chatbots. The integrated functional payment system, including platforms like the EZ-Link card and the Singapore Quick Response Code (SGQR), is also highlighted in some publications for facilitating seamless transactions across various services [43,84]. The mobile app "SingPass" is noted for allowing easy individual access for transactions with the government [77,93].

Smart Economy

Smart economy is identified as a key factor in SC development, focusing on economic competitiveness, productivity, and labor market flexibility, emphasizing integration and transformation within local and national markets [44]. The concept involves the use of digital technologies and human capital, such as skills, creativity, and knowledge, to create competitive products and services based on efficiency and sustainability [11,43]. Additionally, a smart economy may promote a green economy by fostering companies that encourage renewable energy and enhance energy efficiency.

The reviewed documents show that Barcelona has implemented several initiatives under its SC strategy to boost economic progress, focusing on innovation, entrepreneurship, and sustainable economic growth. The role of project 22@Barcelona is highlighted for attracting qualified professionals and knowledge-based businesses linked to the new economy, such as media, design, ICT, and energy [41,52,53]. These conditions have transformed the city into a highly attractive and sustainable business environment [10,44,61]. Noori et al. [50] also emphasized the role of Barcelona Activa, a local development agency focused on promoting entrepreneurship and offering support to companies and startups. The organization of the Smart City Expo World Congress since 2011 is also noted for positioning Barcelona as a leader in SC innovation, attracting global businesses, investors, software companies, and equipment and utility providers [35,60].

Like Barcelona, Helsinki has implemented several initiatives to promote a smart economy to make the city more competitive nationally and globally [73]. The literature indicates that Forum Virium has been committed to creating new businesses, improving services, establishing links to international markets, attracting foreign investment and businesses [64], organizing innovation competitions, and funding [73]. The Smart Kalasatama project is noted in various documents for its role in attracting companies, particularly those focused on sustainable businesses that develop and offer energy savings, and demand management services for city residents [64,66]. The study conducted by Zaman and Hertweck [73] indicated that the city's smart growth heavily relies on Kalasatama for testing solutions to urban problems, including public transportation and energy use. The city's focus on promoting smart economy initiatives is highlighted in various documents. For instance, Nicolas et al. [13] noted that 30% of the interventions and announcements in Helsinki are related to smart economy. This emphasis has significantly contributed to Helsinki's competitive edge, as evidenced by its high rankings in entrepreneurship, startups, and high technology [59].

In the case of Medellin, smart economy issues are explored in only two of the returned documents [32,76]. In their analysis of key aspects of implementing SC initiatives in Medellin, Smith et al. [32] identified socioeconomic inequalities and the lack of technological skills as the most critical challenges for boosting a smart economy. Similarly, Valencia-Arias et al. [76] identified three main socioeconomic challenges for the city: the prevalence of informal labor, the need to redefine industrial and productive styles, and the financial resources required to implement the necessary changes. Recognizing these issues, the city initiated an economic transition process centered on service sectors, with a focus on strategic clusters such as digital businesses, sustainable habitat, health, business tourism, and the reinforcement of ICT infrastructure [32]. Initially driven by the public sector, this process has evolved to include public–private initiatives, exemplified by programs like Medellin Digital City, aimed at enhancing the city's technological infrastructure. Additionally, the innovation and business center Ruta N is noted for its role in fostering entrepreneurship, supporting startups, and attracting international companies [32,76].

Finally, the efforts undertaken in Singapore to create a smart economy are well-documented in the reviewed literature. According to Della Corte et al. [11], Singapore's masterplan emphasizes the use of ICT to enhance economic competitiveness by fostering creativity and enabling innovation among businesses. In Singapore, the private and public sectors have been collaborating to make the city an attractive global living lab destination for smart technologies and improve the industry using ICT [11,84,91]. The government has been actively supporting its living labs, through which several global and national companies, high-tech startups, and Singaporean institutions are collaborating to find innovative technological solutions for various business sectors, including healthcare, autonomous vehicles, security, and tourism, among others [84]. These collaborative initiatives have allowed the city to continuously seek and seize business opportunities and explore the technologies to meet new market needs [11]. In addition, the government launched initiatives such as Build Amazing Startups Here (BASH), an all-in-one startup facility designed to attract and support tech startups [84]. Within the context of the ASCN, Singapore has encouraged the creation of partnerships and positioned itself as a leading global business and technology hub [91]. As a result of these efforts, Singapore has achieved impressive milestones: it has attracted 7000 international companies, boasts one of the highest per capita income levels in the world [56], is one of the most competitive cities in the global economy [84], consistently ranks high in SC rankings [77], and is positioned to build a smart nation ready for the Fourth Industrial Revolution [84].

4. Discussion

This review of smart sustainable practices across Barcelona, Helsinki, Medellin, and Singapore has highlighted significant advancements and persistent challenges within the six SC domains. This discussion critically evaluates the initiatives undertaken, focusing on key insights and identifying areas for improvement in fostering more sustainable urban planning through SC initiatives.

The literature review showed that the emphasis in the four SCs has been on ICT solutions, along with good governance, environment, economic development, and social inclusion as the primary concerns. As shown in Table 3, each city has developed strategies tailored to its local context, yet common themes such as the integration of technology, community engagement, and effective networking have emerged as critical components in enhancing sustainable urban development. Therefore, instead of representing uniformity across the four studied cities, the concept of a SC should be understood as one that interacts with and adapts to local and national contexts. Although there is some overlap in areas like mobility and environmental sustainability, this review revealed that each city has tailored its SC approach to reflect its unique priorities and specific contexts. For example, Singapore has invested heavily in ICT and enhancing the city's global competitiveness through a top-down approach. Medellin has concentrated on reducing urban and social inequalities through transportation, housing, and education initiatives aimed at mitigating socioeconomic disparities. Helsinki emphasizes open data, digital public services, and sustainability, with the Kalasatama district serving as a living lab for testing smart energy solutions, urban mobility, and sustainable living practices. Meanwhile, Barcelona has focused on promoting citizen-centric governance, using digital technologies to improve public services and enhance public participation in urban development through platforms like Decidim.

Table 3. Main smart city initiatives adopted in the four cities.

City	Mobility	Environment	Governance	People	Living	Economy
Barcelona	- Data-driven public transport; - Smart traffic; - Cycling infrastructure.	- Energy efficiency; - Smart grids; - Climate resilience.	- Bottom-up; - Citizen participation; - Digital platforms.	- Educational programs; - Cultural initiatives.	- Digital inclusion; - Public health.	- Innovation hubs; - Support to startups.
Helsinki	- Sustainable transportation; - Electric vehicle infrastructure.	- Carbon neutrality; - Climate monitoring.	- Bottom-up; - Open data; - Citizen empowerment.	- Education on digital literacy; - Living labs and innovation.	- Sustainable building solutions; - Mobile health.	- Smart Kalasatama; - Support for green businesses.
Medellin	- Metrocable; - Public transportation modernization.	- Urban regeneration; - Green corridors.	- Bottom-up; - Social inclusion; - Community-driven projects.	- Citizen engagement; - Tech access inequalities.	- Social urbanism and inclusion.	- Support for local businesses.
Singapore	- Smart traffic; - Autonomous vehicles.	- Biophilic urbanism; - Climate resilience.	- Top-down; - Electronic services; - Global projects.	- Tech education; - Senior and racial inequalities.	- Smart homes; - Digital health; - Fast Internet.	- High-tech industries; - Global business hub.

A key finding of this review is the uneven focus on specific SC domains. Smart governance and smart environment are the most studied dimensions, highlighting an emphasis on governance efficiency and environmental sustainability in the four cities. Conversely, smart economy, smart living, and smart people received significantly less attention, despite their importance in holistic urban sustainability. This trend aligns with prior studies that indicate a lack of emphasis on societal aspects within SCs [22,36]. The limited exploration of these dimensions underscores the need for greater attention to societal factors, particularly economic sustainability and social well-being, which have often been overshadowed by technological and environmental concerns [45].

Regarding the lessons learned, the concept of "smart", particularly in Barcelona, Helsinki, and Singapore, is closely linked to using ICT to enhance urban operations and sustainable practices through the extensive data generated. However, this presents two main challenges. First, ICT-driven initiatives often prioritize economic and competitive aspects over addressing social needs [45,57]. Second, the mere availability of digital technologies does not ensure that all individuals can think or act smartly. For instance, the review found that in Singapore, certain social groups—specifically the elderly and some ethnic minorities—are less included in SC initiatives due to lower educational levels and tech skills [79]. The exclusion of these vulnerable populations can create new forms of socio-technical division and exclusion [25], resulting in SC initiatives disconnected from social and cultural contexts, potentially leading to solutions that are technologically advanced but socially inequitable and unsustainable. To address these challenges, social needs must be integrated into the planning and implementation of ICT-driven initiatives, ensuring that economic and competitive priorities do not overshadow inclusivity. Further, as demonstrated in Medellin, cities must develop inclusive digital literacy programs and foster public–private partnerships to enhance ICT and R&D investments, particularly in resource-limited areas [35].

Governance also emerged as a critical domain in the reviewed SC strategies. In the four cities, ICT solutions assist planning authorities in defining and monitoring public policies, enhancing transparency, local democracy, and public participation. Living labs and innovation centers play a pivotal role in engaging citizens and stakeholders in developing e-services for all, while online platforms and apps are essential for disseminating data and facilitating interactions with residents. However, in terms of governance approach, this review indicated a divergence, with Barcelona, Helsinki, and Medellin adopting a bottom-up strategy [13,26], whereas Singapore employs a more top-down, government-controlled approach [26,43,84]. Bottom-up approaches are more effective in fostering community engagement, empowering local voices, and promoting social and urban inclusion, while top-down strategies tend to be less responsive to the population's needs and public interest [35,57,64]. This review has confirmed that Barcelona, committed to citizen-centric governance, demonstrates higher levels of public participation in SC initiatives compared with Singapore [43]. Therefore, a balanced governance model that combines leadership with community-driven initiatives could be more effective in ensuring the success of SC strategies [35,64].

Confirming previous studies [26,76], environmental sustainability also appeared as a key dimension of SC strategies in the four cities. Advanced ICT and big data are utilized to address issues such as GHG emissions, air and noise pollution, energy and water consumption, and waste management. However, this review suggests that while ICT advancements are well-documented, the environmental benefits of these technologies are less explored. In the environmental dimension, some authors have argued that ICT solutions have been used for promotional purposes and for political motivations to face complex urban challenges, rather than to evaluate their benefits and the actions required to increase urban resilience and livability alongside urban smartness [69,76]. This indicates a pressing need for more comprehensive evaluations to understand the extent to which SCs genuinely contribute to environmental sustainability.

Building on the insights gained from this review, it is essential to examine the smart sustainable urban practices adopted in the four cities. The notion of smart sustainable cities represents a relatively new holistic paradigm of urbanism, emphasizing sustainable urban development supported by ICT [10,40]. Sustainable urban development goals are inherently part of several initiatives described in the reviewed documents. Cities like Barcelona and Helsinki are strongly committed to becoming a showcase for sustainable urban development [39]. But out of the 71 papers reviewed, only a few specifically focused on smart and sustainable urban development [3,37,39,61,65,70]. In the revised literature, the integration of sustainable development goals into SC strategies is rarely explored comprehensively. The searched documents do not effectively demonstrate best practices within a smart sustainable development framework, as the primary focus has been on how ICT solutions address specific issues, such as mobility and environmental concerns, often overlooking broader urban challenges. Although the concept of SCs has been consistently utilized in these four cities and in the literature for over a decade, it remains unclear how effectively they have mitigated or resolved some of the most pressing urban issues. For instance, in Barcelona, SC initiatives have centered on projects like public Wi-Fi and smart sensors for managing public spaces, but these efforts have done little to tackle deeper urban concerns, such as housing affordability and wage inequality. Similarly, Helsinki has made progress in open data and digital public services; however, challenges persist in bridging the digital divide and ensuring equitable access to these innovations for all citizens. In Medellin, despite significant advancements in urban mobility solutions like the Metrocable system, issues related to social equity and poverty reduction remain unresolved. Lastly,

Singapore has adopted a top-down approach to SC development, heavily investing in ICT, yet these initiatives have not effectively addressed social inclusiveness, particularly for aging populations and marginalized groups who struggle with digital literacy and access. As observed in Barcelona [58], Helsinki [73], and in other cities like Rio and Tokyo [96,97], urban redevelopment and regeneration projects associated with SC initiatives can enhance the quality of life by improving infrastructure, connectivity, and public services. However, they also contribute to socioeconomic and urban inequalities, including gentrification and the displacement of local communities. It appears that smart urbanism policies often concentrate on specific urban development strategies in which local actors reframe existing targets to unlock funding and resources, rather than addressing the root causes of urban challenges. These findings align with previous studies indicating that smart urbanism frequently lacks a comprehensive consideration of urbanism, prioritizing technological solutions over holistic urban development strategies [29], indicating that urban sustainability has not been a main driver for SCs [23]. To achieve truly smart and sustainable urban development, cities need to transcend a narrow focus on ICT solutions and tackle deeper socioeconomic and urban challenges. Smart urbanism positions society and the urban context as primary parameters in developing digital initiatives. Therefore, there is a need for a more comprehensive understanding of how technology and urbanization intersect in shaping SC approaches and their impacts on contemporary urban life. Holistic approaches that integrate social equity, inclusiveness, and long-term urban planning are essential for ensuring that SC initiatives effectively address the complex challenges cities face today and contribute to improving the quality of life for all citizens.

5. Conclusions

This study offers valuable insights into the implementation of SC initiatives across Barcelona, Helsinki, Medellin, and Singapore, emphasizing the various strategies employed across six domains to integrate sustainable urban practices. Despite notable advancements in governance and environmental management, there remains a concerning gap in how effectively these initiatives engage with broader urban challenges. As observed, while technological solutions have been prioritized, the need for inclusive and holistic practices that address socioeconomic disparities is critical to ensure that smart urbanism truly enhances the quality of life for all citizens. Drawing on the lessons learned from this review, we recommend for future studies to carry out the following:

(i) Develop comprehensive frameworks to assess the performance of smart sustainable cities, namely in addressing socioeconomic disparities and environmental challenges;
(ii) Analyze methods to enhance community involvement and ensure social equity in smart sustainable urban projects, with a focus on the influence of social and cultural factors on public participation and social inclusion;
(iii) Investigate how SC initiatives are successfully translated into urban sustainability practices, with a focus on identifying the challenges and opportunities that influence their effectiveness in real-world applications;
(iv) Provide additional evidence about the benefits of ICT in promoting urban sustainability through quantitative models, indicators, or metrics.

Finally, this review has several limitations that should be acknowledged. First, it focused solely on scientific papers published in English and indexed in two databases (Scopus and Web of Science). Therefore, studies published in other languages, in different repositories, or alternative formats such as conference papers, were not included. Second, the selection of documents was guided by specific search criteria detailed in the methodology, which may have led to the exclusion of other relevant research. Third, due to the large number of papers reviewed, only key findings were highlighted instead of conducting a comprehensive meta-analysis. Lastly, the findings of this study are limited to four cities. Future research could benefit from including additional cities from all five continents to enhance the representativeness of diverse urban contexts and provide a more comprehensive overview of global SC development, particularly comparing developed and developing countries. Despite these limitations, this study significantly contributes to clarifying the concept of smart sustainable cities, improving the design and implementation of sustainable urban planning practices, and promoting more sustainable and livable urban environments.

Author Contributions: Conceptualization, Á.V. and F.F.; methodology, Á.V. and F.F.; formal analysis, Á.V., F.F., and R.R.; investigation, Á.V. and F.F.; writing—original draft preparation, Á.V. and F.F.; writing—review and editing, R.R.; visualization, F.F.; supervision, R.R. All authors have read and agreed to the published version of the manuscript.

Funding: This research received no external funding.

Data Availability Statement: Data are contained within the article.

Conflicts of Interest: The authors declare no conflicts of interest.

References

1. United Nations Habitat. *Envisaging the Future of Cities, World Cities Report 2022*; UN-Habitat: Nairobi, Kenya, 2022.
2. Gunnarsdóttir, I.; Árnadóttir, Á.; Heinonen, J.; Davíðsdóttir, B. Decarbonization of passenger transport in Reykjavík, Iceland—A stakeholder analysis. *Case Stud. Transp. Policy* **2023**, *12*, 101019. [CrossRef]
3. Shmelev, S.; Shmeleva, I. Global urban sustainability assessment: A multidimensional approach. *Sustain. Dev.* **2018**, *26*, 904–920. [CrossRef]
4. Rogers, D. Understanding Urbanism. In *Understanding Urbanism*; Rogers, D., Keane, A., Alizadeh, T., Nelson, J., Eds.; Springer: Singapore, 2020; Volume 1, pp. 1–13. [CrossRef]
5. Nijhuis, S.; Stolk, E.; Hoekstra, M. Teaching urbanism: The Delft approach. *Proc. Inst. Civ. Eng. Urban Des. Plan.* **2017**, *170*, 96–106. [CrossRef]
6. Anthopoulos, L. Understanding the smart city domain: A literature review. In *Transforming City Governments for Successful Smart Cities*; Rodríguez, E., Ed.; Springer: Cham, Switzerland, 2015; Volume 8, pp. 9–21. [CrossRef]
7. Neirotti, P.; De Marco, A.; Cagliano, A.; Mangano, G.; Scorrano, F. Current trends in smart city initiatives: Some stylised facts. *Cities* **2014**, *38*, 25–36. [CrossRef]
8. Gracias, J.S.; Parnell, G.S.; Specking, E.; Pohl, E.A.; Buchanan, R. Smart Cities—A Structured Literature Review. *Smart Cities* **2023**, *6*, 1719–1743. [CrossRef]
9. Rajamohan, K.; Rangasamy, S.; Abreo, A.; Upadhyay, R.; Sabu, R. Smart cities: Redefining urban life through IoT. In *Handbook of Research on Machine Learning-Enabled IoT for Smart Applications Across Industries*; Goel, N., Yadav, R., Eds.; IGI Global: Hershey, PA, USA, 2023; Volume 1, pp. 316–335. [CrossRef]
10. Fernández, C.; Peek, D. Smart and sustainable? Positioning adaptation to climate change in the european smart city. *Smart Cities* **2020**, *3*, 511–526. [CrossRef]
11. Della Corte, V.; D'Andrea, C.; Savastano, I.; Zamparelli, P. Smart cities and destination management: Impacts and opportunities for tourism competitiveness. *Eur. J. Tour. Res.* **2017**, *17*, 7–27. [CrossRef]
12. Crumpton, C.; Wongthanavasu, S.; Kamnuansilpa, P.; Draper, J.; Bialobrzeski, E. Assessing the ASEAN smart cities network (ASCN) via the quintuple helix innovation framework, with special regard to smart city discourse, civil participation, and environmental performance. *Int. J. Urban Sustain. Dev.* **2021**, *13*, 97–116. [CrossRef]
13. Nicolas, C.; Kim, J.; Chi, S. Natural language processing-based characterization of top-down communication in smart cities for enhancing citizen alignment. *Sustain. Cities Soc.* **2021**, *66*, 102674. [CrossRef]
14. Ben Letaifa, S. How to strategize smart cities: Revealing the SMART model. *J. Bus. Res.* **2015**, *68*, 1414–1419. [CrossRef]
15. Lee, J.; Phaal, R.; Lee, S. An integrated service-device-technology roadmap for smart city development. *Technol. Forecast. Soc. Chang.* **2013**, *80*, 286–306. [CrossRef]
16. Gaur, A.; Scotney, B.; Parr, G.; McClean, S. Smart city architecture and its applications based on IoT. *Procedia Comput. Sci.* **2015**, *52*, 1089–1094. [CrossRef]
17. Caragliu, A.; del Bo, C.; Nijkamp, P. Smart cities in Europe. *J. Urban Technol.* **2011**, *18*, 65–82. [CrossRef]
18. Müller, A.R.; Park, J.; Sonn, J.W. Finding the old in the new: Smart cities in the national and local trajectories of urban development. *Int. J. Urban Sci.* **2023**, *27*, 1–9. [CrossRef]
19. De Jong, M.; Joss, S.; Schraven, D.; Zhan, C.; Weijnen, M. Sustainable-smart-resilient-low carbon-eco-knowledge cities; Making sense of a multitude of concepts promoting sustainable urbanization. *J. Clean. Prod.* **2015**, *109*, 25–38. [CrossRef]
20. Manca, F.; Daina, N.; Sivakumar, A.; Yi, J.W.X.; Zavistas, K.; Gemini, G.; Vegetti, I.; Dargan, L.; Marchet, F. Using digital social market applications to incentivise active travel: Empirical analysis of a smart city initiative. *Sustain. Cities Soc.* **2022**, *77*, 103595. [CrossRef]
21. UNECE (United Nations Economic Commission for Europe). *The UNECE-ITU Smart Sustainable Cities Indicators*; United Nations: Geneva, Switzerland, 2015.

22. Müller-Eie, D.; Kosmidis, I. Sustainable mobility in smart cities: A document study of mobility initiatives of mid-sized Nordic smart cities. *Eur. Transp. Res. Rev.* **2023**, *15*, 36. [CrossRef]
23. Haarstad, H. Constructing the sustainable city: Examining the role of sustainability in the 'smart city' discourse. *J. Environ. Policy Plan.* **2017**, *19*, 423–437. [CrossRef]
24. Wathne, M.; Haarstad, H. The smart city as mobile policy: Insights on contemporary urbanism. *Geoforum* **2020**, *108*, 130–138. [CrossRef]
25. Woods, O. Subverting the logics of "smartness" in Singapore: Smart eldercare and parallel regimes of sustainability. *Sustain. Cities Soc.* **2020**, *53*, 101940. [CrossRef]
26. Shamsuzzoha, A.; Niemi, J.; Piya, S.; Rutledge, K. Smart city for sustainable environment: A comparison of participatory strategies from Helsinki, Singapore and London. *Cities* **2021**, *114*, 103194. [CrossRef]
27. Pira, S. The social issues of smart home: A review of four European cities' experiences. *Eur. J. Futures Res.* **2021**, *9*, 3. [CrossRef]
28. Wang, K.; Zhao, Y.; Gangadhari, R.; Li, Z. Analyzing the adoption challenges of the internet of things (Iot) and artificial intelligence (ai) for smart cities in china. *Sustainability* **2021**, *13*, 10983. [CrossRef]
29. Verrest, H.; Pfeffer, K. Elaborating the urbanism in smart urbanism: Distilling relevant dimensions for a comprehensive analysis of Smart City approaches. *Inf. Commun. Soc.* **2019**, *22*, 1328–1342. [CrossRef]
30. World Competitiveness Center. *IMD Smart City Index 2024*; World Competitiveness Center: Lausanne, Switzerland, 2024.
31. Eden Strategy Institute and ONG&ONG Pte Ltd. *Top 50 Smart City Governments*; Eden Strategy Institute and ONG&ONG Pte Ltd.: Singapore, 2018.
32. Smith, H.; Medero, G.; Crane De Narváez, S.; Castro Mera, W. Exploring the relevance of 'smart city' approaches to low-income communities in Medellín, Colombia. *GeoJournal* **2023**, *88*, 17–38. [CrossRef]
33. Zygiaris, S. Smart City Reference Model: Assisting Planners to Conceptualize the Building of Smart City Innovation Ecosystems. *J. Knowl. Econ.* **2013**, *4*, 217–231. [CrossRef]
34. Angelidou, M. The Role of Smart City Characteristics in the Plans of Fifteen Cities. *J. Urban Technol.* **2017**, *24*, 3–28. [CrossRef]
35. Mora, L.; Deakin, M.; Reid, A. Strategic principles for smart city development: A multiple case study analysis of European best practices. *Technol. Forecast. Soc. Chang.* **2019**, *142*, 70–97. [CrossRef]
36. Csukás, M.; Szabó, R. The many faces of the smart city: Differing value propositions in the activity portfolios of nine cities. *Cities* **2021**, *112*, 103116. [CrossRef]
37. Al-Saffar, M. How to enhance the future of urban environments through smart sustainable urban infrastructures? *Disegnarecon* **2019**, *12*, 141–1414. [CrossRef]
38. Sánchez-Vaquerizo, J. Getting Real: The Challenge of Building and Validating a Large-Scale Digital Twin of Barcelona's Traffic with Empirical Data. *ISPRS Int. J. Geo-Inf.* **2022**, *11*, 24. [CrossRef]
39. Bibri, S.; Krogstie, J. Environmentally data-driven smart sustainable cities: Applied innovative solutions for energy efficiency, pollution reduction, and urban metabolism. *Energy Inform.* **2020**, *3*, 29. [CrossRef]
40. Bibri, S.; Krogstie, J. The emerging data–driven Smart City and its innovative applied solutions for sustainability: The cases of London and Barcelona. *Energy Inform.* **2020**, *3*, 5. [CrossRef]
41. Bolici, R.; Mora, L. Urban regeneration in the digital era: How to develop Smart City strategies in large european cities. *TECHNE* **2015**, *10*, 110–119. [CrossRef]
42. Calzada, I. (Smart) citizens from data providers to decision-makers? The case study of Barcelona. *Sustainability* **2018**, *10*, 3252. [CrossRef]
43. Demirel, D. The Impact of Managing Diversity on Building the Smart City A Comparison of Smart City Strategies: Cases From Europe, America, and Asia. *SAGE Open* **2023**, *13*, 21582440231184971. [CrossRef]
44. Demirel, D.; Mülazımoğlu, M. How the smart governance model shapes cities? Cases from Europe. *J. Enterprising Communities People Places Glob. Econ.* **2022**, *16*, 8–25. [CrossRef]
45. Drapalova, E.; Wegrich, K. Who governs 4.0? Varieties of smart cities. *Public Manag. Rev.* **2020**, *22*, 668–686. [CrossRef]
46. Fonseca, D.; Sanchez-Sepulveda, M.; Necchi, S.; Peña, E. Towards smart city governance. Case study: Improving the interpretation of quantitative traffic measurement data through citizen participation. *Sensors* **2021**, *21*, 5321. [CrossRef]
47. Mancebo, F. Smart city strategies: Time to involve people. Comparing Amsterdam, Barcelona and Paris. *J. Urban.* **2020**, *13*, 133–152. [CrossRef]
48. Mann, M.; Mitchell, P.; Foth, M.; Anastasiu, I. #BlockSidewalk to Barcelona: Technological sovereignty and the social license to operate smart cities. *J. Assoc. Inf. Sci. Technol.* **2020**, *71*, 1103–1115. [CrossRef]
49. March, H.; Ribera-Fumaz, R. Smart contradictions: The politics of making Barcelona a Self-sufficient city. *Eur. Urban Reg. Stud.* **2016**, *23*, 816–830. [CrossRef]
50. Noori, N.; Hoppe, T.; de Jong, M. Classifying pathways for smart city development: Comparing design, governance and implementation in Amsterdam, Barcelona, Dubai, and Abu Dhabi. *Sustainability* **2020**, *12*, 4030. [CrossRef]
51. Peyman, M.; Fluechter, T.; Panadero, J.; Serrat, C.; Xhafa, F.; Juan, A.A. Optimization of Vehicular Networks in Smart Cities: From Agile Optimization to Learnheuristics and Simheuristics. *Sensors* **2023**, *23*, 499. [CrossRef]
52. Phumpiu, C.; Rivera-Kuri, J. Monterrey envisioned as a smart city developed through international model examples. *WIT Trans. Ecol. Environ.* **2014**, *191*, 51–64. [CrossRef]

53. Ricci, L.; Mariano, C. The network construction of the "public city". @22Barcelona: A smart neighbourhood in a smart city. *TECHNE* **2018**, *1*, 121–126. [CrossRef]
54. Ringenson, T.; Höjer, M. Smart city planning and environmental aspects: Lessons from seven cities. In *ICT for Sustainability*; Atlantis Press: Amsterdam, The Netherlands, 2016; pp. 159–166.
55. Sanyal, S.; Mukhopadhyay, S.; Xhafa, F. Enhanced incentive schemes for volunteers to distribute surplus food in smart cities. *Sustain. Cities Soc.* **2023**, *91*, 104409. [CrossRef]
56. Shmelev, S.; Shmeleva, I. Multidimensional sustainability benchmarking for smart megacities. *Cities* **2019**, *92*, 134–163. [CrossRef]
57. Smith, A.; Martín, P. Going Beyond the Smart City? Implementing Technopolitical Platforms for Urban Democracy in Madrid and Barcelona. *J. Urban Technol.* **2021**, *28*, 311–330. [CrossRef]
58. Soriano-Gonzalez, R.; Perez-Bernabeu, E.; Ahsini, Y.; Carracedo, P.; Camacho, A.; Juan, A. Analyzing Key Performance Indicators for Mobility Logistics in Smart and Sustainable Cities: A Case Study Centered on Barcelona. *Logistics* **2023**, *7*, 75. [CrossRef]
59. Steenmans, K.; Robin, E.; Acuto, M.; Iwaszuk, E.; Garza, L. Governing the informed city: Examining local government strategies for information production, consumption and knowledge sharing across ten cities. *Urban Gov.* **2023**, *3*, 243–251. [CrossRef]
60. Tomàs, M. The smart city and urban governance: The urban transformation of Barcelona, 2011–2023. *Urban Res. Pract.* **2023**, *17*, 588–605. [CrossRef]
61. Vardopoulos, I.; Papoui-Evangelou, M.; Nosova, B.; Salvati, L. Smart 'Tourist Cities' Revisited: Culture-Led Urban Sustainability and the Global Real Estate Market. *Sustainability* **2023**, *15*, 4313. [CrossRef]
62. Yoo, Y. Toward sustainable governance: Strategic analysis of the smart city seoul portal in korea. *Sustainability* **2021**, *13*, 5886. [CrossRef]
63. Zambrano-Prado, P.; Muñoz-Liesa, J.; Josa, A.; Rieradevall, J.; Alamús, R.; Gasso-Domingo, S.; Gabarrell, X. Assessment of the food-water-energy nexus suitability of rooftops. A methodological remote sensing approach in an urban Mediterranean area. *Sustain. Cities Soc.* **2021**, *75*, 103287. [CrossRef]
64. Anttiroiko, A. City-as-a-platform: The rise of participatory innovation platforms in finnish cities. *Sustainability* **2016**, *8*, 922. [CrossRef]
65. Garau, C.; Nesi, P.; Paoli, I.; Paolucci, M.; Zamperlin, P. A Big Data Platform for Smart and Sustainable Cities: Environmental Monitoring Case Studies in Europe. In *Computational Science and Its Applications—ICCSA 2020*; Lecture Notes in Computer Science; Springer: Cham, Switzerland, 2020; Volume 12255 LNCS, pp. 393–406. [CrossRef]
66. Härkönen, K.; Hannola, L.; Pyrhönen, O. Advancing the smart city objectives of electric demand management and new services to residents by home automation—Learnings from a case. *Energy Effic.* **2022**, *15*, 25. [CrossRef]
67. Hernández, J.; Garcia, R.; Fischer, M.; Schonowski, J.; Atlan, D.; Ruohomaki, T. An interoperable open specifications framework for smart city urban platforms. In Proceedings of the Global IoT Summit, GIoTS 2019—Proceedings, Aarhus, Denmark, 17–21 June 2019.
68. Javed, A.; Kubler, S.; Malhi, A.; Nurminen, A.; Robert, J.; Framling, K. BIoTope: Building an IoT Open Innovation Ecosystem for Smart Cities. *IEEE Access* **2020**, *8*, 224318–224342. [CrossRef]
69. Kutty, A.; Wakjira, T.; Kucukvar, M.; Abdella, G.; Onat, N. Urban resilience and livability performance of European smart cities: A novel machine learning approach. *J. Clean. Prod.* **2022**, *378*, 134203. [CrossRef]
70. Lapinskaitė, I.; Stasytytė, V.; Skvarciany, V. Assessing the European Union capitals in the context of smart sustainable cities. *Open House Int.* **2022**, *47*, 763–785. [CrossRef]
71. Quijano, A.; Hernández, J.; Nouaille, P.; Virtanen, M.; Sánchez-Sarachu, B.; Pardo-Bosch, F.; Knieilng, J. Towards Sustainable and Smart Cities: Replicable and KPI-Driven Evaluation Framework. *Buildings* **2022**, *12*, 233. [CrossRef]
72. Soe, R.; Ruohomäki, T.; Patzig, H. Urban Open Platform for Borderless Smart Cities. *Appl. Sci.* **2022**, *12*, 700. [CrossRef]
73. Zaman, S.; Hertweck, C. Methods for Uncovering Discourses That Shape the Urban Imaginary in Helsinki's Smart City. *Front. Sustain. Cities* **2022**, *4*, 796469. [CrossRef]
74. Lopes, N. Smart governance: A key factor for smart cities implementation. In Proceedings of the 2017 IEEE International Conference on Smart Grid and Smart Cities, Singapore, 23–26 July 2017; pp. 277–282. [CrossRef]
75. Martinez, J.; Mahajan, S. Smart Cities and Access to Nature: A Framework for Evaluating Green Recreation Space Accessibility. *IEEE Access* **2023**, *11*, 102014–102024. [CrossRef]
76. Valencia-Arias, A.; Urrego-Marín, M.; Bran-Piedrahita, L. A methodological model to evaluate smart city sustainability. *Sustainability* **2021**, *13*, 11214. [CrossRef]
77. Ang-Tan, R.; Ang, S. Understanding the smart city race between Hong Kong and Singapore. *Public Money Manag.* **2022**, *42*, 231–240. [CrossRef]
78. Chib, A.; Alvarez, K.; Todorovic, T. Critical Perspectives on the Smart City: Efficiency Objectives vs Inclusion Ideals. *J. Urban Technol.* **2022**, *29*, 83–99. [CrossRef]
79. Echendu, A.; Okafor, P. Smart city technology: A potential solution to Africa's growing population and rapid urbanization? *Dev. Stud. Res.* **2021**, *8*, 82–93. [CrossRef]
80. Górna, A.; Górny, K. Singapore vs. The 'singapore of africa'—Different approaches to managing urban agriculture. *Land* **2021**, *10*, 987. [CrossRef]
81. Hartmann, R.; Garcia-Acevedo, L. Thermodynamic assessment of cities applying exergetic efficiency as evaluation index. *Sustain. Energy Technol. Assess.* **2022**, *50*, 101801. [CrossRef]

82. Icasiano, C.; Taeihagh, A. Governance of the risks of ridesharing in southeast Asia: An in-depth analysis. *Sustainability* **2021**, *13*, 6474. [CrossRef]
83. Iryna, K.; Iryna, U.; Yevhen, P.; Nataliia, D.; Maxym, B. Factors of successful development of smart cities. *Int. J. Comput. Sci. Netw. Secur. IJCSNS* **2022**, *22*, 21–28. [CrossRef]
84. Joo, Y. Developmentalist smart cities? The cases of Singapore and Seoul. *Int. J. Urban Sci.* **2023**, *27*, 164–182. [CrossRef]
85. Kenger, O.; Kenger, Z.; Ozceylan, E.; Mrugalska, B. Clustering of Cities Based on Their Smart Performances: A Comparative Approach of Fuzzy C-Means, K-Means, and K-Medoids. *IEEE Access* **2023**, *11*, 134446–134459. [CrossRef]
86. Lim, H.; Taeihagh, A. Autonomous vehicles for smart and sustainable cities: An in-depth exploration of privacy and cybersecurity implications. *Energies* **2018**, *11*, 1062. [CrossRef]
87. Papyshev, G.; Yarime, M. Exploring city digital twins as policy tools: A task-based approach to generating synthetic data on urban mobility. *Data Policy* **2021**, *3*, e16. [CrossRef]
88. Sethi, M.; Mittal, S. Developing a smart tool for integrated climate action planning (ICLAP 2050) in Asia-Pacific Cities. *Comput. Urban Sci.* **2022**, *2*, 45. [CrossRef]
89. Sotirelis, P.; Nakopoulos, P.; Valvi, T.; Grigoroudis, E.; Carayannis, E. Measuring Smart City Performance: A Multiple Criteria Decision Analysis Approach. *J. Knowl. Econ.* **2022**, *13*, 2957–2985. [CrossRef]
90. Šulyová, D.; Vodák, J. The impact of cultural aspects on building the smart city approach: Managing diversity in Europe (London), North America (New York) and Asia (Singapore). *Sustainability* **2020**, *12*, 9463. [CrossRef]
91. Tan, S.; Taeihagh, A.; Sha, K. How transboundary learning occurs: Case study of the asean smart cities network (ascn). *Sustainability* **2021**, *13*, 6502. [CrossRef]
92. Wey, W.; Peng, T. Study on Building a Smart Sustainable City Assessment Framework Using Big Data and Analytic Network Process. *J. Urban Plan. Dev.* **2021**, *147*, 04021031. [CrossRef]
93. Wirtz, B.; Müller, W.; Schmidt, F. Public Smart Service Provision in Smart Cities: A Case-Study-Based Approach. *Int. J. Public Adm.* **2020**, *43*, 499–516. [CrossRef]
94. Psara, O.; Fonseca, F.; Nisiforou, O.; Ramos, R. Evaluation of urban sustainability based on transportation and green spaces: The case of Limassol, Cyprus. *Sustainability* **2023**, *15*, 10563. [CrossRef]
95. The Innovation in Politics Institute. *DECIDIM: A Platform for Participatory Processes*; The Innovation in Politics Institute: Vienna, Austria, 2024.
96. Baroghi, F.; Ribeiro, P.; Fonseca, F. Towards a Holistic Framework for the Olympic-Led Sustainable Urban Planning Process. *Sustainability* **2024**, *16*, 951. [CrossRef]
97. Kidokoro, T.; Fukuda, R.; Sho, K. Gentrification in Tokyo: Formation of the Tokyo west creative industry cluster. *Int. J. Urban Reg. Res.* **2022**, *46*, 1055–1077. [CrossRef]

smart cities

MDPI

Review

Smart Cities, Digital Inequalities, and the Challenge of Inclusion

Olga Kolotouchkina [1,*], Laura Ripoll González [2,3] and Warda Belabas [2]

[1] Institute of Knowledge Technology, Complutense University of Madrid, 28040 Madrid, Spain
[2] Department of Public Administration and Sociology, Erasmus School of Social and Behavioural Sciences, Erasmus University Rotterdam, 3062 PA Rotterdam, The Netherlands; ripollgonzalez@essb.eur.nl (L.R.G.); belabas@essb.eur.nl (W.B.)
[3] Griffith Centre for Social and Cultural Research, Griffith University, Gold Coast, QLD 4215, Australia
* Correspondence: olga.kolotouchkina@ucm.es

Highlights:

What are the main findings?

- New smart cities have deepened social and digital inequalities among the technologically savvy residents and those who are usually left behind due to their age, lack of abilities, or resources.
- The reflective framework for inclusive smart cities offers a comprehensive and holistic approach from the perspective of who is included or excluded, what is the value of digital inclusion, and what are the key premises of inclusive digital governance.

What is the implication of the main finding?

- The framework is meant to act as a starting point in the design of inclusive digital urban policies and strategies and can also be employed to assess the directions taken to date by city governments towards more inclusive practices.

Abstract: While cities on a global scale embrace smartness as a roadmap for efficient urban governance, disparities persist in the domain of digital accessibility, literacy, and skills, with manifestations of digital exclusion, ageism, and ableism prevalent in most digital urban experiences. Despite their commitment to bridging the digital divide, governments lack comprehensive frameworks to inform policymaking and action for inclusion in smart cities. This review paper aims to deepen the understanding of the multifaceted challenges confronting the governance of inclusion in smart cities. Drawing from a review of research encompassing digital inclusion, digital transitions, smart cities, and urban governance, we discuss who is included and excluded in the governance of smart cities; what are the necessary conditions to be met for smart cities to be considered inclusive; and how can smart city governance deliver public value and equal benefits for all. As a novel contribution, this paper outlines a reflective framework to inform citizen inclusion in the governance of smart cities. This framework is meant to act as a starting point in the design of inclusive digital urban policies, and can also be employed to assess the directions taken to date in public organizations towards more inclusive urban practices.

Keywords: digital divide; digital inequalities; inclusion; smart cities; urban governance

1. Introduction

Since the emergence of the smart city concept in 1994 with the launch of digital city Amsterdam [1,2], the smart city phenomenon has become widespread among both global and prosperous cities striving to increase efficiency of government services through data-driven solutions and developing cities looking to become more competitive [3]. The definitions of smart cities emphasize the synergy between information and communication technology (ICT), logistics, and energy production and their impact on enhancing the well-being of citizens, their inclusion, and engagement [4]. However, critics of the smart city phenomenon have raised concerns regarding the widening inequalities in access to and usage of ICT, drawing attention to the digital divide of 'new urban poor' class of low-paid workers [5] and citizens left behind the digital progress [6–9].

The priority focus of smart cities on profit maximization through the accumulation of data, information, and knowledge as a main source of urban production and consumption [3,10] leads to a new smart mentality of urban governance [11], assuming that citizens should have the knowledge, skills, abilities, and resources to engage in a digital urban experience. Hence, 'younger', more 'digital savvy', 'able', highly educated and highly skilled citizens are actively embraced by the smart city digital economy, creating exclusion and socio-economic segregation of others (i.e., the working class, older people, people with disabilities) [12–14].

Albeit there exist growing concerns about the marginalization of particular groups of citizens from smart cities strategies and discourses [15–19], to date the literature lacks a reflective conceptual framework to guide the development of governance arrangements that will effectively deliver digital inclusion in the smart city. Anttiroiko and De Jong [15] developed a comprehensive framework of inclusive cities positioning inclusion as an instrumental tool of urban value creation and prosperity of urban communities. Enhancing equality of opportunities, capacity building, motivation, and incentivization are some of the proposed actions aimed at inclusion on an urban scale [15]. Sha et al. [16] argue that poor policymaking in the context of technology-led urban governance can hinder the access of large groups of citizens to the benefits of technology adoption. The authors propose a conceptual model connecting the drivers of disruptive urban technology with specific challenges of inclusive governance. Addressing the needs of vulnerable groups is identified as one of the pillars of inclusive policymaking [16].

The focus on specific needs of vulnerable and underrepresented citizens, such as people with disabilities, older people, and migrants, who often face discrimination and marginalization in access, use, or ownership of digital technologies due a lack of skills, abilities, and/or resources, is increasingly getting the attention of scholars [8,9,17]. However, research by Wang et al. [18] highlights that the value of inclusion is still not a priority in the development of smart city strategies, and stresses the need to expand the understanding of digitally vulnerable and underrepresented citizens in policymaking of smart cities. This trend is also observed in place management literature. For instance, Kavaratzis et al. [19] advocate for citizen-centricity, effective stakeholder engagement, and observing citizens' rights in decision-making processes about places. Against this backdrop, the need to develop more inclusive urban governance frameworks emerges, one that can provide a holistic perspective on the determinants of the digital exclusion of specific groups of citizens, the priorities for their digital engagement, and the effective drivers of their visibility and representation in the decision-making process.

Despite several previous attempts to conceptualize inclusion in the context of smart city developments (i.e., social inclusion or citizen participation) [8,10,18,20], to our knowledge, no comprehensive and holistic framework to deliver digital urban inclusion of the vulnerable and underrepresented groups has been developed to date. In order to fill this research gap, we engage in a problematizing review of digital inclusion in smart cities to provide a first attempt to develop a reflecting framework to inform inclusive smart city governance. We base this framework on previously identified barriers to inclusive urban development [16] and previous work addressing the critical challenges in the design of inclusive urban strategies [15,18]. Thus, we draw from theories and empirical research addressing concepts such as inclusive governance, digital accessibility; digital skills, abilities, and motivations; digital citizenship; co-creation and open innovation, among others, and we address the following guiding questions:

1. Who is included and excluded in the governance of smart cities?
2. What are the necessary conditions for smart cities to be considered inclusive?
3. How can smart city governance deliver public value and equal benefits for all?

To that effect, we first identify key pillars of inclusive urban governance, particularly focusing on vulnerable and underrepresented groups of citizens. We then contextualize the main enabling components of inclusion in digital urban policy processes along with their strategic implications on inclusion outcomes. Finally, we highlight critical challenges that signal important barriers to digital inclusion on an urban scale. The proposed reflective framework aims to enhance the governance of inclusive smart cities by offering a comprehensive approach to inclusion, encompassing stakeholder engagement, policy development, and implementation. It also serves as a starting point in the design and/or enhancement of inclusive digital urban policies and strategies aimed at redefining urban power dynamics and representation structures, enabling an effective exercise of digital rights of all citizens through access and participation, and ensuring that no one is left behind in technologically-assisted urban progress.

The aim of this paper is to provide a deep understanding of how inclusion is embedded in the governance and discourses associated with smart city transitions and digital agendas, with the secondary aim to inform decision making towards a more inclusive digital urban governance logic. Hence, rather than providing a systematic and integrative account of the literature, we aim to critically interrogate the literature to problematize [21] how inclusion has been conceptualized, operationalized, and evaluated, and which are the implicit assumptions underscoring current approaches to delivering inclusion.

This paper is organized as follows. We start by discussing how inclusion has been approached in the literatures of smart cities, urban planning, urban studies, digital transitions, and urban governance, and then turn to the literature on citizen engagement, citizen participation, and inclusive policymaking, among others, to reflect more broadly on the challenges of governing and delivering inclusion at the city level. We provide a theoretical review of concepts, theories, and empirical accounts of inclusion in smart cities. We then present a framework for the governance of inclusive smart cities. This paper concludes by discussing the theoretical and practical implications of our reflection for the scholarship on inclusive digital urban governance and for practitioners and policymakers committed to ensuring inclusion in smart city projects and urban digital transitions at large.

2. Theoretical Framework

> *"We believe a city to be smart when investments in human and social capital and traditional (transport) and modern (ICT) communication infrastructure fuel sustainable economic growth and a high quality of life, with a wise management of natural resources, through participatory governance"* [22] (p. 70).

2.1. Why Are Cities Going Smart?

Digital technology has become the main driving force shaping urban development and imaginaries of urban futures on a global scale [23]. Cities embrace digital economies and smart agendas as a roadmap towards efficiency, sustainability, and resiliency [24–26].

The promise of applying technology to the betterment of cities has resulted in industrial urban neighbourhoods transformed into high-tech precincts (see Barcelona's 22@ district's case in [27]). At the other extreme end of the spectrum, a series of entirely new smart cities are built from scratch around the world as beacons of a new digital urban age through private–public partnerships [28]. However, the smart city construct goes beyond physical transformations, it "has evolved to mean almost any form of technology-based innovation in the planning, development, operation, and management of cities" [29] (p. 146), and more conceptual work is needed to develop a comprehensive definition of smart city encompassing its multiple dimensions and implications [4].

2.2. Smart City Discourses

Giffinger et al. [30] (p. 11) define the smart city as "a city well performing in a forward-looking way in economy, people, governance, mobility, environment, and living, built on the smart combination of endowments and activities of self-decisive, independent and aware citizens". However, the authors acknowledge that the concept of smart city has been strongly influenced by city marketing experts who often promote optimistic but vague claims about the benefits of smart development. Wiig [31] (p. 535) adds that a "rhetoric of intelligent, transformative digital change works much more to "sell" a city in the global economy than to actually address urban inequalities". Governments around the world also increasingly engage in smart city branding, primarily as a tool to signal innovative and future-oriented strategies and to attract investment [28] or to tag their cities as belonging to a new digital urban elite [32]. In this context, branding and marketing are often applied to digital transformation projects rather than becoming aligned with an overall smart city brand. As such, they become targeted communication exercises to promote the benefits of certain technological interventions in urban efficiency [33,34] or to attract investment [35].

What this tells us is that, in the face of global competition, governments introduce measures that favour private rather than public interests [36], thus risking converting 'smart city' into a new valuable urban adjective for promotional purposes [37] that does not necessarily contribute to the betterment of citizens' lives. In fact, the concept of 'smart city' took over from that of 'sustainable city' as the most popular tag around 2010, peaking in 2015 [38], and is often tied into the agendas of big corporations that apply the 'smart' label to rebrand creative and eco city projects [39,40], reinforcing utopian approaches to solving urban societal issues with technology in urban planning.

2.3. *Technology, Society, and Urban Governance*

Given the open and fuzzy character of smart cities, the literature is characterized by multidisciplinary approaches to understanding and questioning the change technological advancements bring to the urban fabric. The initial approach puts forward a techno-centric and top-down perspective linked to the disruptive role of technology in optimizing urban efficiency, operability, and sustainability [41]. Such an approach emphasizes a new urban metabolism led by a selective group of corporations and business providers, and fueled through centralized dashboards that monitor advanced mobility systems, smart power grids, urban security applications, and waste management [42,43]. The techno-centric approach is increasingly challenged by an alternative social logic of smart urbanism that demands a critical understanding of the transformational impact of urban technologies on the social fabric of cities [23]. An alternative people-centred approach has since emerged, emphasizing that smart innovation can have positive and negative effects for the wellbeing and the quality of life of residents; and highlighting the key role of citizens in digital transformation processes as they embody the digital city experience [6,37].

Regardless of whether technology or people are put at the centre, scholars have long been concerned with the governance of the smart city, mainly considered as a multi-faceted policy and normative framework that regulates the management of digital public services, defines decision-making processes, roles, responsibilities, and procedures guiding interactions and cooperation between local governments and urban stakeholders in the digital transition [44–46]. This has led to the transformation of local governments' digital communication channels into dynamic platforms combining several roles as data and information sharing platforms providing access to administrative procedures, tools for enabling online public consultation and citizen engagement processes, and open data management platforms [47]. Similarly, a wide range of municipal services such as payment of municipal bills, property taxes, or fines, as well as welfare assistance, are increasingly performed online through ICT platforms [48]. Hence, digital apps and online platforms enable citizens to actively engage in discussions and decision-making processes and urban knowledge production [49]. Increased digital interaction of citizens with their city gives rise to digital citizenship, or the ability to participate in society online [50] and with smart urban governance models [44]. However, the digital experience is only available for those acquainted with and able to access it, which brings about the challenge of ensuring universal digital inclusion and accessibility [8,51,52].

2.4. The Challenge of Inclusion in the Governance of Smart Cities

Despite the potential inclusive benefits of digital innovation, some critical aspects of everyday life in the city are reproduced and reinforced as many are left behind due to their lack of access, skills, resources, or abilities [12,13,53,54]. Hence, notwithstanding transformative dynamics of the new digital urbanism, its polarizing effects on the citizens' wellbeing and its role within the urban ecosystem are under constant scrutiny by scholars. Evans et al. [55], for instance, critique the excessive technological triumphalism of policymakers and corporations dedicated primarily to the economic growth of digital cities rather than striving to support digital social transformations aimed at tackling widening social inequalities and the digital divide of the most vulnerable.

The limited and predominantly passive role of residents in smart city governance is also of concern to scholars. Residents are often regarded as sources of data feeding urban platforms, consumers of digital services, and/or investors in innovative urban technologies, such roles raising concerns about the ethics, the privacy, and the construct of digital citizenship in the smart city [7,11,56,57]. In parallel, key financial, hi-tech, and cultural services of the new digital economies are mainly aimed at highly educated and highly skilled global elites, leading to capital accumulation [3] and a widening of the societal divide as lower income, older, and migrant citizens lack the necessary digital literacy skills or access to engage in the digital city [58]. Furthermore, ageism and ableism are embedded in most digital experiences, perpetuating stigma and discrimination towards those in lower socioeconomic brackets, elderly, migrants, and people with disabilities. The disregard of specific needs, skills, and abilities of the disadvantaged and vulnerable groups leads to their systematic exclusion from smart city policies and discourses [9,14].

For instance, people with disabilities face significant digital barriers (i.e., a gap of 11% in access to Internet use in developing countries; low ownership rates of smart phones and computers; lack of compliance with web content accessibility guidelines of 63% of national governments portals and 98% of top websites [59]). At the same time, despite the meaningful impact of digital technologies on active ageing, personal autonomy, and community engagement of older people [60,61], the likelihood of using the Internet decreases by 8% each additional year after the age of 65 [13,62]. Furthermore, while digital technology has become a key enabler of social inclusion and place engagement of migrants and refugees [63], Alam and Imran [64] report that the digital divide for this group is further compounded by restricted affordability, low digital literacy, and language barriers.

Tackling digital inequalities and therefore fostering a barrier-free, inclusive, and ethical smart city remains one of the major challenges of policymaking and digital urban governance [7]. Inclusion, in this study, is understood as a public value that local authorities actively need to pursue to ensure it. In doing so, inclusion is not merely about guaranteeing that all citizens have the right to benefit from the innovations adopted, but that they also have a voice in deciding what smart city strategies they want to adopt and how to do so [18].

3. A Framework for Inclusive Smart Cities

A review of the extant scholarship on the emerging complexities and critical challenges of smart city development reveals that the application of urban technologies tends to favour the tech-savvy, wealthy, and well-connected, widening inequalities in ICT access, use, and outcomes of vulnerable groups of citizens. On a conceptual level, the digital divide has been conceptualized in three sub-levels: first order (ICT access or connectivity), second order (ICT use or capability), and third order (ICT outcomes/returns or content) [65–67]. Other scholars have rather focused on conceptualizing the determinants of the digital divide, distinguishing between: economic and technological determinants relating to prosperity and modernity; institutional determinants, relating to the political system and rule of law; and social determinants, relating to people-based characteristics, such as demographics and educational levels [68]. What these categorizations have in common is that inclusivity is measured broadly from the availability and accessibility of technology, the citizen groups benefiting from it, and the nature of institutional arrangements supporting collaboration with stakeholders in the digital transition. In terms of economic and technological determinants, one could think about issues of physical and material access to the Internet and digital devices, but also access and skills to use smart devices and upgraded versions of technology [69]. Additionally, in the context of smart city strategies, institutional determinants can be related to how processes of data collection and data ownership are organized between governments and other stakeholders such as big tech companies [26], but also how participation of and communication with citizens is organized (especially with marginalized or vulnerable citizens) [51]. Finally, social determinants of inclusion [64,70] are often used to further express the importance of including certain vulnerable groups in the context of digitalization, focusing on characteristics such as age, disabilities, and migrant background.

In inclusive smart city policy, issues of decision-making power and democratic legitimacy also play a key role [44]. A focus on inclusion at the urban governance level should result in two significant shifts: lead to outcomes in terms of a higher quality of life for all citizens, but also provide an open participatory approach to planning and acting to achieve those outcomes (increased legitimacy [1,9,44]). Hence, inclusiveness is also required in the structures for collaboration and decision-making, and must be a guiding principle on how citizens and governments interact and for how material output (wealth) and post-material output (health and sustainability) are prioritized [20,65,71].

3.1. For Whom Should Be Inclusive Smart Cities?

3.1.1. Vulnerable Groups and Digital Accessibility

Digital means of organizing access to services is becoming the norm in smart cities, as traditional face-to-face and phone-based ways of service delivery are diminishing. The phasing out of alternatives to the digital often leads to frustration among certain groups as they encounter barriers in the configuration of the Internet access, the outdated design of assistive devices, the cost of adaptive technologies, as well as other safety and security concerns [72,73]. Oftentimes, vulnerable groups such as older people, people with disabilities, low-income groups, ethnic minorities, and indigenous communities are most likely to be left behind in the progressive race to digitalization [51,74].

Van Dijk's [75] categorization of accessibility is useful here as a reflective framework as it encompasses four main elements: material access, motivational access, skills access, and usage access. The first element, accessibility of all citizens to physical or material tools and devices, is one of the most emphasized components of inclusive smart city strategies by local governments [76]. Such accessibility is a continuous concern as, on one hand, many cannot subsidize the cost and, on the other hand, the complexity of technology demands digital skills and upgraded versions of digital products [77]. On the other hand, motivational access refers to the mental barriers that prevent people from using ICT, often related to the level of self-esteem. Digital vulnerability and lower levels of self-esteem can lead to withdrawal from digital environments [9]. Furthermore, skills access refers to the lack of digital skills as a main barrier for usage. Finally, usage access refers to differences at the ICT user capability level [78]. This distinction is in line with the broader literature, in which there is an agreement on seeing accessibility as more than the level of possession of ICT or Internet connection, but rather incorporating (rational) digital literacy skills, abilities (i.e., knowledge) to access the digital environment, and finally (emotional) motivation to engage with the digital environment [70,79]. We expand on these non-physical requirements to access in the next section.

3.1.2. Digital Skills, Abilities, and Motivations

Digital skillsets often are limited to skills to navigate the Internet and to access information services [53]. Beyond the possession (or lack thereof) of the necessary skills to master digital devices, scholars have emphasized the importance of education and/or social support networks in enabling digital access [60,79,80]. This partly signals that digital access often depends on both economic capital (monetary assets and other economic possessions) and social capital, that is, human relationships, networks, and social support.

At this point, we wonder, what constitutes digital skills? Helsper and Deursen [81], amongst others, argue for a classification of skills that transcends technicalities. Van Deursen and Van Dijk [54], in this sense, propose a distinction between medium-related skills such as operational skills (or 'button knowledge'), formal skills (such as browsing and navigating), content-related skills (which includes searching, selecting, and evaluating information), but also communication skills (think about mailing and creating online identities), content creation (e.g., contributions to the Internet), and finally strategic skills, which are related to using the digital environment to achieve one's own personal and professional goals (ibid). In effect, this means that digital skills go beyond digital literacy, to also encompass social media and critical thinking skills. In addition, other cognitive and physical abilities are required for engaging digitally [82]. In this paper, we prefer the use of 'digital abilities' to suggest a more comprehensive conceptualization of the needs for digital accessibility, as not only connectivity or literacy, but also limited hearing and sight, for example, may hinder access of older people and people with disabilities to Internet [14].

Equally as importantly, motivation as the emotional status of citizens in engaging with the online world is required for meaningful digital engagement. Deng and Fei [83] highlight how self-efficacy or an individual's judgement about his/her own skills and capability to perform computer-related activities influences engagement and involvement in smart city structures.

What emerges is that policymakers must be aware and actively consider citizens' differences in access, skills, and abilities to enable their full participation in the smart city. Ideally, policymakers would invest in identifying vulnerable individuals and groups; hear and raise awareness of their needs; and provide the necessary infrastructure and policies to overcome all barriers and motivate all citizens to get involved in the making of the smart city.

3.2. *What Is the Value of Digital Inclusion?*

3.2.1. Digital Citizenship

The understanding of the inclusive principles for governing smart or digital cities also requires attention to the construct of digital citizenship. This concept is grounded on digital competences, norms of behaviour with digital technology [50,84], and participation of digitally enabled citizens in social life [69]. "A digital citizen is someone with regular and flexible access to the Internet, the skills to apply this technology, and a regular use of the Internet for participation and functioning in all spheres of the society" [85] (p. 5). This definition highlights the performative nature of citizenship in the digital domain, determined by actions, processes, and outcomes, rather than by a formal given status of belonging [57,86].

The spread of Artificial Intelligence (AI), blockchain, and of the algorithmic data governance in urban management poses important ethical and social challenges to the construct of digital citizenship as mounting concerns over the dominant role of global Information Technology (IT) firms in ownership and management of data, and the lack of a specific digital rights policy framework emerge. As evidence of the pervasive impact of digital technologies on limiting civil liberties and increasing digital vulnerabilities, Calzada [87] highlights that the COVID-19 pandemic has led to the emergence of a new globalized digital pandemic citizenship regime framed by surveillance and cyber control, and that this has materialized through mobile apps, biometric technologies, and vaccine passports.

Digital rights of citizens in the inclusive smart cities should be therefore framed, on one hand, by the informed and enabling process of digital innovation and its transformative outcomes on their social, economic, and political engagement, and, on the other hand, by the ethical and moral implications of the unfolding algorithmic innovation on the human rights and social justice in the cyberspace [85,88].

3.2.2. Rights and Ownership of Technological Innovation and Data

Different authors have raised concerns about the exclusion of citizens and other stakeholders in the city when it comes to questions of rights and ownership of data and technology. Nederhand et al. [89], for example, argue that for technology firms, there is a lot to gain when it comes to ambitions of cities regarding their smartness, "whereby the knowledge and budgets of powerful private players overshadows those of local governments" (p. 3). In practice, the dominance of tech companies and other private players in smart cities leads to exclusion of citizens that are less tech-savvy, and often ends in the (mis)use of private data for profit making and information control (ibid). Lee et al. [90] similarly warn about the increasing involvement and power of big IT and consultancy firms that search for new business opportunities in smart city development. Governments interested in inclusive smart cities need to think not only about inclusion in terms of the technology they select, but also in terms of who is benefiting from the data and the rights that follow from that technology. The question is hence: from whose perspectives are data collected, who has access to the data, and who determines which data need to be collected?

Too often, smart cities focus on using big data to improve urban efficiency, planning, and policy making. Inclusive smart cities should rather focus on the implications of the collection and use of data for citizens. In the next sections, we elaborate on this technology (or top-down) led vs. people (or bottom-up centred) perspectives and their effects in smart urban governance.

3.3. *How to Ensure Inclusive Governance in Smart Cities?*

3.3.1. Commitment to Inclusive Governance

To build inclusive smart cities, where no one is left behind because of age, lack of abilities, or social and economic vulnerability, inclusion should become a backbone of key strategic and functional areas within city governments. Some of the emerging practices of institutionalizing commitment to inclusion in smart cities include the establishment of specific figures or departments focused on inclusion, public–private partnerships to deliver inclusion, specific training on inclusion for public employees, as well as community engagement aimed at identifying and effectively addressing critical challenges of digital inclusion [8].

Furthermore, establishing accessibility by designing a universal standard of urban policies for digital development of all communication channels and administrative tools and resources simplifies and unifies digital access to the city. In fact, the standardization and compliance with the WCAG (Web Content Accessibility Guidelines) levels A and AA make key urban platforms and digital channels accessible, understandable, and operable to all citizens [91].

3.3.2. Co-Creation and Open Innovation

Cities need to make room for people to participate in smart city decisions, as too often the starting point lies not in the existing knowledge and skills of citizens, but rather in technology and the interests of urban entrepreneurs, as stated above [37]. In the development of inclusive smart cities, citizens play a crucial role as they provide their experiences and expertise, help collect data, and enhance the democratic process through their active participation [71]. Jang and Gim [92] argue that "establishing a citizen-centred smart governance system can make a contribution by taking equity, inclusion and social justice into consideration at every stage of the development of smart city initiatives" (p. 140).

Malek et al. [20] similarly state that an inclusive city provides the participatory infrastructure to enhance citizen centricity in its smart decisions. This means that there is not only attention for decision-making through consensus with citizens, but also power delegation in the design of decision-making structures. Inclusive cities hence need to actively invest in a city infrastructure—both online and offline—in which all citizens can attend and give input to decision-making processes regarding the smartness of the city.

3.3.3. Inclusive Communication and Representation

Finally, inclusive governance is also connected to questions regarding communication and representation. Information about participation in smart city processes needs to equally reach all citizens and communities, which means that government websites and other communication channels need to be accessible for all, including people with disabilities, older people, migrants, and so forth [93]. For underrepresented groups to be visible and to make their voices heard in these smart city processes, three elements are required [94]: to be present, to belong, and to take part not only in planning and development phases, but also in the implementation process of urban strategies. However, bringing these processes to all citizens, designing avenues for them to provide feedback through public forums, town halls, information campaigns, and community engagement depends on the will of local authorities [93].

Including a diverse spectrum of citizens in inclusive smart city processes can lead to a better relationship between cities and their people. This holds particularly true when it comes to citizen representation in (or identification with) smart city brands [95] and the legitimacy of such smart city branding efforts (albeit a full elaboration on smart city brands is out of the scope of this paper).

Drawing from the previous theoretical review and the discussion of the complexities facing smart city development, Table 1 outlines an overview of key questions, components, priority targets, and critical challenges for inclusive smart cities in support of a shift towards more open and inclusive digital transition processes.

Table 1. A reflective framework for inclusion of vulnerable and underrepresented citizens in smart cities.

Key Questions	Main Components	Priority Targets	Examples of Critical Challenges
WHO is included/excluded?	Digital accessibility	Affordability of resources, tools, and devices	Lack of resources (i.e., financial)
		Education, social support networks, social relationships	Lack of social capital (relationships, networks)
	Digital skills and abilities	Digital literacy	Lack of skills in a variety of domains, beyond digital literacy
			Lack of screen-skills, information seeking, communication skills, social media, critical thinking
			Misalignment between limited abilities and use of smart devices and digital services
	Motivations to engage with digital platforms	Enhancing self-esteem, self-efficacy, self-confidence	Persisting mental barriers (low self-esteem, anxiety, fear, etc.)
WHAT is the value of digital inclusion?	Digital citizenship	Ensuring informed and enabling process of digital innovation and its main outcomes	Lack of knowledge on the opportunities, risks, and outcomes of the digital innovation

Table 1. *Cont.*

Key Questions	Main Components	Priority Targets	Examples of Critical Challenges
WHAT is the value of digital inclusion?	Digital citizenship	Safeguarding social justice and human rights in the digital environment (digital/non-digital rights)	Neglecting ethical and moral concerns of algorithmic and datafied environments
	Open data	Ensuring open data system	Dominance of global tech companies, focus on economic profit
		Providing regulatory framework on data availability and access	Lack of control and misuse of private data
	Digital knowledge	Enabling effective access to knowledge resources and dissemination platforms	Lack of effective information resources
		Citizen-centricity in smart decision-making process	Technocratic leadership
HOW can smart cities ensure an inclusive governance?	Institutional commitment	Sustainable and standardized administrative procedures in the digital environment	Lack of commitment in aligning design of digital environment
		Specific training of public employees	Lack of knowledge and institutional procedures
		Compliance with web content accessibility guidelines (WCAG)	Lack of digital accessibility protocols
	Inclusive communication and representation	Inclusive city brands and discourses with active engagement of all urban stakeholders	Misalignment between information and reference world of hard-to-reach groups
		Broad scale of communication channels, both offline and online	Language barriers
		Aligning communication with shared interests and values, resonating with multiple audiences	Lack of intercultural competences and sensitivity
	Co-creation/open innovation	Fostering input from citizens in design and decision making	Aligning power structures to citizen-centricity
		Citizen-centricity in smart innovation	Lack of effective citizen engagement and representation

4. Discussion and Conclusions

4.1. Policy Implications

In the last decades, cities have become unique places for advanced digital innovation aimed at ensuring urban efficiency, operability, and sustainability. While the impact of technological innovation has an undeniable impact on capacity building, cost reduction, and global attractiveness and competitiveness of the place, the negative outcomes of the progressive digitization of urban spaces should not be neglected. New smart cities have deepened social and digital inequalities among their technologically savvy residents and those who are usually left behind due to their age, lack of abilities, or resources. Despite several efforts to address the lack of inclusion, governments still lack comprehensive frameworks to guide policy and action toward inclusive governance in smart and digital cities. At the same time, extant research on smart cities has been primarily focused on the techno-centric dimension of digital urban development, lacking knowledge on social intelligence, social justice, and citizen-centred attributes [96].

Similarly, studies on inclusion in smart cities frequently have a one-dimensional approach, highlighting certain elements of an inclusive strategy while neglecting others, and mostly focusing on assessing effectiveness of digital interventions in the delivery of public services and on tackling the big challenges such as climate change and rapid urbanization [97]. In particular, the literature shows a lack of attention to a deeper understanding of how inclusion works within smart city structures [98].

This paper aims to address this gap by proposing a holistic framework for building inclusive smart cities. We engaged in a problematizing review of the literature to identify key challenges and priority targets for digital inclusion on an urban scale. We do so from the lens of digitally vulnerable and underrepresented groups of citizens and with their needs in mind. Our proposed framework is based on the main values of digital inclusion, as well as the fundamental principles of an inclusive urban governance logic.

The application of this framework as an evaluative tool to reflect on the effectiveness of certain strategies to deliver inclusion could similarly provide important insights into future policy and strategic directions by identifying areas of neglect. The framework is meant to act as a starting point to aid the design of inclusive digital urban policies and strategies, but can also be employed to assess the directions taken to date in the view of assisting learning and adapting in public organizations towards more inclusive practices.

We similarly wish to emphasize that the reflective framework introduced above offers a comprehensive, albeit not exhaustive, perspective to observing inclusion in smart city policy, discourses, and strategies. What also becomes apparent from our problematizing review of the literature is that the smart city concept is inextricably linked to other concepts in the literature such as 'vital city', the 'just city', 'sustainable city' or—more recently—'the societal smart city', in which citizen-driven governance, inclusive services, resilient infrastructures, and skills and information literacy are also considered key elements in the digital transition [99,100]. Our framework incorporates insights from research on city vitality, social justice, and sustainability.

4.2. Future Research

From a theoretical standpoint, this framework provides an overview of determinants of inclusion in smart urban governance that can inspire further empirical research on overlooked aspects of inclusion. For instance, on the misalignment between information of reference worlds for certain marginalized groups. A further discussion is also needed on the intersectionality and the exacerbation of existing vulnerabilities in the practice of smart urban governance and smart city developments. The concepts of social justice, equal opportunities, and digital citizenship are essential pillars of further research in the context of smart urban development.

As suggested above, we welcome further empirical research to reflect on and further develop the introduced framework. Of particular importance would be to get a better understanding of the degree to which the different pillars are considered in decision-making for smart city policies. Importantly, also which pillars are not considered and the reasons underlying such choices. In addition, in line with scholarly concerns on the reinforcing of inequalities in the narratives and action surrounding the transition to digitalization, we call for empirical case research focusing on cities that claim to be smart, investigating the discrepancy between their storytelling and actual inclusive policy design and decision-making in smart city practices. This is particularly relevant as we observe that, in more recent years, smart city branding has been criticized for its economic and market-driven perspective, and its neglect of real community problems that are pushed to the back in brand narratives [101,102].

Finally, a limitation of this framework lies in its focus on the structure/process or decision-making-side of inclusive cities, and not on the equally important question of outcomes of smart city policies in terms of a higher quality of life for all citizens, which deserves equal attention. Nevertheless, we believe that our focus offers the preconditions for coming to such an equal distribution of benefits that lead to higher quality of life for all groups of citizens. By testing the framework in real-life contexts, it offers a method of transformation of smart urban governance into a much more participatory and bottom-up process.

Author Contributions: Conceptualization, O.K., L.R.G. and W.B.; methodology, L.R.G.; investigation, O.K., L.R.G. and W.B.; writing—original draft preparation, O.K.; writing—review and editing, L.R.G. and W.B.; visualization, O.K.; supervision, L.R.G.; project administration, W.B.; funding acquisition, O.K. All authors have read and agreed to the published version of the manuscript.

Funding: This research was partially supported by the EDIRE project under the European Union's Horizon Europe program for widening participation and spreading excellence (grant agreement number 101060145), and by the European Union's Horizon 2020 research and innovation programme under the Marie Skłodowska Curie grant agreement number 895260, and the APC was funded by the EDIRE project under the European Union's Horizon Europe program for widening participation and spreading excellence (grant agreement number 101060145).

Conflicts of Interest: The authors declare no conflicts of interest. The funder had no role in the design of this study and in the writing of the manuscript.

References

1. Capra, C.F. The Smart City and its citizens: Governance and citizen participation in Amsterdam Smart City. *Int. J. E-Plan. Res. (IJEPR)* **2016**, *5*, 20–38. [CrossRef]
2. Mora, L.; Bolici, R. How to become a smart city: Learning from Amsterdam. In *Smart and Sustainable Planning for Cities and Regions: Results of SSPCR 2015*; Springer: Berlin/Heidelberg, Germany, 2017; Volume 1, pp. 251–266.
3. De Jong, M.; Joss, S.; Taeihagh, A. Smart cities as spatial manifestations of 21st century capitalism. *Technol. Forecast. Soc. Change* **2024**, *202*, 123299. [CrossRef]
4. Dameri, R.P. Searching for Smart City Definition: A Comprehensive Proposal. *Int. J. Comput. Technol.* **2013**, *11*, 2544–2551. [CrossRef]
5. Caprotti, F. Eco-urbanism and the Eco-city, or, Denying the Right to the City? *Antipode* **2014**, *46*, 1285–1303. [CrossRef]
6. Calzada, I.; Pérez-Batlle, M.; Batlle-Montserrat, J. People-Centered Smart Cities: An Exploratory Action Research on the Cities' Coalition for Digital Rights. *J. Urban Aff.* **2021**, *45*, 1537–1562. [CrossRef]
7. Kitchin, R.; Cardullo, P.; Di Feliciantonio, C. Citizenship, justice, and the right to the smart city. In *The Right to the Smart City*; Cardullo, P., Di Feliciantonio, C., Kitchin, R., Eds.; Emerald Publishing: Bingley, UK, 2019; pp. 1–24.
8. Kolotouchkina, O.; Barroso, C.L.; Sánchez, J.L.M. Smart Cities, the Digital Divide, and People with Disabilities. *Cities* **2022**, *123*, 103613. [CrossRef]
9. Kolotouchkina, O.; Viñarás-Abad, M.; Mañas-Viniegra, L. Digital Ageism: Emerging Challenges and Best Practices of Age-Friendly Digital Urban Governance. *Media Commun.* **2023**, *11*, 6–17. [CrossRef]
10. Cardullo, P.; Kitchin, R. Smart Urbanism and Smart Citizenship: The Neoliberal Logic of "Citizen-Focused" Smart Cities in Europe. *Environ. Plan. C Politics Space* **2018**, *37*, 813–830. [CrossRef]
11. Vanolo, A. Smartmentality: The Smart City as Disciplinary Strategy. *Urban Stud.* **2014**, *51*, 883–898. [CrossRef]
12. Florida, R. *The New Urban Crisis: Gentrification, Housing Bubbles, Growing Inequality, and What We Can Do About It*; Oneworld Publications: London, UK, 2017; ISBN 9781786074010.

13. Friemel, T.N. The Digital Divide Has Grown Old: Determinants of a Digital Divide among Seniors. *New Media Soc.* **2016**, *18*, 313–331. [CrossRef]
14. Rosales, A.; Fernández-Ardèvol, M. Ageism in the Era of Digital Platforms. *Converg. Int. J. Res. Into New Media Technol.* **2020**, *26*, 1074–1087. [CrossRef]
15. Anttiroiko, A.V.; De Jong, M. *The Inclusive City: The Theory and Practice of Creating Shared Urban Prosperity*; Palgrave Macmillan: Cham, Switzerland, 2021; ISBN 9783030613648.
16. Sha, K.; Taeihagh, A.; De Jong, M. Governing Disruptive Technologies for Inclusive Development in Cities: A Systematic Literature Review. *Technol. Forecast. Soc. Chang.* **2024**, *203*, 123382. [CrossRef]
17. Belabas, W.; George, B. Do Inclusive City Branding and Political Othering Affect Migrants' Identification? Experimental Evidence. *Cities* **2023**, *133*, 104119. [CrossRef]
18. Wang, C.H.; Steinfeld, E.; Maisel, J.L.; Kang, B. Is Your Smart City Inclusive? Evaluating Proposals from the U.S. Department of Transportation's Smart City Challenge. *Sustain. Cities Soc.* **2021**, *74*, 103148. [CrossRef]
19. Kavaratzis, M.; Giovanardi, M.; Lichrou, M. *Inclusive Place Branding: Critical Perspectives on Theory and Practice*; Routledge: London, UK, 2017.
20. Malek, J.A.; Lim, S.B.; Yigitcanlar, T. Social Inclusion Indicators for Building Citizen-Centric Smart Cities: A Systematic Literature Review. *Sustainability* **2021**, *13*, 376. [CrossRef]
21. Alvesson, M.; Sandberg, J. The Problematizing Review: A Counterpoint to Elsbach and van Knippenberg's Argument for Integrative Reviews. *J. Manag. Stud.* **2020**, *57*, 1290–1304. [CrossRef]
22. Caragliu, A.; Del Bo, C.; Nijkamp, P. Smart Cities in Europe. *J. Urban Technol.* **2011**, *18*, 65–82. [CrossRef]
23. Luque-Ayala, A.; Marvin, S. Developing a Critical Understanding of Smart Urbanism? *Urban Stud.* **2015**, *52*, 2105–2116. [CrossRef]
24. Kitchin, R. The Timescape of Smart Cities. *Ann. Am. Assoc. Geogr.* **2019**, *109*, 775–790. [CrossRef]
25. Lim, C.; Kim, K.-J.; Maglio, P.P. Smart Cities with Big Data: Reference Models, Challenges, and Considerations. *Cities* **2018**, *82*, 86–99. [CrossRef]
26. Townsend, A.M. *Smart Cities: Big Data, Civic Hackers, and the Quest for a New Utopia*; W.W. Norton & Company: New York, NY, USA, 2013; ISBN 9780393349788.
27. Bakıcı, T.; Almirall, E.; Wareham, J. A Smart City Initiative: The Case of Barcelona. *J. Knowl. Econ.* **2012**, *4*, 135–148. [CrossRef]
28. Kolotouchkina, O.; Seisdedos, G. Place branding strategies in the context of new smart cities: Songdo IBD, Masdar and Skolkovo. *Place Brand. Public Dipl.* **2018**, *14*, 115–124. [CrossRef]
29. Yigitcanlar, T.; Kamruzzaman, M.; Buys, L.; Ioppolo, G.; Sabatini-Marques, J.; da Costa, E.M.; Yun, J.J. Understanding "Smart Cities": Intertwining Development Drivers with Desired Outcomes in a Multidimensional Framework. *Cities* **2018**, *81*, 145–160. [CrossRef]
30. Giffinger, R.; Fertner, C.; Kramar, H.; Kalasek, R.; Pichler-Milanovic, N.; Meijers, E. Smart Cities—Ranking of European Medium-Sized Cities, Vienna University of Technology. Available online: http://www.smart-cities.eu (accessed on 15 July 2024).
31. Wiig, A. The Empty Rhetoric of the Smart City: From Digital Inclusion to Economic Promotion in Philadelphia. *Urban Geogr.* **2015**, *37*, 535–553. [CrossRef]
32. Engelbert, J.; van Zoonen, L.; Hirzalla, F. Excluding citizens from the European smart city: The discourse practices of pursuing and granting smartness. *Technol. Forecast. Soc. Chang.* **2019**, *142*, 347–353. [CrossRef]
33. Huertas, J.I.; Mahlknecht, J.; Lozoya-Santos, J.d.J.; Uribe, S.; López-Guajardo, E.A.; Ramirez-Mendoza, R.A. Campus City Project: Challenge Living Lab for Smart Cities. *Appl. Sci.* **2021**, *11*, 11085. [CrossRef]
34. Vijaygopal, R.; Bennett, R.; Savani, S. Initiation, marketing and branding of smart city projects: A study of decision processes. *J. Mark. Manag.* **2023**, *39*, 1833–1861. [CrossRef]
35. Grebosz-Krawczyk, M. Place Branding (R)Evolution: The Management of the Smart City's Brand. *Place Brand. Public Dipl.* **2021**, *17*, 93–104. [CrossRef]
36. Capdevila, I.; Zarlenga, M.I. Smart City or Smart Citizens? The Barcelona Case. *SSRN Electron. J.* **2015**, *8*, 266–282. [CrossRef]
37. Hollands, R.G. Will the Real Smart City Please Stand Up? *City* **2008**, *12*, 303–320. [CrossRef]
38. Allam, Z.; Newman, P. Redefining the Smart City: Culture, Metabolism and Governance. *Smart Cities* **2018**, *1*, 4–25. [CrossRef]
39. Anthopoulos, L. Smart Utopia vs Smart Reality: Learning by Experience from 10 Smart City Cases. *Cities* **2017**, *63*, 128–148. [CrossRef]
40. Willis, K.S.; Alessandro, A. *The Routledge Companion to Smart Cities*; Routledge: London, UK; New York, NY, USA; Taylor & Francis Group: Abingdon, UK, 2020; ISBN 9781138036673.
41. Picon, A. *Smart Cities: Theory and Criticism of a Self-Fulfilling Ideal—AD Primer*; John Wiley & Sons: Hoboken, NJ, USA, 2015; ISBN 9781119075592.
42. Baykurt, B.; Raetzsch, C. What Smartness Does in the Smart City: From Visions to Policy. *Converg. Int. J. Res. Into New Media Technol.* **2020**, *26*, 775–789. [CrossRef]
43. Kandt, J.; Batty, M. Smart Cities, Big Data and Urban Policy: Towards Urban Analytics for the Long Run. *Cities* **2020**, *109*, 102992. [CrossRef]

44. Meijer, A.; Bolívar, M.P.R. Governing the Smart City: A Review of the Literature on Smart Urban Governance. *Int. Rev. Adm. Sci.* **2015**, *82*, 392–408. [CrossRef]
45. González, L.R.; Klijn, E.H.; Eshuis, J.; Braun, E. Does Participation Predict Support for Place Brands? An Analysis of the Relationship between Stakeholder Involvement and Brand Citizenship Behavior. *PAR Public Adm. Rev.* **2023**. [CrossRef]
46. Ruhlandt, R.W.S. The Governance of Smart Cities: A Systematic Literature Review. *Cities* **2018**, *81*, 1–23. [CrossRef]
47. United Nations Department of Economic and Social Affairs. *United Nations E-Government Survey 2022. The Future of Digital Government*; United Nations: New York, NY, USA, 2022.
48. Hatuka, T.; Zur, H.; Mendoza, J.A. The Urban Digital Lifestyle: An Analytical Framework for Placing Digital Practices in a Spatial Context and for Developing Applicable Policy. *Cities* **2020**, *111*, 102978. [CrossRef]
49. Cardullo, P. *Citizens in the 'Smart City': Participation, Co-Production, Governance*; Routledge: Oxfordshire, UK, 2020.
50. Ribble, M.S.; Bailey, G.D.; Ross, T.W. Digital citizenship: Addressing appropriate technology behavior. *Learn. Lead. Technol.* **2004**, *32*, 6.
51. Pérez-Escolar, M.; Canet, F. Research on Vulnerable People and Digital Inclusion: Toward a Consolidated Taxonomical Framework. *Univers. Access Inf. Soc.* **2022**, *22*, 1059–1072. [CrossRef]
52. Lee, J.; Babcock, J.; Pham, T.S.; Bui, T.H.; Kang, M. Smart city as a social transition towards inclusive development through technology: A tale of four smart cities. *Int. J. Urban Sci.* **2023**, *27* (Suppl. 1), 75–100. [CrossRef]
53. Graham, S. Bridging Urban Digital Divides? Urban Polarisation and Information and Communications Technologies (ICTs). *Urban Stud.* **2002**, *39*, 33–56. [CrossRef]
54. van Deursen, A.J.; van Dijk, J.A. The First-Level Digital Divide Shifts from Inequalities in Physical Access to Inequalities in Material Access. *New Media Soc.* **2019**, *21*, 354–375. [CrossRef]
55. Evans, J.; Karvonen, A.; Luque-Ayala, A.; Martin, C.; McCormick, K.; Raven, R.; Voytenko Palgan, J. *Smart and Sustainable Cities? Pipedreams, Practicalities and Possibilities*; Routledge: London, UK, 2020.
56. Cardullo, P.; Kitchin, R. Being a "Citizen" in the Smart City: Up and down the Scaffold of Smart Citizen Participation in Dublin, Ireland. *GeoJournal* **2018**, *84*, 1–13. [CrossRef]
57. Isin, E.; Ruppert, E. *Being Digital Citizens*; Rowman & Littlefield: Lanham, MD, USA, 2015.
58. Nijman, J.; Wei, J.D. Urban inequalities in the 21st century economy. *Appl. Geogr.* **2020**, *117*, 102188. [CrossRef]
59. United Nations Department of Economic and Social Affairs. Disability and Development Report 2024. Accelerating the Realization of the Sustainable Development Goals by, for and with Persons with Disabilities. Executive Summary. Available online: https://social.desa.un.org/publications/un-flagship-report-on-disability-and-development-2024 (accessed on 27 September 2024).
60. Li, Y.; Alencar, A. A tale of two cities: Digital placemaking and elderly Houniao migration in China. *J. Ethn. Migr. Stud.* **2023**, *49*, 1032–1049. [CrossRef]
61. Llorente-Barroso, C.; Kolotouchkina, O.; Mañas-Viniegra, L. The Enabling Role of ICT to Mitigate the Negative Effects of Emotional and Social Loneliness of the Elderly During COVID-19 Pandemic. *Int. J. Environ. Res. Public Health* **2021**, *18*, 3923. [CrossRef]
62. McDonough, C.C. The Effect of Ageism on the Digital Divide among Older Adults. *Gerontol. Geriatr. Med.* **2016**, *2*, 1–7. [CrossRef]
63. Georgiou, M. City of Refuge or Digital Order? Refugee Recognition and the Digital Governmentality of Migration in the City. *Telev. New Media* **2019**, *20*, 600–616. [CrossRef]
64. Alam, K.; Imran, S. The Digital Divide and Social Inclusion among Refugee Migrants. *Inf. Technol. People* **2015**, *28*, 344–365. [CrossRef]
65. Caragliu, A.; Del Bo, C.F. Smart Cities and the Urban Digital Divide. *npj Urban Sustain.* **2023**, *3*, 43. [CrossRef]
66. Olphert, W.; Damodaran, L. Older People and Digital Disengagement: A Fourth Digital Divide? *Gerontology* **2013**, *59*, 564–570. [CrossRef] [PubMed]
67. Song, Z.; Wang, C.; Bergmann, L. China's Prefectural Digital Divide: Spatial Analysis and Multivariate Determinants of ICT Diffusion. *Int. J. Inf. Manag.* **2020**, *52*, 102072. [CrossRef]
68. Wijers, G.D.M. Determinants of the Digital Divide: A Study on IT Development in Cambodia. *Technol. Soc.* **2010**, *32*, 336–341. [CrossRef]
69. Mossberger, K.; Tolbert, C.J.; Mcneal, R.S. *Digital Citizenship: The Internet, Society, and Participation*; MIT Press: Cambridge, MA, USA, 2010; ISBN 9780262134859.
70. Helsper, E.J. *Digital Inclusion: An Analysis of Social Disadvantage and the Information Society*; Department for Communities and Local Government: London, UK, 2008; ISBN 9781409806141.
71. Chen, T.; Ramon Gil-Garcia, J.; Gasco-Hernandez, M. Understanding Social Sustainability for Smart Cities: The Importance of Inclusion, Equity, and Citizen Participation as Both Inputs and Long-Term Outcomes. *J. Smart Cities Soc.* **2022**, *1*, 135–148. [CrossRef]
72. Dobransky, K.; Hargittai, E. The Disability Divide in Internet Access and Use. *Inf. Commun. Soc.* **2006**, *9*, 313–334. [CrossRef]

73. Lussier-Desrochers, D.; Normand, C.L.; Romero-Torres, A.; Lachapelle, Y.; Godin-Tremblay, V.; Dupont, M.-È.; Roux, J.; Pépin-Beauchesne, L.; Bilodeau, P. Bridging the Digital Divide for People with Intellectual Disability. *Cyberpsychol. J. Psychosoc. Res. Cyberspace* **2017**, *11*, 1. [CrossRef]
74. Tsatsou, P. Vulnerable People's Digital Inclusion: Intersectionality Patterns and Associated Lessons. *Inf. Commun. Soc.* **2021**, *25*, 1475–1494. [CrossRef]
75. van Dijk, J. *The Network Society*, 4th ed.; Sage Publications: Thousand Oaks, CA, USA, 2020.
76. Serrano, W. Digital systems in smart city and infrastructure: Digital as a service. *Smart Cities* **2018**, *1*, 134–154. [CrossRef]
77. Tsatsou, P. Is Digital Inclusion Fighting Disability Stigma? Opportunities, Barriers, and Recommendations. *Disabil. Soc.* **2020**, *36*, 702–729. [CrossRef]
78. Lytras, M.D.; Visvizi, A. Who uses smart city services and what to make of it: Toward interdisciplinary smart cities research. *Sustainability* **2018**, *10*, 1998. [CrossRef]
79. Al-Muwil, A.; Weerakkody, V.; El-haddadeh, R.; Dwivedi, Y. Balancing Digital-By-Default with Inclusion: A Study of the Factors Influencing E-Inclusion in the UK. *Inf. Syst. Front.* **2019**, *21*, 635–659. [CrossRef]
80. Hänninen, R.; Taipale, S.; Luostari, R. Exploring Heterogeneous ICT Use among Older Adults: The Warm Experts' Perspective. *New Media Soc.* **2020**, *23*, 1584–1601. [CrossRef]
81. Helsper, E.J.; van Deursen, A.J.A.M. Digital skills in Europe: Research and policy. In *Digital Divides: The New Challenges and Opportunities of E-Inclusion*; Andreasson, K., Ed.; Routledge: New York, NY, USA, 2015; pp. 126–148.
82. Raja, D.S. Bridging the Disability Divide Through Digital Technologies. Background Paper for the 2016 World Development Report. 2016. Available online: https://thedocs.worldbank.org/en/doc/123481461249337484-0050022016/original/WDR16BPBridgingtheDisabilityDividethroughDigitalTechnologyRAJA.pdf (accessed on 29 September 2024).
83. Deng, G.; Fei, S. Exploring the Factors Influencing Online Civic Engagement in a Smart City: The Mediating Roles of ICT Self-Efficacy and Commitment to Community. *Comput. Hum. Behav.* **2023**, *143*, 107682. [CrossRef]
84. Xu, S.; Yang, H.H.; MacLeod, J.; Zhu, S. Social Media Competence and Digital Citizenship among College Students. *Converg. Int. J. Res. New Media Technol.* **2018**, *25*, 735–752. [CrossRef]
85. Oyedemi, T.D. The Theory of Digital Citizenship. In *Handbook of Communication for Development and Social Change*; Servaes, J., Ed.; Springer: Singapore, 2018; pp. 237–255. [CrossRef]
86. Hintz, A.; Dencik, L.; Wahl-Jorgensen, K. Digital Citizenship and Surveillance Society. *Int. J. Commun.* **2017**, *11*, 731–739.
87. Calzada, I. Emerging Digital Citizenship Regimes: Pandemic, Algorithmic, Liquid, Metropolitan, and Stateless Citizenships. *Citizsh. Stud.* **2022**, *27*, 160–188. [CrossRef]
88. Mossberger, K.; Tolbert, C.J.; Hamilton, A. Measuring Digital Citizenship: Mobile Access and Broadband. *Int. J. Commun.* **2012**, *6*, 2492–2528.
89. Nederhand, J.; Avelino, F.; Awad, I.; De Jong, P.; Duijn, M.; Edelenbos, J.; Engelbert, J.; Fransen, J.; Schiller, M.; Van Stapele, N. Reclaiming the City from an Urban Vitalism Perspective: Critically Reflecting Smart, Inclusive, Resilient and Sustainable Just City Labels. *Cities* **2023**, *137*, 104257. [CrossRef]
90. Lee, J.Y.; Woods, O.; Kong, L. Towards More Inclusive Smart Cities: Reconciling the Divergent Realities of Data and Discourse at the Margins. *Geogr. Compass* **2020**, *14*, e12504. [CrossRef]
91. WCAG 2.1 Understanding Docs. Available online: http://www.w3.org/WAI/WCAG21/Understanding/intro#understanding-the-four-principles-of-accessibility (accessed on 15 July 2024).
92. Jang, S.; Gim, T.-H.T. Considerations for Encouraging Citizen Participation by Information-Disadvantaged Groups in Smart Cities. *Sustain. Cities Soc.* **2022**, *76*, 103437. [CrossRef]
93. Kempin Reuter, T. Human Rights and the City: Including Marginalized Communities in Urban Development and Smart Cities. *J. Hum. Rights* **2019**, *18*, 382–402. [CrossRef]
94. Boucher, N.; Vincent, P.; Fougeyrollas, P.; Geiser, P.; Hazard, D.; Nouvellet, H. Participation of People with Disabilities in Local Governance: Measuring the Effects of Inclusive Local Development Strategies. Centre Interdisciplinaire de Recherche en Réadaotion et Intégration Sociale 2015. Available online: http://www.firah.org/en/the-participation-of-people-with-disabilities-in-local-governance-how-to-measure-the-effects-on-inclusive-local-development.html (accessed on 12 December 2023).
95. Molinillo, S.; Anaya-Sánchez, R.; Morrison, A.M.; Coca-Stefaniak, J.A. Smart City Communication via Social Media: Analysing Residents' and Visitors' Engagement. *Cities* **2019**, *94*, 247–255. [CrossRef]
96. Mora, L.; Bolici, R.; Deakin, M. The First Two Decades of Smart-City Research: A Bibliometric Analysis. *J. Urban Technol.* **2017**, *24*, 3–27. [CrossRef]
97. Sharifi, A. A Critical Review of Selected Smart City Assessment Tools and Indicator Sets. *J. Clean. Prod.* **2019**, *233*, 1269–1283. [CrossRef]
98. Nam, T.; Pardo, T.A. Conceptualizing smart city with dimensions of technology, people, and institutions. In Proceedings of the 12th Annual International Digital Government Research Conference: Digital Government Innovation in Challenging Times, College Park, MD, USA, 12–15 June 2011; pp. 282–291. [CrossRef]
99. Alizadeh, H.; Sharifi, A. Societal Smart City: Definition and Principles for Post-Pandemic Urban Policy and Practice. *Cities* **2023**, *134*, 104207. [CrossRef]
100. Ripoll González, L.; Gale, F. Sustainability as Economic Value Pluralism: Implications for Urban Politics and Policy. *Cities* **2023**, *134*, 104167. [CrossRef]

101. Gotham, K.F. (Re) branding the big easy: Tourism rebuilding in post-Katrina New Orleans. *Urban Aff. Rev.* **2007**, *42*, 823–850. [CrossRef]
102. Hassen, I.; Giovanardi, M. The Difference of "Being Diverse": City Branding and Multiculturalism in the "Leicester Model". *Cities* **2018**, *80*, 45–52. [CrossRef]

smart cities

MDPI

Review

Embedding Circular Economy in the Construction Sector Policy Framework: Experiences from EU, U.S., and Japan for Better Future Cities

Giulia Marzani [1,*], Simona Tondelli [1], Yuko Kuma [2], Fernanda Cruz Rios [3], Rongbo Hu [4,*], Thomas Bock [5] and Thomas Linner [6]

1 Department of Architecture, Alma Mater Studiorum—University of Bologna, 40136 Bologna, Italy; simona.tondelli@unibo.it
2 Department of Architecture, Kyushu Sangyo University, Fukuoka 813-0004, Japan; kuma@mail.kyusan-u.ac.jp
3 Civil, Architectural, and Environmental Engineering, Drexel University, 3141 Chestnut Street, Curtis 251, Philadelphia, PA 19104, USA; fc432@drexel.edu
4 Kajima Corporation, Kajima Technical Research Institute Singapore, Singapore 489690, Singapore
5 School of Engineering and Design, Technical University of Munich, 80333 Munich, Germany; bockrobotics@web.de
6 Faculty of Civil Engineering, Ostbayerische Technische Hochschule Regensburg, 93053 Regensburg, Germany; thomas.linner@oth-regensburg.de
* Correspondence: giulia.marzani3@unibo.it (G.M.); r.hu@kajima.com.sg (R.H.)

Highlights:

What are the main findings?

- Varied integration of Circular Economy in construction sector policies in the EU, U.S., and Japan.
- Limited acknowledgment of digital technologies in CE Policies, despite their potential.

What are the implications of the main finding?

- Need for Policy Harmonization and Cross-Sectoral Collaboration.
- Policymakers should integrate digital tools more systematically into CE strategies, leveraging research-driven insights to enhance urban resilience, optimize resource use, and support a more efficient and inclusive built environment.

Abstract: The transition towards a Circular Economy (CE) in the construction sector is essential to achieving sustainable, inclusive smart cities. This study examines the integration of CE principles into construction policies across four key global contexts: the European Union (focusing on Italy and Germany), the United States, and Japan. Through a comparative policy analysis, the research identifies best practices, implementation barriers, and the role of digitalization in advancing CE strategies. In Europe, CE is embedded in policy frameworks such as the Green Deal and the New Circular Economy Action Plan, driving the shift toward sustainable urban development. The United States, while in the early stages of CE adoption, is fostering circular initiatives at local levels, particularly in waste management and building deconstruction. Japan's policy landscape integrates CE within a broader strategy for resource efficiency, emphasizing technological innovation. The findings highlight the necessity of a research-driven approach to inform policies that leverage digital tools, such as Building Information Modeling and Digital Product Passports, to enhance material traceability and urban circularity. This study contributes to the global effort of designing smart cities that are not only technologically advanced but also environmentally and socially sustainable through the adoption of CE principles in the built environment.

 https://doi.org/10.3390/smartcities8020048

Keywords: circular smart cities; policy scan; digital technologies; Circular Economy; construction sector

1. Introduction

Urban policies and strategies changed over time, influenced by the global challenges that cities are asked to tackle in order to improve inhabitants' quality of life. Resource scarcity and climate change mitigation and adaptation are some of the most contemporary examples. Thus, contemporary cities are increasingly subject to transformations, embracing visions that reflect and react to the various stimuli coming from the global environment. Over time, this influence has been translated by planners and city leaders into the definition of urban paradigms. Most notably, the sustainable development theories and interpretations led to the conceptualization of new development pathways that "meet the needs of the present without compromising the ability of future generations to meet their own needs" [1]. The Brundtland Report and the Earth Summit (United Nations Conference on Environment and Development) of Rio de Janeiro in 1992 contributed to the wide diffusion of the concept that has been introduced in the common lexicon, also shaping urban strategies and policies, pursuing environmental, social, and economic sustainability. These priorities are still key urgencies for many city governments who, fifteen years after the Rio de Janeiro conference, are still enacting policies and setting ambitious targets to meet the sustainable goals of the Agenda 2030.

Since the 2000s, digitalization has been shaping citizens' habits and needs, clearly highlighting that the traditional way of approaching cities would not fit with the new needs of the population, thus contributing to the rise in the smart cities' paradigm [2]. However, the ecological and sustainability ideals have not been concurrent with the rising debate of the smart city. Rather, urban trajectories have expanded to envelop the smart concept around sustainable discourse, and it can be assumed that smart cities are also sustainable [3]. The so-called "smart eco-city" has been interpreted as a way of achieving environmental goals through and with smart urban strategies [4]. Therefore, the notions of sustainable development, ecological modernization, and regenerative development are influencing the different urban paradigms, creating a certain degree of overlapping and cross-cutting exchange [5].

Despite worldwide policies acknowledging the importance of pursuing social, economic, and environmental sustainability, the dichotomy between the achievement of economic growth and the overconsumption of natural resources is difficult to overcome [6,7]. In the last decade, the Circular Economy (CE) began to emerge as one of the most holistic models to achieve that paradigm shift, having the potential to reverse the consumption of natural resources while ensuring economic growth [8–11]. Even in their diversity, the numerous CE definitions share the idea of a model of production and consumption that is based on the creation of values and economic growth without consuming virgin materials and natural resources, which could lead to environmental degradation [12]. Embracing the Ellen MacArthur Foundation's vision, CE is conceived as overcoming the linear economic model based on the take/make/dispose approach, thus eliminating the concept of "waste". The Ellen MacArthur Foundation, together with the McKinsey Institute, also conceptualized the ReSOLVE framework, which is often used to understand which actions lead to circularity principles, namely: Regenerating, Sharing, Optimizing, Looping, Virtualizing, and Exchanging resources [11,13].

As argued by [3], there is a correlation between smartness and circularity in cities, especially revealing a close interconnection between technology and environmental sustainability. For example, the role of digital technologies is embedded in many of the proposed RESOLVE actions and is considered an essential enabler of CE in many industries [14]. Moreover, in the academic debate, the concept of the "Smart Circular Economy" paradigm began to emerge [15], even though previous studies failed to approach the integration of digitalization practices into the CE discourse in a systematic way, mostly focusing on specialized sectors [16]. Research is even more scarce if applied to smart urban circularity [17]. In particular, the built environment is the city's most investigated domain, where digital transformation has been slower but with significant improvements in the last decades [14]. In fact, the construction sector is the largest emitter of greenhouse gas (GHG) emissions, being responsible for at least 37% of emissions globally. As a result, numerous initiatives attempt to improve the performance of existing buildings and set priorities for the construction of new ones [18]. The United Nations has identified three urgent pathways to reduce GHG emissions and decarbonize the sector by design: avoiding the extraction and production of raw materials, shifting to regenerative material practices, and improving methods to decarbonize conventional materials like cement, steel, and aluminum. For those targets to be achieved, building designers need to adopt resource-efficient, data-driven optimization processes, fostering the reuse and recycling of materials [18]. Key CE design strategies include using computer-aided design optimization for reduced material usage, selecting materials that reduce non-renewable material extraction, designing for material and component reuse, and extending the life of buildings and/or materials through proper maintenance [18].

However, although the literature acknowledges the importance of CE for the construction industry, the integration of CE strategies and the industry's digital transformation remains a critical knowledge gap [19]. Moreover, given the rising attention attributed to CE and digital technologies as a means to achieve sustainability in the construction sector, policymakers are currently facing the challenge of designing and implementing strategies and actions to promote circularity. Hence, the integration of CE principles into policies, though understudied, can play a vital role in enabling CE as a driver for sustainable smart urban transition [19].

Based on these considerations, the present contribution aims to analyze to what extent CE is embedded into the current policy framework targeting the construction sector. Europe, the U.S., and Japan were the regions selected in the grant-in-aid for Scientific Research called "Development, Empirical Research and Dissemination of New Theories and System Techniques for Circular Economy (CE) to Meet SDGs Goal 12; Producer and Consumer Responsibility" coordinated by Aoyama Gakuin University. Italy and Germany are the first two best performers in CE practices among the five main European Countries [20]. Based on that, four countries were included in the scope of this research: Italy, Germany, Japan, and the United States (U.S.).

The policy framework analysis in the four countries will focus on understanding to what extent existing policies facilitate the adoption of CE practices in the construction sector. Through a content analysis, this paper provides (i) an overview of the state-of-the-art and adopted approaches to CE-related policies in different countries around the world; (ii) the identification of the most common strategies adopted by national and local governments in the identified countries to achieve a circular built environment; and (iii) the main barriers for the further development of CE practices in the construction sector, also considering whether the policies acknowledge the role of digital technologies.

Following this introduction, Section 2 proposes a background of policies in the analyzed countries. Section 3 reviews the policy sector in each investigated country and the methodology adopted to analyze the most relevant policy documents. Section 4 presents the results, followed by Sections 5 and 6, in which discussion and conclusions are presented.

2. Background

2.1. Circular Economy Policies for the Construction Sector in the EU

According to the European Union, the construction sector in the EU accounts for: 40% of gross final energy consumption; 35% of greenhouse gas emissions; 50% of extracted material resources; 30% of water consumption; and 35% of waste production [21]. In particular, the building sector plays an important role in Germany's energy transition, accounting for 30% of Germany's greenhouse gas emissions. Residential buildings alone are responsible for 26% of Germany's final energy consumption due to electricity usage and heating. Meanwhile, non-residential buildings are responsible for 47% of Germany's greenhouse gas emissions, despite comprising only 13% of the building volume [22]. As for the Italian scenario, the building sector is the largest contributor to energy demand, responsible for 44% of the energy consumption. Residential buildings are the main energy consumers, accounting for 57%, with lower levels of energy performance compared to the European benchmark [23].

Even though the policy framework analysis is performed at a national level for each country, it is worth mentioning that Italy and Germany belong to the European Union and are therefore influenced by the policies enacted at the European level. The reference policy document for CE in Europe is the New Circular Economy Action Plan [24], which was adopted by the commission in March 2020 as one of the main pillars of the European Green Deal, the EU agenda for sustainable growth [25]. However, as its name predicts, the New Circular Economy Action Plan is rooted in previous initiatives promoted by the European Commission, starting from 2015 when the first Circular Economy Action Plan was adopted [26]. This initial plan included measures stimulating Europe's transition towards a CE while fostering sustainable economic growth and the generation of new jobs. The plan included 54 actions, and after three years of implementation, all of them have been achieved, even if, for some actions, the work continues beyond 2019. Actions were related to the whole life cycle, going into detail of different phases of the production and consumption process. The actions were divided according to the topic they mainly refer to: production, consumption, waste management, the market for secondary raw materials, and sectorial actions concerning some strategic sectors (e.g., plastic, food waste, critical raw materials, construction and demolition, biomass and bio-based materials, innovation and investments, monitoring).

Thanks to the implementation of the 54 actions, the EU is recognized as a global leader in CE policymaking, and the action plan encouraged at least 14 Member States, 8 regions, and 11 cities to put forward CE strategies [27]. All the 54 actions included in the first CE action plan were completed within 2019 and, in December of the same year, the European Commission adopted the European Green Deal, which consists of a package of policy initiatives set up to pave the road for a green transition, with the ultimate goal of reaching climate neutrality in Europe by 2050. The depicted vision for Europe enables the imagination of a continent that will rely on renewable energy, curbing emissions throughout all sectors but with a strong commitment to the energy sector. Target values are proposed in transport, job creation, energy consumption and production, and building renovation. It is foreseen that at least 3% of the total floor area of public buildings should be renovated annually, with a target of 49% renewable energy in buildings by 2030. Europe also requires Member States to increase the use of renewable energy for heating and cooling by 1.1% each year until 2030. As one of the main pillars of the European Green Deal, in March 2020, the European Commission adopted the New CE Action Plan [26]. The new action plan comprises 35 actions to be implemented in 2020–2023, focusing on different areas when compared to the previous action plan. The key actions are listed under seven macro-areas that correspond to as many overall goals and targets:

- A sustainable product policy framework;
- Key product value chains;
- Less waste, more value;
- Making the CE work for people, regions, and cities;
- Crosscutting actions;
- Leading efforts at the global level;
- Monitoring the progress.

The construction and building category is included within the key product value chains. In the Action Plan, the Commission proposes to launch a new comprehensive strategy for a Sustainable Built Environment to coordinate and ensure coherence in climate goals, energy and resource efficiency, construction and demolition (C&D) waste, accessibility, digitalization, and necessary skills. Specifically, the strategy promotes circularity principles through the revision of the construction product regulation, promoting the adaptability of the built environment, using the Level(s) framework [28] to integrate life-cycle assessment in public procurement, revise the legislation for C&D waste, and promote initiatives to reduce soil sealing, fostering the reuse of abandoned sites and contaminated brownfields. The strategy for a Sustainable Built Environment has still not been published, even though it was expected in 2021, but some of the actions have been addressed in the "Renovation Wave for Europe", a document that pays particular attention to green buildings [29].

Recently, in July 2024, the new Ecodesign for Sustainable Products Regulation (ESPR) [30] entered into force as a cornerstone of the Commission's approach to more environmentally sustainable and circular products. It replaces the Ecodesign Directive 2009/125/EC, which was more related only to energy-related products [31]. The ESPR extends the scope to cover virtual and physical products and reinforces the range of ecodesign requirements that can be set for products, which can comprise requirements relating to durability, circularity, and the overall reduction of the environmental and climate footprint of products, amongst many others. The ESPR includes a set of other measures, such as the Digital Product Passport (DPP), which might impact the construction sector. DPP is a digital identity card for products, components, and materials that will store relevant information to support products' sustainability, promote their circularity, and strengthen legal compliance.

Regarding the EU's construction sector, the following tools and regulations to foster circularity and sustainability assessment are also worth noting [32]:

- Transversal regulations on Building Sustainability Assessment by CEN TC 350;
- Cradle-to-Cradle certification scheme;
- Environmental Assessment Methodology (BREEAM) circular framework;
- Leadership in Energy and Environmental Design (LEED) certification.

In this paper, the authors selected Italy and Germany to perform a policy framework analysis at a national level. According to the Circular Economy Network, they are the first two best performers in CE practices among the five main European countries included in the analysis [20].

2.2. *Circular Economy Policies for the Construction Sector in the U.S.*

In the U.S. construction sector, residential and commercial buildings are responsible for 31% of GHG emissions when considering electricity end-use [33]. CE implementation in the U.S. is still in its infancy. While the U.S. administration from 2020 to 2024 supported a transition towards a CE, the federal support came in the form of funding, targets, and action plans. Even at a local level (i.e., cities), there are few CE legislation in effect. Perhaps as a result, CE strategies that preserve the value of the materials over their life cycle are still rare in the U.S. construction industry. Three strategies dominate the end-of-life of construction materials in the country: recycling (usually of metals with high market value like steel, copper, and aluminum), downcycling—mostly concrete and brick—into coarse aggregates, and disposal in landfills. For example, according to the U.S. Environmental Protection Agency (EPA), 76% of C&D waste was diverted from landfills in 2018 [34]. Of the 76%, 52% represented materials downcycled as aggregates [34]. EPA's Sustainable Materials Management plan aims to increase the safe reuse and recycling of C&D materials and improve the tracking of C&D data [34]. Efforts like the Inflation Reduction Act and the Bipartisan Infrastructure Law provided unprecedented funding to support state and local waste management infrastructure, including recycling programs through the National Recycling Strategy [35]. Although those efforts focused more heavily on municipal solid waste management, critical materials, and electric vehicle batteries, C&D stakeholders have actively participated in the workshops and taskforces to promote CE education, strengthen secondary materials markets, and measure progress toward CE [36]. In the building sector, the Executive Order on Catalyzing Clean Energy Industry Jobs Through Federal Sustainability, mandated by the U.S. Federal Government in 2021, establishes legally binding targets for federal buildings and operations. The Order promotes using construction materials with lower embodied emissions and includes an ambitious target for a net-zero emissions building portfolio by 2045. However, at the time of writing this paper, the new U.S. administration has paused funding disbursement related to the Inflation Reduction Act and Bipartisan Infrastructure Law, which can significantly affect recycling programs and CE incentives. At the same time, the withdrawal of the U.S. from the Paris Agreement, the administration's explicit support for fossil fuel expansion, and the recent rollbacks in renewable energy policies strongly indicate a shift away from sustainable economic models [37]. These actions suggest that federal support for CE initiatives will likely diminish, potentially stalling progress in waste reduction, material recovery, and climate-conscious industrial practices.

However, although their authority is shaped by federal constraints and state preemption laws, states and municipalities in the U.S. have relative autonomy in policymaking. In fact, despite the changes at a federal level, some municipal governments had taken proactive steps to advance CE principles, driven by the past administration's strong emphasis on sustainability. As a result, in this paper, U.S. policy frameworks are analyzed at a local level (i.e., cities).

2.3. Circular Economy Policies for the Construction Sector in Japan

In Japan, the total GHG emissions for fiscal year 2022 amounted to 1.135 billion tons of CO_2 equivalent, with CO_2 accounting for 91.1% of the total. Of these CO_2 emissions, the manufacturing and construction sectors contributed 22.7%, while the residential sector accounted for 4.8% [38]. In terms of final energy consumption, the agriculture, forestry, fisheries, mining, and construction sectors combined (excluding energy consumption related to transportation) accounted for 2.9%, whereas the residential sector contributed 15.0%. Regarding industrial waste, the construction industry generated 21.5% of total industrial waste in fiscal year 2021 [39].

Japan's CE policies are fundamentally framed by the Basic Environment Law and the Fifth Basic Environment Plan (2018) [40,41], which provides a comprehensive and long-term outline for environmental conservation policies. In accordance with the Basic Environment Law, Japan has enacted the Basic Act on Establishing a Sound Material-Cycle Society (2000) [42,43], which lays out the foundational principles and planning framework for CE. Additionally, the Fundamental Plan for Establishing a Sound Material-Cycle Society specifies concrete measures and targets. In August 2024, the Cabinet approved the Fifth Fundamental Plan for Establishing a Sound Material-Cycle Society [44–46]. Against the backdrop of finite resource availability and increasing environmental burdens, CE has become an essential component for achieving a sustainable society. Japan's policies have progressed incrementally based on these laws and plans.

The Fifth Fundamental Plan for Establishing a Sound Material-Cycle Society positions the transition to a CE as a central focus [44]. In addition to addressing environmental issues such as climate change and biodiversity conservation, it aims to contribute to economic security, industrial competitiveness, regional revitalization, and the realization of a high-quality life. These objectives are outlined as a national strategy to secure a sustainable future for the coming generations. The plan highlights the following five priority areas:

- Building sustainable communities and societies through the transition to a CE;
- Comprehensive resource circulation across the entire lifecycle through enhanced collaboration among businesses;
- Establishing diverse regional circular systems to achieve regional revitalization;
- Strengthening the foundations for resource circulation and waste management to ensure proper waste treatment and environmental restoration;
- Promoting the development of an international resource circulation framework and expanding Japan's circular industries globally.

As part of the comprehensive resource circulation across the entire lifecycle through enhanced collaboration among businesses, the Fifth Fundamental Plan sets specific targets for various materials and products to be achieved by 2030 [44]. Key material categories include plastics and waste oil, biomass and food waste, metals such as base and rare metals, and soil, stone, and construction materials. The plan emphasizes the promotion of an environmentally conscious design for construction materials, the extension of building lifespans, and the expanded use of blended cement to enhance the value of resources. Key product categories include buildings, automobiles, small household appliances and electronics, textile products, and materials or products newly required for climate change mitigation. The plan outlines measures for buildings to form and maintain high-quality social infrastructure, reduce waste generation through extended building lifespans, promote the reuse of construction materials, and facilitate recycling construction-related plastic waste. Besides this, it aims to minimize waste generation during disasters.

The progress of these policies is evaluated approximately every two years by Japan's Ministry of the Environment in the form of a Progress Report on the Basic Plan for Establishing a Sound Material-Cycle Society. In addition to monitoring material flow indicators such as resource productivity, the circular use rate, and the final disposal volume, the CE Roadmap is utilized to track the progress of key initiatives [47,48]. The 2022 progress report shows the following developments and challenges in the construction sector:

- For soil, stone, and construction materials, the representative indicators show steady progress in the desired direction;
- Although the proportion of certified long-life quality housing in newly constructed homes is increasing, achieving the target remains challenging, indicating the need for further promotion.

Furthermore, following the revision of the Act on the Rational Use of Energy [49], Compliance with Energy Efficiency Standards for Newly Constructed Buildings [50] has been made mandatory. This has led to steady progress in reducing energy consumption in buildings. Such achievements play a crucial role in accelerating the construction sector's transition to a CE. Japan's CE policies also recognize challenges such as advancing recycling technologies, addressing regional disparities in recycling infrastructure, and supporting small and medium-sized enterprises in transitioning to a CE. To tackle these issues, efforts are being made to improve resource management efficiency through the active use of digital technologies and to promote the adoption of circular business models.

3. Materials and Methods

As recalled in the background, the policy scan has been conducted for three regions only, namely the EU, the U.S., and Japan. These three geographic areas have been selected as reference contexts for the grant-in-aid for Scientific Research called "Development, Empirical Research and Dissemination of New Theories and System Techniques for Circular Economy (CE) to Meet SDGs Goal 12; Producer and Consumer Responsibility", being key players in CE adoption and representing the strongest economies worldwide. Moreover, the diverse regions have different regulatory frameworks and strategic approaches to implementing policies, thus offering opportunities to identify the challenges and the best practices that might support the CE transition in the construction sector. In the EU, Italy and Germany have been selected for a more detailed analysis since they represent the first two best performers in CE practices among the five main European countries [20]. In the U.S., CE initiatives at the city scale are also investigated.

To conduct the policies' content analysis and identify the most common CE strategies adopted in the different countries, two reference frameworks must be established. On one hand, defining the various phases of the construction value chain is necessary. On the other hand, given the blurriness of the CE concept and the numerous interpretations and operationalization methods developed over time, the most suitable framework for identifying the CE actions tailored to the construction sector had to be identified.

Similar to what has been adopted by [51], the phases of the construction value chain are the following:

1. Planning and design;
2. Manufacturing;
3. Construction;
4. Use and maintenance;
5. End-of-life.

The proposed framework for CE has been adapted, enriching the CE criteria used by [52] to include the ReSOLVE criteria. Indeed, the former study has a strong focus on waste management and waste prevention in the construction sector, while the ReSOLVE framework encompasses more clearly the role of digital technologies throughout the phases of the construction value chain (e.g., the emphasis on the role of digital technologies in fostering CE as expressed in [53]). The chosen CE criteria are summarized and represented in Table 1.

Table 1. CE criteria coupled with enabling digital technologies.

CE Criteria	Examples of CE-Related Actions	Enabling Technology
(a) Reducing the use of resources through reuse and adaptation	a.1 Sharing of facilities/adaptive use of facilities prevents the need for additional buildings; a.2 Increasing the utilization rate of buildings to prevent the need for additional buildings; a.3 Choice of material or product for options requiring less material for the same performance; a.4 Saving of materials in production, optimizing cut-offs; a.5 Reducing material consumption at the construction site by using products pre-cut to size.	Advanced materials for the shift to renewable energy. Techs for traceability (Internet of Things (IoT), blockchain) for shared spaces and by selecting flexible construction solutions; for reusing components.

Table 1. *Cont.*

CE Criteria	Examples of CE-Related Actions	Enabling Technology
(b) Waste prevention	b.1 Extending life span by renovating old buildings instead of building new ones; b.2 Preventing premature demolition by changing the use of a building; b.3 Repair and maintenance to prevent premature demolition; b.4 Use of demountable construction components enabling the reuse of construction components and reconstruction of buildings.	Advanced materials for prioritizing renovation with respect to demolition and new construction.
(c) Increasing recycling rates;	c.1 Sorting, separating, and recycling activities; c.2 Enabling recycling through selective demolition; c.3 Use of waste-derived/recycled materials in new products; c.4 Use of waste-derived materials in construction.	Advanced materials; automated manufacturing systems.
(d) Use of biobased/ renewable materials	d.1 Biobased construction materials or products, such as wood, cellulose, cotton.	Advanced materials.
(e) Use of recyclable materials;	e.1 Using materials that are recyclable at end-of-life; e.2 Enabling clean dismantling and recycling at end-of-life by not mixing materials at the installation phase.	Advanced manufacturing systems; advanced materials and IoT.
(f) Virtualized processes;	f.1 Digitalization of building information and design to share information and optimize each phase; f.2 Use sensors and machine learning to collect data for monitoring and optimization activities.	IoT, Artificial Intelligence (AI), automated manufacturing systems.

In order to conduct the policy analysis, the following information has been requested:

- Policy name;
- Promoter;
- Timeframe of implementation;
- Impacted phase of the construction value chain [from (1) to (5) according to the above-mentioned phases];
- CE focus [from (a) to (f), according to Table 1];
- Binding tool [Yes/No];
- CE relevance [High, Medium, Low]. It expresses if the policies are directly related to CE, or might only have indirect connections.

4. Results

This section provides a detailed analysis of the relevant policies in the four countries selected in this study, following the above-mentioned methodology. Detailed results are presented in Appendix A.

4.1. Italian Circular Economy Policies Scan

As a Member State of the EU, Italy has ratified all the European Directives to establish binding targets for specific sectors. For example, waste and energy management are regulated by specific directives at the European level, and therefore, targets have been transposed into Italian legislation. Five policies have been analyzed, directly and indirectly, targeting CE and the construction sector.

Italy has also adopted the National strategy for sustainable development, outlining the pathway to achieve the Agenda 2030 goals [54]. It indirectly addresses CE since the topic is included in the strategy; an explicit link is made in the section dedicated to "Prosperity", in which the document directly mentions the necessity of creating a new CE model aiming at a more sustainable and efficient use of resources, minimizing the negative impacts on the environment, thus fostering the progress of humankind. In this respect, the circular material use is calculated among the core indicators, intended as the percentage of secondary raw material used in production processes as well as the recycling rate. Other strategies decrease emissions and increase the efficiency of water management systems. Moreover, urban renovation is fostered, thus aiming at prolonging the lifetime of buildings. However, the regional governments are in charge of the application, and the CE is mentioned only when it comes to waste management in the regional strategies of sustainable development annexed to the document, thus confirming the sectorial (and limited) application of the concept currently adopted.

Specifically targeting the construction sector, a new decree specifies the end of waste criteria for C&D waste. It specifies to what extent C&D waste can re-enter the construction process. Moreover, the national legislative decree n.o. 152/2006 promotes waste prevention through a sustainable design, waste recovery, and recycling, as well as a technological system able to track the waste streams [55].

In terms of CE policies, Italy is one of the EU countries that has a national strategy for the CE transition [56], as foreseen in the mission 2 component 1—Sustainable Agriculture and CE—of the National Resilience and Recovery Plan (PNRR) [57]. The roots of the CE Italian national strategy can be traced back to 2017 when, after public consultation, the document "Towards a circular economy model in Italy. Document of framework and strategic positioning" was published. The document aimed to provide an overview of CE and the Italian positioning on the topic, ensuring coherence with the goals of the Paris Agreement, the SDGs, and the European Union target. Among the objectives, the strategy is identifying new administrative tools to strengthen the market of secondary raw materials, making them competitive if compared to virgin raw materials (e.g., through public procurement and Minimum Environmental Criteria, end of waste, extended producer responsibility, fostering the product as a service, and sharing practices). Chapter 8 identifies the macro-objectives and specific objectives of the strategy together with a list of actions to be undertaken by 2035, divided into specific sectors, among which urban areas and territories are listed.

Italy also enacted the National Green Public Procurement Action Plan (2023) [58] in which it is stated that all public procurement must comply with minimum environmental criteria (CAM). The CAM requirements are defined for the various phases of the public administration's purchasing process with the purpose of identifying the best design solution, products, and services from an environmental point of view. Their systematic and homogeneous application makes it possible to prefer environmentally preferable products and produces a leverage effect on the market, inducing less virtuous economic operators to adapt to the new requests of the public administration. In Italy, the effectiveness of the CAM has been ensured by the transposition of the Action Plan in law (Legislative Decree 36/2023 article 57), which made the CAM application mandatory [59]. The objective is to reduce the environmental impacts and promote a more sustainable and circular production and consumption model. At present, CAM have been adopted in 21 categories of supplies and assignments, among which the construction sector is included. The application of the CAM considers and supports the existing regulations, such as the directive related to the energy performance of buildings (2010/31/EU and subsequent modifications [60]), the EU regulation about construction products (305/2011/EU and subsequent modifications [61]), and the waste management directive (2008/98/CE and subsequent modifications [62]). Concerning the integration of CE principles, a paragraph is dedicated to disassembly and end-of-life: it is stated that the project relating to new buildings, including demolition works and building reconstruction or renovation, requires at least 70% in weight of the building components subject to disassembly or deconstruction or other recovery operations. The threshold has to be demonstrated through a disassembly and selective demolition plan drafted by the applicant to the tender. Another requirement is related to the use of construction materials (e.g., iron, bricks, wood, concrete, insulation) produced with at least a pre-defined percentage of recycled materials. In addition to the minimum requirements, in the law, there are some additional criteria that, if guaranteed by the applicant to the tender, allow the collection of extra scores for the evaluation process. As an example, a premium score is attributed to the economic operator who decides to undertake a Life Cycle Assessment (LCA) and Life Cycle Cost (LCC) study to assess the environmental and economic sustainability of the project as well as the use of Building Information Modeling (BIM) for the different phases of the construction.

Another document in force that indirectly targets CE is the National Plan for the Ecological Transition [63]. The Plan was approved in 2022 by the Ministry of the Ecological Transition, and its general aim is to reach climate neutrality by 2050 and reduce intermediate target GHG emissions by 55% by 2030. Regarding CE implications for the construction sector, the Plan fosters the substitution of materials with bio-based ones (e.g., promoting the use of wood in construction). In terms of energy efficiency, the Plan promotes the design of Net Zero Energy Buildings and the integration of renewable energies in the energy demand of residential buildings. Strategy no. 8 is about the promotion of CE, and it includes a focus on the construction sector in which the strong relation between the CE and the energy efficiency is acknowledged. Secondary raw materials, eco-design, light and durable materials, recyclable materials, and design for disassembly are all crucial elements of the strategy. The target for building renovation is set at 2%/year, and another important aspect highlighted is the need to introduce renewable energies. Figure 1 synthesizes the phases of the construction value chains that are mostly impacted by Italian policies and the CE criteria that are considered. End-of-life is the main life cycle stage addressed, while increasing recycling rates is the dominant CE strategy.

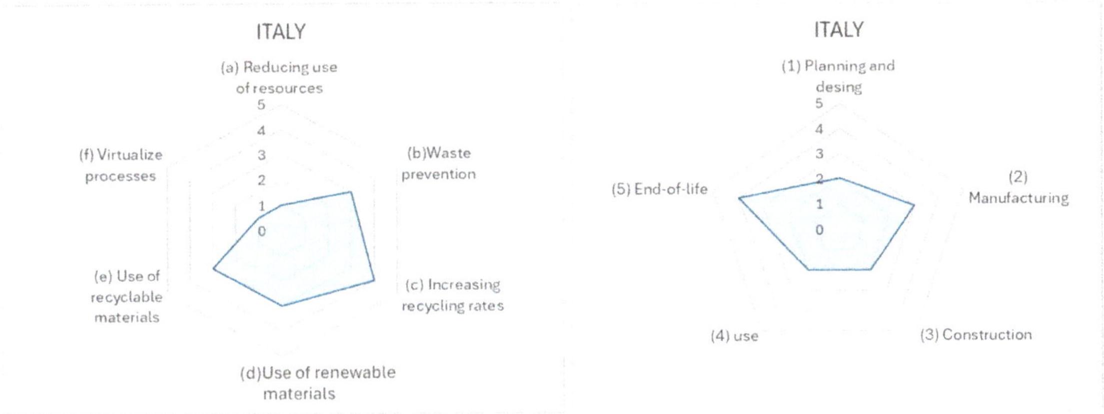

Figure 1. Analysis of CE policies in Italy. On the left, the coverage of the CE processes is shown. On the right, the analysis addresses the most impacted construction process phases.

4.2. Germany Circular Economy Policies Scan

Six policies have been analyzed in detail, two of those highly relevant for CE transition and of a binding nature: the new German Closed Cycle Management Act (Kreislaufwirtschaftsgesetz, KrWG) [64] and the New Buildings Energy Act (Gebäudeenergiegesetz, GEG) [65]. Both impact all the construction value chain phases and address almost the totality of CE actions, except for "virtualize" processes. The waste management aspect, based on the closed-loop concept and disposal responsibilities, is not new in Germany. The relevant policy has been adopted for more than 20 years. In 2013, there were 339.1 million tons of waste produced in Germany with a total recycling rate of 79%, of which 202.7 million tons are C&D waste. The new German Closed Cycle Management Act aiming at transforming waste management in Germany into resource management came into force on 1 June 2012, which has raised public awareness of closed-cycle waste management even more [66].

The New Buildings Energy Act [65] came into force on 1 November 2020, which replaces and unifies the German Energy Saving Act (Energieeinsparungsgesetz, EnEG) [67], the German Energy Saving Ordinance (Energieeinsparverordnung, EnEV) [68], and the German Renewable Energies Heat Act (Erneuerbare-Energien-Wärmegesetz, EEWärmeG) [69]. The new law will also be supported by other existing laws and standards [70].

The other binding tool that is relevant for the CE transition, even though not directly, is the Germany Federal Climate Change Act [71]. All the phases of the construction value chain are impacted by promoting the reduction of resource use, waste prevention, and the use of renewable materials. The same CE actions are fostered by the other two non-binding documents. The Germany Climate Action Plan 2050 [72] and the Germany Energy Concept [73] indirectly foster CE transition of the built environment since the first one aims to cut up to 67% emissions by 2030 compared to 1990 [74], while the second one promoted the reduction of 80–95% of the primary energy demand compared to 2008 [75]. Furthermore, several voluntary certification frameworks have already been established globally to quantify the environmental impact of specific buildings and reduce it over time. These frameworks include world-renowned LEED and BREEAM certification frameworks, and the system particularly made for the German market, the DGNB (Deutsche Gesellschaft für Nachhaltiges Bauen) certification system.

Other incentives in Germany that are worth mentioning include but are not limited to subsidies provided at the federal and state level, such as the following [76]:

- Subsidies provided by the German Credit Institute for Reconstruction with the KfW 55 loan for passive houses;
- Several states have set up/planned a subsidy per ton of biogenic carbon used (e.g., North Rhine-Westphalia, Berlin, Bavaria, Baden-Württemberg).

In Figure 2, it is shown that the most diffused CE practices are the reduction of the use of resources, being implemented in all the analyzed documents, together with the promotion of the use of renewable/renewable materials. Virtualization practices are not fostered by any policy. When it comes to the construction value chain, there is a holistic thinking of the transition, since 5 out of 6 policy documents are considering all the phases of the value chain.

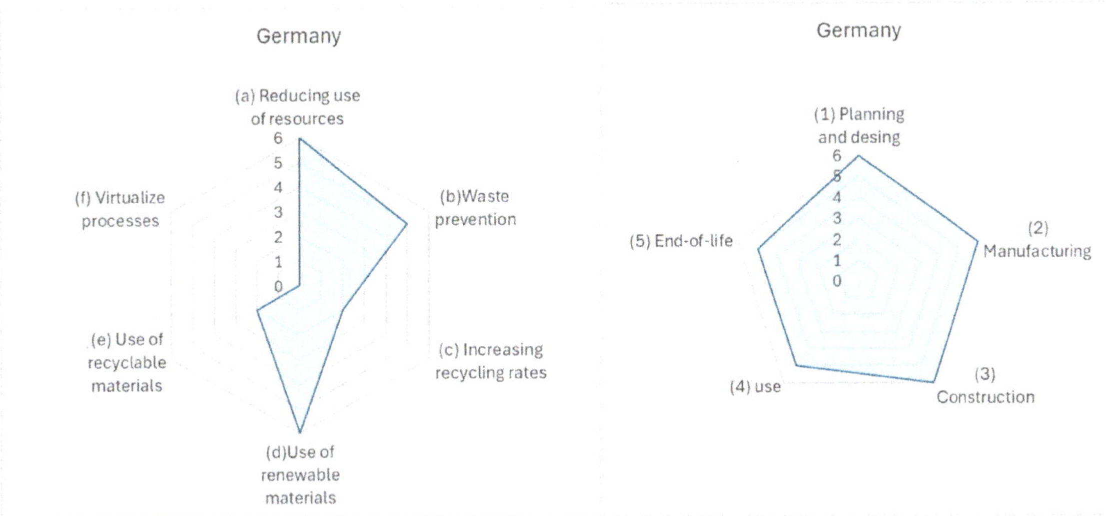

Figure 2. Analysis of CE policies in Germany. On the left, the coverage of the CE processes is shown. On the right, the analysis addresses the most impacted construction process phases.

4.3. U.S. Circular Economy Policies Scan

As mentioned in the background, CE is still in its infancy in the U.S., and federal policy has been largely impacted by the current administration. However, given the relative autonomy of U.S. states and cities and the vast geographical territory of the country (which results in a variation of local infrastructure and different levels of CE implementation), in this paper, we analyze strategies at a city scale. Several U.S. cities have created action plans and targets to foster a CE and decarbonize their industry sectors, including construction. For example, the City of Charlotte, in North Carolina, developed the Sustainable and Resilient Charlotte by 2050 Resolution, which strives to source all the energy use in the local building sector from zero-carbon sources by 2030 [77]. In addition, in 2018, the "Circular Charlotte" report explored how Charlotte could start implementing a strategy to become the first circular city in the U.S. Similarly, the City of Phoenix, in Arizona, created the Phoenix's Zero Waste and Circular Economy roadmap, seeking 50% waste diversion by 2030 and zero waste by 2050 through increasing the reuse and recycling rate of materials [78]. The City of Austin, in Texas, developed the Austin Resource Recovery plan with the goal of reaching zero waste by 2040. The plan includes the Circular City Program, which aims to change citywide policies, create CE pilot projects, and implement zero waste goals for city facilities [79]. San Francisco, in California, highlighted CE strategies for building material management in its San Francisco Climate Action Plan. The Plan includes green building requirements for City facilities, such as a waste limit per square foot of building area, the use of Environmental Product Declarations, and the requirement of whole building life cycle assessments [80].

C&D waste is the largest single waste stream ending up in landfills in many U.S. cities. C&D waste can be prevented through circular building strategies such as design for disassembly and material reuse. Designing for disassembly means planning for future deconstruction and material reuse from early design phases. Existing buildings that were not designed that way can still be deconstructed, but a much higher rate of material reuse and recycling can be achieved through design for disassembly [81]. Although very few buildings in the U.S. (and worldwide) have been intentionally designed for disassembly, some U.S. cities have been implementing deconstruction ordinances as part of their CE policies. San Francisco and Los Angeles are part of the Clean Construction Declaration shared by the C40 cities, a group of large cities worldwide that share a commitment to acting on climate change. The commitments include requiring the deconstruction of buildings by 2025 and advancing design for disassembly and Building as Material Banks by 2030 [82].

Portland, in Oregon, pioneered deconstruction regulations by requiring certain buildings to be fully deconstructed [83]. The policy started in 2016 and is limited to single-family residential units built in 1940 or earlier. In Milwaukee (Wisconsin), a deconstruction ordinance dating from 2018 mandates deconstruction for historic buildings or structures built in 1929 or earlier [84]. The City of San Antonio (Texas) adopted a similar ordinance in 2022, requiring all small-scale residential units built prior to 1920 to be deconstructed [85]. The City of Palo Alto (California) went one step further and started requiring the deconstruction of all residential and commercial buildings in 2020 [86].

The results show that all eight city-level policies included in the analysis address waste prevention and recycling rates, while none directly acknowledge the digital technologies needed for virtualizing CE processes. Due to the increasing number of deconstruction ordinances, U.S.-based CE policies focus heavily on the end-of-life phase of the construction value chain (Figure 3).

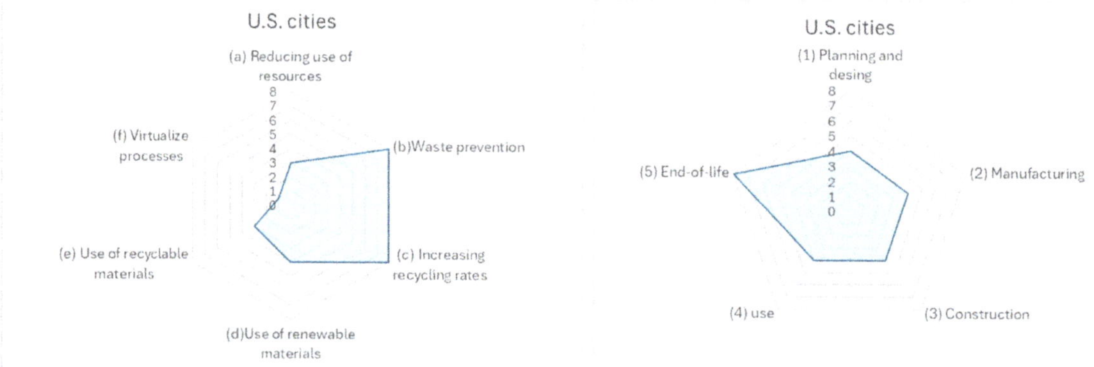

Figure 3. Analysis of CE policies in U.S. On the left, the coverage of the CE processes is shown. On the right, the analysis addresses the most impacted construction process phases.

4.4. *Japanese Circular Economy Policies Scan*

In Japan, while there have been laws aimed at promoting recycling and reuse, the Basic Act on Establishing a Sound Material-Cycle Society serves as the central legal framework that incorporates the comprehensive concept of a CE [42,43]. This law aims to shift away from a mass production, mass consumption, and mass disposal economic model. It promotes the efficient use of resources and recycling throughout the entire lifecycle, from production and distribution to consumption and disposal. Through these measures, it seeks to reduce resource consumption and establish a material-cycle society with minimal environmental impact.

Building upon this framework, this section analyzes five key laws and plans closely related to promoting CE in Japan's construction sector, as well as four additional laws and policies that have a significant impact.

Japan's CE policies in the construction sector have been progressively strengthened, as outlined in the Fourth Fundamental Plan for Establishing a Sound Material-Cycle Society (2018) [87] and the CE Roadmap (2022) [48], which was developed to monitor progress. The CE Roadmap, in particular, sets forth specific measures aimed at achieving a circular society by 2050. These include expanding the use of construction waste, improving the efficiency of raw material usage, promoting an environmentally conscious design, extending building lifespans, forming and maintaining high-quality social infrastructure, and utilizing reusable construction materials. Subsequently, in August 2024, the Fifth Fundamental Plan for Establishing a Sound Material-Cycle Society was approved by the Cabinet, positioning the transition to a CE as a national strategy and setting targets for 2030 [44]. This plan specifies improving the quality and expanding the use of recycled construction materials, promoting compact and resilient urban development, and reducing disaster-related building waste. A new CE Roadmap, based on the Fifth Fundamental Plan, is expected to be developed in the near future.

Key laws that serve as the foundation for implementing CE in the construction sector include the Act on Recycling of Construction Materials [88], the Act on the Promotion of Effective Utilization of Resources [89], the Act on the Rational Use of Energy (Energy Conservation Act) [49], and the Act on Promoting Green Procurement (Green Purchasing Act) [90]. These laws play a crucial role in promoting resource circulation and reducing environmental impact within the construction sector, thereby supporting the formation of a sound material-cycle society.

The Act on Recycling of Construction Materials [88] mandates that construction contractors properly sort waste materials such as concrete, asphalt, and wood generated from demolition and construction activities. The law aims to achieve high recycling rates and reduce environmental impact through the reuse and recycling of these materials.

Land, Infrastructure, Transport, and Tourism has committed to the steady implementation of the Construction Recycling Promotion Plan 2020 [91,92] as part of its Environmental Action Plan [93]. The plan notes that the recycling rate for construction waste, which was around 60% in the 1990s, had reached approximately 97% by fiscal year 2018, marking a transition from a growth phase to a maintenance and stabilization phase. Moving forward, the focus will shift toward improving the quality of recycling, specifically by enhancing the value of recycled materials and promoting the use of high-value-added recycled products.

The Act on the Promotion of Effective Utilization of Resources aims to promote the recycling and reuse of construction materials and recycled resources [89]. It mandates construction contractors to utilize recycled resources and recycle by-products for certain specified construction materials. Specifically, it encourages the use of recycled materials such as crushed concrete and asphalt concrete, as well as the recycling of soil and crushed concrete generated during construction activities. This law seeks to enhance the efficient use of resources in the construction sector.

The Act on the Rational Use of Energy [49] was revised in April 2023, mandating compliance with energy efficiency standards for nearly all newly constructed buildings. This revision contributes to reducing energy consumption in buildings and lowering greenhouse gas emissions.

The Act on Promoting Green Procurement mandates national and other public institutions to procure environmentally friendly goods and imposes a duty of effort on local governments [90]. This promotes the use of eco-materials and recycled resources in public sector projects, contributing to the reduction in environmental impact.

Looking more broadly at laws and policies related to the construction sector, the Waste Management and Public Cleansing Act (Waste Management Act) [94] and the Government's comprehensive plan based on the Law Concerning the Promotion of Measures to Cope with Global Warming (Global Warming Countermeasures Plan) [95] are noteworthy. The Waste Management Act establishes a management framework for construction waste from generation to final disposal, forming the basis for proper waste management within the construction sector's CE efforts.

The Global Warming Countermeasures Plan [90], formulated in 2021, aims to reduce greenhouse gas emissions by 46% from 2013 levels by fiscal year 2030, with a commitment to further pursue a 50% reduction. The plan outlines the direction of CE policies aimed at achieving carbon neutrality, including the expansion of construction waste utilization, improvement in the efficiency of raw material use, promotion of an environmentally conscious design, and extension of building lifespans.

The Green Growth Strategy [96] outlines action plans and concrete projections for 14 key sectors expected to drive growth from both industrial and energy policy perspectives. These sectors include next-generation renewable energy, hydrogen industries, automotive and thermal storage battery industries, semiconductor and Information and Communication Technologies (ICTs) industries, civil infrastructure, housing and building industries, and resource recycling-related industries. For housing and buildings, the strategy emphasizes strengthened regulatory measures, such as mandating compliance with energy efficiency standards, and the promotion of zero-energy homes and buildings. Additionally, the government supports corporate efforts through various means, including budgetary allocations, tax incentives, financial support, regulatory reforms, international cooperation, the promotion of initiatives in universities, the hosting of the 2025 World Expo, and the establishment of a youth working group.

The Circular Economy Vision 2020 [97] outlines the challenges and strategic directions for transitioning to a CE. Against the backdrop of the limitations of the linear economic model, advancements in digital technologies, and increasing demands from markets and society for environmental considerations, the vision aims to shift from the 3Rs (Reduce, Reuse, Recycle) as environmental activities to CE as an integral part of economic activities. It also promotes the development and global deployment of circular products and business models, advocating for companies to voluntarily advance their management and business strategies. Additionally, the vision emphasizes the long-term goal of reconstructing a resilient circular system.

The Growth-Oriented Resource Autonomy Strategy [98] presents a comprehensive policy package aimed at reconstructing resource circulation industrial policies to achieve autonomy and resilience in Japan's domestic resource circulation system while capturing international markets. In the context of the construction sector, the strategy highlights a resource-efficient design, including the use of recycled and renewable resources (such as wood and bio-based materials). Instead of setting detailed standards at the national level, the government supports private-sector standardization efforts, encouraging companies to voluntarily establish and disclose quantitative targets.

Thus, the promotion of CE in Japan's construction sector is progressively supported by a framework centered on the Basic Act on Establishing a Sound Material-Cycle Society and reinforced by related laws and policies. This framework enables the steady implementation of concrete practices across various stages of the construction value chain, including the proper disposal and improved recycling rates of construction waste, the promotion of recycled material use, enhancements in energy efficiency, an environmentally conscious design, and the extension of building lifespans.

Figure 4 synthesizes the phases of the construction value chains that are mostly impacted by the above-mentioned Japanese policies and the CE criteria that are considered.

Figure 4. Analysis of CE policies in Japan. On the left, the coverage of the CE processes is shown. On the right, the analysis addresses the most impacted construction process phases.

5. Discussion

The results show that 23 out of 30 analyzed policies concern waste management, and 20 of those deal with increasing recycling rates. In contrast, virtualized processes are considered in five policies only, thus demonstrating that the major attention is still devoted to waste management and prevention while processes that deal with the digitalization of building information and design to optimize each life cycle stage or the use of sensors and machine learning to collect data for monitoring and optimizing activities are still less addressed by policies. Moreover, in terms of circular construction, Japanese and German policies had the highest emphasis on reducing resource use. U.S. cities' policies mainly consider waste prevention and increasing recycling rates due to the high number of deconstruction ordinances included in the study, which specifically target those aspects.

The analysis of the construction process phases shows that the end-of-life is the most addressed by policies. This can be explained also by the fact that at least in Europe and Japan there is a regulation specifically targeting waste management, thus including the C&D waste in the category. The other two most investigated construction phases are manufacturing and construction, while the least addressed is the use and maintenance phase, considered by 19 policies across the four countries. This is important especially for the design of better future smart cities, since there is the need to interpret the already-built environment according to a circular perspective, thus trying to leverage retrofitting processes inspired by CE principles. For the use and maintenance phase, there is a need for policies not only focusing on the material and environmental dimensions but also acknowledging the importance of the social dimension and residents' behaviour to achieve a more holistic and systemic transition.

As far as the role of digital technologies is concerned, there has been a growing number of publications considering them as enablers of CE strategies [99–102]. However, not all policies reviewed in this study yet incorporate or acknowledge digital tools in their frameworks. Digital technologies can support CE in the construction sector in several ways, and the literature emphasizes that the design phase is the most crucial step in the lifetime of buildings to encompass circularity. Still, designers are not aware of CE importance [19,103]. For example, the integration of BIM with IoT and Blockchain can allow material traceability throughout the construction value chain [104–106], a concept called "material passports" [107]. Spatial data acquisition technologies such as geographic information systems (GISs), light detection and ranging (LiDAR), and unmanned aerial vehicles (UMVs) can help collect data on existing building material stocks, and, if integrated into BIM and material passports, can create digital twins of existing building infrastructure.

Digital twins are virtual replicas of existing objects (in this case, buildings) that utilize real-time data to monitor and track building performance, maintenance needs, and material data [14,108]. Digital marketplaces can help connect supply and demand for reusable building materials and close loops in the construction industry [109,110]. Artificial intelligence and machine learning technologies can predict material flows and C&D waste with accuracy, helping cities plan CE infrastructure [111,112].

However, outside the realms of academic research, numerous barriers hinder the application of digital technologies in the construction sector; this includes, for example, problems with data reliability and interpretation [113,114], technology integration challenges when different digital tools collect, store, and process data differently [115,116], and the lack of collaboration and transparency in the construction industry [117]. These challenges have been slowing down the process of digitalization in the building sector, which may explain why digital technologies are not usually acknowledged in CE policy in the U.S. and Europe. For example, in the Italian scenario, the National Circular Economy Plan [56] acknowledges the role of digitalization, especially for waste management, but does not specifically target the construction sector. Similarly, the National Action Plan on Green Public Procurement [58] mentions the BIM technology, but only for public buildings and only as a premium criterion. The same is valid for the German scenario, in which digital technologies were never emphasized or even mentioned in any of the previous circular-related regulations, including the German Closed Cycle Management Act [64], New Buildings Energy Act [65], Germany's Federal Climate Change Act [71], and Germany Energy Concept [73].

However, for Europe, a new stimulus is coming from the recent adoption of the new eco-design directive [30], which will foster the adoption of the Digital Material Passport putting more emphasis on the role of digital technologies in the CE transition. The first result can already be seen in Germany, where at the end of 2024, the National Circular Economy Strategy (NCES) has been envisaged to further mitigate climate change, protect the environment, and create opportunities for value creation and competitiveness [118]. This national strategy particularly highlights the importance of fully leveraging the potential of digital technologies in the CE. To achieve these goals, key measures include promoting the Digital Product Passport (DPP) as a central tool, supporting DPP lighthouse projects in critical sectors including the construction sector, informing consumers about CE options, and developing digital services to encourage repair, reuse, and sustainable consumption. For the construction sector in particular, the strategy emphasizes establishing new CE regulations and promoting the design of resource-efficient, circular buildings and materials [118].

Conversely, many of the Japanese policies already promote the adoption and utilization of digital technologies. The Fifth Basic Environment Plan highlights the development and dissemination of advanced technologies as one of its key strategies for achieving a sustainable society [40]. It emphasizes optimizing production through the use of ICT, such as AI and IoT, and promoting low-carbon logistics through the adoption of autonomous vehicles and drones. The Fifth Basic Plan for Establishing a Sound Material-Cycle Society [46] emphasizes the importance of leveraging ICT and digital transformation, fostering human resource development, and enhancing collaboration among stakeholders as part of efforts to strengthen resource circulation and waste management infrastructure and ensure proper disposal and environmental restoration. It also references the European Digital Product Passport, highlighting the need to establish a quality database for recycled materials. In addition, the plan advocates for the use of digital technologies and robotics to enhance traceability and improve efficiency, thereby reinforcing resource circulation infrastructure and promoting decarbonization. The Environmental Action Plan [93] identifies the utilization of digital technologies and data as a key cross-sectoral and public-private partnership project. It calls for the promotion of digital transformation in urban infrastructure, transportation, and logistics, as well as the use of GIS and the development of open data platforms to enhance productivity and optimize green initiatives. Additionally, it highlights the importance of leveraging open data to foster new industries and services, advance civic technology, and enhance environmental data and monitoring systems to strengthen the PDCA cycle. The Construction Recycling Promotion Plan [91] sets the improvement of productivity in the construction recycling sector as a medium- to long-term goal and promotes the use of ICT through initiatives such as iConstruction [119]. With the realization of Society 5.0 [120], the utilization of data on materials and soil through BIM/City Information Modelling (CIM) has become increasingly important. Furthermore, efficient monitoring of by-product logistics, as well as the streamlining of administrative procedures and statistical surveys, are also emphasized. The Global Warming Countermeasures Plan [95] promotes comprehensive energy management in the residential sector through the adoption of Home Energy Management Systems, smart meters, and smart home devices. It aims for the widespread implementation of these technologies by 2030, facilitating the visualization of energy usage and promoting efficient energy management. The Green Growth Strategy [96] outlines support for the creation of new business models in energy management within the housing and building sector by leveraging AI, IoT, electric vehicles, and storage batteries. It emphasizes the importance of establishing relevant legal frameworks and providing demonstration support to facilitate these developments.

6. Conclusions

The paper analyzed CE direct and indirect policies in Italy, Germany, U.S. cities, and Japan. An overview of the policy instruments has been given, resulting in 18 binding regulations out of 30 policies analyzed in the study. The research reveals significant differences in how CE principles are embedded in construction policies across the EU, the U.S., and Japan. While the EU has a well-structured framework (e.g., Green Deal, Circular Economy Action Plan), the U.S. focuses more on local initiatives, and Japan integrates CE within a broader resource efficiency strategy.

The most common strategies adopted by national and local governments in the identified countries to achieve a circular built environment resulted in waste prevention and increased recycling rates, thus mainly addressing the end-of-life stage. This survey also confirms that Japan's CE promotion policies in the construction sector emphasize using digital technologies across key CE processes, including production, distribution, operation, waste management, and recycling. Specifically, ICT technologies such as AI and IoT contribute to optimizing production processes and reducing carbon emissions in logistics, while data utilization tools like BIM/CIM play a critical role in resource traceability and improving the efficiency of by-product logistics. In the operation phase, energy management is advanced through the use of HEMS and smart meters. Additionally, initiatives such as the development of quality databases for recycled materials and the enhancement of energy management systems promote the recycling of construction materials and energy efficiency in residential buildings. Conversely, more effort seems needed in Europe and the U.S. to promote a more systematic use of digital technologies through the all-life cycle phases of buildings, even though new stimuli will probably come in the near future thanks to the recent adoption of the new eco-design regulation by the European Commission. This is especially true for the use and operation phase, in which the social dimension should be considered while planning policies for the development of better future smart and circular cities.

While the study has offered valuable insights, the findings may not fully capture the diversity of policy approaches in other global contexts, such as emerging economies, since it focuses on selected regions only (EU, U.S., Japan). Moreover, even though this study identifies barriers to integrating CE in construction policies, it does not conduct an empirical assessment of real-world implementation. Further studies could explore case studies or stakeholder perspectives to assess policy effectiveness. Future research could also expand the geographic scope of the study, exploring how CE policies are shaped by different socio-economic and regulatory contexts. Moreover, investigating how different actors (e.g., governments, private sector, and communities) collaborate in CE policy implementation could provide insights into governance models that enhance sustainable and circular urban transitions in the construction sector.

Author Contributions: Conceptualization, G.M. and S.T.; methodology, formal analysis, S.T.; investigation G.M., R.H., Y.K. and F.C.R.; resources, G.M., R.H., Y.K., F.C.R., T.B. and T.L.; data curation, G.M.; writing—original draft preparation, G.M., R.H., F.C.R. and Y.K.; writing—review and editing, G.M. and S.T.; visualization, G.M.; supervision, S.T., T.B. and T.L. All authors have read and agreed to the published version of the manuscript.

Funding: This research Grant-in-aid for Scientific Research, Basic research (B), Japan "Development, Empirical Research and Dissemination of New Theories and System Techniques for Circular Economy (CE) to Meet SDGs Goal 12; Producer and Consumer Responsibility", grant number 22H01717 is funded by Japan Society for the Promotion of Science.

Data Availability Statement: Analyzed data are publicly available, even though many documents might be in local languages.

Conflicts of Interest: Author Rongbo Hu was employed by the company Kajima Corporation. The remaining authors declare that the research was conducted in the absence of any commercial or financial relationships that could be construed as potential conflicts of interest.

Appendix A

Policy Name	Promoter	Timeframe of Implementation	Impacted Phases of the Value Chain	CE Criteria	Binding Tool (Yes/No)	CE Relevance
[IT] Ecological transition plan	Ministry of Ecological Transition	2022	All	(a), (b), (c), (d), (e)	Long-term strategy	Medium
[IT] End of Waste—C&D	Ministry of the environment and energy safety	2024	(5)	(c)	Yes	High
[IT] National Circular Economy plan	Ministry of Ecological Transition	2022	(1), (2), (5)	(b), (c), (d), (e)	Long-term strategy	High
[IT] National Action Plan on the Green Public Procurement (PANGPP)	Legislative Decree 36/2023	2022	(2), (3), (5)	(c), (d), (e), (f)	Yes	Medium
[IT] National Strategy for Sustainable Development	Ministry of the environment and energy safety	2022	(4)	(b)	Long term strategy	Low
[DE] KfW Efficiency House 55, Efficiency House 40, and Efficiency House 40 Plus Programs.	German government, KfW Banking Group (Credit Institute for Reconstruction)	2009	(1)			Low

Policy Name	Promoter	Timeframe of Implementation	Impacted Phases of the Value Chain	CE Criteria	Binding Tool (Yes/No)	CE Relevance
[U.S.] Sustainable and Resilient Charlotte by 2050 Resolution	City of Charlotte, North Carolina	2030–2050	All	(a), (b), (c), (d)	No (long-term strategy)	High
[U.S.] Phoenix's Zero Waste and Circular Economy roadmap	City of Phoenix, Arizona	2030–2050	All	(b), (c), (d), (e)	No (long-term strategy)	High
[U.S.] Austin Resource Recovery Comprehensive plan	City of Austin, Texas	2040	All	All	No (long-term strategy)	Medium
[U.S.] San Francisco Climate Action Plan	City of San Francisco, California	2030–2040	All	(a), (b), (c), (d), (e)	No (long-term strategy)	Medium
[U.S.] Deconstruction Ordinance	Portland, Oregon	Current	(5)	(b), (c)	Yes	High
[U.S.] Deconstruction Ordinance	Milwaukee (Wisconsin)	Current	(5)	(b), (c)	Yes	High
[U.S.] Deconstruction Ordinance	City of San Antonio (Texas)	Current	(5)	(b), (c)	Yes	High
[U.S.] Deconstruction Ordinance	Palo Alto (California)	Current	(5)	(b), (c)	Yes	High
[JP] Basic Act on Establishing a Sound Material-Cycle Society	Government of Japan	Ongoing	All	(a), (b)	Yes	High
[JP] Fundamental Plan for Establishing a Sound Material-Cycle Society (including CE Roadmap)	Ministry of the Environment	Ongoing, reviewed every five years	All	(a), (b), (c)	Yes	High
[JP] Act on Recycling of Construction Materials	Ministry of Land, Infrastructure, Transport and Tourism	Ongoing	(2), (3), (4), (5)	(c), (e)	Yes	High
[JP] Act on the Promotion of Effective Utilization of Resources	Ministry of Economy, Trade and Industry	Ongoing	(2), (3), (5)	(a), (c)	Yes	High
[JP] Act on the Rational Use of Energy	Ministry of Land, Infrastructure, Transport and Tourism	Ongoing	(1), (3), (4)	(a)	Yes	Low
[JP] Act on Promoting Green Procurement	Ministry of the Environment	Ongoing	(1), (2), (3)	(a), (d)	Yes	Low
[JP] Waste Management and Public Cleansing Act	Ministry of the Environment	Ongoing	(5)	(b)	Yes	High

Policy Name	Promoter	Timeframe of Implementation	Impacted Phases of the Value Chain	CE Criteria	Binding Tool (Yes/No)	CE Relevance
[JP] Government's comprehensive plan based on the Law Concerning the Promotion of Measures to Cope with Global Warming	Government of Japan	Ongoing	All	(a), (b), (d)	Yes (the plan is implemented through legally binding regulations)	Medium
[JP] Green Growth Strategy Through Achieving Carbon Neutrality in 2050	Ministry of Economy, Trade and Industry	Ongoing	All	All	Yes (the plan is implemented through legally binding regulations)	High
[JP] Circular Economy Vision	Ministry of Economy, Trade and Industry	Ongoing	All	All	No (guides legally binding regulations)	High
[JP] Growth-Oriented Resource Autonomy Strategy	Ministry of Economy, Trade and Industry	Ongoing	All	All	No (guides legally binding regulations)	High

References

1. Brundtland, G.H. *Our Common Future: Report of the World Commission on Environment and Development*; United Nations: Geneva, Switzerland, 1987.
2. Batty, M.; Axhausen, K.W.; Giannotti, F.; Pozdnoukhov, A.; Bazzani, A.; Wachowicz, M.; Ouzounis, G.; Portugali, Y. Smart Cities of the Future. *Eur. Phys. J. Spec. Top.* **2012**, *214*, 481–518. [CrossRef]
3. Formisano, V.; Iannucci, E.; Fedele, M.; Bonab, A.B. City in the Loop: Assessing the Relationship between Circular Economy and Smart Sustainable Cities. *Sinergie Ital. J. Manag.* **2022**, *40*, 147–168. [CrossRef]
4. Caprotti, F. Smart to Green: Smart Eco-Cities in the Green Economy. In *The Routledge Companion to Smart Cities*; Willis, K.S., Aurigi, A., Eds.; Routledge: New York, NY, USA, 2020; ISBN 9781315178387.
5. De Jong, M.; Joss, S.; Schraven, D.; Zhan, C.; Weijnen, M. Sustainable-Smart-Resilient-Low Carbon-Eco-Knowledge Cities; Making Sense of a Multitude of Concepts Promoting Sustainable Urbanization. *J. Clean. Prod.* **2015**, *109*, 25–38. [CrossRef]
6. Schulz, C.; Bailey, I. The Green Economy and Post-growth Regimes: Opportunities and Challenges for Economic Geography. *Geogr. Ann. Ser. B* **2014**, *96*, 277–291. [CrossRef]
7. Arslan, H.M.; Khan, I.; Latif, M.I.; Komal, B.; Chen, S. Understanding the Dynamics of Natural Resources Rents, Environmental Sustainability, and Sustainable Economic Growth: New Insights from China. *Environ. Sci. Pollut. Res.* **2022**, *29*, 58746–58761. [CrossRef] [PubMed]
8. Circle Economy. Circularity Gap Report. 2023. Available online: https://www.circularity-gap.world/2023 (accessed on 14 February 2025).
9. Ellen MacArthur Foundation. *How the Circular Economy Tackles Climate Change*; Ellen MacArthur Foundation: Isle of Wight, UK, 2019.
10. Ellen MacArthur Foundation. *Towards the Circular Economy Vol. 1: An Economic and Business Rationale for an Accelerated Transition*; Ellen MacArthur Foundation: Isle of Wight, UK, 2013.
11. Ellen MacArthur Foundation. *Growth Within: A Circular Economy Vision for a Competitive Europe*; Ellen MacArthur Foundation: Isle of Wight, UK, 2015.
12. Jackson, T. *Prosperity Without Growth*; Routledge: London, UK, 2009; ISBN 9781136546785.
13. Rossé, M.; Stuchtey, M.; Vanthournout, H. *Finding Growth Within: A new Framework for Europe in The Circular Economy: Moving from Theory to Practice*; McKinsey Center for Business and Environment; McKinsey & Company: Washington, DC, USA, 2016; pp. 4–10.
14. Çetin, S.; De Wolf, C.; Bocken, N. Circular Digital Built Environment: An Emerging Framework. *Sustainability* **2021**, *13*, 6348. [CrossRef]
15. Bressanelli, G.; Adrodegari, F.; Pigosso, D.C.A.; Parida, V. Circular Economy in the Digital Age. *Sustainability* **2022**, *14*, 5565. [CrossRef]

16. De Felice, F.; Petrillo, A. Green Transition: The Frontier of the Digicircular Economy Evidenced from a Systematic Literature Review. *Sustainability* **2021**, *13*, 11068. [CrossRef]
17. Anttiroiko, A.V. Smart Circular Cities: Governing the Relationality, Spatiality, and Digitality in the Promotion of Circular Economy in an Urban Region. *Sustainability* **2023**, *15*, 12680. [CrossRef]
18. United Nations Environment Programme. *Building Materials and the Climate: Constructing a New Future*; United Nations Environment Programme: Nairobi, Kenya, 2023.
19. Banihashemi, S.; Meskin, S.; Sheikhkhoshkar, M.; Mohandes, S.R.; Hajirasouli, A.; LeNguyen, K. Circular Economy in Construction: The Digital Transformation Perspective. *Clean. Eng. Technol.* **2024**, *18*, 100715. [CrossRef]
20. Circular Economy Network. 6th Report on Circular Economy in Italy—2024. Available online: https://circulareconomynetwork.it/wp-content/uploads/2024/05/CEN2024_SINTESI-ENG_DEF.pdf (accessed on 14 February 2025).
21. European Circular Economy Stakeholder Platform. *Circular Buildings and Infrastructure. State of Play Report*; European Union: Brussels, Belgium, 2022.
22. Novikova, A.; Stelmakh, K.; Emmrich, J. *Climate Finance Landscape of the German Building Sector*; Institute for Climate Protection, Energy and Mobility (IKEM): Berlin, Germany, 2018.
23. Barbarella, A.; Montanini, C.; Ferruccio, V. Italy Climate Report 2022. La Performance dell'Italia Sui Temi del Clima e dell'Energia; Italy for Climate, Italy. Available online: https://coilink.org/20.500.12592/nbcf9t (accessed on 11 March 2025).
24. European Commission. *Directorate-General for Communication Circular Economy Action Plan—For a Cleaner and More Competitive Europe*; Publications Office of the European Union: Luxembourg, 2020.
25. European Commission. *Directorate-General for Communication European Green Deal—Delivering on Our Targets*; Publications Office of the European Union: Luxembourg, 2021.
26. European Commission. Circular Economy Action Plan. Available online: https://environment.ec.europa.eu/strategy/circular-economy-action-plan_en (accessed on 14 February 2025).
27. Ellen MacArthur Foundation. The EU's Circular Economy Action Plan. Available online: https://www.ellenmacarthurfoundation.org/circular-examples/the-eus-circular-economy-action-plan (accessed on 14 February 2025).
28. European Commission. Level(s) Framework. Available online: https://environment.ec.europa.eu/topics/circular-economy/levels_en (accessed on 14 February 2025).
29. European Commission; Directorate-General for Energy (European Commission). A Renovation Wave for Europe—GreeA Renovation Wave for Europe—Greening Our Buildings, Creating Jobs, Improving Livesning Our Buildings, Creating Jobs, Improving Lives. Available online: https://eur-lex.europa.eu/legal-content/EN/TXT/?qid=1603122220757&uri=CELEX:52020DC0662 (accessed on 14 February 2025).
30. European Commission. Ecodesign for Sustainable Products Regulation. 2024. Available online: https://eur-lex.europa.eu/legal-content/EN/TXT/?uri=CELEX:32024R1781&qid=1719580391746 (accessed on 14 February 2025).
31. European Parliament and the Council Directive 2005/32/EC—Establishing a Framework for the Setting of Ecodesign Requirements for Energy-Related Products. 2009. Available online: https://eur-lex.europa.eu/eli/dir/2009/125/oj/eng (accessed on 14 February 2025).
32. González, A.; Sendra, C.; Herena, A.; Rosquillas, M.; Vaz, D. Methodology to Assess the Circularity in Building Construction and Refurbishment Activities. *Resour. Conserv. Recycl. Adv.* **2021**, *12*, 200051. [CrossRef] [PubMed]
33. U.S. Government. *Inventory of U.S. Greenhouse Gas Emissions and Sinks: 1990–2021*; United States Environmental Protection Agency: Washington, DC, USA, 2023.
34. United States Environmental Protection Agency; Office of Land and Emergency Management. *Advancing Sustainable Materials Management: 2018 Fact Sheet Assessing Trends in Materials Generation and Management in the United States*; United States Environmental Protection Agency: Washington, DC, USA, 2020.
35. Toward, P.; Change, T. *Building a Circular Economy for All: Progress Toward Transformative Change*; United States Environmental Protection Agency: Washington, DC, USA, 2022.
36. United States Environmental Protection Agency. *National Framework for Advancing the U.S. Recycling System*; United States Environmental Protection Agency: Washington, DC, USA, 2019.
37. Milman, O. How Trump Is Targeting Wind and Solar Energy—And Delighting Big Oil. *The Guardian*, 3 February 2025.
38. National Institute for Environmental. *Studies Greenhouse Gas Inventory Report of Japan*; Ministry of the Environment: Tokyo, Japan, 2024.
39. Ministry of the Environment. Status of Industrial Waste Emissions and Treatment (FY2021 Results). 2024. Available online: https://www.env.go.jp/content/000211701.pdf (accessed on 9 December 2024).
40. Ministry of the Environment. Outline of the 5th Basic Environmental Plan. Available online: https://www.env.go.jp/content/900511403.pdf (accessed on 14 February 2025).
41. Ministry of the Environment. The Basic Environmental Plan. Available online: https://www.env.go.jp/content/900511404.pdf (accessed on 14 February 2025).

42. Ministry of the Environment. Purpose of the Basic Act for Establishing a Sound Material-Cycle Society. Available online: https://www.env.go.jp/recycle/circul/kihonho/shushi.html (accessed on 9 December 2024).
43. Ministry of the Environment. Overview of the Basic Act for Establishing a Sound Material-Cycle Society. Available online: https://www.env.go.jp/recycle/circul/kihonho/gaiyo.html (accessed on 9 December 2024).
44. Ministry of the Environment. The 5th Fundamental Plan for Establishing a Sound Material-Cycle Society. Available online: https://www.env.go.jp/content/000264244.pdf (accessed on 14 February 2025).
45. Ministry of Environment. The Basic Plan for Establishing a Sound Material-Cycle Society. Available online: https://www.env.go.jp/content/000242999.pdf (accessed on 14 February 2025).
46. Ministry of the Environment. Outline of the 5th Fundamental Plan for Establishing a Sound Material-Cycle Society. Available online: https://www.env.go.jp/content/000243000.pdf (accessed on 14 February 2025).
47. Ministry of the Environment. Overview of the 4th Fundamental Plan for Establishing a Sound Material-Cycle Society and the Circular Economy Roadmap. Available online: https://www.env.go.jp/content/000138209.pdf (accessed on 14 February 2025).
48. Ministry of the Environment. Second Progress Report on the 4th Fundamental Plan for Establishing a Sound Material-Cycle Society (Circular Economy Roadmap). Available online: https://www.env.go.jp/content/000071596.pdf (accessed on 14 February 2025).
49. Ministry of Land Infrastructure Transport and Tourism. Overview of the 2022 Revision of the Act on the Rational Use of Energy for Buildings. Available online: https://www.mlit.go.jp/jutakukentiku/house/shouenehou_r4.html (accessed on 9 December 2024).
50. Ministry of Land Infrastructure Transport and Tourism. Explanation of the Energy Efficiency Standards Compliance Obligation System (Second Edition). Available online: https://www.mlit.go.jp/common/001627105.pdf (accessed on 14 February 2025).
51. Sáez-de-Guinoa, A.; Zambrana-Vasquez, D.; Fernández, V.; Bartolomé, C. Circular Economy in the European Construction Sector: A Review of Strategies for Implementation in Building Renovation. *Energies* **2022**, *15*, 4747. [CrossRef]
52. Zu Castell-Rüdenhausen, M.; Wahlström, M.; Fruergaard Astrup, T.; Jensen, C.; Oberender, A.; Johansson, P.; Waerner, E.R. Policies as Drivers for Circular Economy in the Construction Sector in the Nordics. *Sustainability* **2021**, *13*, 9350. [CrossRef]
53. Pimponi, D.; Porcari, A. Circular Economy in the Building and Construction Sector in Italy—Towards Sustainable Production and Consumption; 2020. Available online: https://www.airi.it/airi2020/wp-content/uploads/2021/06/SocKETs_D1.1_Report_Airi_Final_Website.pdf (accessed on 13 February 2025).
54. Ministry of the Environment and energy safety. National Strategy for Sustainable Development 2022. 2022. Available online: https://www.mase.gov.it/pagina/national-sustainable-development-strategy (accessed on 13 February 2025).
55. President of the Italian Republic. Legislative Decree 3rd April 2006, No 152—Norme in Materia Ambientale. 2006. Available online: https://www.normattiva.it/uri-res/N2Ls?urn:nir:stato:decreto.legislativo:2006-04-03;152 (accessed on 13 February 2025).
56. Ministry of the Ecological Transition. *National Strategy for the Circular Economy*; Ministry of the Ecological Transition: Paris, France, 2022.
57. Italian Government. Italia Domani, Il Piano Nazionale Di Ripresa e Resilienza. Available online: https://www.italiadomani.gov.it/it/strumenti/documenti/archivio-documenti/piano-nazionale-di-ripresa-e-resilienza.html (accessed on 13 February 2025).
58. Ministry of the Environment and Energy Safety. Piano d'Azione Nazionale Green Public Procurement. Available online: https://gpp.mase.gov.it/Home/PianoAzioneNazionaleGPP (accessed on 13 February 2025).
59. President of the Italian Republic Legislative Decree 31st March 2023, No 36—Codice dei Contratti Pubblici; Italy. 2023. Available online: https://www.normattiva.it/uri-res/N2Ls?urn:nir:stato:decreto.legislativo:2023;036 (accessed on 13 February 2025).
60. European Parliament and the Council Directive 2010/31/EU of the European Parliament and of the Council of 19 May 2010 on the Energy Performance of Buildings (Recast). 2010. Available online: https://eur-lex.europa.eu/eli/dir/2010/31/oj/eng (accessed on 13 February 2025).
61. European Parliament and the Council Consolidated Text: Regulation (EU) No 305/2011 of the European Parliament and of the Council of 9 March 2011 Laying down Harmonised Conditions for the Marketing of Construction Products and Repealing Council Directive 89/106/EEC. 2011. Available online: https://eur-lex.europa.eu/legal-content/EN/TXT/?uri=CELEX:02011R0305-20241117 (accessed on 13 February 2025).
62. European Parliament and the Council Consolidated Text: Directive 2008/98/EC of the European Parliament and of the Council of 19 November 2008 on Waste and Repealing Certain Directives. 2008. Available online: https://eur-lex.europa.eu/legal-content/EN/TXT/?uri=CELEX:02008L0098-20240218 (accessed on 13 February 2025).
63. Ministry of the Ecological Transition. *Piano per la Transizione Ecologica*; Ministry of the Ecological Transition: Rome, Italy, 2022. Available online: https://www.mase.gov.it/sites/default/files/archivio/allegati/PTE/PTE-definitivo.pdf (accessed on 13 February 2025).
64. New German Closed Cycle Management Act. 2023. Available online: https://www.gesetze-im-internet.de/krwg/BJNR021210012.html (accessed on 13 February 2025).
65. New Buildings Energy Act. Available online: https://www.gesetze-im-internet.de/geg/BJNR172810020.html (accessed on 6 March 2025).

66. Nelles, M.; Grünes, J.; Morscheck, G. Waste Management in Germany—Development to a Sustainable Circular Economy? *Procedia Environ. Sci.* **2016**, *35*, 6–14. [CrossRef]
67. German Energy Saving Act. Available online: https://climate-laws.org/documents/energy-saving-act-eneg_f81a?id=energy-saving-act-eneg_acdd (accessed on 6 March 2025).
68. German Energy Saving Ordinance. Available online: https://www.bmwk.de/Redaktion/DE/Gesetze/Energie/EnEV.html (accessed on 6 March 2025).
69. German Renewable Energies Heat Act. Available online: https://www.bmwk.de/Redaktion/DE/Gesetze/Energie/EEWaermeG.html (accessed on 6 March 2025).
70. German Energy Agency. Energy-Efficient Building and Refurbishment the Right Way. 2020. Available online: https://www.dena.de/en/topics-projects/energy-efficiency/buildings/consulting-and-planning/german-buildings-energy-act-geg-standards-and-laws/ (accessed on 13 February 2025).
71. Federal Ministry for Economic Affairs and Climate Action Germany's Federal Climate Change Act. Available online: https://www.gesetze-im-internet.de/englisch_ksg/englisch_ksg.html (accessed on 6 March 2025).
72. Federal Ministry for the Environment. Nature Conservation Building and Nuclear Safety (BMUB) Germany Climate Action Plan 2050. Available online: https://www.bmuv.de/en/publication/climate-action-plan-2050-en (accessed on 6 March 2025).
73. German Federal Government. Germany Energy Concept. Available online: https://www.gtai.de/en/invest/industries/healthcare-market-germany/germany-s-energy-concept-105260#:~:text=Central%20to%20Germany%E2%80%99s%20Energy%20Concept,supply-oriented%20energy%20supply%20structure (accessed on 6 March 2025).
74. BMUB. *Climate Action Plan 2050*; BMUB: Bonn, Germany, 2016.
75. BMWK. *Der Weg zur Energie der Zukunft—Sicher, Bezahlbar und Umweltfreundlich*; BMWK: Berlin, Germany, 2011.
76. Rheude, F.; Röder, H. Estimating the Use of Materials and Their GHG Emissions in the German Building Sector. *Clean. Environ. Syst.* **2022**, *7*, 100095. [CrossRef]
77. City of Charlotte. Strategic Energy Action Plan. Available online: https://tinyurl.com/4xcdn9au (accessed on 6 March 2025).
78. City of Phoenix. Phoenix's Zero Waste and Circular Economy Roadmap. Available online: https://www.phoenix.gov/publicworks/zerowasteplan (accessed on 6 March 2025).
79. City of Austin. 2023 Austin Resource Recovery Comprehensive Plan. 2023. Available online: https://www.austintexas.gov/sites/default/files/files/Resource_Recovery/2023%20Austin%20Resource%20Recovery%20Comprehensive%20Plan.pdf (accessed on 13 February 2025).
80. San Francisco Department of Environment. *San Francisco's Climate Action Plan 2021*; San Francisco Department of Environment: San Francisco, CA, USA, 2020.
81. Rios, F.C.; Chong, W.K.; Grau, D. Design for Disassembly and Deconstruction—Challenges and Opportunities. *Procedia Eng.* **2015**, *118*, 1296–1304. [CrossRef]
82. C40 Cities. Clean Construction Declaration Planned Actions to Deliver Commitments. Available online: https://www.c40.org/wp-content/uploads/2022/11/Clean-Construction_planned-actions-to-deliver-commitments.pdf (accessed on 13 February 2025).
83. City of Portland Deconstruction Permit Requirements. Available online: https://www.portland.gov/bps/garbage-recycling/decon/deconstruction-requirements (accessed on 6 March 2025).
84. City of Milwaukee Deconstruction Ordinance. Available online: https://city.milwaukee.gov/DNS/Inspections_Sections/Decon (accessed on 6 March 2025).
85. San Antonio City Council Deconstruction Requirements. Available online: https://www.sareuse.com/deconstruction (accessed on 6 March 2025).
86. City of Palo Alto Deconstruction Ordinance. Available online: https://www.cityofpaloalto.org/Departments/Public-Works/Zero-Waste/Zero-Waste-Requirements-Guidelines/Deconstruction-Ordinance#section-3 (accessed on 6 March 2025).
87. Ministry of the Environment. The 4th Fundamental Plan for Establishing a Sound Material-Cycle Society. Available online: https://www.env.go.jp/content/900535436.pdf (accessed on 14 February 2025).
88. Ministry of Land Infrastructure Transport and Tourism. Construction Material Recycling Act. Available online: https://www.mlit.go.jp/sogoseisaku/region/recycle/d03project/d0301/page_030110recyclelaw.htm (accessed on 14 February 2025).
89. Ministry of Economy Trade and Industry. Act on the Promotion of Effective Utilization of Resources. Available online: https://www.meti.go.jp/policy/recycle/main/admin_info/law/02/ (accessed on 14 February 2025).
90. Ministry of the Environment. Green Purchasing Act. Available online: https://www.env.go.jp/policy/hozen/green/g-law/index.html (accessed on 14 February 2025).
91. Ministry of Land Infrastructure Transport and Tourism. Construction Recycling Promotion Plan 2020. Available online: https://www.mlit.go.jp/report/press/content/001365044.pdf (accessed on 14 February 2025).

92. Ministry of Land Infrastructure Transport and Tourism. Recent Changes in Social Conditions Surrounding Construction Recycling and Past Efforts. Available online: https://www.mlit.go.jp/policy/shingikai/content/05shiryou2.pdf (accessed on 14 February 2025).
93. Ministry of Land Infrastructure Transport and Tourism. Environmental Action Plan. Available online: https://www.mlit.go.jp/common/001448170.pdf (accessed on 14 February 2025).
94. Ministry of the Environment. Waste Management and Public Cleansing Act. Available online: https://www.env.go.jp/recycle/waste/laws.html (accessed on 14 February 2025).
95. Ministry of the Environment. Global Warming Countermeasure Plan. Available online: https://www.env.go.jp/content/900440195.pdf (accessed on 14 February 2025).
96. Ministry of Economy Trade and Industry. Green Growth Strategy Through Achieving Carbon Neutrality by 2050. Available online: https://www.meti.go.jp/policy/energy_environment/global_warming/ggs/pdf/green_honbun.pdf (accessed on 14 February 2025).
97. Ministry of Economy Trade and Industry. Circular Economy Vision 2020. Available online: https://www.meti.go.jp/shingikai/energy_environment/junkai_keizai/pdf/20200522_02.pdf (accessed on 14 February 2025).
98. Ministry of Economy Trade and Industry. Outline of the Growth-Oriented Autonomous Resource Economy Strategy. Available online: https://www.meti.go.jp/press/2022/03/20230331010/20230331010-1.pdf (accessed on 14 February 2025).
99. Caldas, L.R.; Silva, M.V.; Silva, V.P.; Carvalho, M.T.M.; Toledo Filho, R.D. How Different Tools Contribute to Climate Change Mitigation in a Circular Building Environment?—A Systematic Literature Review. *Sustainability* **2022**, *14*, 3759. [CrossRef]
100. Çetin, S.; Gruis, V.; Straub, A. Digitalization for a Circular Economy in the Building Industry: Multiple-Case Study of Dutch Social Housing Organizations. *Resour. Conserv. Recycl. Adv.* **2022**, *15*, 200110. [CrossRef]
101. Yu, Y.; Yazan, D.M.; Junjan, V.; Iacob, M.-E. Circular Economy in the Construction Industry: A Review of Decision Support Tools Based on Information & Communication Technologies. *J. Clean. Prod.* **2022**, *349*, 131335. [CrossRef]
102. Setaki, F.; van Timmeren, A. Disruptive Technologies for a Circular Building Industry. *Build. Environ.* **2022**, *223*, 109394. [CrossRef]
103. Anastasiades, K.; Blom, J.; Buyle, M.; Audenaert, A. Translating the Circular Economy to Bridge Construction: Lessons Learnt from a Critical Literature Review. *Renew. Sustain. Energy Rev.* **2020**, *117*, 109522. [CrossRef]
104. Cai, G.; Waldmann, D. A Material and Component Bank to Facilitate Material Recycling and Component Reuse for a Sustainable Construction: Concept and Preliminary Study. *Clean Technol. Environ. Policy* **2019**, *21*, 2015–2032. [CrossRef]
105. Honic, M.; Ferschin, P.; Breitfuss, D.; Cencic, O.; Gourlis, G.; Kovacic, I.; De Wolf, C. Framework for the Assessment of the Existing Building Stock through BIM and GIS. *Dev. Built Environ.* **2023**, *13*, 100110. [CrossRef]
106. Copeland, S.; Bilec, M. Buildings as Material Banks Using RFID and Building Information Modeling in a Circular Economy. *Procedia CIRP* **2020**, *90*, 143–147. [CrossRef]
107. BAMB Project. Building as Material Banks (BAMB). 2023. Available online: https://www.bamb2020.eu/ (accessed on 14 February 2025).
108. Jemal, K.M.; Kabzhassarova, M.; Shaimkhanov, R.; Dikhanbayeva, D.; Turkyilmaz, A.; Durdyev, S.; Karaca, F. Facilitating Circular Economy Strategies Using Digital Construction Tools: Framework Development. *Sustainability* **2023**, *15*, 877. [CrossRef]
109. Kovacic, I.; Honic, M.; Sreckovic, M. Digital Platform for Circular Economy in AEC Industry. *Eng. Proj. Organ. J.* **2020**, *9*, 16. [CrossRef]
110. Xing, K.; Kim, K.P.; Ness, D. Cloud-BIM Enabled Cyber-Physical Data and Service Platforms for Building Component Reuse. *Sustainability* **2020**, *12*, 10329. [CrossRef]
111. Davis, P.; Aziz, F.; Newaz, M.T.; Sher, W.; Simon, L. The Classification of Construction Waste Material Using a Deep Convolutional Neural Network. *Autom. Constr.* **2021**, *122*, 103481. [CrossRef]
112. Oluleye, B.I.; Chan, D.W.M.; Antwi-Afari, P. Adopting Artificial Intelligence for Enhancing the Implementation of Systemic Circularity in the Construction Industry: A Critical Review. *Sustain. Prod. Consum.* **2023**, *35*, 509–524. [CrossRef]
113. Yuan, L.; Lu, W.; Xue, F.; Li, M. Building Feature-based Machine Learning Regression to Quantify Urban Material Stocks: A Hong Kong Study. *J. Ind. Ecol.* **2023**, *27*, 336–349. [CrossRef]
114. Giovanardi, M.; Konstantinou, T.; Pollo, R.; Klein, T. Internet of Things for Building Façade Traceability: A Theoretical Framework to Enable Circular Economy through Life-Cycle Information Flows. *J. Clean. Prod.* **2023**, *382*, 135261. [CrossRef]
115. Kleemann, F.; Lehner, H.; Szczypińska, A.; Lederer, J.; Fellner, J. Using Change Detection Data to Assess Amount and Composition of Demolition Waste from Buildings in Vienna. *Resour. Conserv. Recycl.* **2017**, *123*, 37–46. [CrossRef]
116. Gordon, M.; Batallé, A.; De Wolf, C.; Sollazzo, A.; Dubor, A.; Wang, T. Automating Building Element Detection for Deconstruction Planning and Material Reuse: A Case Study. *Autom. Constr.* **2023**, *146*, 104697. [CrossRef]
117. Costa, M.M.; Barreto Neto, J.F.; Varela Alberte, E.P.; Carneiro, A.P. Blockchain-Based Framework for Improving Waste Management and Circular Economy in Construction. *IOP Conf. Ser. Earth Environ. Sci.* **2022**, *1101*, 062009. [CrossRef]
118. BMUV. *The National Circular Economy Strategy*; BMUV: Bonn, Germany, 2024.

119. Ministry of Land Infrastructure Transport and Tourism. I-Construction. Available online: https://www.mlit.go.jp/common/001137123.pdf (accessed on 14 February 2025).
120. Cabinet Office. Society 5.0. Available online: https://www8.cao.go.jp/cstp/society5_0/ (accessed on 14 February 2025).

smart cities

MDPI

Article

The Use of the Smart Technology for Creating an Inclusive Urban Public Space

Mohammed Itair [1,*], Isam Shahrour [1] and Ihab Hijazi [2]

1 Civil and Geo-Environmental Laboratory (LGCgE), Lille University, 59650 Lille, France; ishahrour@gmail.com
2 Urban Planning Engineering Department, An-Najah National University, Nablus P400, Palestine; eehab@najah.edu
* Correspondence: mohammeditair@gmail.com; Tel.: +33-7495-8796

Abstract: This paper strives to enhance the inclusivity of urban public spaces, which play a crucial role in providing essential services for all citizens, including community building, physical and mental well-being, social interaction, civic engagement, citizen participation, and economic vitality. Despite the importance of these spaces, as recognized by the UN's 2030 sustainability goals, the 2023 UN sustainable development report and scholars have drawn attention to their low availability, particularly for low-income individuals, women, children, and people with disabilities. To improve the inclusivity of public spaces, this paper offers the following contributions. (i) The establishment of a comprehensive framework for assessing public space inclusivity. This framework incorporates eight indicators: spatial distribution, typology, facilities and services, green and humid areas, governance and management, safety, user categories, and user satisfaction. (ii) The utilization of the framework to assess the inclusivity of public spaces in Nablus, a major Palestinian city. This assessment confirms the observations made by the UN and scholars regarding the low inclusivity of public spaces; in particular, a lack of public space, poor spatial distribution, and user dissatisfaction with safety conditions and services. (iii) The introduction of the concept of smart public space, which involves citizens in the governance of this space and leverages smart technology for monitoring, providing real-time information and services to citizens, improving facility efficiency, and creating an eco-friendly environment that preserves resources and biodiversity. By addressing these aspects, this paper enhances inclusivity. It promotes the development of an urban public space that caters to the diverse needs of the community, fostering a sense of belonging and well-being for all.

Keywords: inclusive; public space; SDGs; smart; citizens' participation; comprehensive; governance; monitoring

1. Introduction

This paper discusses the implementation of UN target 11.7 concerning inclusive urban public space and proposes the smart public space concept to help achieve this goal. The significance of this research is related to the crucial role of public space in community building, inclusivity, democracy, physical and mental health, social interaction, civic engagement, citizens' participation, and economic vitality [1,2]. In addition, this space has a vital role in improving the urban environment, particularly the air quality and biodiversity [3]. The UN 2030 Agenda for Sustainable Development highlighted through target 11.7 the necessity of providing a safe and inclusive public space for all, particularly for women, children, and people with disabilities [4,5].

Based on data from 1072 cities, the 2023 UN Sustainable Development Goals Report [6] alerts that access to open public spaces remains low globally, particularly in developing countries. This report reveals that over 75% of the participating cities have less than 20% of their area dedicated to public spaces and streets. This ratio is much lower than the UN goal, which targets a 45–50% ratio of public spaces to streets. The situation for public space is more critical because this space accounts for only 3.2% of the urban land in 2020, about four times less than that occupied by streets. Scholars highlighted significant disparities in the accessibility to public space [7]. For example, women, children, and people with disabilities face substantial difficulties accessing public space due to safety concerns and inadequate infrastructure. According to the World Health Organization (WHO), women face a higher risk of experiencing violence in public spaces than men, with an estimated one in three women experiencing physical or sexual violence in their lifetime [8]. Furthermore, women are more likely to be excluded from the public space due to gender norms. Similarly, children and persons with disabilities face challenges accessing and utilizing public spaces due to a lack of appropriate infrastructure and facilities, especially in developing countries [9].

Various approaches have been proposed to improve accessibility to public spaces. Esfandfard et al. [10] proposed the "universal design" approach that facilitates access of all to public spaces, regardless of people's abilities or disabilities. Kent [11] proposed the "placemaking" concept to involve the community in designing and developing public spaces. Ellery and Ellery [12] used this concept to create a public space that is not only functional but also meaningful and enjoyable for all citizens. Community-led initiatives have also been proposed as a practical approach to creating inclusive public spaces. These initiatives engage community members in designing and implementing public spaces, promoting ownership, and ensuring that the public spaces meet the needs of the diverse groups of the community [13]. These initiatives also foster a sense of community and encourage responsible use of public spaces. Zhou [14] highlighted the role of citizens' participation in creating an inclusive urban public space.

However, initiatives proposed to improve public spaces are still fragmented and have not yet been allowed to achieve the target of creating an inclusive public space, as stated by the last UN Sustainable Development Goals Reports [15,16]. Considering the crucial role of this space in improving the quality of life for all, there is a pressing need for an innovative concept for an inclusive public space that guarantees accessibility for all and improves services to the community. This paper contributes to this objective by (i) establishing a comprehensive framework for an inclusive public space, which is based on both social and environmental indicators, (ii) the application of this framework for the assessment of a public space in the city of Nablus in Palestine to understand the challenges of this space and the expectations of citizens, and (iii) the elaboration of the smart public space concept to address the challenges of an inclusive public space. This concept is inspired by the smart city concept, which offers several opportunities, particularly citizens' engagement [17–20] and the use of digital technologies such as the IoT, crowdsourcing, and artificial intelligence for real-time monitoring of public spaces and assessment of public space services. It also provides tools to improve citizens' safety and accessibility. This concept integrates recent developments in mobile applications and online platforms that offer valuable information on accessibility features, empowering individuals to navigate urban environments [21,22]. Additionally, immersive technologies like virtual and augmented reality offer innovative ways to simulate and design inclusive spaces [23,24].

Smart governance has also a crucial role in the smart public space concept as confirmed by Glebova and Desbordes [25], who used this concept to improve the accessibility and usability of sports facilities and promote tourism.

As highlighted by Dudek et al. [26], a smart intelligent urban system should go beyond simply incorporating advanced technology; it should also play a central role in promoting social responsibility by nurturing a strong sense of community and shared responsibility.

This research aims to establish a framework for the construction of inclusive public spaces, which have a major role in building a socio- and eco-friendly city. The framework includes a set of social and environmental indicators for the assessment of the inclusivity of public spaces. It is used for evaluating public spaces in a complex urban context in the city of Nablus in Palestine to understand the functioning of this place and the expectations of citizens. Based on this evaluation, the research establishes the concept of smart public spaces that improve the functioning and services of spaces to meet citizens' expectations.

This paper is structured as follows. First, it presents the research methodology and material and emphasizes the indicators for inclusive public spaces. Then, it discusses applying this concept to assessing the public space in the city of Nablus in Palestine. The assessment includes data collection through site visits and users' opinions and data analysis to determine the public space's main challenges and the users' expectations. Finally, this paper presents smart public spaces and how they could help address public space challenges.

2. Methodology and Materials

2.1. Overview

This research focuses on developing an inclusive public space that enables equal access for citizens, regardless of gender, ethnicity, age, and ability. This research is based on a comprehensive literature review and analysis of collected data. This research starts with an extensive literature review about inclusive public spaces to explore previous works about the inclusivity of public spaces and the indicators used for constructing these spaces and their assessments. The indicators are then used to assess the public space in Nablus, a major city in Palestine. The evaluation includes collecting data about the spatial distribution of the public space and its significant features. It also includes interviews with the users of public spaces about their perception of and expectations of these spaces. Data analysis resulted in identifying the social and environmental challenges of public spaces. Smart public space is proposed to address the challenges of public spaces. This concept ensures citizens' effective participation in public space governance to make it more inclusive. It also uses the capacity of smart technology to enhance the public space's safety and provide a set of digital services to users.

The following sections provide an extensive description of the research methodology's first stages, including establishing indicators for inclusive public spaces and data collection for assessing the public space in Nablus.

Sections 3 and 4 will present the analysis of the collected data to understand the challenges of the public space and the concept of the smart public space, which was established to address these challenges.

2.2. Indicators of Inclusive Public Spaces

An inclusive public space should ensure accessibility to all citizens regardless of their home location, gender, ethnicity, age, and abilities [27,28]. It also should provide adequate facilities and services to citizens. Both public authorities and communities need indicators to establish and assess strategies for inclusive public spaces. These indicators could be classified into the following categories (Table 1): (i) spatial distribution, (ii) typology, (iii) facilities and services, (iv) green and humid areas, (v) governance and management, (vi) safety, (vii) users' category, and (viii) users' satisfaction.

The spatial distribution measures the proximity of the public space to citizens regardless of their residence or working location. The typology informs about public spaces, such as plazas, linear markets, community yards, sidewalks, recreation areas and parks, and playgrounds. The facilities and services indicators concern the public space equipment such as public toilets, seat shading, water taps, garbage bins, lights, baby services, parking areas, and catering. The green and humid area indicators indicate the ratio and distribution of green and humid areas in the public space. The management of the public space concerns decision-making, citizens' participation, and maintenance of resources. The safety indicators concern safety equipment, vandalism, and security measurements. Finally, the category of public space users indicates the classifying of the population using the public space according to their gender, age, and abilities, while the users' satisfaction informs about their degree of satisfaction with the public space.

Table 1. Indicators for inclusive public spaces.

No.	Indicator Category	Description
1	Spatial distribution	Proximity of the public space to citizens, regardless of their residence or working location.
2	Typology	Type of public spaces: plazas, linear markets, community yards, sidewalks, recreation areas and parks, and playgrounds.
3	Facilities and services	Public space equipment and services: public toilets, seat shading, water taps, garbage bins, public lights, baby services, parking areas, and catering.
4	Green and humid areas	Ratio and distribution of green and humid areas in the public space.
5	Governance and management	Decision-making, citizens' participation, and maintenance of resources.
6	Safety	Safety equipment, vandalism, and security measurements.
7	Users' category	Category of the population using the public space (gender, age, disabilities).
8	Users' satisfaction	The user's degree of satisfaction with the public space services.

2.3. Presentation of the Case Study: Nablus City, Palestine

The set of indicators presented in the previous section was used for the assessment of the public space in the city of Nablus in Palestine. This city constitutes a regional urban center in the northern part of the West Bank [29]. This city is recognized as a vibrant commercial and cultural center, housing An-Najah National University and the Palestinian Stock Exchange. With a population of approximately 200,000 residents [30], it plays a crucial role as an administrative, educational, commercial, and industrial hub in the region. Its historical development reflects the influences of the Ottoman and British mandate periods, visible in the old and modern neighborhoods that emerged around the city [31].

The city of Nablus is experiencing significant population growth and density [29]. According to the Palestinian Central Bureau of Statistics [30], Nablus has an annual population growth rate of 2.5% and an average family size of 4.8. This rapid population increase has resulted in high population density, surpassing other regional cities. The expansion of urban areas has been limited, leading to a notable difference in built-up density, which reaches 733 inhabitants/km^2 in some neighborhoods [32].

2.4. Data Collection

Based on discussions with the municipality and the community, 63 spaces were selected for this study. Figure 1 illustrates the location of these spaces in the city.

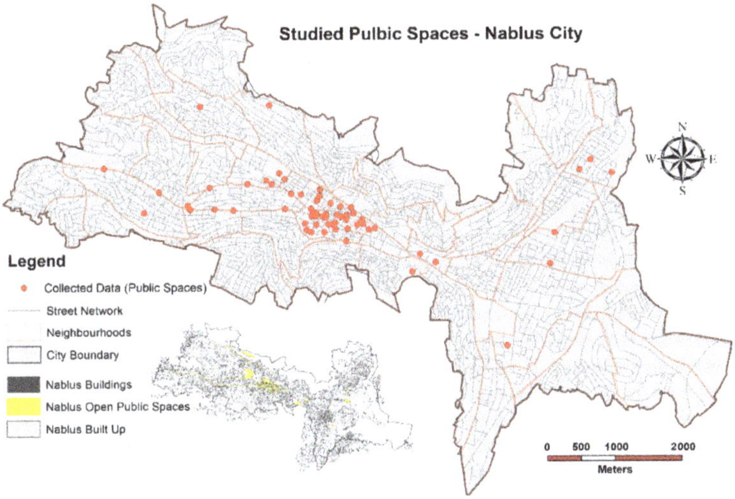

Figure 1. Localization of the surveyed public spaces in the city of Nablus.

Ten field surveyors collected data using the open-source Kobo Toolbox App [33]. The use of Kobo Toolbox streamlined the data collection process by facilitating efficient and accurate data entry. This toolbox allowed for the collection of various data types, including images and videos. For a comprehensive understanding of the public spaces' inclusivity, interviews were conducted with different categories of individuals during both daytime and nighttime. This approach enables capturing the perspectives of diverse citizens, including individuals from different age groups, genders, and people with disabilities.

The collected data concerned the indicators presented in Table 1. Data concerning the first two indicators (spatial distribution and typology) were collected by the survey team and confirmed by the municipality. Data concerning other indicators were collected from the users of the public spaces. Around 630 users participated in this survey. The participants encompassed a diverse cross-section of the community, representative of the broader population's characteristics. The survey protocol dictated that five individuals be interviewed in each public space: two males, two females, and one individual with a disability or a child. This approach aimed to capture a well-rounded understanding of users' experiences across different demographics. Notably, the female respondents exhibited greater responsiveness in completing the survey forms, which can be attributed to factors such as the form's length and the women's engagement with the subject matter.

3. Results

3.1. Spatial Distribution of Public Spaces

The total area of the public spaces within the city was found to be 496,359 m^2, which includes market areas, community yards, and streets, such as linear market streets. The per capita ratio of public spaces is 2.6 m^2/inhabitant, which falls below local and international standards that vary between 5 m^2 and 9 m^2 per person [34,35]. The allocation of public spaces in the master plan is below the standards; at least an additional one km^2 of new public space is required to meet the standards.

Figure 2 illustrates the distribution of public spaces in the city. It shows that public spaces are concentrated in the city center and the old city. The eastern part has reduced public spaces, including two camps and a low-income population. The results show that only 30% of the city's neighborhoods have access to public spaces within a five min walking distance, while over 41% need more than ten minutes of walking to access public spaces. The open spaces better serve the western part than the eastern part. The results also show that 41% of the public spaces' entrances are unclear, posing difficulty in accessing these spaces.

When explicitly considering parks and playgrounds, a more nuanced perspective emerges regarding walking distances. In this context, merely 7% of these recreational facilities are conveniently accessible within a 5 min walking radius. Conversely, 73% of parks and playgrounds demand traversing distances exceeding ten minutes on foot. This differentiation highlights a more pronounced challenge in accessing recreational spaces designed for leisure and outdoor activities. The disparities between accessibility to public spaces at large and specifically to parks and playgrounds underscore the complex spatial distribution within the city.

Moreover, the survey revealed that most public spaces operate on a neighborhood scale, typically ranging from 500 to 2000 m^2 in size. Pocket parks spanning fewer than 500 m^2 are conspicuously limited in number despite their noteworthy impact on urban aesthetics. These small green areas contain a distinctive capacity to increase the aesthetic attractiveness of the urban landscape. Regrettably, their limited presence could curtail small-scale relaxation, leisure, and visual delight opportunities.

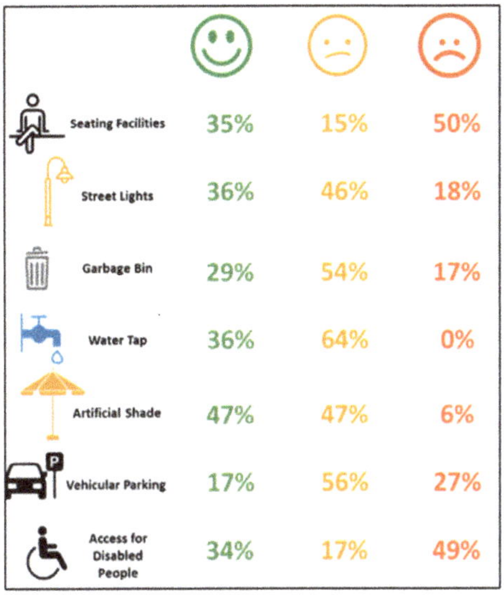

Figure 9. Satisfaction levels with physical features in public spaces.

These results highlight a high level of user dissatisfaction, particularly with seating facilities, waste disposal, and facilities for individuals with disabilities. In turn, this significantly affects the aesthetic value of the public spaces. Specifically, around 40% of the public spaces are regarded as uncomfortable, while over 60% of the reviewed open public spaces have garbage management issues.

4. Discussion

The analysis conducted in the preceding section reveals a significant shortfall in the quality of public spaces within the city of Nablus, highlighting a clear lack of inclusivity. This assessment has exposed a multitude of shortcomings, including but not limited to inadequate spatial coverage, inequitable distribution, notable absences in key areas of the city, security concerns, particularly affecting women, low levels of civic engagement, and a dearth of essential facilities such as public toilets, access to clean tap water, seating, artificial shade, clear navigation signage, and facilities catering to individuals with disabilities. These collective deficiencies severely restrict citizens' accessibility to public spaces, underscoring the urgent need for comprehensive improvement measures.

The following sections provide an extensive discussion of a significant issue of inclusive public spaces: their accessibility and smart public spaces as solutions to improve the inclusivity of public spaces.

4.1. Accessibility of Public Spaces

The distribution and accessibility of public spaces emerge as a significant concern, with the total area falling below recommended standards. The spatial concentration of public spaces in certain areas exacerbates the disparities between neighborhoods. This is particularly evident in the city's east, where reduced public spaces are available for low-income populations and two refugee camps. Consequently, urban planning efforts should prioritize the equitable distribution of public spaces across the city, considering the diverse needs of neighborhoods and population groups [36,37]. Engaging community stakeholders, including residents from marginalized groups, can provide valuable insights and ensure that public spaces reflect the needs and expectations of the entire community [38,39].

Furthermore, urban planning should focus on improving the accessibility and visibility of public spaces through clear entrances and wayfinding signage [40]. This will enhance the overall user experience and encourage more people, including those with disabilities, to utilize the city's public spaces.

The results also indicate a gender disparity in the use of public spaces, particularly during nighttime. Women reported feeling less safe than men, which was likely influenced by insufficient lighting and social problems occurring in these spaces. Moreover, people with disabilities have a reduced presence in public spaces, reflecting the lack of inclusive design and facilities that address their needs.

Strategies should be implemented to encourage the understanding and diverse participation of different age groups, genders, and abilities in public spaces. This should inform the design and management of public spaces to ensure their inclusivity, particularly safety for all community members [41]. Integrating the voices and perspectives of women, people with disabilities, and other marginalized groups in decision-making will foster a sense of ownership and belonging among users [42].

The physical features of public spaces are generally unsatisfactory, in particular seating, lighting, garbage disposal, green coverage, and facilities for women and people with disabilities. Seating was identified as a critical area for improvement. Moreover, the lack of necessary features for specific user groups, such as women's needs and accessibility considerations for people with disabilities, highlights a significant gap in the design and construction of these spaces. Enhancing the quality and diversity of physical features, along with ensuring the provision of inclusive amenities, can contribute to creating more attractive and user-friendly public spaces that meet the diverse needs and preferences of the community.

4.2. Smart Public Spaces

Analysis of the data collected regarding the public spaces in Nablus highlights the imperative for proactive measures to enhance their inclusivity and ensure alignment with users' expectations. More precisely, the analysis showed a high user dissatisfaction with public space safety, facilities, services, and accessibility. It also highlighted a lack of citizen participation in the governance and management of public spaces.

Considering the vital role of public spaces in cities and the Palestinian regulations concerning the crucial role of public spaces [43–46], a new strategy is required in Palestine for building smart spaces that involve citizens in their government and use the latest technology to improve the quality of services to citizens and the environment.

Smart public spaces should involve citizens in public space governance because they are the main ones concerned with this space. Also, they can provide users with valuable information about the public space location, design, facilities, services, and social environment. These spaces should also be monitored by the latest technology to improve services to users, in particular accessibility, safety, navigation, access to tap water, toilets, seating, shaded areas, and green and humid spaces. They should also provide specific services for people with disabilities. These spaces should also be eco-friendly, use advanced technology for water saving, irrigation, renewable energy, smart public lighting, and biodiversity survey and protection.

Smart public spaces are inspired by the smart city concept [47–49], which aims to improve the quality of life in cities by involving citizens in city affairs and improve the efficiency of urban systems and functions.

Figure 10 depicts the architecture of the innovative smart public space concept, as outlined by Shahrour and Xie [50]. This architectural framework comprises six layers, each vital in shaping the concept's functionality and efficacy. These layers encompass governance, physical infrastructure, data collection mechanisms, communication protocols, server infrastructure, and smart services.

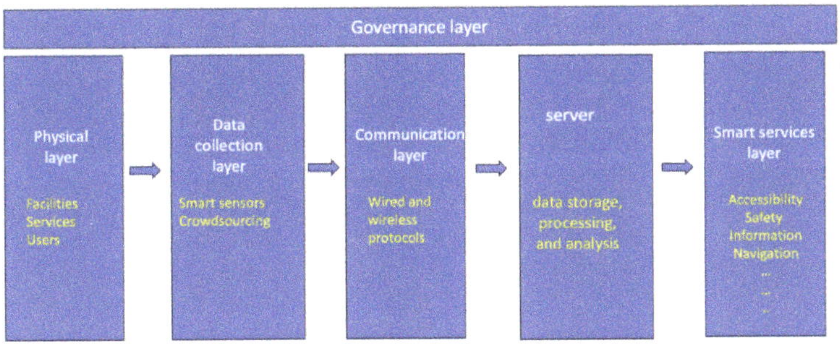

Figure 10. The smart public space layers.

The governance layer, positioned horizontally, serves as the linchpin for orchestrating the organizational dynamics and decision-making processes inherent to the design, amenities, services, and overall operation of the public space. Central to its role is ensuring a robust and inclusive engagement of critical stakeholders—from citizens who utilize the space to policymakers who shape its policies and the administrative entities overseeing its execution.

The physical layer constitutes the tangible foundation of the concept, encompassing the spatial elements, equipment installations, facilities, and the diverse services offered within the public space. It also extends its scope to encompass citizens, who not only act as end-users of the space but also serve as a wellspring of valuable data and critical evaluations that contribute to the enhancement of the public space experience.

The data collection layer includes tools and resources meticulously orchestrated to amass comprehensive data about the public space. This layer encompasses an array of Internet of Things (IoT) components, such as intelligent sensors endowed with the capability to meticulously monitor a spectrum of vital parameters within the public space ecosystem. These parameters encompass aspects ranging from spatial occupancy, ambient air quality, and accessibility metrics to the intricate functionality of lighting installations, facilities' operational status, and overall safety benchmarks. The data layer also includes interactive tools like mobile apps that transmit users' observations and suggestions. The data collection process must adhere to a stringent commitment to transparency while dutifully respecting the overarching regulations concerning data privacy.

The communication layer connects the physical layer to the server infrastructure. Employing both wired and wireless communication protocols, this layer orchestrates the seamless transmission and exchange of data. Advanced communication mechanisms, such as 5G technology and the innovative LoRaWan protocols, are at the forefront of its capabilities and contribute to the swift and robust transmission of data across the interconnected network. Through these sophisticated communication avenues, the public space concept attains an elevated level of cohesion, ensuring that the insights gleaned from the data-rich physical layer are channeled effectively to and from the central server hub.

The server layer is pivotal in executing multiple functions, including data storage, processing, and analysis. The latter employs various mathematical and information technology tools, including statistics and artificial intelligence, to metamorphose collected data into valuable insights, actions, and services. It can furnish statistical insights regarding public space usage and operational efficacy. These insights, in turn, contribute to enhancing service quality, optimizing equipment efficiency, and rationalizing resource allocation. Additionally, the system excels in real-time anomaly detection, swiftly implementing corrective measures when necessary. Furthermore, its contributions extend to fortifying safety through enhancing lighting systems, utilizing security cameras, and deploying a mobile app alert system.

The smart services layer encompasses a broad spectrum of actions that elevate citizen experiences and environmental conditions. Services include real-time updates on public space dynamics, facility operations, service availability, navigation aids, social connections, Wi-Fi accessibility, efficient irrigation practices, intelligent public lighting, and streamlined parking management.

5. Conclusions

This paper examined the improvement of implementing the UN Target 11.7 concerning inclusive urban public spaces. It contributes to this goal by elaborating the concept of the inclusive public space. The significance of this concept lies in recognizing the crucial role that public spaces play in community building, physical and mental well-being, social interaction, civic engagement, citizen participation, and economic vitality. However, the 2023 UN Sustainable Development Goals Report and scholars have raised concerns about the limited access to public spaces, particularly for low-income individuals, women, children, and people with disabilities.

Considering the significance of inclusive public spaces, this paper proposed a comprehensive framework for assessing this inclusivity. The framework employs eight indicators: spatial distribution, typology, facilities, green areas, governance, safety, user categories, and user satisfaction. These indicators were employed to evaluate the inclusivity of public spaces in Nablus, a major Palestinian city. The assessment validated the observations made by the UN and scholars, highlighting the low inclusivity of public spaces in Nablus, particularly due to inadequate availability and poor spatial distribution. Additionally, the assessment revealed significant citizens' dissatisfaction with safety conditions and services. It underscores the significance of involving citizens in the governance of public spaces, as they are the primary stakeholders who can provide valuable data and insights to enhance the quality and performance of these spaces.

Enhancing the inclusivity of public spaces necessitates a comprehensive approach that engages citizens in their governance and leverages the latest technology, particularly smart technology. By employing such technology, real-time information and services can be provided to citizens, enabling the monitoring of public spaces, improving facility efficiency, and creating an eco-friendly environment that preserves resources and biodiversity. This paper outlines how smart public spaces contribute to achieving this inclusivity and proposes an architectural framework for its implementation.

This paper suffered from some limitations. The main limitation concerned data collection because authorities were not able to provide basic data about the public space policy strategy, planning, facilities, and functions. Field visits and interviews were then used to collect data. In the future, data could be collected more frequently and involve more people through mobile applications and smart monitoring. The interviews focused on the citizens' perception of public space facilities, safety, and governance. In the future, interviews should be extended to the use of digital tools to understand citizens' expectations and fears about these tools, which will help in the development of smart tools for public spaces.

Author Contributions: Conceptualization, M.I. and I.S.; methodology, M.I. and I.S.; software, M.I.; validation, I.S. and I.H.; formal analysis, M.I. and I.H.; writing review and editing, M.I. and I.S.; supervision, I.S. All authors have read and agreed to the published version of the manuscript.

Funding: This research received no external funding.

Data Availability Statement: The data presented in this study are available on request from the corresponding author. The data are not publicly available due to privacy and proprietary concerns.

Conflicts of Interest: The authors declare no conflict of interest.

References

1. World Health Organization, Regional Office for Europe. *Urban Green Spaces and Health*; WHO/EURO:2016-3352-43111-60341; World Health Organization, Regional Office for Europe: Geneva, Switzerland, 2016; Available online: https://apps.who.int/iris/handle/10665/345751 (accessed on 18 August 2023).
2. UN. General Assembly Transforming Our World: The 2030 Agenda for Sustainable Development | Department of Economic and Social Affairs. 2015. Available online: https://sdgs.un.org/2030agenda (accessed on 14 March 2023).

3. Yang, J.; Zhao, Y.; Zou, Y.; Xia, D.; Lou, S.; Liu, W.; Ji, K. Effects of Tree Species and Layout on the Outdoor Thermal Environment of Squares in Hot-Humid Areas of China. *Buildings* **2022**, *12*, 1867. [CrossRef]
4. UN DESA. Goal 11 | Department of Economic and Social Affairs. 2015. Available online: https://sdgs.un.org/goals/goal11 (accessed on 14 March 2023).
5. OHCHR. *Policy Guidelines for Inclusive Sustainable Development Goals: Sustainable Cities and Communities*; OHCHR: Geneva, Switzerland, 2020.
6. United Nations Department for Economic and Social Affairs. *Sustainable Development Goals Report 2023: Special Edition*; United Nations: San Francisco, CA, USA, 2023. Available online: https://unstats.un.org/sdgs/report/2023/ (accessed on 18 August 2023).
7. Ahmimed, C. *Inclusive Cities: Trends and New Initiatives*; Department for Economic and Social Affairs (UNDESA) Division for Social Policy and Development: New York, NY, USA, 2018. Available online: https://www.un.org/development/desa/family/wp-content/uploads/sites/23/2018/05/Inclusivecities-DESAEGM16May2018CAHMIMED.pdf (accessed on 20 July 2023).
8. WHO. Violence against Women. 2021. Available online: https://www.who.int/news-room/fact-sheets/detail/violence-against-women (accessed on 14 March 2023).
9. Wellenstein, A.; García Mora, M.E. Breaking Barriers and Reconstructing a Disability-Inclusive Society. 3 December 2021. Available online: https://blogs.worldbank.org/latinamerica/breaking-barriers-and-reconstructing-disability-inclusive-society (accessed on 20 June 2023).
10. Esfandfard, E.; Wahab, M.H.; Amat, R.C. Universal design in urban public spaces for people with disability. Case study of Tehran, Iran. *Plan. Malays.* **2018**, *16*. [CrossRef]
11. Kent, E. Leading urban change with people powered public spaces. The history, and new directions, of the Placemaking movement. *J. Public Space* **2019**, *4*, 127–134. [CrossRef]
12. Ellery, P.J.; Ellery, J. Strengthening Community Sense of Place through Placemaking. *Urban Plan.* **2019**, *4*, 237–248. [CrossRef]
13. Ni, M.; Cattaneo, T. Design for Urban Resilience: A Case of Community-led Placemaking Approach in Shanghai China. In *Cross-Cultural Design. Culture and Society*; Lecture Notes in Computer Science; Rau, P.-L.P., Ed.; Springer International Publishing: Cham, Switzerland, 2019; pp. 207–222.
14. Zhou, S. Understanding "Inclusiveness" in Public Space: Learning from Existing Approaches. sustain.ubc.ca. 20 September 2019. Available online: https://sustain.ubc.ca/about/resources/understanding-inclusiveness-public-space-learning-existing-approaches (accessed on 12 June 2023).
15. Ahvenniemi, H.; Huovila, A.; Pinto-Seppä, I.; Airaksinen, M. What are the differences between sustainable and smart cities? *Cities* **2017**, *60*, 234–245. [CrossRef]
16. Angelidou, M.; Psaltoglou, A.; Komninos, N.; Kakderi, C.; Tsarchopoulos, P.; Panori, A. Enhancing sustainable urban development through smart city applications. *J. Sci. Technol. Policy Manag.* **2017**, *9*, 146–169. [CrossRef]
17. Granier, B.; Kudo, H. How are citizens involved in smart cities? Analysing citizen participation in Japanese "Smart Communities". *Inf. Polity* **2016**, *21*, 61–76. [CrossRef]
18. Cardullo, P.; Kitchin, R. Being a 'citizen' in the smart city: Up and down the scaffold of smart citizen participation in Dublin, Ireland. *GeoJournal* **2019**, *84*, 1–13. [CrossRef]
19. Prandi, C.; Barricelli, B.R.; Mirri, S.; Fogli, D. Accessible wayfinding and navigation: A systematic mapping study. *Univ. Access Inf. Soc.* **2023**, *22*, 185–212. [CrossRef]
20. Bastos, D.; Fernández-Caballero, A.; Pereira, A.; Rocha, N.P. Smart City Applications to Promote Citizen Participation in City Management and Governance: A Systematic Review. *Informatics* **2022**, *9*, 89. [CrossRef]
21. Chiu, C.-C.; Lee, L.-C. System satisfaction survey for the App to integrate search and augmented reality with geographical information technology. *Microsyst. Technol.* **2018**, *24*, 319–341. [CrossRef]
22. Lau, B.P.L.; Wijerathne, N.; Ng, B.K.K.; Yuen, C. Sensor Fusion for Public Space Utilization Monitoring in a Smart City. *IEEE Internet Things J.* **2018**, *5*, 473–481. [CrossRef]
23. Dick, E. Current and Potential Uses of AR/VR for Equity and Inclusion. Available online: https://itif.org/publications/2021/06/01/current-and-potential-uses-arvr-equity-and-inclusion/ (accessed on 20 July 2023).
24. McKenna, H.P. Adaptive Reuse of Cultural Heritage Elements and Fragments in Public Spaces: The Internet of Cultural Things and Applications as Infrastructures for Learning in Smart Cities. In Proceedings of the 2017 13th International Conference on Signal-Image Technology & Internet-Based Systems (SITIS), Jaipur, India, 4–7 December 2017; pp. 479–484. [CrossRef]
25. Glebova, E.; Desbordes, M. Smart Sports in Smart Cities. In *Smart Cities and Tourism: Co-Creating Experiences, Challenges and Opportunities: Co-Creating Experiences, Challenges and Opportunities*; Goodfellow Publishers Ltd.: Oxford, UK, 2021; pp. 60–73. Available online: https://books.google.fr/books?hl=en&lr=&id=TaWaEAAAQBAJ&oi=fnd&pg=PA60&dq=Smart+Sports+in+Smart+Cities&ots=MBCL6s3szX&sig=VvVe_j7RmSSDU8MhATW6XJYohMU&redir_esc=y#v=onepage&q=Smart%20Sports%20in%20Smart%20Cities&f=false (accessed on 8 September 2023).
26. Dudek, M.; Bashynska, I.; Filyppova, S.; Yermak, S.; Cichoń, D. Methodology for assessment of inclusive social responsibility of the energy industry enterprises. *J. Clean. Prod.* **2023**, *394*, 136317. [CrossRef]
27. Landman, K. Inclusive public space: Rethinking practices of mitigation, adaptation and transformation. *Urban Des. Int.* **2020**, *25*, 211–214. [CrossRef]
28. Pineda, V.S. What is Inclusive and Accessible Public Space? *J. Public Space* **2022**, *7*, 5–8. [CrossRef]

29. The World Bank. *Palestinian Cities Get a Boost through Integrated Urban Planning and Innovation*; World Bank: Washington, DC, USA, 2021. Available online: https://www.worldbank.org/en/news/feature/2021/10/14/palestinian-cities-get-a-boost-through-integrated-urban-planning-and-innovation (accessed on 24 June 2023).
30. Palestinian Central Bureau of Statistics. Preliminary Results of the Population, Housing and Establishments Census 2017. 2018. Available online: https://www.pcbs.gov.ps/Downloads/book2364-1.pdf (accessed on 8 September 2023).
31. Ghadban, S.S. Architecture in Palestine. In *Encyclopaedia of the History of Science, Technology, and Medicine in Non-Western Cultures*; Selin, H., Ed.; Springer: Dordrecht, The Netherlands, 2008; pp. 225–238. [CrossRef]
32. Hadba, L.; Ashqar, H.I.; Tuqan, N.A.S.; Lourenço, J. Effects of Public Participation in Urban Planning: The Case of Nablus City. 2014. Available online: http://repositorium.sdum.uminho.pt/ (accessed on 24 June 2023).
33. KoboToolbox. Available online: https://www.kobotoolbox.org/ (accessed on 23 June 2023).
34. Ministry of Municipal Rural Affairs and Housing. المعايير التخطيطية للخدمات العامة والاقليمية المحلية ومستوياتها المختلفة(Planning Standards for Local Public and Regional Services and Their Different Levels. 2015. Available online: https://momrah.gov.sa/ar/node/13020 (accessed on 5 August 2023).
35. European Commission. *How Can Public Space in a City Help to Address Future Urban Challenges?* European Commission: Brussels, Belgium, 2019; Available online: https://urban.jrc.ec.europa.eu/thefutureofcities/ (accessed on 5 August 2023).
36. UN Habitat. Background Note by UN Habitat-Proposal for a Public Open Space Indicator. 2015. Available online: https://unstats.un.org/unsd/post-2015/activities/egm-on-indicator-framework/docs/Background%20note%20by%20UN%20Habitat-%20Proposal%20for%20a%20public%20open%20space%20indicator-EGM_Feb2015.pdf (accessed on 8 September 2023).
37. Wolch, J.R.; Byrne, J.; Newell, J.P. Urban green space, public health, and environmental justice: The challenge of making cities 'just green enough'. *Landsc. Urban Plan.* **2014**, *125*, 234–244. [CrossRef]
38. Abusaada, H.; Elshater, A. Effect of people on placemaking and affective atmospheres in city streets. *Ain Shams Eng. J.* **2021**, *12*, 3389–3403. [CrossRef]
39. Waddington, H.; Sonnenfeld, A.; Finetti, J.; Gaarder, M.; John, D.; Stevenson, J. Citizen engagement in public services in low- and middle-income countries: A mixed-methods systematic review of participation, inclusion, transparency and accountability (PITA) initiatives. *Campbell Syst. Rev.* **2019**, *15*, e1025. [CrossRef]
40. Project for Public Spaces. What Role Can Design Play in Creating Safer Parks? 31 December 2018. Available online: https://www.pps.org/article/what-role-can-design-play-in-creating-safer-parks (accessed on 5 August 2023).
41. Wojnowska-Heciak, M.; Suchocka, M.; Błaszczyk, M.; Muszyńska, M. Urban Parks as Perceived by City Residents with Mobility Difficulties: A Qualitative Study with In-Depth Interviews. *Int. J. Environ. Res. Public Health* **2022**, *19*, 2018. [CrossRef]
42. Polko, P.; Kimic, K. Gender as a factor differentiating the perceptions of safety in urban parks. *Ain Shams Eng. J.* **2022**, *13*, 101608. [CrossRef]
43. MoLG. التخطيط التنموي المحلي(*Strategic Development & Investment Plan)*; 3rd ed.; Ramallah, Palestine, 2018. Available online: https://www.mdlf.org.ps/Document/SDIP%20manual%202018(1).pdf (accessed on 8 September 2023).
44. MoLG. دليل التخطيط العمراني دليل إجراءات وأدوات إعداد المخططات الهيكلية في الضفة الغربية وقطاع غزة(*Urban Planning Manual—A Guide to Procedures and Tools for Preparing Master Plans in the West Bank and Gaza Strip)*; Ramallah, Palestine, 2013. Available online: https://procurement-notices.undp.org/view_file.cfm?doc_id=139183 (accessed on 8 September 2023).
45. MoLG. دليل تخطيط الطرق والمواصلات في المناطق الحضرية بفلسطين (Guide to Planning Roads and Transportation in Urban Areas in Palestine). 2013. Available online: https://urbanplanninglibrary.blogspot.com/2022/01/ROADS-MANUAL.html (accessed on 24 June 2023).
46. Itair, M.; Hijazi, I.; Qanazi, S.; Zaidalkilani, S.; Issa, A. Localizing sustainable development goals in strategic development planning procedures. Case study: Local development planning manual for the Palestinian cities and towns. *Najah Univ. J. Res.-B* **2022**, *37*, 1491–1526. [CrossRef]
47. Szczepańska, A.; Kaźmierczak, R.; Myszkowska, M. Smart City Solutions from a Societal Perspective—A Case Study. *Int. J. Environ. Res. Public Health* **2023**, *20*, 5136. [CrossRef] [PubMed]
48. Gracias, J.S.; Parnell, G.S.; Specking, E.; Pohl, E.A.; Buchanan, R. Smart Cities—A Structured Literature Review. *Smart Cities* **2023**, *6*, 1719–1743. [CrossRef]
49. Albrecht, F.; Moser, J.; Hijazi, I. Assessing façade visibility in 3D city models for city marketing. *Int. Arch. Photogramm. Remote Sens. Spat. Inf. Sci.* **2013**, *XL-2/W2*, 1–5. [CrossRef]
50. Shahrour, I.; Xie, X. Role of Internet of Things (IoT) and Crowdsourcing in Smart City Projects. *Smart Cities* **2021**, *4*, 1276–1292. [CrossRef]

smart cities

MDPI

Article

Fit Islands: Designing a Multifunctional Virtual Urban Community to Promote Healthy Aging for Chinese Older Adults

Zixin Shen [1], Rongbo Hu [2,3,*], Dong Wan [4] and Thomas Bock [1,3]

1 School of Engineering and Design, Technical University of Munich, 80333 Munich, Germany; zixin.shen@tum.de (Z.S.); bockrobotics@web.de (T.B.)
2 Department of System Design Engineering, Keio University, Yokohama 223-8522, Japan
3 CREDO Robotics GmbH, 94315 Straubing, Germany
4 Civil Design Department, Sinoma International Engineering Co., Ltd., Nanjing 211111, China; wandong@sinoma-ncdri.cn
* Correspondence: rh@credorobotics.com; Tel.: +49-176-579-66382

Abstract: Within the context of an aging global population, the demographic structure of emerging economies is undergoing a dramatic transformation. Emerging economies have a large population base and rapid economic development, but they are ill-prepared to deal with population aging. Limited resources force many older adults to face health issues such as chronic diseases and loss of physical independence, exacerbating the burden of traditional family and societal elderly care. Uncontrollable events such as the COVID-19 pandemic and regional conflicts have exacerbated the plight of older adults. Improving the quality of life and health of older adults has become a development priority in emerging economies in the face of a rapidly aging population. The development of smart cities has brought with it many available digital technologies, and the consequent development of smart aging offers endless possibilities for improving the quality of life and health of older people, making cities more inclusive of older people. Researchers from developed economies have attempted to address the health issues of older adults through a technology that combines physical exercise and digital technology called Exergame. However, existing projects are not suitable for older adults in emerging economies due to differences in national conditions. The aim of this project is therefore to propose a universal approach to designing a health-promoting Exergame system in the format of a virtual urban community to help emerging economies cope with aging populations, making cities more inclusive. To verify the feasibility of this approach, the authors designed an expandable Exergame called "Fit Islands", using China as a case study. Based on the initial demonstration, the authors conducted functional tests. The result is that Fit Islands can meet the development objective of motivating Chinese older people to increase their physical activity, providing initial evidence of the feasibility of an Exergame system to promote healthy aging in emerging economies. The application of Fit Islands demonstrates the feasibility of the universal Exergame development method, which can, in principle, provide comprehensive and practical guidance for other countries.

Keywords: smart aging; inclusive smart cities; aging in China; health; COVID-19 pandemic; digital technology; Exergame; population aging

1. Introduction

Population aging is one of the most significant trends of the 21st century [1]. Compared to developed countries, the growth rate and magnitude of the elderly population in developing regions are much higher [2]. By 2050, the proportion of the global population aged 60 and above living in developing countries is expected to increase from the current 62% to 80% [3]. This trend is particularly evident in emerging economies, such as China and India [4]. Improvements in healthcare conditions and economic growth have allowed more people to live longer in emerging economies. The large population base, strong economic growth, abundant resources, and enormous market, present both opportunities and challenges [5]. With the rapid aging of the population, problems have arisen such as a huge gap between the rich and the poor in urban and rural areas, an unfriendly society for older adults, empty nesters, a rapid increase in the number of chronically ill patients, and an underprepared healthcare system [6–8]. Emerging economies have less time and face greater difficulties in preparing for aging compared to developed economies.

The COVID-19 pandemic and regional conflicts have undoubtedly exacerbated the physical and mental ill health of older adults. With weaker immune systems and a higher likelihood of developing chronic diseases such as heart disease, diabetes, lung disease and cancer, older adults are the most susceptible population in the COVID-19 epidemic [9,10]. Lockdowns and other restrictions on travel have been widely adopted by countries to combat the spread of COVID-19 [11]. Spatial isolation makes older adults, who are already socially isolated due to family relationships, social connections, marital status, digital gap, etc., more likely to develop a sense of loneliness, which in turn triggers psychological problems such as depression and anxiety. These psychological problems can easily lead to a decline in the quality of sleep, increased incidence of cardiovascular and cerebrovascular diseases, and can lead to cognitive decline in older adults, which directly and negatively affects their quality of life [12–14].

With the development of technologies such as the Internet of Things (IoT), big data, and artificial intelligence (AI), smart aging, as an important part of smart cities, offers new possibilities to deal with the problem of an aging population. By utilizing advanced Internet, cloud computing, wearable and other new-generation information technology tools, it is possible to build urban IoT systems and urban information platforms for home-based elderly care, community-based elderly care, and institutional elderly care, to maximize the satisfaction of the needs of older adults in terms of social participation, social inclusion, medical care, communication, leisure, and culture [15]. Smart aging makes smart urban environments more age-friendly. The experience of older persons is fundamentally enhanced by addressing the risks of old age, improving the efficiency of aging, and empowering older persons with equal rights and capabilities [16]. Uncontrollable events such as the COVID-19 pandemic have even advanced the development of the digital world, and smart aging is gradually becoming an inevitable trend. In order to cope with the problems brought about by the rapid aging of the population mentioned above, emerging economies can make rational use of existing digital technologies to enrich city support for older adults and improve their quality of life.

Video games are one of the most popular digital interactive entertainment products nowadays, capable of carrying core values such as education, health, culture, and art. In smart cities, digital entertainment interactive applications such as age-friendly games have great potential to enhance the quality of life and the physical and mental health of older adults. In recent years, a new digital technology has been introduced to motivate people to engage in physical exercise. This technology combines technology, playfulness, and physical activity, requiring physical exertion to play the games, and is known as Exergame [17]. Exercise has potential benefits for improving the physical health, cognitive abilities, mood, personality and overall well-being of older adults [18–20]. In the context of aging societies, Exergame is no longer just for teenagers or game enthusiasts, but offers new ways of coping with the problems of aging as a tool for disease prevention and health education for older people. Many attempts have been made in countries with developed economies. However, older adults in emerging economies have complex and challenging requirements for smart products or services because of their specificities, and existing research projects are not entirely appropriate.

The aim of this study is therefore to try to incorporate Exergame's game format in the design of a virtual community that encourages physical activity, breaks down spatial isolation in urban areas, crosses the digital gap for older adults, and reduces the impact of uncontrollable events on the city in emerging economies. By proposing a design methodology for the health-promoting Exergame system, new solution ideas are proposed to help different emerging economies cope with the problem of population aging. An Exergame system called Fit Islands has been developed using China as an example. To adapt to different scenarios in the city and meet the needs of more Chinese older adults, the system integrates sports, social, educational and entertainment functions and is universal, affordable, adaptable, and age-friendly. Fit Islands provides a platform for older adults who are in nursing homes, hospitals, community centers and residences, and even older adults in senior colleges and in employee lounges at companies, creating a virtual urban community to link them together (see Figure 1). Functional testing of Fit Islands has confirmed the positive effects of the health-promoting Exergame system in addressing the issue of aging in China, and provides guidance for other emerging economies.

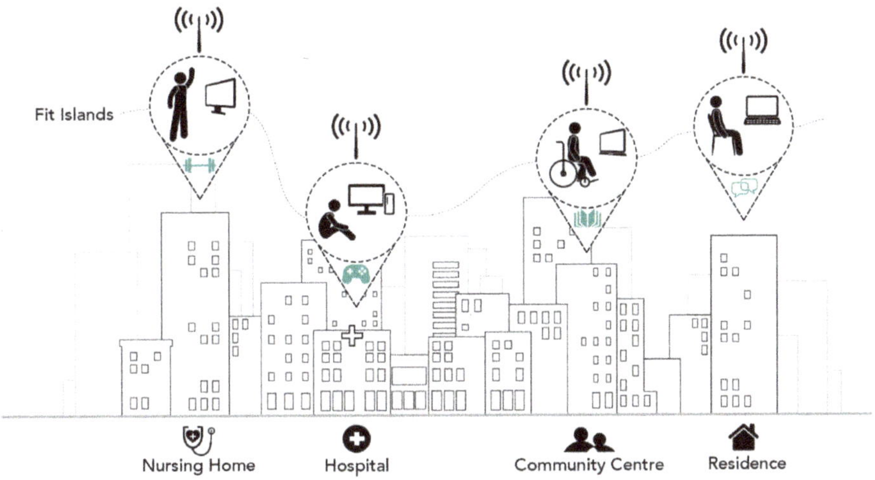

Figure 1. Fit Islands breaks spatial isolation and provides a health-promoting virtual platform for Chinese older adults.

2. Background Research

Exergames typically utilize various types of sensors as input devices, allowing players to interact with the game through bodily movements in specific gameplay scenarios. The most widely used sensors in Exergame input devices include accelerometers, gyroscopes, infrared (IR) and RGB optical sensors/cameras, and pressure sensors. Optical sensors/cameras, along with corresponding algorithms, can enable motion capture functionality, allowing the possibility of playing Exergames without holding or wearing input devices [21–26]. As the gaming industry has evolved and expanded, the variety of Exergames has also grown, incorporating different game interfaces and devices. For example, in *Dance Dance Revolution*, players need to step on pressure-sensitive panels following arrow indicators on the screen in sync with different musical rhythms [27]. In *Kick Ass Kung-Fu*, real-time image capture through cameras is used to enable player interaction with the virtual world [28]. In *Nautilus*, floor sensors are employed to complete tasks such as rescuing trapped dolphins at the bottom of a lake [29].

As a tool for health improvement and education, a growing number of such games are beginning to be developed with older adults as their primary users. For example, Gerling et al. in 2010 explored design guidelines for sports game design for older adults with a game called *SliverBalance* [30]. Chartomatsidis et al. in 2019 developed an Exergame called *Fruit Collector*, based on the Kinect sensor and developed using Unity for 3D game world design, demonstrating the positive impact of Exergames on improving physical health in older adults [31]. In 2017, a virtual therapeutic gaming system called *SilverFit 3D* further demonstrated that treating older adults through Exergames is an accepted, effective, and potentially cost-efficient approach, through a system based on scientific research and clinical practice that can develop rehabilitation protocols and targeted exercise programs for older adults [32–34]. The REACH research project conducted by Hu et al. included an Exergame system called the *ActivLife Gaming Platform*, which senses and assists with a smart device called the iStander, designed to prevent falls in older adults through physical activity and cognitive training [35]. For example, *SliverBalance* takes into account the physical condition of older adults, both standing and sitting, and the cognitive limitations of older adults by presenting a simple user interface (UI) and gameplay that allows them to focus more on the game. *Fruit Collector*'s creators interviewed orthopedic therapists and physiotherapists to assess the game's movements, and focused on building older adults' confidence in the game. *SilverFit 3D* can create over 1.2 million exercises by combining over 30 different visual effects and hundreds of game variations with 40 unique movements based on scientific research and clinical practice. The flexibility and adaptability of the iStander smart device in REACH also ensure the safety of older adults during play. However, these games also have some drawbacks. For example, *SliverBalance* and *Fruit Collector* lack variety in content and do not take into account the long-term motivation of the players. The *SilverFit 3D* and REACH systems are expensive, require bulky medical or smart devices, and are not widely used in everyday life.

Older adults, as individuals, experience the psychological and physiological changes associated with aging to different degrees. For example, cognitive impairments may prevent older adults from engaging in complex game tasks for extended periods, while physical limitations can reduce their interaction with the surrounding environment. It is important to consider the changes that come with age, ability, language, culture, and gender when designing digital games and interaction modes for older adults, taking into account the principles of inclusive design [36].

In the 1980s and 1990s, scholars (e.g., Weisman and Whitcomb) began to make recommendations after studying the effects of age, suggesting that gameplay should include adjustable game speed and difficulty, and easy-to-use interface design [37,38]. Ijsselsteijn et al. also suggested in their study of digital games in the early 21st century that visually adjustable game interfaces should be considered. For example, fonts, colors, and contrast, providing a presentation that incorporates both dynamic resources and multimodal feedback in addition to text [39]. Flores et al. expanded on this by arguing that it makes sense to combine appropriate cognitive challenges, simple UIs, and motivational feedback in the design [40]. They also found that older players preferred to play simple puzzle and quiz games. Gerling et al. also suggest that older players should be given the possibility to adjust the difficulty of the game, the speed of the game, and the sensitivity of the input device [30]. A large proportion of today's older adults have limited experience with digital games, therefore Exergames should be designed to minimize the number of steps required to complete tasks and reduce cognitive load, thus enhancing the user adaptability of the gaming system. The following design principles also need to be considered:

- When designing specific types of movement control for older players in Exergames, the movement capabilities of individual body parts should be fully considered. The interaction mechanism of the game should allow the user to move through multiple pathways, whether sitting or standing.
- Considering the age-related physical limitations that older adults may have, such as a lack of balance, the game's movement design should not include too many dangerous movements, such as leaning backwards. The movements in the game should also be supported by some purpose and scientific theory.
- The motivation of older users to play Exergames should be fully considered due to their unfamiliarity with digital technology. The game should simplify the processes and present the interfaces in a way that meets the psychological needs of older adults, and should also give reasonable hardware requirements to avoid the possibility of difficulty in getting started or frustration due to the operation of the game, thus fostering a positive player experience.

- Consider the practicality and comprehensiveness of the game. Besides the game itself, attention should be paid to the extension and playability of the game, depending on the gender, culture, and preferences of older adults of different ages.

3. A universal Exergame Development Method

The process of Exergame development for emerging economies in particular is proposed as shown in Figure 2. This development method was applied in the development process of the Exergame in this study.

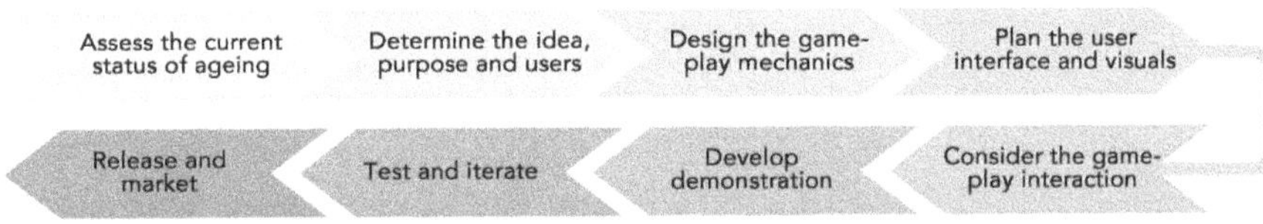

Figure 2. A universal health-promoting Exergame system method for emerging economies.

3.1. Assess the Current Status of Aging

Assessing the current status of aging in a country involves collecting relevant data and analyzing various aspects related to population aging. The specific activities include: collecting demographic data, examining the age structure of the population, studying life expectancy and mortality rates, assessing health indicators, exploring socio-economic factors, examining social support systems, analyzing education and employment opportunities, researching social integration and well-being, identifying policy implications, and exploring the process of smart aging development.

3.2. Determine the Main Idea, Objectives and Users of the Game

Based on the knowledge and assessment of the target market, the next step is to identify the type of sport to incorporate and the target users of the game. The specific activities include: considering whether the game will be created using an existing gaming platform (e.g., handheld game consoles, mobile devices) or develop a customized hardware setup, determining the target users (e.g., specifically for older people who cannot walk or can be adopted by all older people), and determining the goal of the game (e.g., to motivate older adults to exercise, to participate in a rehabilitation program, to improve balance, etc.).

3.3. Design the Gameplay Mechanics

In this step, the core game mechanics that encourage physical activity will be created. This may include activities such as running, jumping, dancing, or using specialized controllers. This can be designed according to the interests and characteristics of relevant countries. The rules, objectives, and challenges that players will encounter will also be defined. A balance between exercise and engaging gameplay also needs to be found to keep the users motivated.

3.4. Plan the UI and Visuals

In this step, the UI elements and visual style of the game will be designed. Consider using vibrant and colorful images to make the game more visually appealing. Based on the background research in Section 2, game design for older people needs to follow inclusive design principles. It is important to ensure that the UI is intuitive and easy to understand, providing clear instructions to players.

3.5. Consider the Gameplay Interaction

The game needs to use simple interaction mechanisms wherever possible. To make the game more effective, hardware and software components are implemented according to the current state of digital development in the country to accurately measure and track the player's movement. This can be done using motion sensors, cameras, or specialized controllers. These data can then be used to influence the gameplay or provide feedback to players about their exercise performance. To add an extra layer of motivation, it is worth considering the incorporation of multiplayer and social features into the game. This could involve online leaderboards, cooperative or competitive gameplay modes, or the ability to connect with friends and challenge each other.

3.6. Develop Demonstration

In this step, the appropriate platform for the game development needs to be chosen. If the game is being developed for console or PC, programming languages such as C++, C# or Java may be required. For mobile platforms, languages such as Swift (for iOS) or Java/Kotlin (for Android) can be considered. There are also game development engines, such as Unity or Unreal Engine, that provide powerful tools and libraries to simplify the development process.

3.7. Test and Iterate

In this step, it is necessary to conduct extensive playtesting to identify any bugs, gameplay issues, or areas for improvement. Feedback from players will be gathered and necessary adjustments can be made to enhance the game's experience. Then, we need to continuously iterate and refine the game based on user feedback to create a more engaging and enjoyable Exergame experience.

3.8. Release and Market

Once the game's mechanics, visuals, and overall experience are satisfactory, the process of optimizing the final product and preparing for release begins. Depending on the platform, specific submission guidelines and requirements need to be followed. A marketing strategy needs to be developed to make the game known to the target audience. Social media, gaming communities, and other relevant channels can be used to promote the game. It is worth considering collaborating with fitness influencers or partnering with fitness-related organizations to increase exposure.

4. Fit Islands

The authors developed an Exergame using China as an example, following the proposed general system development approach described above. By analyzing the data of Chinese older adults, they explored in detail the current aging challenges faced by Chinese older adults and the causes of those challenges. After determining the purpose of the design, the project team designed an expandable virtual community called Fit Islands.

4.1. The Current Status of Aging in China

China, as a representative emerging economy, has a huge population base. The improvement in medical conditions and the one-child policy have directly contributed to population aging in China [41–43]. Generally, individuals aged 65 and above are considered older adults. However, in China, based on the retirement age of 60 for men (55 for women), individuals aged 60 and above are classified as older adults [44]. Data shows that even using the age of 65 as the threshold, China's population aging exhibits significant scale, depth, and speed. In 2020, the elderly population aged 65 and above in China reached 191 million, accounting for 13.5% of the total population. Globally, one in four older adults is Chinese [45]. By the end of 2022, China's population had decreased by 850,000 compared to the previous year, marking the first year of negative growth. The rapid decline in the number of women of childbearing age and the fertility rate suggests that China is entering a period of normalized population decline [46]. It is projected that by 2057, China's elderly population aged 65 and above will peak at 425 million, accounting for 32.9–37.6% of the total population. The speed of population aging in China has surpassed that of any other country in modern history [47]. From 2001 to 2021, China transitioned into deep aging in just 20 years, which is shorter than the 126 years it took France, 46 years for the United Kingdom, and 40 years for Germany [48]. China has had much less time to prepare for aging compared to other countries, resulting in increasingly prominent issues such as empty nest syndrome, a significant increase in the old-age dependency ratio, increased burden of elderly care, premature aging, and a decline in the willingness of young people to have children [47,49–53].

4.1.1. Age Structure and Information Technology Proficiency

Since the early 2000s, the level of information technology in China has developed rapidly. Nationwide, Internet penetration has increased from 16% in 2007 to 70% by 2020. During the same period, the number of Internet users increased from 210 million to 989 million. In 2007, the Internet penetration rate was 26% in urban areas and only 7% in rural areas. By 2020, Internet penetration in urban areas reached 80%, while in rural areas it reached 56% during the same period, which is eight times higher than in 2007 [54]. In terms of age structure, the elderly population in China has a relatively young age structure [55,56]. The cohort of individuals born in the 1960s and later, who are now entering old age in China, are referred to as the "new older adults" [57]. These "new" older adults embrace fashion, technology, and life, and are more adaptable to various changes in life. Social, entertainment, and information-based applications are commonly used by older adults [58]. An increasing number of older adults are participating in the production of content such as text, audio, and video [59]. In terms of education, the education level of older adults in China has greatly improved [56]. Older adults with higher levels of education exhibit higher adoption rates of smart home technologies and have higher expectations for smart furniture designed for older adults [60]. However, consistent with research in other countries, the adaptation of Chinese older adults to digital technology shows a polarization phenomenon [61]. For a considerable portion of the elderly population, rapid changes in lifestyle can lead to confusion, decreased sense of security and belonging, and even exclusion from the digital world, known as the digital divide [62]. With the widespread outbreak of the COVID-19 pandemic in 2019, digital transformation has further swept across the globe. During the COVID-19 epidemic prevention and control period, the widespread promotion of "health codes" and contactless payment has further highlighted the issue of the digital divide.

4.1.2. Health Status

Self-rated health assessments among older adults show overall improvement but still remain at a relatively low level, with significant disparities between urban and rural areas [56]. The elderly population is highly susceptible to chronic diseases, and the number of older adults with chronic diseases is rapidly increasing due to population aging. According to statistics from 2022, over 180 million older adults in China suffer from chronic (non-communicable) diseases such as ischemic heart disease, cancer, stroke, arthritis, and dementia, with a prevalence rate of at least one chronic disease reaching 75% [63]. The number of disabled and partially disabled older adults exceeds 40 million. Furthermore, the prevalence of hypertension among people aged 60 and above is as high as 58.3%, and the prevalence of diabetes is 19.4% [64]. The health status of the elderly population is not optimistic.

4.1.3. Psychological Condition

As China's social and economic development has brought significant improvements to the lives of older adults in both urban and rural areas, the quality of life demanded by older adults in China has increased, and the abundance of leisure time after retirement has further strengthened their need for spiritual communication and care. In China, family relationships and support are the most important part of social support, with Chinese cultural traditions emphasizing the institution of the family and collectivism. Family support is the most important factor influencing older adults' social isolation and loneliness [65]. For most older adults, children are the most important social relationship, and reduced communication with them can exacerbate loneliness [66]. In the context of the one-child policy, more than two million older adults have lost their only child [67]. The proportion of older Chinese adults living with their children is decreasing as the size of the family decreases. The proportion of older adults living independently (alone or with a partner) is 65.4%, with only 26.8% of older adults currently living with their children [68,69]. The proportion of empty nesters among China's elderly population is now more than half [70]. Rural areas, which have 70% of the elderly population, are aging far more rapidly than urban areas, and the empty nest phenomenon is even more serious [71]. In a survey by Qiang Li et al., most older adults do not have close ties with their relatives [68]. Meanwhile, in a survey of older adults aged 60 and over living alone in Chongming District, Shanghai, 44.68% of older adults said they sometimes felt lonely and 15.56% said they often felt lonely [72]. In response to the sudden global outbreak of the COVID-19 epidemic, many countries, including China, have introduced lockdowns and home confinement measures to limit the spread of infection. Older adults, a vulnerable population, are the main target of isolation [11]. This uncontrollable adverse event makes older adults, who are already socially isolated due to family ties, social connections, and marital status, more susceptible to induced psychological problems such as depression and anxiety.

4.1.4. Leisure and Residence

Overall, leisure activities of older adults are mostly sedentary [73]. In terms of physical activity, square dance is the first choice of middle-aged and older adults in China [74]. But in recent years, conflicts around music noise and music venues have been common in the city. China is promoting a three-tiered elderly care system, with plans for 90% of older people to stay at home, 7% in community centers, and 3% in institutional care centers. Most older Chinese now own their own private living space. The average rural older person has 1.93 rooms, while the average urban older person has 2.79 rooms [75]. However, most of the existing living environments for Chinese seniors do not meet the needs of seniors in terms of usage. There is also a lack of community facilities for different functions, such as activity centers for older adults, physical exercise centers, barrier-free facilities, and other appropriate community facilities, which limits the space available to older adults and reduces their quality of life. China's traditional "family pension" is facing increasingly serious challenges. At the same time, China's elderly care products and aging services such as "institutional pension" and "community nursing service" [76] have not yet developed to the extent that they can meet the needs of the aging population, and are not yet an alternative to "family pension" [69].

4.1.5. Economic Status and Social Welfare

China's older adults are growing old before they get rich, and their ability to pay is insufficient. GDP per capita in 2022 will be only USD 12,741, while the standard GDP per capita in developed countries reaches over USD 20,000 [77]. Despite China's rapid development over the past few decades, the gap with developed countries is still huge, and the disparity between rich and poor in various regions is very large. China's pension insurance system started late and relies mainly on the first pillar, basic pension insurance [78]. 58% of older adults do not have a pension, and older adults are not yet well protected financially. There is a marked difference between the guaranteed income of older adults in urban and rural areas [56]. In 2020, the average per capita receipt of rural old-age pension insurance was less than USD 28 per month, which is about twenty times less than the average per capita receipt of basic old-age pension insurance for urban workers [48]. In 2014 data, the per capita annual income of urban older adults was USD 3348, while the per capita annual income of rural older adults was USD 1066 [55]. Older adults rely mainly on savings, children, continued work, and property for their old age. Although a significant portion of Chinese seniors have insufficient ability to pay, China's smart retirement is still a very promising market as aging increases. The market consumption demand of China's senior recreation industry was above USD 700 billion in 2016, and is expected to reach USD 2.8 trillion in 2030 [79,80].

4.1.6. Smart Aging

In China, research on smart aging can be traced back to 2007 [81]. Since then, it has developed rapidly from its nascent stage, and various cities have gradually implemented smart aging practices. However, studies have shown that China still lags behind other developed countries in terms of research capabilities in the field of elderly welfare technology [82]. Currently, there is no unified and standardized national system or set of standards for smart aging in China, resulting in fragmented and disordered development. As a result, regional disparities, variations in service content, and technological differences are observed in smart aging across the country [83]. Furthermore, the lack of industry standards and management regulations has led to variations in the quality and technological level of smart aging products produced by different companies. The interfaces, data formats, and interoperability between different products and systems are not standardized, resulting in issues such as compromised quality and safety, low utilization of related resources, and difficulties in data sharing [84]. These factors hinder the widespread adoption and promotion of smart aging. The number of categories of smart aging products and services, as well as their user-friendliness for older adults, are insufficient to meet the rapidly growing demands of the elderly population in China. Currently, most smart aging products and services in the Chinese market focus on meeting the basic living needs of older adults. There is a scarcity of products and services that cater to the social, emotional, cultural, entertainment, and lifelong learning needs of older adults. Furthermore, products and services that address more advanced aspects of smart aging are lacking. Existing smart aging products and services have not adequately considered the capabilities and physiological characteristics of older adults. Issues such as small fonts on the UI, complex operations, and excessive use of technical jargon impact the user experience for older adults, making it challenging to promote and popularize smart aging effectively.

4.2. Determine the Main Idea, Objectives, and Users of the Game

"Fit" represents health, symbolizing that this game will promote physical and mental well-being for older adults. Fit Islands will primarily focus on Exergaming and will create a 3D gaming world for older adults in China. In Fit Islands, players will assume the role of lively animated characters living on their own floating village. These villages are like individual islands, which also represent lonely older adults in the city. There are different scenarios on the island, allowing players to freely explore and wander the map through simple operations. On the islands, players can visit shops, cultivate crops, interact with friends on their contact lists at the post office, and play games at the game center. Elderly players will fully immerse themselves in their own gaming world, enjoying different music and game experiences. The game will be targeted at all Chinese older people, to give fairness and accessibility to older people with different abilities. Therefore, the game is designed to include two forms of exercise, full-body standing exercise and upper-body exercise. Considering the cost and accessibility of the game, Fit Islands will be available for public download and installation in the form of an online game (see Figure 3).

To achieve the desired impact on the target users, the game design will need to meet the following objectives:

- Adaptable: The game will fully consider the physical condition of the target users, ensuring adaptability and fairness for different players. Membership will be set up to provide extended games for various numbers of players, ensuring that the games can be adapted to different scenarios, such as homes, community centers, offices, nursing homes.

- Affordable: The game will be played without external wearable devices. It will only require a computer with a camera to play. The aim is to minimize the cost of the game and make it accessible to more older adults.
- Elderly-friendly: The game design will strictly adhere to the principles of inclusive design. Preferences and characteristics of elderly users will be considered, ensuring friendly interface colors, fonts, usability, and interactions that cater to older adults. The design will be inclusive of different levels of cultural, linguistic, visual, and auditory abilities, creating a user-friendly interface and game experience for the older adults.
- Practicality: The focus should not only be on the game itself but also on providing practical assistance to the target users. For example, valuable health knowledge may be incorporated into the game. The game's adaptability to different scenarios allows older adults' physical activities, such as square dance, to no longer be limited by urban facilities.
- Enjoyability: The game will draw inspiration from popular games, creating an enjoyable gaming world that integrates socializing, physical activity, cognitive training, and other functionalities for older adults.
- Promote initiative: The game will simplify the process of playing games and allow older adults to actively engage in physical activity, making friends and learning through games and sports, rather than being passively assisted by a number of assistive devices.

Figure 3. Fit Islands poster.

4.3. Design the Gameplay Mechanics

The process of Fit Islands will fully consider the needs of older adults and strive to simplify the game as much as possible. The specific process can be summarized as follows: Initial game installation—Game instructions—Account creation—Main interface—Game map—Gameplay—Points and health knowledge (see Figure 4).

Elderly players who may be less familiar with information technology can receive assistance from family members or caregivers during the initial game installation. After running the game, they will enter the initial interface where they can click on a "Start" button, and the game will provide instructions and gameplay explanations. Once players have a general understanding of the game content and gameplay, they can proceed to create an account. For first-time players, a series of questions and options will be used to gather basic user information. After successfully creating a new game account, players can choose an avatar representing different genders to enter the main interface. The main interface will contain all the information about the individual user's gameplay, including game duration, news, friends list, as well as features such as starting the game, membership, shopping cart, wardrobe, and settings. Upon starting the game, players will enter the game map, which consists of a floating island with various scenes, including a game center, shops, a post office, and a farm. The game center will not only feature physical activity games but also include two games specifically designed for cognitive training and knowledge learning for older adults. The game will utilize a points-based incentive system, allowing players to use points to redeem various items in the game store. Additionally, at the end of each gaming session, players will receive different health knowledge, further motivating older adults to complete the game successfully. It is worth noting that to simplify the preference settings for older adults, the game will employ individual user profiles and automatically match game content. The game can accommodate two users in total, and after each game session, the progress and user data will be automatically saved. The content of physical activity games will be determined based on the user's initial basic information and settings. For example, if an older adult has difficulty standing due to conditions like arthritis, they can choose to play seated games during the initial information setup, and the game will automatically match them with appropriate seated physical activity games. In the settings, older adults can freely adjust game language, color brightness and contrast, font size, and more. If they wish to change the timing of physical activities, they can access the personal information section in the settings interface for adjustments. Furthermore, Fit Islands is expandable, allowing different membership levels to enjoy larger game maps, more game varieties, and accommodate additional independent user profiles and services.

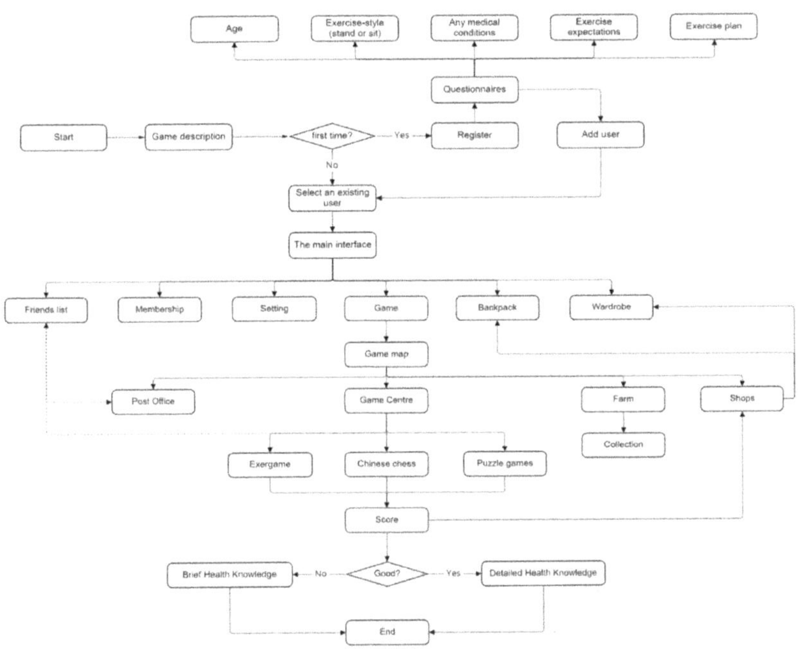

Figure 4. Functional flowchart of Fit Islands.

4.3.1. Exergame Design

In Section 4.1, based on a survey and study of the interests and hobbies of older adults, square dance emerged as the most popular form of physical activity among Chinese senior citizens. However, conflicts have arisen in recent years due to issues related to music noise and venue availability. The root cause of these problems lies in the incomplete public infrastructure catering to the needs of older adults in China. Furthermore, the outbreak of the COVID-19 pandemic in recent years has further restricted the gathering of large crowds, imposing additional limitations on square dance. Therefore, Fit Islands aims to provide older adults with the opportunity to engage in rhythmic exercise at home through the medium of gaming and digital technology. The game is therefore designed to include two forms of exercise, namely full body standing and upper body exercise (see Figure 4). The Exergame in Fit Islands draws inspiration from the VR game *OhShape* [85]. However, in Fit Islands, a different approach is taken to visually demonstrate exercise movements in a more intuitive manner for elderly players. Shadows representing the movements are used to directly display the poses, providing visual cues for the actions. These shadows will appear in a semi-transparent form, allowing players to see through the previous shadow and anticipate what comes next. Through motion capture, the game character located in the center-bottom of the screen will mirror the actions performed by the player. The shadow representing the action will gradually move closer to the game character from behind. The degree of overlap between the shadow and the game character will determine the player's score (see Figure 5).

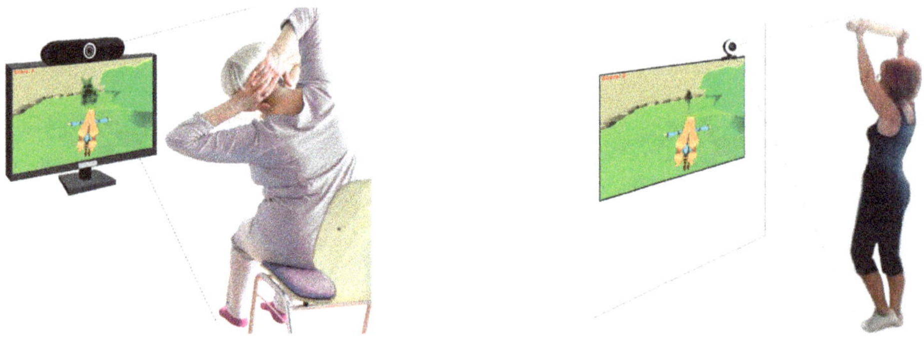

(**a**) (**b**)

Figure 5. Two game modes: (**a**) upper body movement; (**b**) full body standing movement.

4.3.2. Exercise Content and Exercise Programs

The movements in Fit Islands will primarily focus on preventing cardiovascular diseases [86,87], enhancing muscle strength [88–90], and improving flexibility and balance among older adults [91–94]. It will provide suitable exercise plans for older adults through a simple interactive interface. High-intensity exercise modes may not always yield significant results and can pose risks to older adults. Therefore, when designing exercise plans, it is important to start with low-intensity and short-duration exercises and allow the older adults to control the difficulty of the game themselves. Fit Islands will categorize and assess the difficulty of all expert-evaluated movements, and these movements will be combined in a default sequence to form a set of progressively more-challenging game levels. Each level will last approximately 10–15 min by default, and follow the sequence of warm-up, exercise, and stretching for the movements [95]. After completing each level, the game will automatically save the progress, ensuring that the individual exercise plans of each user are not disrupted. Furthermore, the exercise content will be adjustable. In Fit Islands, apart from setting limitations on the exercise time and specific movements based on initial basic questions, an option to adjust the difficulty level will be added at the end of each game. This design allows the game to maintain a basic level of challenge while providing inclusivity, ensuring the autonomy and safety of older adults.

4.3.3. Health Knowledge

In addition to exercising cognitive levels and increasing physical activity through the game, health knowledge appears on the interface after each round of the game. Research studies have shown that older adults are generally at a low level of interest in exercise, but are very active in collecting health knowledge, so health knowledge can also be used as a means to increase older adults' motivation to exercise in Fit Islands.

The Fit Islands system contains a rich health knowledge base. We have collected knowledge on diet, exercise, living habits, self-monitoring, disease prevention, and psychology, such as: knowledge of nutrients and proper dietary combinations [96,97], different types of exercise and exercise precautions [98], how to maintain good sleep [99], how to prevent cardiovascular and cerebral vascular diseases [100,101], and how to maintain a good state of mind and self-psychological guidance [102].

The presentation of this health knowledge needs to be tailored to the target audience of older adults. When designing applications specifically for older adults, it is important to use clear, objective, and educational language. Using a condescending tone or complex language will only hinder the efficiency of game interaction and increase older adults' fear of digital technology. Therefore, when presenting health knowledge to older adults, it is important not to excessively emphasize the authority of the knowledge but to consider ways to make it easier and quicker for them to comprehend the information. For example, if the health knowledge aims to convey information related to diet, as shown in Figure 6, the image on the left may appear too authoritative and monotonous. The image on the right uses a friendly tone and presents the health knowledge through visuals and audio.

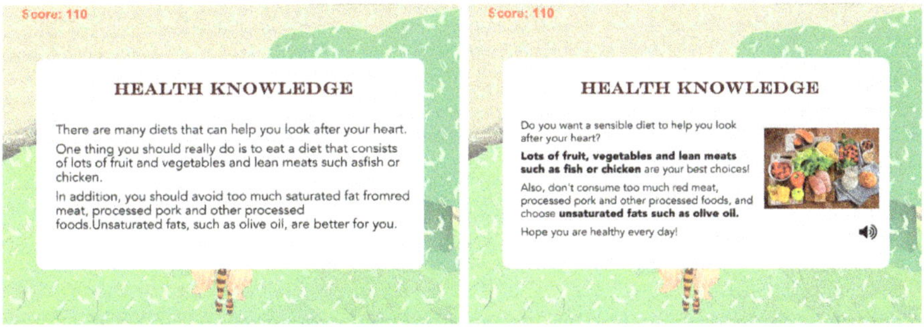

Figure 6. Comparison of two presentations of health knowledge.

4.3.4. Business Strategy

Fit Islands has added a product point exchange and membership system, where points can be earned through farm cultivation, sports games, cognitive games, and interaction with friends. These points can be exchanged in the shop for virtual game character skins as well as some real elderly goods, such as electronic blood pressure monitors, etc. As a free digital product targeting older adults, appropriate adverts or videos are played at the end of the game. This will attract brands, communities, nursing homes, etc. with the same target audience to place commercial ads or offer real products. Using this advertising revenue wisely, the system could be further improved and upgraded to attract more older adults to play Fit Islands. Fit Islands will also strictly control the quality of these commercial products and the correct orientation of the advertising content. The products will be delivered to the seniors by post, and Fit Islands will have an agreement with the brand to keep the address of the user private to ensure the security of the seniors' information. To adapt to different scenarios in smart cities, Fit Islands offers three levels of membership for different tiers of players (see Figure 7).

4.4. Plan the UI and Visuals

The design of UI for older adults needs to take into account their habits and preferences in terms of style and aesthetics, to compensate for actual physical and cognitive changes. Fit Islands has been designed to meet the design principles of Section 2. For example, ensuring that the UI has both large and easily viewable key elements, using bold and age-friendly colors [103], having an adaptable interface and settings, incorporating dynamic resources such as images and voice and multimodal feedback, and using blank areas to emphasize key points (see Figure 8). After entering the game map, the game not only gives players full freedom, but also creates a natural scene through ambient sounds and dynamic effects, so players will have an immersive experience of running freely and exploring everywhere on the island.

Figure 7. Fit Islands' gaming membership interface.

(a) (b) (c)

(d) (e) (f)

Figure 8. *Cont.*

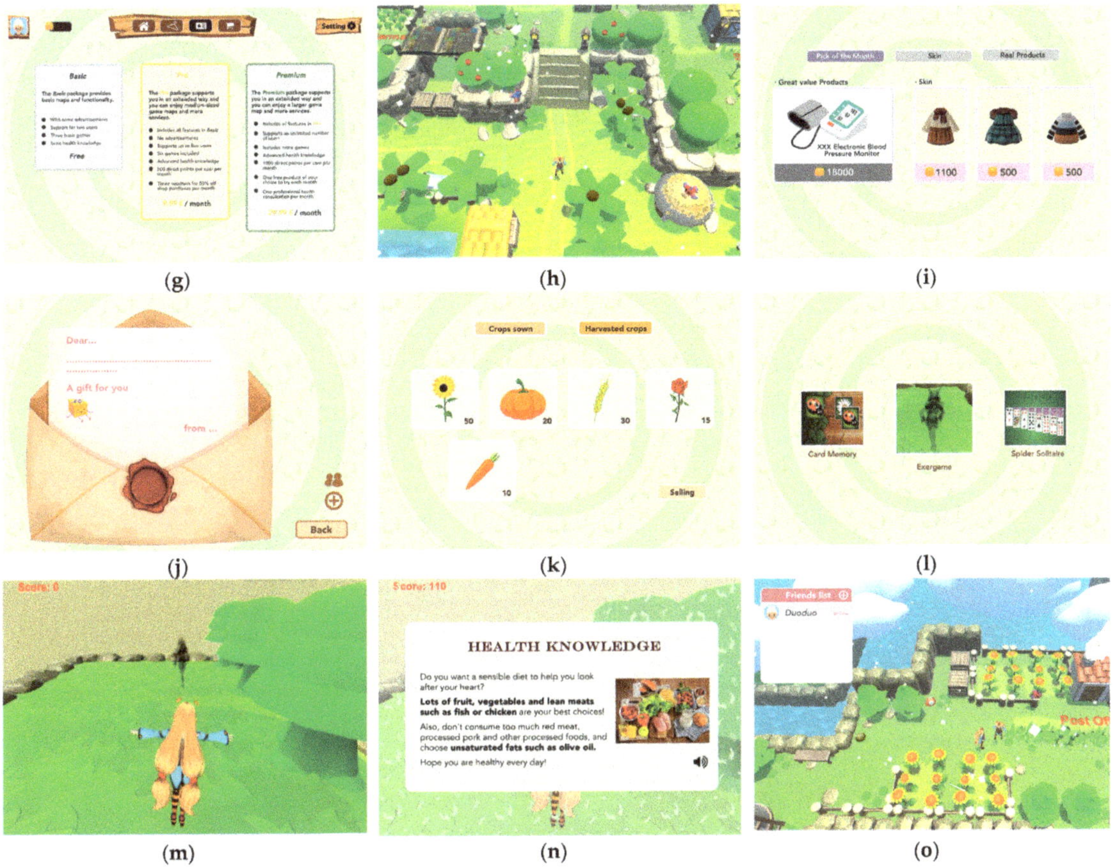

Figure 8. The main interface in the demonstration. (**a**) Initial interface; (**b**) game description; (**c**) user registration; (**d**) interface for selecting gender; (**e**) main interface; (**f**) settings interface; (**g**) membership interface; (**h**) game map; (**i**) shop interface; (**j**) post office interface; (**k**) farm interface; (**l**) game center; (**m**) Exergame interface; (**n**) health knowledge interface; (**o**) multiplayer interface.

4.5. Consider the Gameplay Interaction

In order to cater for older people who are illiterate and slow to process information, Fit Islands employs an audio-visual sensory experience to simplify data and information feedback, making it easier for older people to take in different information through voice-reading and pictures. Voice recognition powered by existing voice input methods has been added so that in the Post Office players can use speech to draft and send "letters" to their friends. The automatic matching of game content based on questionnaire information mentioned in Section 4.3 is also an optimization of age-appropriate game interaction. In terms of operation, Fit Islands fully avoids complex operations and gestures. The interface switching throughout the game mainly relies on direct mouse clicks, while the motion game is performed through motion capture. Motion capture in Fit Islands avoids any wearable devices, and the game can be played simply by standing in front of a computer with a camera. Therefore, it can avoid the possibility that older people may resist or even be afraid of the game because of poor interaction experiences. Fit Islands also adds online interaction, with the aim of creating a platform for older people to connect with each other, breaking down the isolation caused by geography, beliefs, and perceptions of older people. Once players are online, they will be able to see each other's images on their respective islands. By clicking on each other's characters, they can engage in dialogue and perform actions such as shaking hands. Players can play against each other in cognitive games or complete challenges in two-player mode in Exergame.

4.6. Develop Demonstration

The development of the Fit Islands demonstration consisted of interface design, a motion capture program, motion pose image recognition, and the creation of a 3D game world. For the motion capture part, the demonstration uses the ThreeD Pose Tracker algorithm [104]. This algorithm has several advantages: it does not require any wearable device, it can capture the upper body alone, it does not require much hardware, and it is very fast to process real-time data. However, it also has some disadvantages, such as the need for the player to stand completely within camera range to be recognized, and the lack of accuracy in motion capture. Motion pose recognition uses a skeleton recognition algorithm to identify imported motion pose images and display these skeletons in the image of the game character [105]. The 3D game world for the Fit Islands demonstration was created using an open-source resource called 'Top down island' from the Unity shop [106]. The game is played from a bird's eye view and the character is controlled by running around with a right click of the mouse.

4.7. Test and Iterate

The initial testing of the demonstration for Fit Islands was split into two phases. In the first phase, a 62-year-old German woman (test subject A) was invited to test the functionality and operation of the game after the initial demonstration was completed. After receiving feedback, the initial game demonstration was refined. The second stage was to invite nine Chinese older adults (test subjects B–J) to test the functionality and usability of the game after the completion of the refined game demonstration. The nine seniors, aged 51–77, included three women and six men. These older adults were asked to operate and use the game and to provide feedback on the corresponding issues (see Figure 9).

Figure 9. Field photos of nine older Chinese adult test subjects (test subjects B–J).

Once individuals' data is involved, protecting their data privacy becomes critical. Best practice for protecting individual privacy in the test was applied as follows. First and foremost, the test was conducted in a fully anonymous manner, meaning that information such as names, birth dates, addresses, and resident identity card numbers, were not recorded. This approach excludes any possibility of identifying any individual test participant. Furthermore, in order to participate in the test, all the participants were informed of the transparent purpose of the study and needed to give consent to provide their basic demographic information such as age, gender, province of residence, and level of education. These measures also go in line with the Guide on Good Data Protection Practice in Research [107].

The initial demonstration design for the first phase involved an older adult (test subject A) living alone in a retirement flat in Munich, Germany. After being introducing to the characters and the game, she showed great interest. The test showed that the initial demonstration still contained many problems. Because the motion recognition needed to recognize the distance from head to toe, A had to stand far away, and was not able to see the action on the screen. Scoring was also difficult, as performance was only scored if done to an exceptional standard. However, A said she liked the colors of the interface in Fit Islands and the health knowledge provided after the game.

After debugging and refining the demonstration based on A's feedback, nine older Chinese adults were invited to test Fit Islands further. Five of them, B–F, were tested in the computer room of a community center in Yancheng, Jiangsu Province, China; G and H were tested in the activity room of a community center in Yancheng, Jiangsu Province, China; and the other two, I and J, were tested in their homes. The results of the tests are shown in Table 1.

From the results of these tests, it can be concluded that the expandable Fit Islands is an adaptable, affordable, age-friendly, practical, and enjoyable Exergame system that contains a wide range of functions (see Figure 10). Specifically, the following conclusions can be drawn:

- This system can effectively motivate older Chinese people to exercise and increase their amount of physical activity, thus becoming more energetic. The initial test participants were able to complete the exercises relatively easily, and the music and scoring format allowed them to concentrate on the exercises.
- This system provides a free platform for Chinese seniors to communicate and play multiplayer games. The affordability of gaming device requirements for this system lowers the barrier for older adults to use it. This anonymous and fun way can be used as a complement to family pensions, allowing older adults to socialize more with people of their age, mentally reducing over-dependence on their children and alleviating feelings of social isolation and loneliness. Several older adults in the test expressed their enjoyment of the post office function and their expectations of the online multiplayer function.
- The age-friendly and inclusive design of this system can attract older adults to actively use the technology. Most of the older adults responded positively to the game's functions and interface, found the game relatively easy to operate and use, and expressed acceptance of the game's format. They all expressed a desire to use the game in the future.
- The system is very adaptable to different devices and different building environments. The adaptability of Fit Islands was demonstrated using different equipment in different test locations; for example, desktop computers were used in the computer room, while laptops were used in the activity room and the testers' homes. The system's Exergame is not too strict in terms of site requirements and older adults can play the game in different building scenarios using existing equipment.

Table 1. Information and feedback for the initial test of the demonstration.

	Age	Gender	Country of Residence	Occupation	Retired or Not	Living Condition	Evaluation of the Game Content and Features	Evaluation of the Game Interface	Whether the Game Will Improve Your Health	Whether You Want to Play This Game in the Future
A	62	Female	Germany	Housing Management	Yes	Living alone in nursing home	In addition to the exercise function, I think the health knowledge function is very meaningful.	I liked the overall style and color scheme of the game, which gave it a very warm and welcoming feel.	Yes	Yes
B	76	Male	China	Freelance	Yes	Living with his partner	I've never played a game like this before and I think the point system is great	The words are big and the interface is easy to use	Yes	Yes
C	64	Female	China	Junior school teacher	Yes	Living with her son	I think some of the movements are a bit confusing, the way the shadows are maybe not as good as the colored form	It's a novel approach	Yes	Yes
D	53	Female	China	Community staff	No	Living with her partner	Very useful game, perfect for a job like mine that requires a lot of sitting. The news feature is very helpful to me.	The game's interface looks comfortable and the buttons are large.	Yes	Yes
E	68	Male	China	Engineer	No	Living alone in a 60 sqm apartment	Games that allow people to exercise make me feel healthy, and the movements are not too difficult.	Very good	Yes	Very enjoyable
F	51	Male	China	PE teacher	No	Living with his partner	The game characters are rather monotonous, and it would be nice to have the freedom to decide the gender and image of the game characters.	The map interface is very realistic and I like the feeling of exploration.	Yes	Yes
G	56	Male	China	Community staff	No	Living with his partner	The music and rhythm of the game is very good and I will try my best to complete each move to get a higher score.	The game characters are cute	Yes	Yes
H	66	Male	China	Freelance	Yes	Living alone	I'm interested in being able to exchange real products, which will motivate me to complete the game more and better.	Everything else is fine, except that the game character's hair covers up the action a bit.	Yes	Yes
I	75	Female	China	Housewife	/	Living with her partner in a 90 sqm apartment	I can't type, so writing letters is a bit difficult for me	I would have been afraid that it would be difficult to play the game because I can't read, but the game uses more pictures, audio and voice recognition, which has helped me a lot.	Yes	Yes
J	77	Male	China	Civil Servant	Yes	Living with his partner in a 90 sqm apartment	It's good that my partner and I can take turns playing the game and we have our own progress	I like the feeling of being able to control the game character and run around the island.	Yes	Yes, I think the health knowledge at the end of the game will be the motivation to keep using this game for me

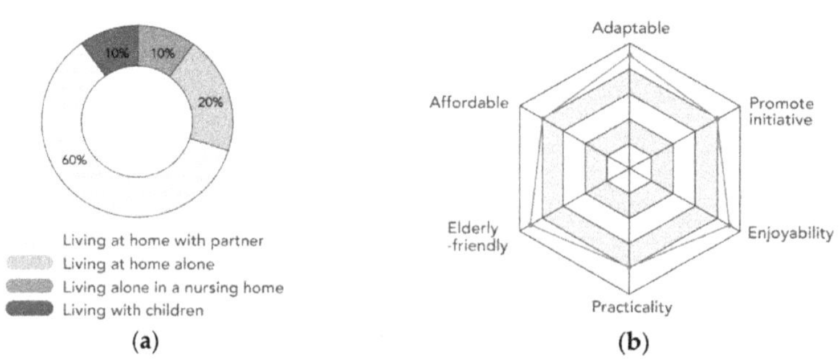

Figure 10. Analysis of test results. (**a**) Residency (**b**) Evaluation scores of the game.

5. Limitations and Future Work

Fit Islands' demonstration has implemented most of the game's design and functionality. Further development of the demonstration will allow for more accurate motion capture, refinement of the entire game's library of motion levels, and the implementation of online connectivity. Further testing and evaluation will require larger numbers of participants and long-term follow-up observations to ensure the long-term sustainability and effectiveness of Fit Islands. Given the unique nature of older adults, it is necessary to further refine and evaluate the efficacy and safety of Exergame's specific movements to meet the specific needs and preferences of older adults in different countries and regions.

After further refinement of the game demonstration, several older adults with experience of digital technology will be invited for further testing ($n \geq 30$). A uniform site will be chosen and the test subjects will be brought together at that site for testing. The future test will be divided into four main phases. The first phase of the test will be the distribution of information about Fit Islands' research objectives and planned tasks to the test participants, who will be introduced to the Exergames equipment and procedures. The second phase will divide the test subjects into a control group ($n \geq 10$) and an intervention group ($n \geq 20$). The physical and psychological conditions of the two groups will be assessed separately to ensure that the subjects are not statistically significantly different. There are many methods of testing specifically for older adults. For example, a research team from Tianjin Medical University used data from the China Health and Retirement Longitudinal Study (CHARLS) to assess the functional status of older adults using the Activities of Daily Living Scale (ADL) and the Instrumental Activities of Daily Living Scale (IADL) [108]. A condensed version of the Social Network Scale developed by Lubben et al. (Lubben Social Network Scale 6, LSNS-6) can be used to assess the social isolation of older adults [109]. Mental health status can be tested using the UCLA Loneliness Scale or subjective questions [110,111]. The third phase will have the intervention group ($n \geq 20$) play the game three times a week for five weeks, either at home or at this uniform venue of their choice. Various parameters will be recorded during the game, including the time of entry into the game, the time of user creation, the time of exercise, the score of the game and whether the participant completed the game in its entirety or exited early, etc. The control group ($n \geq 10$) will not receive any interventions and will live their lives according to their usual habits. In the fourth phase, the physical and psychological well-being of the two groups will be assessed again after five weeks in the same way as in the second phase. A comparison of the data from the control group ($n \geq 10$) and the intervention group ($n \geq 20$) in Phase 4 will provide evidence to address the hypothesis (see Figure 11). As mentioned in Section 4.1.1, Chinese older adults' adaptation to digital technology shows polarization. In order to verify Fit Islands' elderly-friendly and adaptable characteristics, some older adults with no experience in digital technology will also be invited to conduct the above test ($n \geq 30$), thus providing data from control group 2 ($n \geq 10$) and intervention 2 ($n \geq 20$). By comparing the four groups of data, evidence can be provided to support a better promotion of Fit Islands.

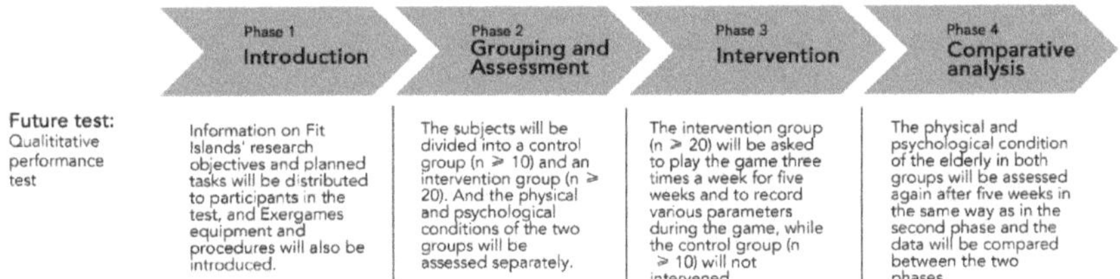

Figure 11. Initial and planned future tests.

After the validation, the project team plans to initiate an internet-based start-up to further promote the project and its products to as many people as possible through multiple channels in order to make a real impact. Furthermore, the start-up (i.e., online store) can potentially create an economy around elderly-focused educational and health games, where profits can in turn support future product iterations and new product development, and potentially create new jobs around the business (e.g., customer service, manufacturing, etc.).

6. Conclusions

Aging will inevitably become an important feature of the population in the coming decades. Compared with developed countries, which have already entered into aging societies, emerging economies need to address their own unique challenges due to cultural, economic and demographic differences. Based on the experiences of developed economies, emerging economies need to consider their own realities and implement innovative solutions within a limited time to ensure the well-being and inclusiveness of their aging populations. Smart city development cannot ignore older adults as a group. Exergame is a technology that allows older people to engage in regular physical activity while having fun in their daily lives. It improves the quality of life and health of older adults by actively increasing physical activity, exercising cognition, and improving cardiovascular health, muscle strength, flexibility, coordination, and balance.

Fit Islands strictly follows the generic development methodology presented in Section 3. In the context of China's rapidly aging population, Fit Islands is an attempt to improve the well-being of older adults, and is a very promising virtual world for smart aging in smart cities. The main characteristics are summarized as follows:

- Fit Islands is fun, practical, inclusive, and age-friendly. By providing a free digital platform for the older population, a special group, it allows older adults to be motivated to be physically active, make friends and learn through games and sports, rather than being passively assisted by some assistive devices.
- Rather than a single game design, Fit Islands aims to create an expandable game world for older adults. Using a 3D virtual community format, Fit Islands combines socializing, exercise, and cognitive training. Different membership levels also allow for larger island maps and a wider variety of games. In addition to the lively game interface and immersive game interaction experience, it also allows players to learn various aspects of health knowledge. Fit Islands uses an individual player archive, where each player has his/her own virtual community and game progress, increasing the player's sense of personal belonging. Fit Islands provides a place for geographically distant seniors and their friends to socialize. Especially during the COVID-19 epidemic, it allowed older adults to overcome spatial isolation by providing contact with their friends and loved ones, and to exercise and work on cognitive awareness in order to stay or become healthier.
- Fit Islands takes full account of age-related design requirements and inclusive design principles. The interface, content, and format of the game are designed based on the characteristics and preferences of older persons. The game's provision of text, pictures, and audio is intended to be inclusive of the literacy levels, language, and visual and auditory abilities of different older adults. Exergame also takes into account the physical condition of older people by providing both standing and sitting modes. In terms of operation, Fit Islands is played with no external devices to minimize the cost of the game, simplify the game, and make it less difficult for older people to use. The most basic computer equipment with a camera is sufficient to play the game.

Furthermore, due to its advantages (such as high flexibility, affordability, and ease of use), the proposed system can be easily integrated into larger smart city frameworks such as Dynamic Vertical Urbanism [112] to generate a comprehensive smart city experience for the aging population and beyond. With the problem of aging increasing in emerging economies, this study not only provides guidance for policy makers, developers, and researchers in emerging economies, but also suggests new possibilities for further promoting smart aging, and facilitating the development of smart cities.

Author Contributions: Conceptualization, Z.S., R.H. and T.B.; methodology, R.H.; software, Z.S.; validation, Z.S. and D.W.; formal analysis, Z.S. and R.H.; investigation, Z.S.; resources, R.H. and T.B.; data curation, D.W.; writing—original draft preparation, Z.S.; writing—review and editing, R.H. and Z.S.; visualization, Z.S. and R.H.; supervision, R.H. and T.B.; project administration, T.B. All authors have read and agreed to the published version of the manuscript.

Funding: This research received no external funding.

Informed Consent Statement: Informed consent was obtained from all subjects involved in the study.

Data Availability Statement: Data are contained within the article.

Acknowledgments: The authors are grateful to the older adults who participated in the prototype testing. Zixin Shen and Rongbo Hu contributed equally to this manuscript.

Conflicts of Interest: Authors Rongbo Hu and Thomas Bock were employed by the company CREDO Robotics GmbH. Author Dong Wan was employed by the company Sinoma International Engineering Co., Ltd. The remaining authors declare that the research was conducted in the absence of any commercial or financial relationships that could be construed as a potential conflict of interest.

References

1. United Nations. World Population Prospects: The 2017 Revision, Key Findings and Advance Tables. Available online: https://population.un.org/wpp/Publications/Files/WPP2017_KeyFindings.pdf (accessed on 26 December 2022).
2. United Nations. *World Population Aging, 2017 Highlights*; United Nations: New York, NY, USA, 2017; ISBN 978-92-1-151551-0.
3. Schwarz, A. *Old Age Security and Social Pensions*; Working Paper; World Bank Human Development (Social Protection) Hub: Washington, DC, USA, 2003.
4. Gietel-Basten, S.; Scherbov, S.; Sanderson, W. Towards a reconceptualising of population aging in emerging markets. *Populationyearbook* **2017**, *1*, 41–66. [CrossRef]
5. Hoskisson, R.E.; Eden, L.; Lau, C.M.; Wright, M. Strategy in Emerging Economies. *Acad. Manag. J.* **2017**, *43*, 249–267. [CrossRef]
6. Green Templeton College. *Aging in Emerging Markets*; Green Templeton College: Oxford, UK, 2015.
7. Shetty, P. Grey Matter: Aging in Developing Countries. *Lancet* **2012**, *379*, 1285–1287. [CrossRef]
8. World Health Organization. *The Global Burden of Disease: 2004 Update*; World Health Organization: Geneva, Switzerland, 2008.
9. Lebrasseur, A.; Fortin-Bédard, N.; Lettre, J.; Raymond, E.; Bussières, E.-L.; Lapierre, N.; Faieta, J.; Vincent, C.; Duchesne, L.; Ouellet, M.-C.; et al. Impact of the COVID-19 Pandemic on Older Adults: Rapid Review. *JMIR Aging* **2021**, *4*, e26474. [CrossRef] [PubMed]
10. Shahid, Z.; Kalayanamitra, R.; McClafferty, B.; Kepko, D.; Ramgobin, D.; Patel, R.; Aggarwal, C.S.; Vunnam, R.; Sahu, N.; Bhatt, D.; et al. COVID-19 and Older Adults: What We Know. *J. Am. Geriatr. Soc.* **2020**, *68*, 926–929. [CrossRef]
11. Peeri, N.C.; Shrestha, N.; Rahman, M.S.; Zaki, R.; Tan, Z.; Bibi, S.; Baghbanzadeh, M.; Aghamohammadi, N.; Zhang, W.; Haque, U. The SARS, MERS and Novel Coronavirus (COVID-19) Epidemics, the Newest and Biggest Global Health Threats: What Lessons Have We Learned? *Int. J. Epidemiol.* **2020**, *49*, 717–726. [CrossRef] [PubMed]
12. Cacioppo, J.T.; Hawkley, L.C.; Thisted, R.A. Perceived Social Isolation Makes Me Sad: 5-Year Cross-Lagged Analyses of Loneliness and Depressive Symptomatology in the Chicago Health, Aging, and Social Relations Study. *Psychol. Aging* **2010**, *25*, 453–463. [CrossRef] [PubMed]
13. Luanaigh, C.Ó.; Lawlor, B.A. Loneliness and the Health of Older People. *Int. J. Geriatr. Psychiatry* **2008**, *23*, 1213–1221. [CrossRef]
14. Steptoe, A.; Owen, N.; Kunz-Ebrecht, S.R.; Brydon, L. Loneliness and Neuroendocrine, Cardiovascular, and Inflammatory Stress Responses in Middle-Aged Men and Women. *Psychoneuroendocrinology* **2004**, *29*, 593–611. [CrossRef]
15. Skouby, K.E.; Kivimäki, A.; Haukipuro, L.; Lynggaard, P.; Windekilde, I. Smart Cities and the Aging Population. In Proceedings of the 32nd Meeting of WWRF 2014, Marrakech, Morocco, 20–22 May 2014.
16. Zhu, H. Smart Aging: Innovations and Reflections on China's Elderly Care Model. *J. Soc. Sci. Hunan Norm. Univ.* **2016**, *45*, 68–73.
17. Larsen, L.H.; Schou, L.; Lund, H.H.; Langberg, H. The Physical Effect of Exergames in Healthy Elderly—A Systematic Review. *Games Health* **2013**, *2*, 205–212. [CrossRef] [PubMed]
18. Chen, A.; Yin, H.; Yan, J. The Relationship between Physical Activity and Well-Being in Older Adults: The Mediating Role of Loneliness. *China Sports Technol.* **2010**, 135–139. Available online: http://www.cqvip.com/qk/83653x/201001/32632328.html (accessed on 27 December 2022).
19. Li, J. *A Study of the Effects of Physical Activity on the Mental Health, Well-Being and Quality of Life of Older People and Its Associated Influencing Factors*; Soochow University: Suzhou, China, 2005. [CrossRef]
20. Zhao, Y. A Study on the Effect of Physical Exercise on Loneliness in Middle-Aged and Elderly People. *Sci. Educ.* **2009**, 287–288. Available online: http://www.cqvip.com/qk/87947x/200903/29258430.html (accessed on 14 January 2024).
21. Gil-Gómez, J.-A.; Lloréns, R.; Alcañiz, M.; Colomer, C. Effectiveness of a Wii Balance Board-Based System (eBaViR) for Balance Rehabilitation: A Pilot Randomized Clinical Trial in Patients with Acquired Brain Injury. *J. Neuroeng. Rehabil.* **2011**, *8*, 30. [CrossRef]

22. Kosse, N.M.; Caljouw, S.R.; Vuijk, P.-J.; Lamoth, C.J. Exergaming: Interactive Balance Training in Healthy Community-Dwelling Older Adults. *J. Cyber Ther. Rehabil.* **2011**, *4*, 399–407.
23. Saposnik, G.; Teasell, R.; Mamdani, M.; Hall, J.; McIlroy, W.; Cheung, D.; Thorpe, K.E.; Cohen, L.G.; Bayley, M. Effectiveness of Virtual Reality Using Wii Gaming Technology in Stroke Rehabilitation: A Pilot Randomized Clinical Trial and Proof of Principle. *Stroke* **2010**, *41*, 1477–1484. [CrossRef] [PubMed]
24. Smith, S.T.; Sherrington, C.; Studenski, S.; Schoene, D.; Lord, S.R. A Novel Dance Dance Revolution (DDR) System for in-Home Training of Stepping Ability: Basic Parameters of System Use by Older Adults. *Br. J. Sports Med.* **2011**, *45*, 441–445. [CrossRef]
25. Studenski, S.; Perera, S.; Hile, E.; Keller, V.; Spadola-Bogard, J.; Garcia, J. Interactive Video Dance Games for Healthy Older Adults. *J. Nutr. Health Aging* **2010**, *14*, 850–852. [CrossRef]
26. Yavuzer, G.; Senel, A.; Atay, M.B.; Stam, H.J. "Playstation Eyetoy Games" Improve Upper Extremity-Related Motor Functioning in Subacute Stroke: A Randomized Controlled Clinical Trial. *Eur. J. Phys. Rehabil. Med.* **2008**, *44*, 237–244.
27. Dance Dance Revolution (1998 Video Game). Wikipedia. 2022. Available online: https://en.wikipedia.org/w/index.php?title=Dance_Dance_Revolution_(1998_video_game)&oldid=1123044439 (accessed on 27 December 2022).
28. Hämäläinen, P.; Ilmonen, T.; Höysniemi, J.; Lindholm, M.; Nykänen, A. Martial Arts in Artificial Reality. In Proceedings of the SIGCHI Conference on Human Factors in Computing Systems, Portland, OR, USA, 2–7 April 2005; pp. 781–790.
29. Strömberg, H.; Väätänen, A.; Räty, V.-P. A Group Game Played in Interactive Virtual Space: Design and Evaluation. In Proceedings of the 4th Conference on Designing Interactive Systems: Processes, Practices, Methods, and Techniques, London, UK, 25–28 June 2002; pp. 56–63.
30. Gerling, K.M.; Schild, J.; Masuch, M. Exergame Design for Elderly Users: The Case Study of SilverBalance. In Proceedings of the 7th International Conference on Advances in Computer Entertainment Technology, Taipei, Taiwan, 17–19 November 2010; pp. 66–69.
31. Chartomatsidis, M.; Goumopoulos, C. A Balance Training Game Tool for Seniors Using Microsoft Kinect and 3D Worlds. In Proceedings of the ICT4AWE, Crete, Greece, 2–4 May 2019; pp. 135–145.
32. Overview SilverFit 3D—SilverFit. Available online: https://www.silverfit.com/en/products/silverfit-3d-camera/817-silverfit-3d-overview (accessed on 27 December 2022).
33. SilverFit 3D Interactive Rehabilitation Training System. Available online: https://gies.hk/zh-hk/expo/exhibition-products/detail/1223 (accessed on 27 December 2022).
34. SilverFit Literature and Research Review. Available online: https://docplayer.nl/203504175-Silverfit-literature-and-research-review.html (accessed on 27 December 2022).
35. Hu, R.; Linner, T.; Trummer, J.; Güttler, J.; Kabouteh, A.; Langosch, K.; Bock, T. Developing a Smart Home Solution Based on Personalized Intelligent Interior Units to Promote Activity and Customized Healthcare for Aging Society. *J. Popul. Aging* **2020**, *13*, 257–280. [CrossRef]
36. Czaja, S.J.; Lee, C.C. Information Technology and Older Adults. In *Human-Computer Interaction*; CRC Press: Boca Raton, FL, USA, 2009; pp. 35–50, ISBN 0-429-13938-1.
37. Weisman, S. Computer Games for the Frail Elderly. *Gerontologist* **1983**, *23*, 361–363. [CrossRef] [PubMed]
38. Weisman, S. Computer Games for the Frail Elderly. *Comput. Hum. Serv.* **1995**, *11*, 229–234. Available online: https://www.tandfonline.com/doi/abs/10.1300/J407v11n01_21 (accessed on 27 December 2022). [CrossRef] [PubMed]
39. Ijsselsteijn, W.; Nap, H.H.; de Kort, Y.; Poels, K. Digital Game Design for Elderly Users. In Proceedings of the 2007 Conference on Future Play, Toronto, ON, Canada, 15–17 November 2007; pp. 17–22.
40. Flores, E.; Tobon, G.; Cavallaro, E.; Cavallaro, F.I.; Perry, J.C.; Keller, T. Improving Patient Motivation in Game Development for Motor Deficit Rehabilitation. In Proceedings of the 2008 International Conference on Advances in Computer Entertainment Technology, Yokohama, Japan, 3–5 December 2008; pp. 381–384.
41. China Suffers Aging Population Nearly 40 Years after Introduction of One-Child Policy. Available online: https://www.worldfinance.com/strategy/china-suffers-aging-population-nearly-40-years-after-introduction-of-one-child-policy (accessed on 27 December 2022).
42. Mao, G.; Lu, F.; Fan, X.; Wu, D. China's Aging Population: The Present Situation and Prospects. In *Population Change and Impacts in Asia and the Pacific*; Poot, J., Roskruge, M., Eds.; New Frontiers in Regional Science: Asian Perspectives; Springer: Singapore, 2020; pp. 269–287, ISBN 978-981-10-0230-4.
43. Woo, J.; Kwok, T.; Sze, F.; Yuan, H. Aging in China: Health and Social Consequences and Responses. *Int. J. Epidemiol.* **2002**, *31*, 772–775. [CrossRef] [PubMed]
44. Law of the People's Republic of China on the Protection of the Rights and Interests of the Elderly. Available online: http://www.npc.gov.cn/npc/c30834/201901/47231a5b9cf94527a4a995bd5ae827f0.shtml (accessed on 16 February 2023).
45. National Bureau of Statistics. National Bureau of Statistics of China Bulletin of the Seventh National Census (No.5). Available online: http://www.stats.gov.cn/ztjc/zdtjgz/zgrkpc/dqcrkpc/ggl/202105/t20210519_1817698.html (accessed on 26 December 2022).
46. Zhang, A.; Master, F. China's First Population Drop in Six Decades Sounds Alarm on Demographic Crisis. Reuters. Available online: https://www.reuters.com/world/china/chinas-population-shrinks-first-time-since-1961-2023-01-17/ (accessed on 18 February 2023).

47. Does China Have an Aging Problem? ChinaPower Project. 2016. Available online: https://chinapower.csis.org/aging-problem/ (accessed on 27 December 2022).
48. Ren Zeping's Team China Aging Research Report 2022. Available online: https://www.caoss.org.cn/UploadFile/pic/20229281791192316.pdf (accessed on 26 December 2022).
49. Banister, J.; Bloom, D.E.; Rosenberg, L. Population Aging and Economic Growth in China. In *The Chinese Economy*; Aoki, M., Wu, J., Eds.; Palgrave Macmillan: London, UK, 2012; pp. 114–149, ISBN 978-1-137-03428-1.
50. China Could Scrap Two-Child Policy, Ending Nearly 40 Years of Limits. China. The Guardian. Available online: https://www.theguardian.com/world/2018/aug/28/china-could-scrap-two-child-policy-ending-nearly-40-years-of-limits (accessed on 27 December 2022).
51. Lv, X.-L.; Jiang, Y.-H.; Sun, Y.-H.; Ren, C.-Z.; Sun, C.-Y.; Sun, L.; Wu, Z.-Q.; Zhao, X. Short Form 36-Item Health Survey Test Result on the Empty Nest Elderly in China: A Meta-Analysis. *Arch. Gerontol. Geriatr.* **2013**, *56*, 291–297. [CrossRef] [PubMed]
52. Moran, A.; Zhao, D.; Gu, D.; Coxson, P.; Chen, C.-S.; Cheng, J.; Liu, J.; He, J.; Goldman, L. The Future Impact of Population Growth and Aging on Coronary Heart Disease in China: Projections from the Coronary Heart Disease Policy Model-China. *BMC Public Health* **2008**, *8*, 394. [CrossRef] [PubMed]
53. Su, D.; Wu, X.-N.; Zhang, Y.-X.; Li, H.-P.; Wang, W.-L.; Zhang, J.-P.; Zhou, L.-S. Depression and Social Support between China' Rural and Urban Empty-Nest Elderly. *Arch. Gerontol. Geriatr.* **2012**, *55*, 564–569. [CrossRef] [PubMed]
54. China Internet Network Information Center (CNNIC). The 46th China Statistical Report on Internet Development. Office of the Central Leading Group for Cyberspace Affairs. 2020. Available online: https://www.cnnic.com.cn/IDR/ReportDownloads/202012/P020201201530023411644.pdf (accessed on 27 December 2023).
55. China Sample Survey on Living Conditions of the Elderly in Urban and Rural Areas Data Brief. 2015. Available online: http://www.crca.cn/index.php/19-life/27-2015.html (accessed on 27 December 2022).
56. Fang, Y.; Wang, H.; Ouyang, Z. An Overview Report on the Situation of Elderly People in Urban and Rural China: 2000–2015. *China Popul. Dev. Stud.* **2019**, *2*, 323–345. [CrossRef]
57. The "New Older Adults": The Changing Face of Retirement after 60. Available online: http://www.crca.cn/index.php/13-agednews/711-60.html (accessed on 27 December 2022).
58. 2019 Silver-Haired Digital Users' Entertainment Behaviour Analysis. Available online: https://www.analysys.cn/article/detail/20019305 (accessed on 27 December 2022).
59. Cotten, S.R.; Anderson, W.A.; McCullough, B.M. Impact of Internet Use on Loneliness and Contact with Others among Older Adults: Cross-Sectional Analysis. *J. Med. Internet Res.* **2013**, *15*, e2306. [CrossRef]
60. Hu, R.; Linner, T.; Wang, S.; Cheng, W.; Liu, X.S.; Güttler, J.; Zhao, C.; Lu, Y.; Bock, T. Towards a Distributed Intelligent Home Based on Smart Furniture for China's Aging Population: A Survey. In Proceedings of the 38th International Symposium on Automation and Robotics in Construction (ISARC), Dubai, United Arab Emirates, 2–4 November 2021; pp. 41–48. [CrossRef]
61. Benesch, K. Causes of Social Isolation in Elderly Adults. Available online: https://www.griswoldhomecare.com/blog/2020/march/causes-of-social-isolation-in-elderly-adults/ (accessed on 27 December 2022).
62. Guo, J.; Xu, Y.; Chen, S.; Zhu, L. Study on the Influence of Internet Use on Loneliness of the Elderly: The Moderating Role of Age. *Chin. J. Health Policy* **2021**, *14*, 29–36. [CrossRef]
63. World Health Organization. *China Country Assessment Report on Aging and Health*; World Health Organization: Geneva, Switzerland, 2015.
64. Transcript of the NHS Commission's Special Press Conference on 1 November 2019. Available online: http://www.nhc.gov.cn/xcs/s7847/201911/b01a5ca22bef4570ab44605c12940f97.shtml (accessed on 16 February 2023).
65. Li, Z.; Zhou, L. The Epidemiologic Studies on Loneliness among Elderly: A Review. *Chin. Ment. Health J.* **2012**, *26*, 658–663.
66. de Jong-Gierveld, J. Developing and Testing a Model of Loneliness. *J. Personal. Soc. Psychol.* **1987**, *53*, 119. [CrossRef] [PubMed]
67. Li, Y. A Perspective on Health Care for the Elderly Who Lose Their Only Child in China. *Scand. J. Public Health* **2013**, *41*, 550–552. [CrossRef] [PubMed]
68. Qiang, L.I.; Gang, X.U.; Zhen, Z. Loneliness and Its Correlates Among the Aged People Living Alone in Urban Areas. *J. East China Norm. Univ. (Philos. Soc. Sci.)* **2019**, *51*, 160. [CrossRef]
69. The Current Situation of Aging in Place in the Community: Over 60% of the Elderly Live Alone. Available online: https://www.pishu.cn/zxzx/xwdt/582639.shtml (accessed on 27 December 2022).
70. Ministry of Civil Affairs: The Proportion of Empty Nesters among China's Elderly Population Has Exceeded Half, with over 70% in Some Big Cities and Rural Areas. Available online: https://www.caoss.org.cn/1article.asp?id=6818 (accessed on 27 December 2022).
71. China National Statistical Bureau. Seventh National Population Census Bulletin. Available online: http://www.gov.cn/guoqing/2021-05/13/content_5606149.htm (accessed on 27 December 2022).
72. Chen, Y.; Hicks, A.; While, A.E. Loneliness and Social Support of Older People in C Hina: A Systematic Literature Review. *Health Soc. Care Community* **2014**, *22*, 113–123. [CrossRef] [PubMed]
73. 5% of Chinese Seniors Regularly Go Online More than 80% Watch TV or Listen to Radio. Available online: http://world.people.com.cn/n1/2016/1010/c57506-28765430.html (accessed on 27 December 2022).
74. Wang, X.M.; Luo, W.J.; Chen, X.Y. Influence of Square Dance on Physical and Psychological Health of the Elderly. *Chin. J. Gerontol.* **2014**, *34*, 477–478.

75. Li, B. Diversification and Characteristics: The Living Arrangements and Housing of the Elderly in China. *Chin. J. Popul. Sci.* **2010**, *30*, 101–110. Available online: http://www.zgrkkx.com/Files/ywfxzz/MagazinePDF/1534.pdf (accessed on 27 December 2022). [CrossRef]
76. National Aging Development Bulletin. 2020. Available online: http://www.nhc.gov.cn/cms-search/xxgk/getManuscriptXxgk.htm?id=c794a6b1a2084964a7ef45f69bef5423 (accessed on 27 December 2022).
77. World Economic Outlook Database. April 2022. Available online: https://www.imf.org/en/Publications/WEO/weo-database/2022/April (accessed on 16 February 2023).
78. China Pension Schemes: Private Plan Opens New Opportunities. Available online: https://www.china-briefing.com/news/china-pension-schemes-private-plan-opens-investment-opportunities/ (accessed on 27 December 2022).
79. Dong, B.; Ding, Q. Aging in China: A Challenge or an Opportunity? *J. Am. Med. Dir. Assoc.* **2009**, *10*, 456–458. [CrossRef]
80. Si, Y. China's Elderly Recreation Consumption Demand to Reach 20 Trillion Yuan by 2030. *China Food* **2018**, 173–174.
81. Li, B.; Sun, X. Analysis of the Current Situation of "Internet+" Home Wisdom Elderly Care in China. *Soft Sci. Health* **2019**, 8–11. Available online: http://www.wsrkx.net/CN/article/downloadArticleFile.do?attachType=PDF&id=296 (accessed on 14 January 2024).
82. Huang, L.; Zhang, L.; Wu, F.; Miao, H.; Li, X. International Comparison of Technology Research on the Welfare of the Elderly. *J. Intell.* **2015**, *34*, 22–27.
83. Zhang, L.; Song, X. Study on the Application and Countermeasures of Information Technology in the Elderly Service Industry. *Sci. Technol. Manag. Res.* **2015**, *35*, 5.
84. Ying, Z.; Chen, X.; Wu, Z. Research on the Construction Path of Smart Pension Community in Zhejiang: Based on SWOT-CLPV Analysis. *Constr. Econ.* **2017**, *38*, 71–75. [CrossRef]
85. OhShape. Wikipedia. 2022. Available online: https://en.wikipedia.org/w/index.php?title=OhShape&oldid=1102598709 (accessed on 28 December 2023).
86. Printable. 10 Best Chair Gym Exercises. Available online: https://www.printablee.com/post_chair-gym-exercises-printable_178691/ (accessed on 28 December 2022).
87. Hollmann, W.; Strüder, H.K.; Tagarakis, C.V.; King, G. Physical Activity and the Elderly. *Eur. J. Prev. Cardiol.* **2007**, *14*, 730–739. [CrossRef] [PubMed]
88. 21 Chair Exercises for Seniors: Complete Visual Guide. Available online: https://californiamobility.com/21-chair-exercises-for-seniors-visual-guide/ (accessed on 28 December 2022).
89. Marciano, C.; Galderisi, M.; Gargiulo, P.; Acampa, W.; D'Amore, C.; Esposito, R.; Capasso, E.; Savarese, G.; Casaretti, L.; Iudice, F.L. Effects of Type 2 Diabetes Mellitus on Coronary Microvascular Function and Myocardial Perfusion in Patients without Obstructive Coronary Artery Disease. *Eur. J. Nucl. Med. Mol. Imaging* **2012**, *39*, 1199–1206. [CrossRef] [PubMed]
90. Paolillo, S.; Rengo, G.; Pagano, G.; Pellegrino, T.; Savarese, G.; Femminella, G.D.; Tuccillo, M.; Boemio, A.; Attena, E.; Formisano, R. Impact of Diabetes on Cardiac Sympathetic Innervation in Patients with Heart Failure: A 123I Meta-Iodobenzylguanidine (123I MIBG) Scintigraphic Study. *Diabetes Care* **2013**, *36*, 2395–2401. [CrossRef]
91. Printablee. 10 Best Printable Chair Yoga Exercises for Seniors. Available online: https://www.printablee.com/post_printable-chair-yoga-exercises-for-seniors_178538/ (accessed on 28 December 2022).
92. Patel, N.K.; Newstead, A.H.; Ferrer, R.L. The Effects of Yoga on Physical Functioning and Health Related Quality of Life in Older Adults: A Systematic Review and Meta-Analysis. *J. Altern. Complement. Med.* **2012**, *18*, 902–917. [CrossRef]
93. Robinson, T.N.; Eiseman, B.; Wallace, J.I.; Church, S.D.; McFann, K.K.; Pfister, S.M.; Sharp, T.J.; Moss, M. Redefining Geriatric Preoperative Assessment Using Frailty, Disability and Co-Morbidity. *Ann. Surg.* **2009**, *250*, 449–455. [CrossRef] [PubMed]
94. Schmid, A.A.; Van Puymbroeck, M.; Koceja, D.M. Effect of a 12-Week Yoga Intervention on Fear of Falling and Balance in Older Adults: A Pilot Study. *Arch. Phys. Med. Rehabil.* **2010**, *91*, 576–583. [CrossRef]
95. What Is Cardiovascular Exercise? Available online: https://www.fyzical.com/lakewoodranch/lakewoodranch/blog/What-Is-a-Cardiovascular-Exercise (accessed on 28 December 2022).
96. Alzheimer's Association. Available online: https://act.alz.org/site/SPageServer/;jsessionid=00000000.app20115b?NONCE_TOKEN=9C725D55D38175E0DC73CF979C87911E (accessed on 5 November 2023).
97. Dietary Guidelines for Americans 2015–2020. Available online: https://www.google.com.hk/books/edition/Dietary_Guidelines_for_Americans_2015_20/HGeCDwAAQBAJ?hl=en&gbpv=0 (accessed on 5 November 2023).
98. Physical Activity Guidelines for Older Adults. Available online: https://www.nhs.uk/live-well/exercise/exercise-guidelines/physical-activity-guidelines-older-adults/ (accessed on 5 November 2023).
99. Aging and Sleep: How Does Growing Old Affect Sleep? Available online: https://www.sleepfoundation.org/aging-and-sleep (accessed on 5 November 2023).
100. Cardiovascular Diseases (CVDs). Available online: https://www.who.int/news-room/fact-sheets/detail/cardiovascular-diseases-(cvds) (accessed on 5 November 2023).
101. Herrschaft, H. Prevention of Cerebrovascular Circulatory Disorders. *Fortschritte Der Neurol. Psychiatr.* **1985**, *53*, 337–343. [CrossRef]
102. Fuller, H.R.; Huseth-Zosel, A. Lessons in Resilience: Initial Coping Among Older Adults During the COVID-19 Pandemic. *Gerontologist* **2021**, *61*, 114–125. [CrossRef]

103. Lee, W.-Y.; Gong, S.-M.; Leung, C.-Y. Is Color Preference Affected by Age Difference. *Int. Assoc. Soc. Des. Res.* **2009**, 1837–1846. Available online: https://www.researchgate.net/publication/267857549_Is_Color_Preference_Affected_by_Age_Difference (accessed on 14 January 2024).
104. ThreeDPoseTracker. Available online: https://digital-standard.com/tdpt/ (accessed on 28 December 2022).
105. 3D Ghost Effect. Particles/Effects. Unity Asset Store. Available online: https://assetstore.unity.com/packages/tools/particles-effects/3d-ghost-effect-159182?locale=zh-CN (accessed on 28 December 2022).
106. Top Down Island. Available online: https://assetstore.unity.com/packages/3d/environments/top-down-island-222313 (accessed on 28 December 2022).
107. Wilms, G. Guide on good data protection practice in research. European University Institute. 2019. Available online: https://www.eui.eu/documents/servicesadmin/deanofstudies/researchethics/guide-data-protection-research.pdf (accessed on 5 November 2023).
108. Guo, L.; An, L.; Luo, F.; Yu, B. Social Isolation, Loneliness and Functional Disability in Chinese Older Women and Men: A Longitudinal Study. *Age Aging* **2021**, *50*, 1222–1228. [CrossRef] [PubMed]
109. Roppolo, M.; Kunnen, E.S.; van Geert, P.L.; Mulasso, A.; Rabaglietti, E. A Quantitative Dynamic Systems Model of Health-Related Quality of Life among Older Adults. *Clin. Interv. Aging* **2015**, *10*, 1755–1770. [CrossRef] [PubMed]
110. Andersson, L. Loneliness Research and Interventions: A Review of the Literature. *Aging Ment. Health* **1998**, *2*, 264–274. [CrossRef]
111. Russell, D.; Peplau, L.A.; Cutrona, C.E. The Revised UCLA Loneliness Scale: Concurrent and Discriminant Validity Evidence. *J. Personal. Soc. Psychol.* **1980**, *39*, 472–480. [CrossRef]
112. Hu, R.; Pan, W.; Bock, T. Towards dynamic vertical urbanism: A novel conceptual framework to develop vertical city based on construction automation, open building principles, and industrialized prefabrication. *Int. J. Ind. Constr.* **2020**, *1*, 34–47. [CrossRef]

smart cities

MDPI

Article

Smart Cities for All? Bridging Digital Divides for Socially Sustainable and Inclusive Cities

Johan Colding [1,2,*], Caroline Nilsson [1] and Stefan Sjöberg [3]

1 Department of Building Engineering, Energy Systems and Sustainability Science, University of Gävle, Kungsbäcksvägen 47, 80176 Gävle, Sweden; caroline.nilsson@hig.se
2 The Beijer Institute of Ecological Economics, Royal Swedish Academy of Sciences, 104 05 Stockholm, Sweden
3 Department of Social Work, Criminology and Public Health, University of Gävle, 80176 Gävle, Sweden; stefan.sjoberg@hig.se
* Correspondence: johan.colding@hig.se

Abstract: This paper aims to emphasize the need for enhancing inclusivity and accessibility within smart-city societies. It represents the first attempt to apply Amartya Sen's capability approach by exploring the implications of digital divides for promoting inclusive and climate-friendly cities that prioritize well-being, equity, and societal participation. Sen's framework recognizes individual variations in converting resources into valuable 'functionings', and herein emphasizes the importance of aligning personal, social, and environmental conversion factors for individuals to fully navigate, participate in, and enjoy the benefits provided by smart cities. Adopting the capability approach and employing a cross-disciplinary analysis of the scientific literature, the primary objective is to broaden understanding of how to improve inclusivity and accessibility within smart-city societies, with a specific focus on marginalized community members facing first- and second-level digital divides. This paper underscores the importance of adopting a systemic perspective on climate-smart city navigation and stresses the importance of establishing a unified governing body responsible for monitoring, evaluating, and enhancing smart-city functionality. The paper concludes by summarizing some policy recommendations to boost social inclusion and address climate change in smart cities, such as creating capability-enhancing institutions, safeguarding redundancy in public-choice options, empowering citizens, and leveraging academic knowledge in smart-city policy formulation.

Keywords: smart cities; digital divides; social sustainability; climate-proofing; capability approach; conversion factors; redundancy; institutions; social inclusion

1. Introduction

While the digitalization of cities and society can contribute to climate-proofing, the governance of smart cities often overlooks crucial social sustainability aspects. The neglect of this issue is evident in the gap between people's views of smart cities. While they are assumed to be climate-friendly and inclusive, they often fall short of these ideals. With few exceptions (see, e.g., [1,2]), there have been limited large-scale scientific surveys conducted on citizens' perceptions of smart cities. In a systematic analysis, van Twist et al. [1] found that citizens' appreciation varies from active discontent (expressed dissatisfaction with the technology, democratic process, and societal impact of the smart city) to passive discontent (citizen dissatisfaction that remains unexpressed due to citizens' lack of awareness and skills, coupled with the absence of channels to express their discontentment). In the absence of channels for individuals to express passive discontent, it is challenging for governing entities, if they exist at all, to monitor, evaluate, and improve the functioning of smart cities. In essence, this issue poses a fundamental democratic question that needs to be handled appropriately.

The notion of 'sustainability' has been at the heart of the international research and policy forefront ever since the Brundtland Report was first published in 1987 [3]. In general, research and practice have given insufficient attention to aspects of social sustainability when compared to economic and ecological sustainability [4–8]. This discrepancy is also evident when it comes to smart-city developments where social sustainability aspects remain inadequately explored. Despite a substantial surge in research on economic development over the past decade, social sustainability factors have not progressed at a similar pace. Remarkably, 1.6% of all studies on smart cities have focused on social sustainability factors, while 92.8% primarily address economic considerations. The remaining studies primarily concentrate on environmental considerations [9]. This disparity underscores the need for a more comprehensive exploration of social sustainability within the context of sustainability research.

Due to the nature of technological innovations and their rapid development, primarily driven by the high-tech industry [10], it is crucial for cities and society at large to establish governing entities with a legal mandate and responsibility to regulate this progress. City development today is affected by the general societal development characterized by increasing inequalities, social exclusion, and marginalization of disadvantaged communities [11,12]. Social exclusion is, therefore, a multi-dimensional process with mutually influencing components such as lack of resources and services, weak social capital, lack of participation and influence, and negative neighborhood effects [13,14]. While the notion of social sustainability has been lacking consensus and clarity [11], there exists a whole cadre of scholars arguing that in smart-city development, it is essential to pay increased attention to the existing social inequalities and increase efforts to address them [9,10].

The importance of considering social inclusion, equality, and justice has received wide recognition among scholars in the academic circles that conduct research on social sustainability. Nilsson et al. [11] discovered that 85% of the literature defining social sustainability connects the concept to various facets of 'equity', encompassing the fair allocation of societal resources, opportunities, and costs. The notion of social sustainability emphasizes, along with the imperative of equity and societal and individual well-being, the facilitation of active participation in and influence over societal development and the promotion of social capital [11]. Smart-city initiatives often promise to improve urban life by using innovative technology to create inclusive environments that ensure healthy, safe, and efficient living for everyone [12].

Scholars define 'equity' as the fair distribution of resources, especially to vulnerable groups. It is often tied to social justice and fair distribution of costs, benefits, and participation in community matters [13–15]. Turker and Ozdemir [16] argue that a socially sustainable society is characterized by justice and equality, ensuring that no one is excluded and by empowering citizens to live fulfilling lives in freedom.

The proclamations listed above align well with Amartya Sen's capability approach, which emphasizes freedom as a fundamental component in enhancing equity, well-being, and inclusion. Well-being can be explained as both physical (e.g., health, access to water and food) and social/emotional well-being, where the latter connects to social inclusion and relationships, trust, education, and social equity [17]. In addition to involvement in decision-making and community influence, participation is a vital aspect of civil, democratic, and sustainable societies [18], and it is key for people living in smart cities to fully take part in all the benefits offered by digital resources and services, get access to information, and be part of urban decision-making processes.

1.1. Paper Objectives

This paper emphasizes the need for enhancing inclusivity and accessibility in smart-city societies. Its novelty lies in being the first scientific publication to apply Amartya Sen's capability approach to smart city research. With a background in urban planning and design, sociology, social work, sustainability science, and institutional economics, the authors of this paper employ a cross-disciplinary analysis of the scientific literature [19] related to how digital disparities inhibit the potential of digital technologies to work for the climate-proofing of smart cities. Adopting a social sustainability framework that emphasizes equity, well-being, participation and influence, and social capital [11], the major objective of this paper is to highlight the imperative of improving inclusivity and accessibility in the smart-city society by way of the capability approach, with a specific focus on marginalized community members who face first- and second-level digital divides. The former pertains to the accessibility to the internet, and the latter to the skills, participation, and efficacy of users in handling digital technologies [20]. By the term 'disadvantaged groups', we here mean segments of the population that face socio-economic, educational, or other challenges and who lack access to or skills in digital technologies that impede their ability to fully participate in and benefit from digitalization more generally. Experiencing digital barriers hinders their social and economic advancement, resulting in so-called digital divides compared to more digitally skilled and advanced segments of the population. As argued here, these barriers have implications for smart cities in dealing with climate change. We acknowledge that there are more aspects that affect the digital divide and social sustainability than those addressed in this paper, like language barriers, disability access, and other aspects. However, our focus here is limited to those factors that restrict access to full participation in a smart-city society and some of the repercussions that may result from this limitation.

Applying ideas from the economist and philosopher Amartya Sen's capability theory [21], we delve into key characteristics essential for advancing the promotion of equitable and inclusive smart cities. These characteristics aim to foster well-being and facilitate active societal participation. The capability approach was first developed by Sen in the early 1980s and has been further refined and expanded upon in various writings related to Sen's development studies and poverty analyses [22]. It posits that a socio-economic system encompasses numerous roles and 'functionings'. When appropriately employed, these functionings have the potential to enhance socio-economic performance and overall happiness of a country, region, or specific population group. Sen emphasizes the importance of focusing not only on material resources but also on individuals' capabilities and opportunities to lead the lives they value (Figure 1). While critics of Sen have voiced concerns about his failure to provide concrete policy recommendations [23], the United Nations Development Programme (UNDP) has adopted the capability approach in constructing the Human Development Index (HDI), which includes indicators like life expectancy, education, and per capita income [24].

Studies on smart cities have previously focused on various types of social inequalities and exclusion. For instance, Chen et al. 2022 provide examples of such studies, e.g., how transport apps take account of the needs of marginalized groups of people [25]. Moreover, Rebernik et al. [26] address the question of whether smart and sustainable cities consider the inclusiveness of all their inhabitants, assessing disability inclusion based on twenty indicators. Furthermore, Nápoles et al. [27] have elaborated on social inclusion in smart cities, and Nesti [28] has elaborated on gender equality in smart cities. The capability approach, however, has only rudimentarily been applied to urban research; but see Basta [29] and Kourtit et al. [30]. Except for a conference proceeding authored by Baldascino and Mosca [31], the capability approach has never before been employed in a scientifically grounded analysis of smart cities.

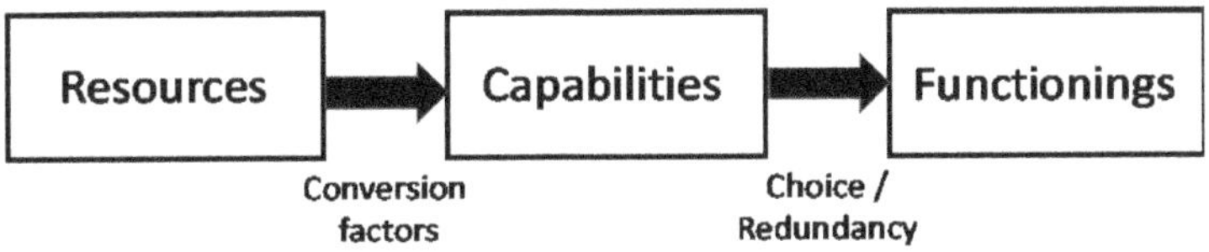

Figure 1. In the capability approach, 'functionings' are the ultimate focus, as they reflect individuals' real freedoms and opportunities to lead the lives they value. 'Resources' encompass the various assets, opportunities, and entitlements that individuals can access to enhance their well-being and capabilities. Certain so-called 'conversion factors' encompass personal, social, and environmental elements that collectively influence individuals' capacity to transform resources into 'capabilities' that represent the primary freedoms that people possess to choose and achieve various valuable 'functionings'. The latter represent different ways of being and doing that people consider important for their well-being and life fulfillment. By recognizing and enabling a broad range of functionings (the various things individuals can do or be), a society can foster a more comprehensive and meaningful measure of well-being beyond traditional economic indicators. In Sen's framework, the emphasis is on promoting human agency, freedom, and the removal of barriers that hinder people from achieving their desired functionings [22]. Adapted and modified from Bonvin 2008 [32].

1.2. Paper Outline

This paper is structured as follows: We commence by framing the key argument presented in this paper by outlining some essential benefits of digitalization for climate-proofing cities. While digital technologies have a real potential to provide huge climate-proofing benefits, they function as a double-edged sword. On the one hand, they broaden access to online services in the smart-city context for a large portion of citizens, arguably improving their use and benefits of these services. For instance, they provide digital access and inclusion of people with disabilities [33–35]. On the other hand, they simultaneously deny this advantage to a subset of disadvantaged groups, giving rise to social exclusion and 'digital divides'. This study concentrates on these divides, exemplified by an exploration of digital technology accessibility and the impact of human behavioral traits. Following this, we adopt Amartya Sen's capability approach to delve into the significance of individuals' navigation within the smart-city society. We underscore a systemic perspective on climate-smart city navigation and stress the importance of establishing a unified governing body responsible for monitoring, evaluating, and enhancing the functionality of smart cities. We conclude by summarizing policy recommendations for such a governing body.

2. Climate Change, Smart Cities, and Social Sustainability

Due to ongoing climate change and the necessity of addressing it successfully, Obringer and Nateghi [36] contend that the link between climate justice and smart cities is unclear and argue that more research is needed on smart cities coupled with social equity and sustainability. While not explicitly focused on climate justice, this paper aims to underscore the impact of digital divides on climate justice within smart-city societies. Addressing these disparities is important for several reasons: First, in the pursuit of establishing a smart city accessible to all, it is crucial to elucidate the various constraints that individuals encounter while navigating in the smart-city landscape. Second, enhanced planning and design of smart-city policies are not realizable unless efforts are directed towards overcoming these specific constraints within the smart-city society. Third, it is crucial to address democracy-related issues and citizenship to further improve the smart-city paradigm [37,38]. Fourthly, capability constraints often carry systemic implications. As argued herein and drawing inspiration from Sen's work, the freedom of agency that individuals hold in the smart-city society is ultimately contingent and constrained by the social, political, and economic resources and opportunities available to them [22]. These opportunities, in turn, we argue, are never offered by the Information and Communication Technology (ICT) itself but from the ruling body of the state or local government of which the prevailing institutions are paramount, as they "enable or restrict the operation of political and economic activities, and in so doing they have an important influence on the achievement of capabilities" [39] (p. 222).

While it is difficult to assess the environmental impact of ICTs on a city level, it is often argued that they hold key value for climate-proofing [40]. ICTs and the Internet of Things can, for instance, aid in reducing greenhouse gas emissions through reduced energy use by substituting resource-intensive activities such as conference travel through virtual means or by making existing services more efficient [41]. Energy savings through ICT technology in smart cities have been shown, for instance, for the operation of transport and heating of buildings [42], as well as numerous other benefits [43].

Estimating the full impact and potential of ICTs in smart-city developments remains a challenge and will probably remain so in the future as well, as novel technologies based on, for example, Artificial Intelligence (AI) are emerging, leading to quite a substantial electricity consumption [44].

While the potential of ICTs in combating climate change is seemingly huge, accessibility to this technology is paramount. Currently, though, one-third of the human population does not have access to the internet. Hence, not everyone can take part in all the benefits that digitalization provides; there are still significant differences between and within countries in accessing and using ICTs. At the same time, there are huge differences in the potential for electrification of transport among developed and developing countries and megacities [45]. There are also significant disparities in internet access and usage between countries in the global north and south [46]. North America recorded the highest internet usage rate, reaching 87.65% in 2017. This figure was more than double the percentages observed in Sub-Saharan Africa and South Asia, as reported by Yang et al. [20].

Some groups of humans are also having more difficulties in accessing and using ICTs than other groups, expressed in digital divides related to gender, geographical location, and income The term 'digital divide' is commonly employed to describe the inequality in the access to and/or utilization of digital technology among households, regions, and countries [20]. The concept emerged in the early 1990s in the context of discussions about disparities in access to information and communication technologies, highlighting the unequal distribution of digital resources and skills among different populations. The disparities that are most referred to are first- and second-level digital divides, with the former pertaining to accessibility to the internet and the latter to skills, participation, and efficacy of users in handling digital technologies such as smartphones, computers, and other gadgets [47]. Third-level digital divides refer to the skewed benefits among individuals being online and where certain groups get more out of their ICT skills than others [48,49]. Gender inequality related to ICTs is a significant challenge, particularly in developing countries, where it is both a consequence and cause of human rights violations [50]. Acilar and Sæbø [50] (p. 234) define the 'gender digital divide' as "the unequal access and use of ICT between genders." This inequality results, among other issues, in the exclusion of women from education [51]. Moreover, the ramifications of digital exclusion are amplified by the fact that education serves as a foundational requirement for personal fulfillment, societal engagement, political representation, and employment opportunities [3]. Since ICT skills are crucial for success in both work and academia, any systematic inequality in these skills becomes a social sustainability issue—a problem also affecting developing countries.

Digital technology for remote work is considered a key factor in reducing climate impacts, such as through decreased commuting [52]. Nevertheless, not all employee groups have access to this option. For example, nine out of ten workers within the Swedish Trade Union Confederation cannot work from home using digital technology due to their work situation [53]. Currently, white-collar workers and knowledge workers are significantly overrepresented in remote work, with a substantial increase during and after the COVID-19 pandemic. Therefore, a 'temporal digital divide' might also be observed, referring to a situation in which a person has access to ICT and possesses the cognitive skills and capacity to use it but cannot use it during working hours. It is also important to emphasize that some persons make an active choice to abstain from using ICTs for various reasons, such as having a negative attitude towards technological equipment and services [54].

2.1. Limiting Behavioral Traits

Despite the climate-friendly nature of a technology, its impact can be compromised by the way humans engage with it. For example, it is widely known that even as a technology becomes increasingly energy efficient, 'rebound effects' often occur with increased use of the technology and the overall increase in energy use [55,56]. Rebound effects refer to the phenomenon wherein efficiency measures aimed at reducing the use of a resource (e.g., energy, money, time) are counteracted by undesired secondary effects that result in consequences in the opposite direction [57]. This phenomenon was already observed during the early stages of industrialization when the increased efficiency of coal-powered steam engines, expected to reduce coal dependency, instead led to a faster depletion of the coal reserves. In market economies, when the production and use of a resource are made more efficient, the price usually decreases, creating a situation that can lead to increased demand and consumption of the resource in question [58].

Rebounding effects are but one classic example of wicked problems that can obscure the climate-proofing of smart cities [58]. Suffice it to say, psychological obstacles often impact human environmentally friendly behavior [59]. Among others, human decision-making often depends on simple rules of thumb known as heuristics. These heuristics can introduce systematic biases into the human cognitive system, stemming from both automatic and rapid intuitive processes, as well as conscious, analytical, and reflective processes [60]. In addition, the design of the physical environment can establish limits on behavior, shaping habits and triggering action responses [61]. In the field of environmental psychology, this spectrum of potential behaviors is framed as the connections between environmental features and the capabilities of the individuals, commonly referred to as 'affordances' [62,63].

A tool related to affordances to boost climate-friendly behavioral change is the use of nudging and defaults. In a decision theoretical context, a default is the outcome that users are left with when they refrain from making an explicit choice [58]. In the affordance literature, various characteristics of the physical environment frequently act as default scenarios, subtly guiding individuals toward specific behaviors [64]. For example, programmable thermostats can be set with energy-saving default settings, encouraging users to maintain more moderate temperatures and reduce energy consumption without having to manually adjust the settings. Also, manufacturers can set energy-saving modes as the default settings on appliances, encouraging users to use energy-efficient options without having to actively select them. A third example of such defaults constitutes public transportation defaults, such as automated ticketing systems, in which integrated and user-friendly automated ticketing systems can be implemented to streamline the boarding process for public transportation. This may include contactless payment methods, mobile apps, or smart cards, making it convenient for individuals to use public transit without the hassle of purchasing tickets every time.

2.2. Second- and Third-Level Digital Disparities

Disparities in digital skills, participation, and efficacy, known as second-level digital divides, have wide-ranging effects on a significant portion of the human population, including individuals possessing technical expertise. For instance, the ongoing updates of applications and the introduction of new smartphone models require cognitive abilities like memory and problem-solving skills, which typically diminish with age [65,66]. While smartphones and other mobile technologies can potentially enhance cognitive functions, the physical aspects, such as navigating touchscreen interfaces, become more challenging as individuals age [67].

With age, people inevitably encounter generational time lags in skill loss and cognitive decline due to the constant evolution of digital technology and the introduction of new smart gadgets. Consequently, as ICT increasingly integrates with traditional urban infrastructure, there is a risk of further marginalizing those facing challenges with digital technologies, either due to difficulty or a refusal to use them [68].

Barbosa and Vetere [69] (p. 2) note that ageing is a heterogenous experience, marked by "both structural (e.g., social class and education) and agentic dimensions (e.g., individual action, attitudes, and dispositions)." Nevertheless, it is essential to recognize that the swift progression of technological advancements has left numerous older adults struggling to adjust, resulting in their growing marginalization within an information-based society [70].

Digital disparities also exist at the third level, with varying benefits from ICT skills highlighted in studies by Wei et al. [48], Van Deursen and Helsper [71], and Van Deursen and van Dijk [72]. These disparities, especially in internet outcomes, can reinforce existing social inequalities. For instance, Public Participation GIS (PPGIS) [73], which can enhance community participation in planning through citizen involvement, may pose democratic challenges. Townsend [38] suggests that PPGIS could lead to a permanent offloading of government responsibilities, disallowing less digitally skilled groups to have a voice in city planning. Prior to the digital era, participatory events faced challenges in adequately representing disadvantaged groups, such as people living in marginalized neighborhoods and low-income earners. The introduction of digital tools has facilitated broader participation, enabling individuals to contribute their opinions regardless of time and location. However, even if there is a potential for digital tools to facilitate the inclusion of a more diverse array of participants in decision-making processes and civic engagement, participation might be skewed towards more privileged groups with greater internet access and digital skills [74].

3. Capabilities for Navigation in the Smart-City Society

It should be noted that ICTs provide important technological tools and solutions to disabled individuals who have challenges, limitations, and restrictions in their ability to perform everyday activities due to physical, sensory, cognitive, or other types of disabilities [33–35]. Hence, the merits of ICTs for the inclusion of people in the smart-city society are indisputable. Still, and as earlier pointed out, ICTs often act as a double-edged sword by denying the advantage of having access to and making use of digital technologies to a subset of disadvantaged individuals in the form of various digital divides. These disparities create challenges for individuals, groups, and communities in the smart-city society, making navigation within this environment problematic. Such navigation involves utilizing digital technologies to access services and critical resources electronically. It also encompasses the process of efficiently moving through the urban landscape with the aid of digital tools. For instance, individuals can navigate smart-city systems to access public transportation schedules, find real-time traffic information, or locate nearby healthcare services through mobile technology applications. The seamless integration of digital navigation enhances accessibility and streamlines the utilization of essential resources within the city environment.

From a sustainability standpoint, the task of leveling out disparities and achieving social justice in the smart-city society involves establishing conditions that foster and elevate people's skills, abilities, and participation. Otherwise, the inclusion of the prefix 'smart' before cities becomes devoid of any meaningful purpose.

To enhance social sustainability and to attain a more nuanced comprehension of social well-being within the realm of smart cities, we contend that it is essential for a more extensive exploration of the capabilities and freedoms crucial for individuals and groups to lead liberated and fully functional lives [75]. This holistic perspective is inspired by the 'capability approach' of Amartya Sen, underscoring the importance of individuals having the freedom to choose diverse ways of living that align with skills and values. In this perspective, development is perceived as the broadening of these tangible freedoms.

In alignment with the insights posed by Sen [22] and Nussbaum [76], capabilities refer to the actual opportunities that individuals possess to lead functioning lives. By adopting this perspective, we argue that smart-city initiatives can better address the diverse needs and aspirations of their citizens, fostering a more inclusive and sustainable urban environment. Unlike traditional economic approaches as indicators of well-being, the capability approach recognizes that income is only a means to an end—the end being the ability to lead a life that one values. Sen highlights the importance of diversity in options and means for reaching human goals and the fact that individuals have different priorities and values. Hence, facilitating a range of different options for navigation in the smart-city society is an important governance aim. The one-size-fits-all approach that currently circumscribes many smart-city developments should be avoided.

Sen's ideas about the diversity of options align well with the resilience thinking principle of 'redundancy', which plays a central role in establishing resilient social-ecological systems [77]. The concept of redundancy, in this context, emphasizes the importance of having a variety of responses available to navigate and address both anticipated and unforeseen changes, as well as to seize new opportunities [78].

In the smart-city development, the redundancy principle echoes Sen's emphasis on expanding real freedoms by fostering a diverse set of options and responses through which people can enhance their adaptive capacity, ensuring they are better equipped to handle the complexities and uncertainties inherent in urban environments. This approach not only promotes resilience but also aligns with Sen's vision of expanding individuals' capabilities to choose and pursue ways of living that resonate with their values and preferences.

3.1. Navigation in the Smart-City Society: A System of Capabilities

So-called 'commodities' represent a key concept in the capability approach. Sen distinguishes between 'functioning' and 'instrumental' commodities, with the former constituting the actual goods and services that individuals use or consume to achieve a certain functioning or capability. The latter represents commodities that serve as instruments or tools to achieve other commodities or functionings. Navigation in the smart-city society greatly depends on the possession of instrumental commodities, such as various gadgets of ICT technology, and the proficiency in using these tools, which together can be regarded as capabilities [79]. Technical artifacts can enhance social justice, well-being, and equality, yet they can also hinder full participation in activities of society, as various so-called 'conversion factors' may impede this process [80].

Conversion factors, as outlined by Sen [22], encompass 'personal', 'social', and 'environmental' elements that collectively influence individuals' capacity to transform resources into valuable functionings. Personal factors are those characteristics that a person is endowed with and which affect the bodily operation as well as her or his psychological make-up and operation. Examples include a person's intelligence, psycho-motor skills, metabolism, physical or mental handicaps, and height. Social factors include institutions, social norms, social hierarchies, and government policies that can promote or inhibit the conversion of a commodity into useful functionings. Environmental factors refer to the external conditions and circumstances that impact an individual's capability conversion. This includes the physical environment, socioeconomic conditions, and access to resources such as public goods, technological infrastructure, and natural resources.

In accordance with the capability approach, to fully navigate, participate, and enjoy the full benefits of society, the three conversion factors need to be aligned; if one is failing, it is impacting the other two, limiting individuals' chances to pursue well-being. It is important to recognize that in Sen's argument, the diversity of options and the fact that people have different capacities to convert resources into valuable functionings are important determinants for reducing inequalities and promoting social sustainability.

Disruptive technologies, such as the internet, smartphones, blockchain technology, and AI, play pivotal roles in shaping smart cities [81]. The integration of these technologies into various applications contributes to the development of smarter cities, fostering improved living conditions and granting residents enhanced access to products and services. Nevertheless, and as previously highlighted, realizing the benefits of these technologies, or what Sen terms "functionings", necessitates the harmonization of the three conversion factors.

In the context of smart cities and drawing an analogy to Sen's capability approach, digital divides emerge as critical obstacles to individuals' capabilities. For example, first-level digital divides can be seen as the exclusion from essential environmental factors required to navigate within a smart-city society, stemming from an individual's incapability to access the internet and other ICTs. On the other hand, second-level digital divides can be construed as the absence of personal factors necessary to adeptly manage instrumental commodities like digital gadgets. This includes both cognitive skills and motor skills, such as the handling of touchscreen interfaces, updating apps and gadgets, and so forth—skills that often decline with age.

While smartphones and other mobile technologies have the potential to enhance human cognition, it is important to note that they may also impede an individual's ability to fully participate in all the functionings offered by the smart-city society. The dual nature of these technologies, serving as both facilitators and barriers, underscores the importance of addressing digital divides to ensure equitable access to and sharing all the benefits of smart-city developments.

3.2. Non-Choice Default Technologies

It has been argued that the prevalence of one-size-fits-all thinking in smart-city societies limits the availability of diverse options for navigation [58]. An illustrative example of this phenomenon is the widespread reliance on smartphones and inexpensive tablet computers, which increasingly perform various functions within smart cities. For instance, a growing number of remotely controlled electronic home devices, which require financial resources to acquire, are now managed through smartphones or other online gadgets. Certain public services are exclusively accessible through these devices, including cashless transaction apps for everyday payments, smartphone-operated parking meters, various medicare services, and so forth. Furthermore, as the use of smartphone- and online-operated services experiences a significant surge, there is a corresponding decline in alternative methods such as credit card or cash payments. This shift has notable implications for public choice and accessibility.

When disruptive technologies become the exclusive means for carrying out daily tasks, leaving little room for alternative or supplementary options, they represent 'non-choice default technologies' (NDT) [58]. A pertinent illustration of an NDT is the complete reliance of farmers on a single type of genetically modified crop and a tailor-made pesticide designed for it. Such tightly interwoven dependencies can result in severe social consequences [32]. In urban settings, the widespread adoption of a specific ride-sharing platform to the exclusion of other options could also be considered an NDT. In other words, if most people use and rely on a specific ride-sharing service to the exclusion of other options, it limits the choices available to individuals in that context. This can happen if a particular platform becomes so popular and widely adopted that other alternatives are less practical or less commonly used. This process is driven by market forces, but local governments can regulate this imbalance by devising adequate institutions that safeguard analog alternatives or by substituting alternative ride-sharing platforms.

A development where smartphones become the only default option to access various kinds of services and operative functions in a city is inherently problematic from a resilience perspective. Maintaining traditional, analog, manual back-up systems that co-exist with digital ICT-driven systems is critical in times of disturbance, for example, hacker attacks or extreme weather events that are predicted to increase in the near future [82]. During crises, analog technologies can step in and perform functions in case of digital system malfunctions. For instance, manual traffic surveillance systems should supplement automatic ones, alternate payment methods should support smartphone-operated parking meters in the event of digital failures, and mechanical locks should complement digital locks [83].

Another illustrative instance highlighting the impact of a default non-choice scenario occurred during the COVID-19 pandemic when certain groups of people faced challenges in scheduling screening appointments and vaccination slots. Despite studies indicating that smart technologies contributed to improving community well-being and enhancing urban resilience during the COVID-19 pandemic [84], the intended goals of these technologies and screening programs were hindered because a significant number of patients could not access the online system [85]. Thus, smart gadgets function as a double-edged sword. On the one hand, they broaden access to online services for a large portion of the human population. On the other hand, they concurrently exclude this benefit for other individuals and already marginalized groups.

NDTs challenge Sen's concept of the freedom to exercise free choice, a fundamental aspect of capabilities [39]. In instances where they dominate the management of daily affairs in a city, individuals find themselves compelled to adopt the technology, regardless of whether they want it or not. Consequently, an individual's agency to engage in the life of a smart-city society is constrained by digital divides, leading to a missed opportunity to benefit from digitalization. Participation limitations arise not only from personal characteristics but also from a combination of social and environmental factors, including the state of ICT infrastructure.

4. Governing the Smart-City Society for Social Sustainability

The three levels of digital divides are intricately linked to social conditions and the evolution of socially sustainable societies. The first level of the digital division concerns the distribution of material resources, crucial for gaining access to the digital sphere. The second and third levels of digital divides emphasize the need for discourse surrounding the equitable distribution of intangible resources and opportunities. Digital development, in connection with social sustainability, extends beyond mere access to the internet and material resources. While the digital divide may appear to be a relatively minor issue affecting only a minority of the human population, the global landscape of the digital divide shows significant variations. Obtaining real-time statistics about the size of the digital gaps is a challenge. However, Jan van Dijk, in his book "The Digital Divide", emphasizes that this divide is not a simplistic binary concept but rather a continuum spanning an entire population. As he notes, it ranges "from people having no access and use at all to those with full access and using several applications every day" [86] (p. 3). This spectrum spans from an information elite at one extreme to a digitally illiterate segment at the other, fully excluded from the digital landscape. Suffice it to say that those in the digital elite will have a much greater capacity to fully realize their functionings, sensu Sen, than those who are less skilled or completely unskilled. This makes inequalities almost inevitable, risking increasing already existing inequalities and social exclusion.

To completely close the digital gap may not be entirely possible. Still, much can be done, and it is going in the right direction. For one thing, access to the internet through broadband development has improved, particularly in developing countries. Nevertheless, bridging the gap in skills and usage remains a considerably greater challenge [86]. Van Dijk alerts that the development and spread of digital technology could unexpectedly worsen social and digital inequality at both national and international levels, contending that addressing this issue requires a focus on social development rather than solely technological progress [86].

The connection between social and digital inequalities emphasizes the significance of prioritizing social sustainability aspects in the planning, design, and management of smart cities. Therefore, it is crucial to integrate considerations of equity, inclusion, and social justice at every stage of smart-city development [9,87]. The architect and urban designer Jan Gehl argues that physical changes in the urban landscape may entail, in addition to altered patterns of movement, mental and social transformations [88]. This can affect how the city is used, determine the representation of diverse groups within its confines, and dictate the extent to which various communities feel a sense of ownership, engage in decision-making processes, and establish a sense of place in the urban environment. As the urban landscape has evolved beyond its conventional physical boundaries and now incorporates a digital sphere, we also need to discuss how the digitized and smart cityscape fosters changes in movement patterns and impacts mental as well as social reconfigurations. Furthermore, interaction with the digital sphere serves as a prerequisite for accessing various facets of the physical environment, as exemplified by parking and public transportation applications. What societal consequences might follow if a part of human diversity is lost within these domains because of digital barriers? Addressing these issues is also pivotal in the context of climate-proofing cities, as the climate crisis is not solely a technical problem but rather a challenge rooted in social conditions and human behavior where climate justice constitutes a precondition for implementing economic and institutional policy instruments aimed at mitigating climate change [89].

Research also indicates that minority populations are more negatively affected by these divides than majority populations [49,71], hindering their access to crucial information, including information about climate issues. For instance, the substantial energy demand of the transport sector and the rapid increase in emissions underscore the importance of promoting commuting through public transportation. This is crucial to conserve energy and reduce emissions compared to other commuting options [45]. However, as smart navigation apps become more prevalent for activities like ticketing and fare collection, there is a risk of excluding individuals from various generations who are not digitally connected or skilled enough to commute by public transport without considerable personal costs. From a sustainable governance perspective, this necessitates the creation of instrumental commodities that offer a blend of digital and analog technological options. Additionally, services that involve manual operation by human beings should be considered to support the development of a green job market and to improve service-mindedness. To encourage the preference for public transportation, like commuting by train instead of flying, cities should carefully plan services to overcome both physical and cultural obstacles, avoiding NDTs and the resultant digital divides. This guarantees the existence of an attractive alternative for residents, fostering both socially and environmentally sustainable transportation decisions.

5. Concluding Remarks

While ICTs provide important technological tools and solutions to disabled individuals, the general expectation is that online services are accessible to everyone and always positive [72]. However, as discussed in this paper, a smart city for all will remain more of an ideal than a reality unless more firm action is taken to avoid the various types of digital disparities that exist in the smart-city society. As clearly elaborated on here, digital development is clearly related to social sustainability issues and to overall societal development. Currently, the development of smart cities is predominantly led by a conglomerate of ICT developers, high-tech companies, and urban prospectors actively seeking investment opportunities in urban development, real estate, and high-tech ventures [90]. This conglomerate may differ substantially from the public authorities, responsible for governing the smart-city network and managing its diverse operation and maintenance. Due to the demand for specialized technological expertise, governance tends to lean toward technocracy, leaving limited space for social expertise to address sustainability aspects [91]. As a result, a participation gap arises in the governance of smart-city development, posing a risk of further diluting existing social exclusion. This situation is likely to contribute to inertia in addressing inequalities associated with the growing digitization of services in cities.

To conclude, it is imperative to establish a unified governing body tasked with monitoring, evaluating, and enhancing the functionality of smart cities. This necessitates the creation of new institutions that facilitate active participation from the bottom up, with significant involvement from the public sector and civil society actors. Given the bureaucratic challenges prevalent in many European countries, such a body would ideally be integrated with existing pan-European initiatives and programs, leveraging their resources and expertise. One such initiative is the European Innovation Partnership on Smart Cities and Communities (EIP-SCC), initiated by the European Commission in 2012 [92]. The EIP-SCC aims to promote collaboration among cities, industry stakeholders, and other relevant parties to develop and deploy smart-city solutions. The EIP-SCC also includes various working groups dedicated to topics such as ICT, citizen participation, and engagement, further facilitating the advancement of smart-city initiatives.

In Table 1 and based on the major issues brought forward in this paper with its focus on Amartya Sen's capability approach, we conclude by providing some more explicit policy and research advice to such a governing body for the promotion of more inclusive and socially sustainable smart cities. This is with the hope that all individuals, including disadvantaged groups, can reach their desired functionings in a smart-city society.

Table 1. Policy recommendations for inclusive and socially sustainable smart cities.

Devise Capability-Enhancing Institutions	Develop policies and institutions that align with Amartya Sen's capability approach to enhance people's capabilities to participate and navigate in the smart-city society
Prioritize Conversion Factors	Focus on conversion factors to maximize the benefits of digital technology, ensuring a comprehensive approach to societal well-being
Safeguard Redundancy in Public-Choice Options	Maintain diverse choices, including analog options, to enhance public-choice options and bolster resilience
Empowering Citizen Voices	Establish a smart city feedback system linked to the capability-enhancing institutions, fostering channels for ordinary citizens and marginalized groups to voice their experiences, including dissatisfaction. Continuous monitoring of satisfaction evolution over time for informed decision-making, enhancing sustainability perspectives
Promote Research on Smart Cities Aligned with Social Equity and Sustainability	Prioritize research on smart cities intertwined with social equity and sustainability, mirroring, for example, the UNDP's incorporation of capability approach indicators into the Human Development Index
Leverage Scientific Knowledge in Policy Formulation	Encourage the integration of more peer-reviewed scientific knowledge into smart-city policy documents, moving away from reliance on other non-reviewed policy documents
Avoid One-Size-Fits-All Approaches	Integrate Amartya Sen's diversity of options into smart-city frameworks, creating smart cities that are both technologically advanced and socially sustainable
Prioritize Social and Organizational Aspects in Digitalization Policies	Highlight the need for policies to place greater emphasis on the social and organizational aspects of digitalization, considering social exclusion, inequalities, and divides in society

Author Contributions: Conceptualization, J.C. and C.N.; methodology, J.C., C.N. and S.S.; formal analysis, J.C., C.N. and S.S..; writing—original draft preparation, J.C.; writing—review and editing, J.C., C.N. and S.S.; project administration, J.C., C.N. and S.S. All authors have read and agreed to the published version of the manuscript.

Funding: Johan Colding's and Stefan Sjöberg's research was partly funded by a joint grant from Mistra [DIA 2019/28] and Formas via the national research programme on climate (2021–00416). Colding's work was also funded by the University of Gävle, and the Beijer Institute of Ecological Economics, the Royal Academy of Swedish Sciences. Caroline Nilsson's research has been supported through financial support from the University of Gävle via the program Future Proof Cities (Grant No. 2019–0129), which is financed by the Knowledge Foundation of Sweden. Stefan Sjöberg's research was also partly funded by the strategic research area Sustainable Cities and the program Urban Commons, University of Gävle.

Data Availability Statement: Data are contained within the article.

Acknowledgments: The authors wish to thank the research program Fairtrans and the University of Gävle for their support of this work.

Conflicts of Interest: The authors declare no conflicts of interest. The funders had no role in the design of the study; in the collection, analyses, or interpretation of data; in the writing of the manuscript; or in the decision to publish the results.

References

1. Van Twist, A.; Ruijer, E.; Meijer, A. Smart Cities & Citizen Discontent: A Systematic Review of the Literature. *Gov. Inf. Q.* **2023**, *40*, 101799. [CrossRef]
2. Lehtiö, A.; Hartikainen, M.; Ala-Luopa, S.; Olsson, T.; Väänänen, K. Understanding Citizen Perceptions of AI in the Smart City. *AI Soc.* **2023**, *38*, 1123–1134. [CrossRef]
3. Åhman, H. Social Sustainability—Society at the Intersection of Development and Maintenance. *Local Environ.* **2013**, *18*, 1153–1166. [CrossRef]
4. Boström, M. A Missing Pillar? Challenges in Theorizing and Practicing Social Sustainability: Introduction to the Special Issue. *Sustain. Sci. Pract. Policy* **2012**, *8*, 3–14. [CrossRef]
5. Colantonio, A. Urban Social Sustainability Themes and Assessment Methods. *Proc. Inst. Civ. Eng. Urban Des. Plan.* **2010**, *163*, 79–88. [CrossRef]
6. Cuthill, M. Strengthening the 'Social' in Sustainable Development: Developing a Conceptual Framework for Social Sustainability in a Rapid Urban Growth Region in Australia. *Sustain. Dev.* **2010**, *18*, 362–373. [CrossRef]
7. Nilipour, A. (Azi) Introduction to Social Sustainability. In *Social Sustainability in the Global Wine Industry: Concepts and Cases*; Forbes, S.L., De Silva, T.-A., Gilinsky, A., Jr., Eds.; Springer International Publishing: Cham, Switzerland, 2020; pp. 1–14; ISBN 978-3-030-30413-3.
8. Vallance, S.; Perkins, H.C.; Dixon, J.E. What Is Social Sustainability? A Clarification of Concepts. *Geoforum* **2011**, *42*, 342–348. [CrossRef]
9. Chen, T.; Ramon Gil-Garcia, J.; Gasco-Hernandez, M. Understanding Social Sustainability for Smart Cities: The Importance of Inclusion, Equity, and Citizen Participation as Both Inputs and Long-Term Outcomes. *J. Smart Cities Soc.* **2022**, *1*, 135–148. [CrossRef]
10. Hollands, R.G. Will the Real Smart City Please Stand up?: Intelligent, Progressive or Entrepreneurial? In *The Routledge Companion to Smart Cities*; Routledge: London, UK, 2020; ISBN 978-1-315-17838-7.
11. Nilsson, C.; Levin, T.; Colding, J.; Sjöberg, S.; Barthel, S. Navigating Complexity with the Four Pillars of Social Sustainability. *Sustain. Dev.* 2024; *accepted*. [CrossRef]
12. Pasolini, G.; Buratti, C.; Feltrin, L.; Zabini, F.; De Castro, C.; Verdone, R.; Andrisano, O. Smart City Pilot Projects Using LoRa and IEEE802.15.4 Technologies. *Sensors* **2018**, *18*, 1118. [CrossRef]
13. McKenzie, K.B.; Scheurich, J.J. Equity Traps: A Useful Construct for Preparing Principals to Lead Schools That Are Successful With Racially Diverse Students. *Educ. Adm. Q.* **2004**, *40*, 601–632. [CrossRef]
14. Dempsey, N.; Bramley, G.; Power, S.; Brown, C. The Social Dimension of Sustainable Development: Defining Urban Social Sustainability. *Sustain. Dev.* **2011**, *19*, 289–300. [CrossRef]
15. Hemani, S.; Das, A.K. Humanising Urban Development in India: Call for a More Comprehensive Approach to Social Sustainability in the Urban Policy and Design Context. *Int. J. Urban Sustain. Dev.* **2016**, *8*, 144–173. [CrossRef]
16. Turker, D.; Ozdemir, G. Modeling Social Sustainability: Analysis of Hospitality e-Distributors. *Sustain. Account. Manag. Policy J.* **2019**, *11*, 799–824. [CrossRef]
17. Rogers, D.S.; Duraiappah, A.K.; Antons, D.C.; Munoz, P.; Bai, X.; Fragkias, M.; Gutscher, H. A Vision for Human Well-Being: Transition to Social Sustainability. *Curr. Opin. Environ. Sustain.* **2012**, *4*, 61–73. [CrossRef]
18. Rashidfarokhi, A.; Yrjänä, L.; Wallenius, M.; Toivonen, S.; Ekroos, A.; Viitanen, K. Social Sustainability Tool for Assessing Land Use Planning Processes. *Eur. Plan. Stud.* **2018**, *26*, 1269–1296. [CrossRef]
19. Conrad, D.; Sinner, A. *Creating Together: Participatory, Community-Based, and Collaborative Arts Practices and Scholarship across Canada*; Wilfrid Laurier University Press: Waterloo, ON, Canada, 2015; ISBN 978-1-77112-024-1.
20. Yang, H.; Wang, S.; Zheng, Y. Spatial-Temporal Variations and Trends of Internet Users: Assessment from Global Perspective. *Inf. Dev.* **2023**, *39*, 136–146. [CrossRef]
21. Comim, F.; Qizilbash, M.; Alkire, S. *The Capability Approach: Concepts, Measures and Applications*; Cambridge University Press: Cambridge, UK, 2008; ISBN 978-1-139-46986-9.
22. Sen, A. *Development as Freedom*, 1st ed.; Oxford University Press: New York, NY, USA, 1999.
23. Miletzki, J.; Broten, N. *An Analysis of Amartya Sen's Development as Freedom*; Macat Library: London, UK, 2017; ISBN 978-1-912281-27-5.

24. Kuhumba, S. Amartya Sen's Capability Approach as Theoretical Foundation of Human Development. *J. Sociol. Dev.* **2018**, *1*, 127–145.
25. Pérez-delHoyo, R.; Andújar-Montoya, M.D.; Mora, H.; Gilart-Iglesias, V. Urban and Building Accessibility Diagnosis Using 'Accessibility App' in Smart Cities—A Case Study. In Proceedings of the 6th International Conference on Smart Cities and Green ICT Systems, Porto, Portugal, 22–24 April 2017; SCITEPRESS—Science and Technology Publications: Porto, Portugal, 2017; pp. 99–108.
26. Rebernik, N.; Szajczyk, M.; Bahillo, A.; Goličnik Marušić, B. Measuring Disability Inclusion Performance in Cities Using Disability Inclusion Evaluation Tool (DIETool). *Sustainability* **2020**, *12*, 1378. [CrossRef]
27. Padrón Nápoles, V.M.; Gachet Páez, D.; Esteban Penelas, J.L.; García Pérez, O.; Martín de Pablos, F.; Muñoz Gil, R. Social Inclusion in Smart Cities. In *Handbook of Smart Cities*; Augusto, J.C., Ed.; Springer International Publishing: Cham, Switzerland, 2021; pp. 469–514. ISBN 978-3-030-69698-6.
28. Nesti, G. Mainstreaming Gender Equality in Smart Cities: Theoretical, Methodological and Empirical Challenges. *Inf. Polity* **2019**, *24*, 289–304. [CrossRef]
29. Basta, C. From Justice in Planning toward Planning for Justice: A Capability Approach. *Plan. Theory* **2016**, *15*, 190–212. [CrossRef]
30. Kourtit, K.; Nijkamp, P.; Suzuki, S. Quantitative Performance Assessment of Asian Stellar Cities by a DEA Cascade System: A Capability Interpretation. *Ann. Reg. Sci.* **2023**, *70*, 259–286. [CrossRef]
31. Baldascino, M.; Mosca, M. The Capability Approach and the Tools of Economic Policies for Smart City. *Procedia Soc. Behav. Sci.* **2016**, *223*, 884–889. [CrossRef]
32. Bonvin, J.-M. Capacités et démocratie. In *La liberté au Prisme des Capacités: Amartya Sen au-delà du Libéralisme*; De Munck, J., Zimmermann, B., Eds.; Raisons Pratiques; Éditions de l'École des Hautes Études en Sciences Sociales: Paris, France, 2008; pp. 237–261. ISBN 978-2-7132-3096-7.
33. Baumgartner, A.; Rohrbach, T.; Schönhagen, P. 'If the Phone Were Broken, I'd Be Screwed': Media Use of People with Disabilities in the Digital Era. *Disabil. Soc.* **2023**, *38*, 73–97. [CrossRef]
34. Goggin, G.; Zhuang, K.V. Disability as Smart Equality: Inclusive Technology in a Digitally Advanced Nation. In *Vulnerable People and Digital Inclusion: Theoretical and Applied Perspectives*; Tsatsou, P., Ed.; Springer International Publishing: Cham, Switzerland, 2022; pp. 257–275. ISBN 978-3-030-94122-2.
35. Manzoor, M.; Vimarlund, V. Digital Technologies for Social Inclusion of Individuals with Disabilities. *Health Technol.* **2018**, *8*, 377–390. [CrossRef] [PubMed]
36. Obringer, R.; Nateghi, R. What Makes a City 'Smart' in the Anthropocene? A Critical Review of Smart Cities under Climate Change. *Sustain. Cities Soc.* **2021**, *75*, 103278. [CrossRef]
37. Luque-Ayala, A.; Marvin, S. Developing a Critical Understanding of Smart Urbanism? *Urban Stud.* **2015**, *52*, 2105–2116. [CrossRef]
38. Townsend, A.M. *Smart Cities: Big Data, Civic Hackers, and the Quest for a New Utopia*; W. W. Norton & Company: New York, NY, USA, 2013; ISBN 978-0-393-08287-6.
39. Nambiar, S. Capabilities, Conversion Factors and Institutions. *Prog. Dev. Stud.* **2013**, *13*, 221–230. [CrossRef]
40. George, G.; Merrill, R.K.; Schillebeeckx, S.J.D. Digital Sustainability and Entrepreneurship: How Digital Innovations Are Helping Tackle Climate Change and Sustainable Development. *Entrep. Theory Pract.* **2021**, *45*, 999–1027. [CrossRef]
41. Coroamă, V.C.; Bergmark, P.; Höjer, M.; Malmodin, J. A Methodology for Assessing the Environmental Effects Induced by ICT Services: Part I: Single Services. In Proceedings of the 7th International Conference on ICT for Sustainability, Bristol, UK, 21–26 June 2020; Association for Computing Machinery: New York, NY, USA, 2020; pp. 36–45.
42. Kramers, A.; Höjer, M.; Lövehagen, N.; Wangel, J. Smart Sustainable Cities—Exploring ICT Solutions for Reduced Energy Use in Cities. *Environ. Model. Softw.* **2014**, *56*, 52–62. [CrossRef]
43. Dwivedi, Y.K.; Hughes, L.; Kar, A.K.; Baabdullah, A.M.; Grover, P.; Abbas, R.; Andreini, D.; Abumoghli, I.; Barlette, Y.; Bunker, D.; et al. Climate Change and COP26: Are Digital Technologies and Information Management Part of the Problem or the Solution? An Editorial Reflection and Call to Action. *Int. J. Inf. Manag.* **2022**, *63*, 102456. [CrossRef]
44. The Impact of GenAI on Electricity: How GenAI Is Fueling the Data Center Boom in the U.S. Available online: https://www.linkedin.com/pulse/impact-genai-electricity-how-fueling-data-center-boom-vivian-lee/ (accessed on 10 February 2024).
45. Fan, J.-L.; Wang, J.-X.; Li, F.; Yu, H.; Zhang, X. Energy Demand and Greenhouse Gas Emissions of Urban Passenger Transport in the Internet Era: A Case Study of Beijing. *J. Clean. Prod.* **2017**, *165*, 177–189. [CrossRef]
46. Mammen, J.T.; Rugmini, D.M.; Girish, K.R. North–South Digital Divide: A Comparative Study of Personal and Positional Inequalities in USA and India. *Afr. J. Sci. Technol. Innov. Dev.* **2023**, *15*, 482–495. [CrossRef]
47. Robinson, L.; Cotten, S.R.; Ono, H.; Quan-Haase, A.; Mesch, G.; Chen, W.; Schulz, J.; Hale, T.M.; Stern, M.J. Digital Inequalities and Why They Matter. *Inf. Commun. Soc.* **2015**, *18*, 569–582. [CrossRef]
48. Wei, K.-K.; Teo, H.-H.; Chan, H.C.; Tan, B.C.Y. Conceptualizing and Testing a Social Cognitive Model of the Digital Divide. *Inf. Syst. Res.* **2011**, *22*, 170–187. [CrossRef]
49. Van Deursen, A.J.; Van Dijk, J.A. The First-Level Digital Divide Shifts from Inequalities in Physical Access to Inequalities in Material Access. *New Media Soc.* **2019**, *21*, 354–375. [CrossRef]
50. Acilar, A.; Sæbø, Ø. Towards Understanding the Gender Digital Divide: A Systematic Literature Review. *Glob. Knowl. Mem. Commun.* **2021**, *72*, 233–249. [CrossRef]
51. Cooper, J. The Digital Divide: The Special Case of Gender. *J. Comput. Assist. Learn.* **2006**, *22*, 320–334. [CrossRef]

52. Colding, J.; Barthel, S.; Ljung, R.; Eriksson, F.; Sjöberg, S. Urban Commons and Collective Action to Address Climate Change. *Soc. Incl.* **2022**, *10*, 103–114. [CrossRef]
53. Nilsson, M. 9 av 10 i LO-yrken Kaninte Jobba Hemifrån. 2021. Available online: https://ka.se/2021/03/04/9-av-10-i-lo-yrken-kan-inte-jobba-hemifran/ (accessed on 10 February 2024).
54. Findahl, O. *En Miljon Svenskar Vill Inte Använda Internet: En Rapport Om Digital Delaktighet*; SE: Stockholm, Sweden, 2013; ISBN 978-91-87437-06-9.
55. Hilty, L.M.; Aebischer, B.; Rizzoli, A.E. Modeling and Evaluating the Sustainability of Smart Solutions. *Environ. Model. Softw.* **2014**, *56*, 1–5. [CrossRef]
56. Galvin, R. The ICT/Electronics Question: Structural Change and the Rebound Effect. *Ecol. Econ.* **2015**, *120*, 23–31. [CrossRef]
57. Malmaeus, M.; Nyblom, Å.; Mellin, A. *Rekyleffekter och Utformning av Styrmedel*; IVL Svenska Miljöinstitutet: Stockholm, Sweden, 2021.
58. Colding, J.; Barthel, S.; Sörqvist, P. Wicked Problems of Smart Cities. *Smart Cities* **2019**, *2*, 512–521. [CrossRef]
59. Holmgren, M.; Kabanshi, A.; Langeborg, L.; Barthel, S.; Colding, J.; Eriksson, O.; Sörqvist, P. Deceptive Sustainability: Cognitive Bias in People's Judgment of the Benefits of CO_2 Emission Cuts. *J. Environ. Psychol.* **2019**, *64*, 48–55. [CrossRef]
60. Evans, J.S.B.T. Reflections on Reflection: The Nature and Function of Type 2 Processes in Dual-Process Theories of Reasoning. *Think. Reason.* **2019**, *25*, 383–415. [CrossRef]
61. Linder, N.; Giusti, M.; Samuelsson, K.; Barthel, S. Pro-Environmental Habits: An Underexplored Research Agenda in Sustainability Science. *Ambio* **2022**, *51*, 546–556. [CrossRef] [PubMed]
62. Chemero: An Outline of a Theory of Affordances—Google Scholar. Available online: https://scholar.google.com/scholar_lookup?&title=An%20outline%20of%20a%20theory%20of%20affordances&journal=Ecological%20Psychology&doi=10.1207/S15326969ECO1502_5&volume=15&pages=181-195&publication_year=2003&author=Chemero,A (accessed on 9 January 2024).
63. Kaaronen, R.O. Affording Sustainability: Adopting a Theory of Affordances as a Guiding Heuristic for Environmental Policy. *Front. Psychol.* **2017**, *8*, 1974. [CrossRef] [PubMed]
64. Huh, Y.E.; Vosgerau, J.; Morewedge, C.K. Social Defaults: Observed Choices Become Choice Defaults. *J. Consum. Res.* **2014**, *41*, 746–760. [CrossRef]
65. Reichman, W.E.; Fiocco, A.J.; Rose, N.S. Exercising the Brain to Avoid Cognitive Decline: Examining the Evidence. *Aging Health* **2010**, *6*, 565–584. [CrossRef]
66. Harada, C.N.; Natelson Love, M.C.; Triebel, K.L. Normal Cognitive Aging. *Clin. Geriatr. Med.* **2013**, *29*, 737–752. [CrossRef] [PubMed]
67. Hwangbo, H.; Yoon, S.H.; Jin, B.S.; Han, Y.S.; Ji, Y.G. A Study of Pointing Performance of Elderly Users on Smartphones. *Int. J. Hum. Comput. Interact.* **2013**, *29*, 604–618. [CrossRef]
68. Colding, J.; Barthel, S. An Urban Ecology Critique on the "Smart City" Model. *J. Clean. Prod.* **2017**, *164*, 95–101. [CrossRef]
69. Barbosa, B.; Vetere, F. *Ageing and Digital Technology: Designing and Evaluating Emerging Technologies for Older Adults*; Neves, B.B., Vetere, F., Eds.; Springer: Singapore, 2019; ISBN 9789811336928.
70. Li, Y.; Yang, Y.; Shi, S.; Wang, B.; Chen, G. Seniors' Knowledge-Based Digital Marginalization in the Era of Information Technology Advancements. *J. Knowl. Econ.* **2023**. [CrossRef]
71. Van Deursen, A.J.; Helsper, E.J. A Nuanced Understanding of Internet Use and Non-Use among the Elderly. *Eur. J. Commun.* **2015**, *30*, 171–187. [CrossRef]
72. Van Deursen, A.J.; van Dijk, J.A. The Digital Divide Shifts to Differences in Usage. *New Media Soc.* **2014**, *16*, 507–526. [CrossRef]
73. Kahila-Tani, M.; Broberg, A.; Kyttä, M.; Tyger, T. Let the Citizens Map—Public Participation GIS as a Planning Support System in the Helsinki Master Plan Process. *Plan. Pract. Res.* **2016**, *31*, 195–214. [CrossRef]
74. Tilly, C.; Wood, L.J. *Social Movements, 1768–2008*; Paradigm Publishers: Boulder, CO, USA, 2009; ISBN 978-1-59451-610-8.
75. Wenta, J.; McDonald, J.; McGee, J.S. Enhancing Resilience and Justice in Climate Adaptation Laws. *Transnatl. Environ. Law* **2019**, *8*, 89–118. [CrossRef]
76. Nussbaum, M.C. Capabilities as Fundamental Entitlements: Sen and Social Justice. In *Capabilities Equality*; Routledge: London, UK, 2005; ISBN 978-0-203-79944-4.
77. Berkes, F.; Colding, J.; Folke, C. *Navigating Social-Ecological Systems*; Berkes, F., Colding, J., Folke, C., Eds.; Cambridge University Press: Cambridge, UK, 2001; ISBN 978-0-521-81592-5.
78. Elmqvist, T.; Folke, C.; Nyström, M.; Peterson, G.; Bengtsson, J.; Walker, B.; Norberg, J. Response Diversity, Ecosystem Change, and Resilience. *Front. Ecol. Environ.* **2003**, *1*, 488–494. [CrossRef]
79. Loh, Y.A.-C.; Chib, A. Tackling Social Inequality in Development: Beyond Access to Appropriation of ICTs for Employability. *Inf. Technol. Dev.* **2019**, *25*, 532–551. [CrossRef]
80. Oosterlaken, I.; van den Hoven, J. Editorial: ICT and the Capability Approach. *Ethics Inf. Technol.* **2011**, *13*, 65–67. [CrossRef]
81. Radu, L.-D. Disruptive Technologies in Smart Cities: A Survey on Current Trends and Challenges. *Smart Cities* **2020**, *3*, 1022–1038. [CrossRef]
82. Lee, H.; Calvin, K.; Dasgupta, D.; Krinner, G.; Mukherji, A.; Thorne, P.; Trisos, C.; Romero, J.; Aldunce, P.; Barret, K.; et al. *IPCC, 2023: Climate Change 2023: Synthesis Report, Summary for Policymakers. Contribution of Working Groups I, II and III to the Sixth Assessment Report of the Intergovernmental Panel on Climate Change*; Core Writing Team, Lee, H., Romero, J., Eds.; IPCC: Geneva, Switzerland, 2023.

83. Colding, J.; Colding, M.; Barthel, S. Applying Seven Resilience Principles on the Vision of the Digital City. *Cities* **2020**, *103*, 102761. [CrossRef]
84. Hassankhani, M.; Alidadi, M.; Sharifi, A.; Azhdari, A. Smart City and Crisis Management: Lessons for the COVID-19 Pandemic. *Int. J. Environ. Res. Public Health* **2021**, *18*, 7736. [CrossRef] [PubMed]
85. Ramsetty, A.; Adams, C. Impact of the Digital Divide in the Age of COVID-19. *J. Am. Med. Inform. Assoc.* **2020**, *27*, 1147–1148. [CrossRef] [PubMed]
86. Van Dijk, J. *The Digital Divide*; John Wiley & Sons: Hoboken, NJ, USA, 2020; ISBN 978-1-5095-3446-3.
87. Michalec, A.; Hayes, E.; Longhurst, J. Building Smart Cities, the Just Way. A Critical Review of "Smart" and "Just" Initiatives in Bristol, UK. *Sustain. Cities Soc.* **2019**, *47*, 101510. [CrossRef]
88. Gehl, J. *Life Between Buildings: Using Public Space*; Island Press: Washington, DC, USA, 2011; ISBN 978-1-59726-827-1.
89. Bergquist, M.; Nilsson, A.; Harring, N.; Jagers, S.C. Meta-Analyses of Fifteen Determinants of Public Opinion about Climate Change Taxes and Laws. *Nat. Clim. Chang.* **2022**, *12*, 235–240. [CrossRef]
90. De Jong, M.; Joss, S.; Schraven, D.; Zhan, C.; Weijnen, M. Sustainable-Smart-Resilient-Low Carbon-Eco-Knowledge Cities; Making Sense of a Multitude of Concepts Promoting Sustainable Urba. *J. Clean. Prod.* **2015**, *109*, 25–38. [CrossRef]
91. Hollands, R.G. Critical Interventions into the Corporate Smart City. *Camb. J. Reg. Econ. Soc.* **2015**, *8*, 61–77. [CrossRef]
92. Maschio, I. European Innovation Partnership on Smart Cities and Communities. Available online: https://e3p.jrc.ec.europa.eu/articles/european-innovation-partnership-smart-cities-and-communities (accessed on 15 April 2024).

smart cities

MDPI

Article

Redesigning Municipal Waste Collection for Aging and Shrinking Communities

Andante Hadi Pandyaswargo [1,*], Chaoxia Shan [2], Akihisa Ogawa [2], Ryota Tsubouchi [3] and Hiroshi Onoda [2]

1 Environmental Research Institute, Waseda University, 1-104 Totsukamachi, Shinjuku-ku, Tokyo 169-8050, Japan
2 Graduate School of Environment and Energy Engineering, Waseda University, 513 Waseda Tsurumaki-cho, Shinjuku-ku, Tokyo 162-0041, Japan; shanchaoxia9816@ruri.waseda.jp (C.S.); ogajanchan0317@akane.waseda.jp (A.O.); onoda@waseda.jp (H.O.)
3 Daiei Kankyo Research Institute Co., Ltd., H¹O Kanda, Kanda Higashimatsushita-cho, Chiyoda-ku, Tokyo 101-0042, Japan; tsubouchi11925@dinsgr.co.jp
* Correspondence: andante.hadi@aoni.waseda.jp

Abstract: Due to aging and depopulation, cities in Japan struggle to maintain their municipal waste collection services. These challenges were exacerbated by the pandemic. To overcome these challenges, a prototype of collective and contactless waste collection technology has been developed. However, its acceptance by society is unknown. In this study, we surveyed Japanese people's preferences regarding household waste disposal. The results showed that older adults (older than 60) are willing to walk longer (more than 2 min) to carry their waste to the disposal site than younger adults. They are also less concerned about the risk of disease infection from touching other people's garbage than younger respondents (at a 0.24 count ratio). Other significant findings are that people who live alone prefer the temporary disposal site to be placed more than one minute away from their house (at a 0.19 count ratio). People living alone also produce less plastic and packaging waste than larger households. With more Japanese older adults living alone because of the scarcity of older-adult care facilities, we proposed two waste collection strategies that can allow for the implementation of more collective and automatized contactless waste pickup technology. Each design poses different challenges, such as the need for residents' cooperation and a higher energy supply. However, they also open new opportunities, such as encouraging active aging and using renewable energy.

Keywords: depopulation; aging society; smart cities; municipal waste collection; smart technology

1. Introduction

Demographic shifts such as population aging and shrinking cities have significant implications for sustainability in rapidly urbanizing contexts [1]. While they pose challenges, they also present opportunities; the growing population of skilled, healthy, and active older adults can engage in sustainability initiatives at the local level, contributing to sustainable development. The rate of demographic aging in Japan is faster and more intense than that in many other countries around the world [2]. In 2021, the Japanese population aged 65 and older reached 30% of its total population [3]. This percentage is much greater than that of the world's other largest economies, such as the US at 16.8% [4] and China at 18.1% [5]. Strategies that Japan adopts to address issues arising from demographic time bombs could become a reference for other countries [6]. It is projected that, eventually, the world will face a similar aging demographic transition [7,8]. Although an aging demographic has been referred to as a time bomb [6], Ref. [9] saw the positive side of it; the trend of the aging demographic has opened up opportunities to take action to reap the benefits of longevity.

The use of digital technologies in smart cities can support the livelihood of an aging society. Japan's know-how in developing urban city infrastructures with digital technologies has been obtained not only through its domestic experiences; Japan has also performed a significant number of international urban development projects. Some major active international examples are Vietnam, Thailand, Indonesia, the United States, Central and South America, and Australia [10]. A new term for smart cities was created to distinguish the shift made in the Japanese smart city paradigm: Society 5.0. Fundamentally, the concept of Society 5.0 is similar to other terminologies used by various scholars and institutions, such as the "super city" [10], "super smart society" [11], "smart city 2.0" [12], and "responsive city" [13,14]. Society 5.0 aims to balance economic advancement with the resolution of social problems through Artificial Intelligence adoption, which can help identify the short-term and long-term effects of different services [15].

Waste management services are an important aspect of Society 5.0. The sector has been determined to improve recycling efficiency and energy use through technological intervention [10]. Recycling efficiency and energy use depend largely on the quality of the source-separated waste [16]. Therefore, waste collection plays a significant role in ensuring the achievement of waste management in Society 5.0. This study explored the social problems Japanese cities face in the municipal waste collection sector and developed the technology intervention needed.

Over the past several decades, many Japanese cities have experienced shrinkage. A total of 81.6% of the municipalities in Japan are shrinking in population. This phenomenon has been experienced in rural cities and the suburbs of metropolitan areas in Japan [17]. A case study in Kitakyushu, Japan, showed that the shrinking phenomenon is correlated with the duration of residence, underage population ratio, and aging population ratio [18].

The phenomenon of shrinking cities is occurring in all municipalities in Japan, with Central Tokyo serving as the only exception [19]. In 2022, very few municipalities in Japan had a density of more than 2800 people/km^2 (Figure 1), which is the average density of cities in high-income countries [20]. This phenomenon has led to the deflation of housing assets and has become a new barrier to promoting compact cities [19], let alone smart cities, which are often associated with high initial costs [21]. During the latter half of the 1980s, asset price inflation in Japan, specifically for high-rise condominiums, was caused by the myth that the price would rise [22]. Because the expectations were unmet, the number of deserted dwellings is increasing yearly. Previous studies have reported that the ratio of desertified houses to dwelling houses is greater than 10% for all municipalities in Japan [23]. By 2018, the number of desertified dwellings had reached 8.5 million units, or approximately 14% of Japan's total housing stock [24].

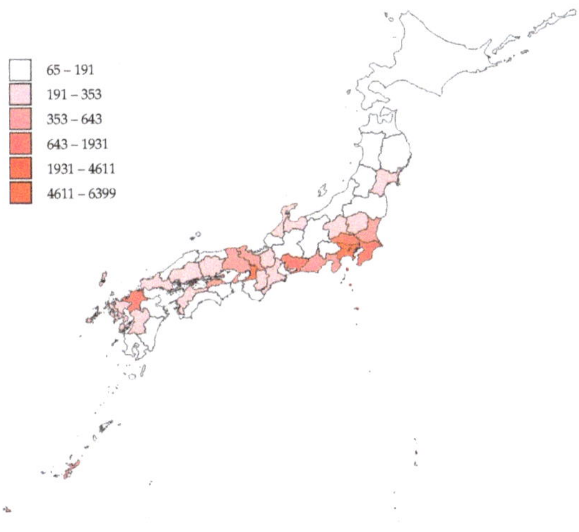

Figure 1. Population density in Japanese prefectures (people/km^2) (adapted by the authors from [25] data).

By adjusting to the current situation in aging and shrinking communities, this study sought to redesign a municipal waste collection system that is more appropriate and maintainable. To achieve these objectives, this study is organized as follows: (1) we collected and analyzed survey data from 1000 Japanese people to understand their perspectives on contactless and more collective waste collection technology through the use of exploratory statistical tools; based on the findings, (2) we proposed two new design alternatives for smart waste collection systems that are more appropriate from the perspectives of waste collection workers and civil society in the current situation.

2. Literature Review

2.1. Municipal Waste Collection in Japanese Aging and Declining Communities

The waste collection rate in Japan is currently greater than 90% [26], and it is managed by municipal governments. Such a near-universal collection rate is expected in high-income countries [27]. However, as the country struggles to maintain city infrastructure related to shrinkage [23], the waste management sector also faces several problems. Japan is currently facing severe labor shortages. The most severe shortage occurs in the face-to-face service sector but is expected to spread to a wide range of industries in the country [28]. The pandemic has triggered worsening conditions where strict immigration measures were imposed during the pandemic and reduced immigrant intake [28,29], preventing sufficient labor intake from abroad. The workforce potential of women is also limited because they are still commonly expected to handle the job of caregiving for both children [30] and elderly family members [31].

A previous study thoroughly discussed how the pandemic changed the dynamics of waste management in Japanese smart cities [32]. Given the speed of depopulation and aging, the need for smart waste management technologies is inevitable and urgent. To seek a concrete solution, we performed participant observation in a previous study by riding a city's waste collection truck with a waste collection worker on their normal operating day. This observation resulted in an understanding of the demand for smart contactless waste collection technology from the worker's perspective [33].

Thus, there is a need to design a smart waste collection technology that responds not only to the challenges faced by collection workers but also to the characteristics and preferences of aging Japanese civil society.

2.2. Development of Contactless Waste Collection Technology in Japan

In a previous participant observation study in a Japanese city, we revealed daily waste management activities and identified the need for smart technology from the perspective of waste collection workers. The study resulted in a pilot development project of a contactless and automated waste collection technology [33]. The study also called for a survey to reveal whether such a technology can be implemented in Japanese society. The developed technology allowed for (1) the reduction of physical human workers; and (2) a contactless manner of lifting, transporting, and dumping waste containers. Due to the more collective nature of waste pickup using this technology, large waste containers must be placed slightly farther than the currently used house-front temporary disposal locations. The experts consulted in the previous study raised a concern about whether the technology would be acceptable considering that aging Japanese people are presumably less fit than younger people. Additionally, detached houses represented approximately 80% of the surveyed case area, and many were small alleys [33]. This situation would mean that a significant number of people could carry their waste bags to the prepared waste containers that are likely to be placed on larger roads for waste packer trucks to have better access (Figure 2). To ensure that the technology is suitable for implementation, clarification is required to determine whether the developed technology needs to be equipped with smaller autonomous mobility devices to pick up waste from house fronts if people are unwilling or unable to carry it to temporary disposal sites.

Figure 2. Contactless collection device that requires residents to dispose of waste in collective (green) waste containers. source: [33].

3. Materials and Methods

3.1. Data Collection

We prepared questionnaires to gather information on Japanese people's perspectives. For the questionnaire distribution, we commissioned Rakuten Insight, who programmed it into a digital survey and sent them online to its panel respondents between 2 and 7 August 2023.

Rakuten Insight was chosen to support the survey collection because of the wide reach of the panel, particularly for the Japanese people. As of July 2022, Rakuten Insight had 15.1 million global panel members, 2.2 million of whom are from the Japanese population [34]. Rakuten Insight performed an annual survey of its panels using survey content similar to that used in the Japanese national census to ensure that the quality of the sample panel was consistent [35]. In this study, Rakuten Insight sent a stratified random sampling survey on our behalf to its Japanese panel. We requested equal proportions of (1) gender, (2) age, and (3) location in Japan (Northern, Central, and Southern Japan) to ensure that we avoid bias and to allow us to derive meaningful insights from those respondent attributes. Furthermore, the particular interest of this research was to observe the inclusivity of the aging society and shrinking cities in Japan. Therefore, we considered stratified random sampling to be the most appropriate method to ensure impartial and equal opportunity in the sample selection process [36]. Rakuten Insight contacted 16,000 individuals and received approximately 1300 responses, resulting in an 8.13% response rate. One thousand responses were received after incomprehensible and ineligible responses were eliminated. Figure 3 shows the locations of the respondents on Japan's map. Table 1 presents the descriptive statistics of the basic social attributes of the respondents. The largest share of respondents is living in a household of 2 people (33%). This group is largely made up of adults older than 60 years old. Very few people are living in households of 5 people and above. It is also very rare to see older adults living in a large household. This may indicate that multi-generational living is no longer practiced. Meanwhile, those in their 60s, 70s, 80s and older also make up a large share of those unemployed.

Table 2 summarizes the respondents' responses to the following characteristics: (1) acceptable time to carry waste to the temporary disposal site, (2) weekly household plastic and packaging waste volume, and (3) need to avoid touching other people's waste to reduce the risk of disease infection. The actual questions and the close-ended answer options are presented in Appendix A. For reference, 45 L trash bags are commonly used in Japanese municipalities, along with other smaller trash bags, such as 20 L trash bags [33].

Figure 3. Location of survey respondents.

Table 1. Summary of respondents' basic social attributes.

Variable	by Age	Obs.								Percentage (%)
		10s	20s	30s	40s	50s	60s	70s	More than 80s	
Gender										
	Female	86	78	56	68	58	53	46	55	50
	Male	39	47	69	57	67	72	79	70	50
Household size										
	1 person	31	46	33	19	32	28	22	22	23
	2 persons	12	19	25	24	36	61	74	80	33
	3 persons	20	30	29	39	32	24	18	14	21
	4 persons	38	23	27	27	15	6	6	7	15
	5 persons	19	5	9	10	8	3	1	1	6
	6 persons	4	2	2	1	1	0	2	1	1
	7 persons	0	0	0	3	0	2	2	0	1
	8 persons	0	0	0	0	1	0	0	0	0.1
	9 persons and more	1	0	0	2	0	1	0	0	0.4

Table 1. *Cont.*

Variable	by Age	Obs.								Percentage (%)
		10s	20s	30s	40s	50s	60s	70s	More than 80s	
Occupations										
	Company employee—general employee	3	41	41	40	35	20	3	1	18
	Company employee—comprehensive work	0	12	11	8	7	6	0	0	4
	Company employee—board member)	0	1	5	2	9	0	3	3	2
	Civil servant/university teacher/non-profit organization	0	12	11	8	11	6	0	1	5
	Temporary employe-es/contract employees	0	1	4	10	5	7	5	1	3
	Self-employed	0	2	4	8	7	3	6	6	4
	Agriculture/forestry/fisheries	0	0	0	2	0	0	1	0	0.3
	Legal professional	0	0	2	0	0	1	0	0	0.3
	Medical professional	0	12	12	6	5	4	1	0	4
	Part-time worker	4	11	10	21	17	17	12	2	9
	Homemaker	1	4	12	10	17	28	22	23	12
	Student	115	22	0	2	0	0	0	0	14
	Unemployed	2	7	12	6	9	32	68	84	22
	Other (specify)	0	0	1	2	3	1	4	4	2

Table 2. Summary of respondents' responses on their household waste disposal.

Acceptable Time to Carry Waste to Temporary Disposal Site	Frequency	Percentage (%)
Less than 1 min	401	40.1
1 to 2 min	306	30.6
2 to 5 min	228	22.8
5 to 10 min	31	3.1
More than 10 min	10	1
I have never experienced carrying trash to the temporary disposal site, so I do not know	11	1.1
I do not know	13	1.3
Weekly household plastic and packaging waste volume in 45 L waste bags		
Less than 1 bag	591	59.1
1 to 2 bags	311	31.1
3 to 5 bags	57	5.7
6 bags or more	4	0.4
Since the plastic waste is not sorted, it is not possible to calculate the conversion to a 45 L garbage bag.	37	3.7
I do not know		
Needs to avoid other people's waste to reduce risk of disease infection		
Very necessary	483	48.3
Necessary	341	34.1
Neutral	128	12.8
Unnecessary	38	3.8
Not necessary at all	10	1

3.2. Methodological Framework

The methodological framework of this study is summarized in Figure 4. The formulas employed in this study follows those that were elaborated in [37,38]. The detailed statistical tools employed are described in Appendix B.

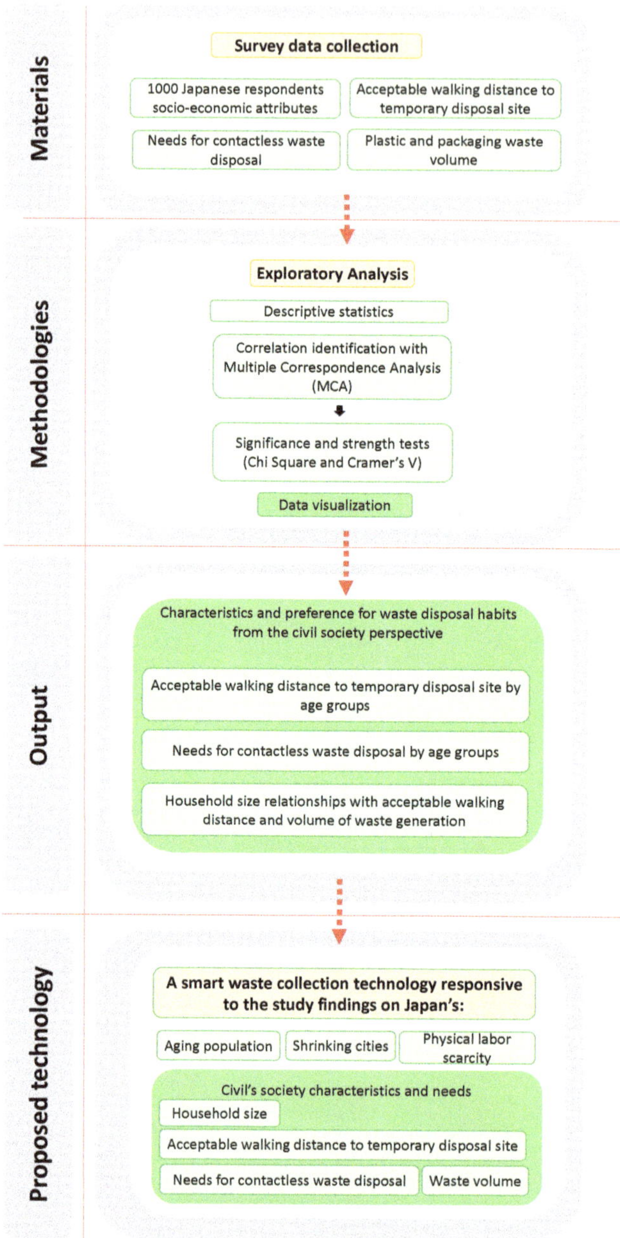

Figure 4. Methodological framework. This study analyzed the literature, collected survey data, and identified respondents' characteristics and needs to develop an appropriate smart waste collection technology through statistical approaches.

3.3. Multiple Correspondence Analysis (MCA) Preparation

Although the multiple correspondence analysis (MCA) does not require model construction, the data must be prepared into groups of classifications for easier interpretation. The collected data are organized into the categories presented in Table 3. Data categorization was guided by the frequency quartiles of the data, where applicable. The detailed walkthrough of the MCA employed in this study was described previously [38]. The functions and packages employed are from the R programming language FactoMineR. We followed the guidelines and principles of MCA from [39,40]. The codes for the graphics were adapted from [41,42].

Table 3. Response categories.

Attributes and Responses	Obs	Unit	Categories (Ranges), Frequency			NA (Excluded from MCA)
			1st (Lo)	2nd (Med)	3rd (Hi)	
Age	1000	Year	Low (≤29), 250	Med (≥30, ≤59), 375	High (≥60), 375	
Gender	1000		Female, 500	Male, 500		
Occupation	985		Low (students and unemployed), 359	Med (part-time worker and homemaker *), 211	High (full-time worker **), 415	Other, 15
Household size	1000	person	Low (=1), 233	Med (=2), 331	High (≥3), 436	
Acceptable time to carry waste	976	Minute	Low (<1), 401	Med (≥1, ≤2), 306	High (>2), 269	I do not know, 24
Weekly waste volume	963	Bag	Low (<1), 591	Med (≥1, ≤2), 311	High (≥3), 61	Unknown due to nonseparation, 37
Needs for contactless technology	1000		Low (not necessary at all and unnecessary), 48	Med (Neutral), 128	High (necessary and very Necessary), 824	

* A homemaker is a person who manages a household and does not work outside the house to gain financial incentives. ** A full-time worker refers to a person working a minimum of 5 days a week, 7 h a day.

4. Results

4.1. Relationships Identified by the MCA

The graphical output of the MCA is shown in Figure 5. The lighter areas indicate a stronger relationship between the attributes. The black dots represent observations. Two clusters of light-colored clouds were identified. The first cluster centers on the lower acceptable time to carry waste to the temporary disposal site (AT_Lo), and the second cluster centers on the lower need for contactless technology to reduce the risk of disease infection (NC_Lo).

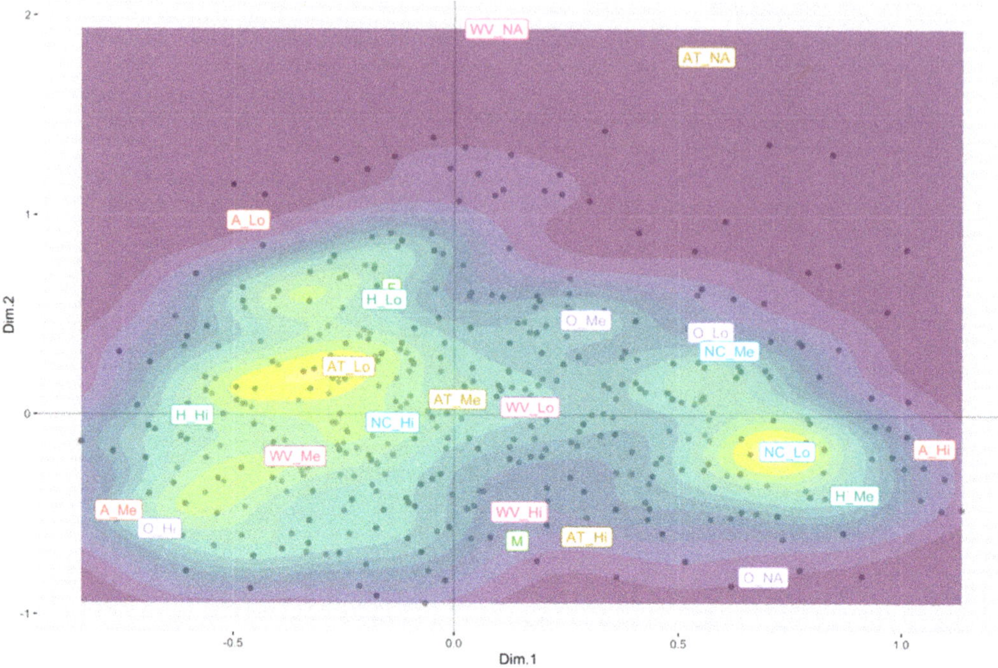

Figure 5. MCA plot. A, age; G, gender; O, occupation; H, household size; F, female; M, male; AT, acceptable time (walking distance to temporary disposal site); WV, waste volume (plastic and packaging waste volume generated by the household); NC, need for contactless technology (necessity of a technology to prevent the chance of touching other people's waste to prevent risk of disease infection); Hi, high; Me, medium; Lo, low.

4.2. Significant Relationships Identified by Nonparametric Tests

Chi-square and Cramer's V tests were performed to (1) test which respondent's attributes or variables were related to the identified clusters, (2) determine the significance and strength of the relationships, and (3) identify the direction of the relationships. The identified variables and attributes of the significant relationships (as indicated by the value $p < 0.001$) are summarized in Appendix C.

Four pairs of variables showed significant relationships (Appendix C): (1) the younger group of respondents (younger than 30 years) was the most hesitant to walk longer (more than 1 min) to carry waste to the temporary disposal sites; (2) the oldest group of respondents (older than 60 years) tended to express the unnecessity of a technology to reduce the possibility of touching other people's waste to minimize the risk of disease infection; (3) the smallest group of households (people who live alone) tended to be more willing to walk a medium distance to the waste disposal site (more than 1 min but less than 2 min); and (4) household size was positively related to waste volume, as larger households produced more waste than smaller households.

Although the four pairs of relationships are only weakly significant (Cramer's V lower than 0.25) [43], they are all deemed significantly correlated ($p < 0.001$). The four pairs of correlated attributes are further detailed in Figures 6–10. As age groups are evenly distributed, meaningful insights can be derived from observing the frequency of age-related responses (Figures 6 and 7). On the other hand, as household size is not evenly distributed among respondents, Figures 8–10 present the relationships related to counts and expected count ratios. The following implications can be made from observing the figures: (1) older people are more willing to walk longer distances to temporary disposal sites (up to 5 min) than younger people are, (2) older people are less concerned about touching other people's waste than younger people are (at a 0.24 count ratio), (3) the smallest household (consisting of just one person) tends to be willing to walk more than the minimum distance (1 to 2 min) but not longer distances (at a 0.19 count ratio), and (4) larger households produce more waste than smaller households.

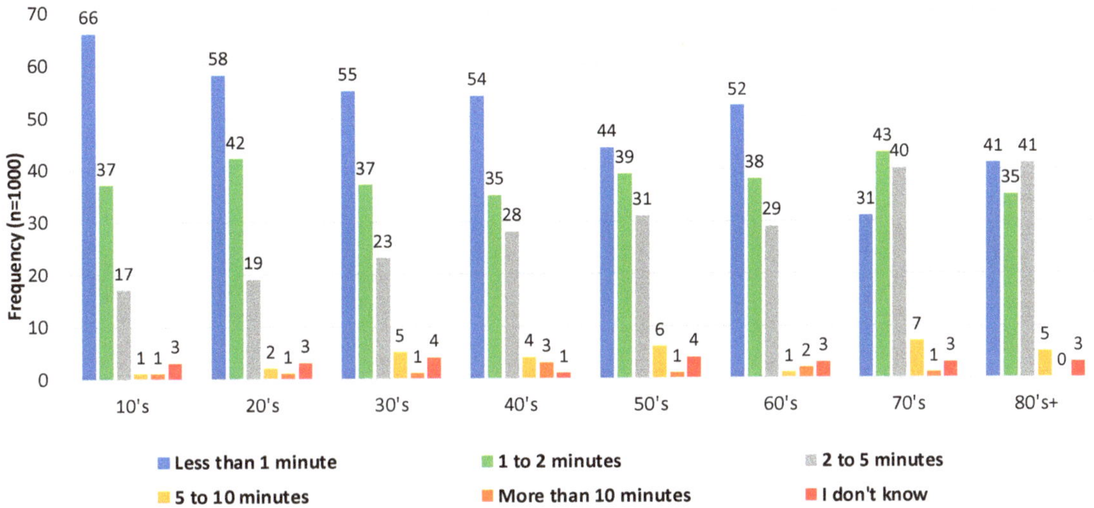

Figure 6. Responses to acceptable time to carry waste to the disposal site by age.

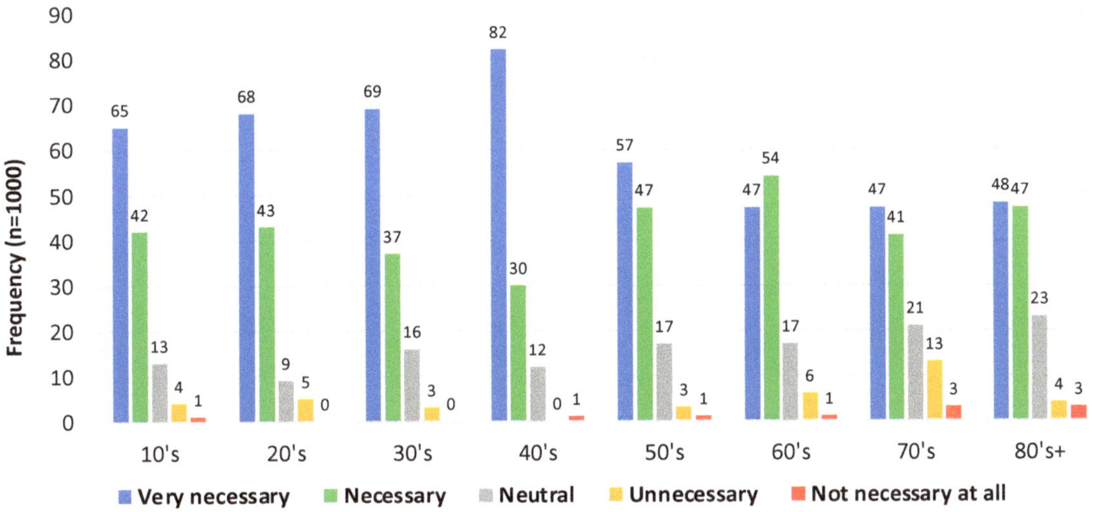

Figure 7. Responses to the need for a technology that reduces the possibility of touching other people's waste to lower the risk of disease infection by age.

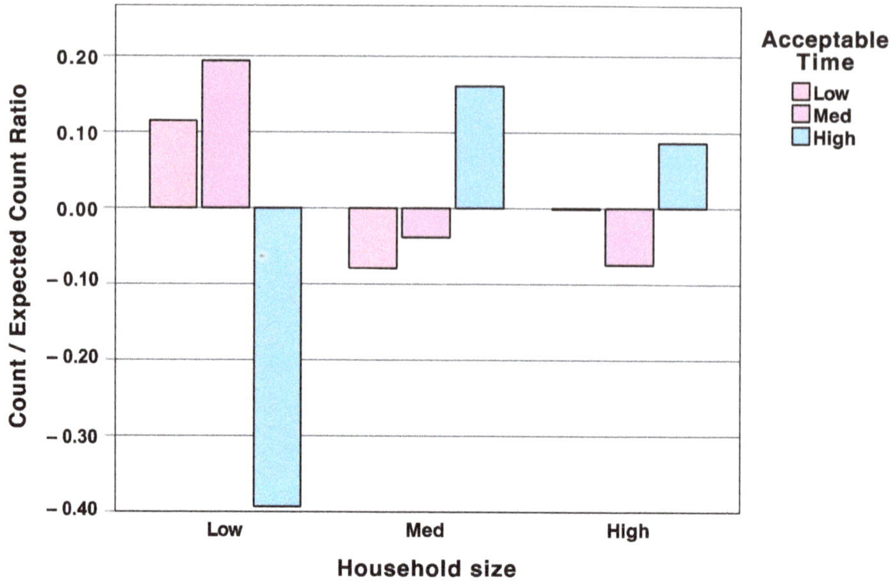

Figure 8. The count and expected count ratio of responses on acceptable time to carry waste to the temporary disposal site by household size.

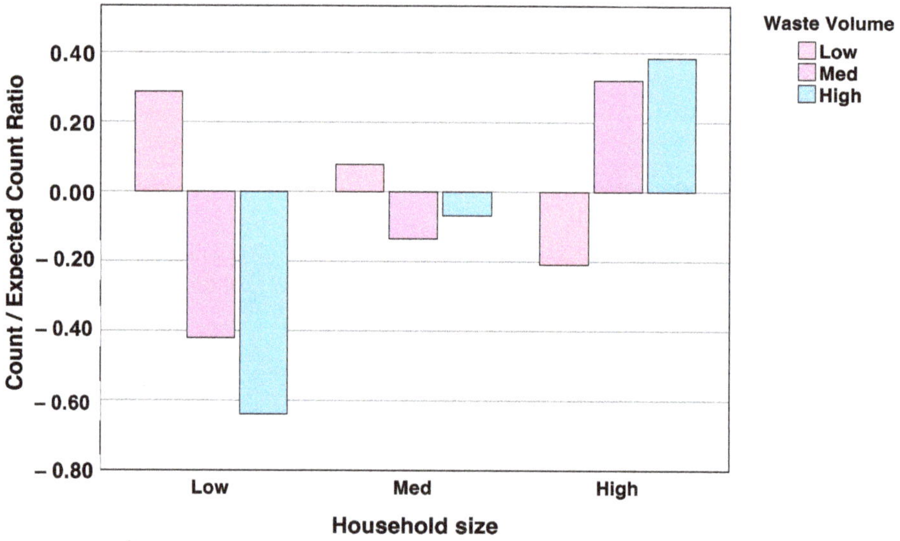

Figure 9. Count and expected count ratios of the responses to waste volume by household size.

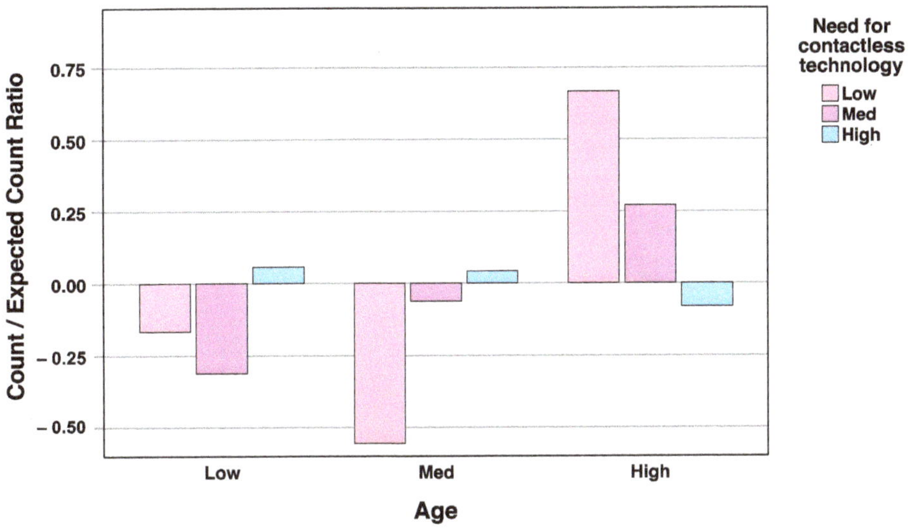

Figure 10. Count and expected count ratios of the response on need for contactless technology by age.

5. Discussions

Aging is often associated with a decline in physical fitness. However, the findings of this study revealed that older people are willing to walk longer distances when carrying their waste to temporary disposal sites than younger people. Although Japan faces the challenge of maintaining the availability of physical workers [28], including the workforce carrying and lifting waste containers that weigh hundreds of kilograms [33], this does not mean that older adult residents are unwilling to walk a certain distance. This study revealed that the oldest group (in their 70s and 80 s) showed greater willingness to walk longer distances (up to 5 min) than did the younger age group (Figure 6). Conversely, the youngest groups of respondents (those in their 10s, 20s, 30s, 40s, and 50s) showed the lowest tolerance for the duration of walking to carry waste, with a nearly negative linear relationship. This finding is surprising because strength and physical perseverance are often expected of young people. On the other hand, it is reassuring that Japan's older adults are willing to walk longer because walking is associated with mental health and active living maintenance [44,45].

The Japanese older adults are known for their longevity [46]. The Japanese society also encourages active aging as a way to extend healthy life expectancy and quality of life in older persons [47]. Perhaps, these are the reasons behind the older adults' greater willingness to walk longer distances than the younger group. However, this result should not discourage the development of autonomous and micro-mobility technologies for waste collection. It is also known that the very advanced age population (75 years old and above) have challenges in their mobility, such as when using public transportation [38]. Therefore, the results of this study should be used to consider the most cost-effective and sustainable technological solution where necessary, including possible acceptance of all age groups.

Another surprising finding is that, although there is a popular belief that older people are at greater risk of contracting diseases, the analysis results of this study revealed that this group of people is less concerned about the risk of disease infection from unintended contact with other people (Figure 7). Conversely, the youngest group of respondents (in their 40s and below) were the most worried and showed a strong need for a technology that could allow for contactless waste disposal. The COVID-19 pandemic has triggered various concerns regarding disease infection possibilities from household waste. These concerns are particularly related to handling single-use masks, testing kits, and other products used by an infected person. Studies have highlighted the need for proper disposal methods, such as incineration to kill pathogens [48], appropriate resting time to decrease the risk of spread of the disease [49], and the possibility of contactless waste pickup [33]. The present study is the first to highlight what the general public thought about this matter, especially among people of various age groups and after the vaccine was widely implemented in society. By the time the survey in this study was taken, most Japanese people were already in their fifth or sixth round of COVID-19 shots. While the virus is currently seen at a similar level of threat as influenza in Japan, there are, of course, future possibilities of a pandemic occurrence similar to that of COVID-19.

In terms of household size, those who live alone are willing to walk more than respondents with larger households (Figure 8). There are many possible reasons why more people who live alone choose the medium walking distance (1 to 2 min) rather than the shortest distance (<1 min). Several possibilities are (1) because no one is expecting them at home to return soon, (2) because of a preference to have their waste disposed farther from their living place to prevent unpleasant smells or other hygiene issues, or (3) because the weight to carry is lighter than that of larger households. However, many Japanese older adults are living alone, and the prevalence of this condition is increasing due to the insufficient capacity of older-adult care facilities [50]. Therefore, to anticipate rapid demographic and social changes in Japan, responses from older age groups and the responses of the single-person households in this study should be seriously considered by waste collection service providers when redesigning waste collection strategies.

Proposed New Designs Based on Findings

In a previous study, the opinions and concerns of city waste collection service operators were incorporated into a pilot demonstration project of a smart waste collection device [33]. This study collected information on the characteristics and preferences of the aging society of Japan for waste disposal activities to determine whether such devices are socially acceptable. The results of both studies, supported by the literature review, provided insights into redesigning Japan's waste collection strategy.

Japan's waste collection practice is facing maintenance difficulties due to labor shortages and aging. We propose two designs to solve this problem, as presented in Figure 11a,b. In the old strategy, people who live in detached houses typically place their waste bags in front of their houses, on sidewalks and corners. The manual collection of these waste bags has physically strained scarce and aging waste collection workers [33]. Since the present study revealed that most older residents are willing to walk up to 5 min to carry their waste to a temporary disposal site, we suggest design (a), which involves placing larger waste containers (the one in Figure 11a is 700 L in capacity) between 1- and 5-min walking distances to allow for a more collective waste collection strategy using the developed contactless waste container dumping device. This new strategy responds to (1) waste collection worker scarcity and physical strains, (2) truck access to containers that can be placed only in larger areas, (3) aging residents' willingness to walk slightly farther, and (4) single-person households' preference for having waste bins placed farther from their living place (more than a 1-min walk). To anticipate the declining physical strength of very advanced age population and to accommodate the needs of people with limited mobility, we also propose design (b), which involves developing autonomous pickup, dumping, and returning of waste bins to the house fronts and corners of residential blocks.

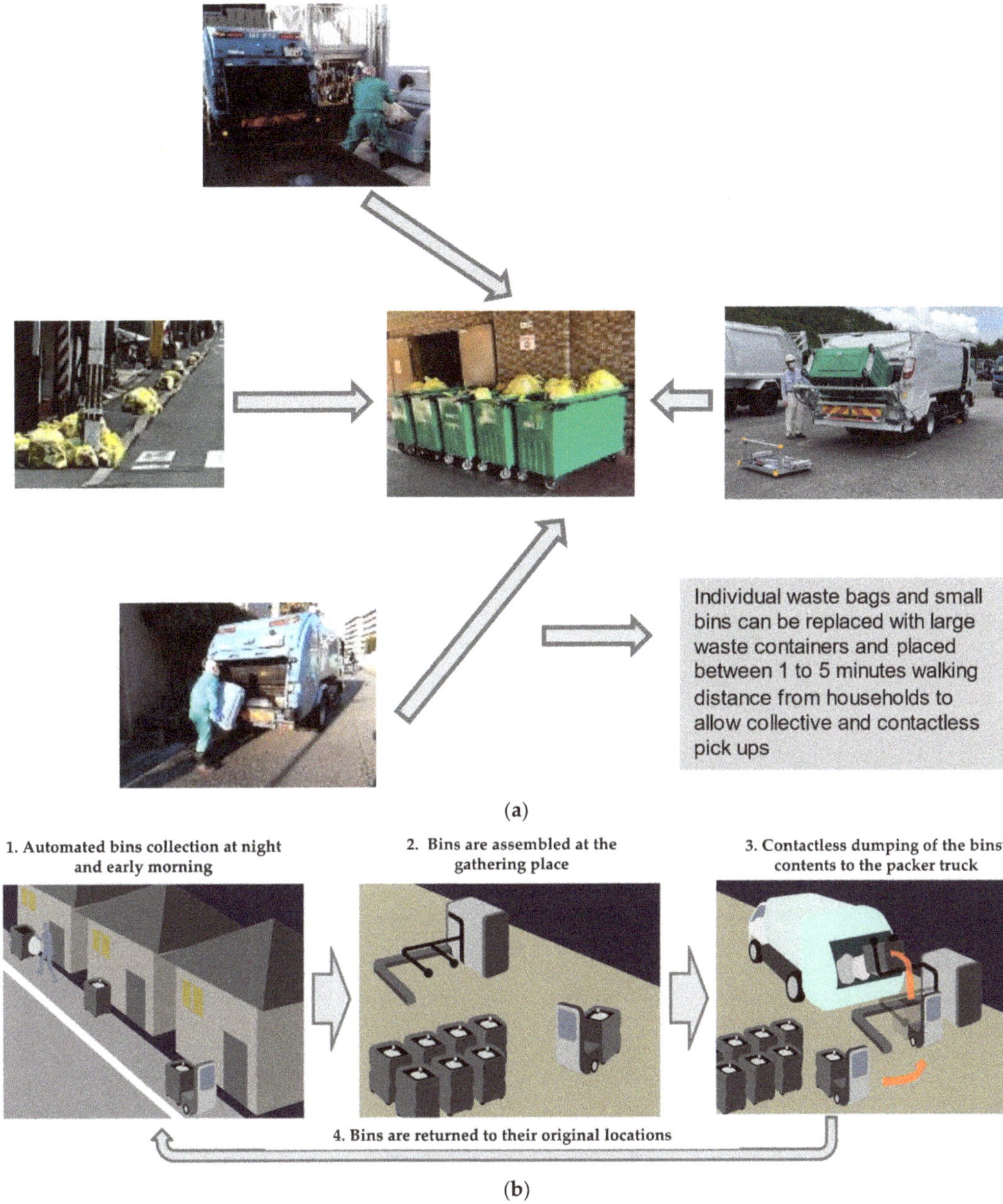

Figure 11. Redesigned household waste collection to reduce collection workers with (**a**) increased walking distance and (**b**) the use of automatization technologies [51].

Both strategies may be faced with various challenges, opportunities, and policy implications. Table 4 provides a summary of these factors to be considered by decision makers and practitioners. While design (a) may bring opportunities of encouraging active lifestyle and communal monitoring, it may be faced with resistance by those unwilling to walk longer to the temporary disposal site. Furthermore, people's cooperation and understanding should be gained through a possibly lengthy process of socialization. Meanwhile, design (b) may be quicker in terms of implementation due to changes that are mainly required at the management levels, but it may imply a higher demand for energy supply. Ensuring that the energy supply comes from renewable resources may improve the sustainability of the strategy.

Table 4. Challenges, opportunities, and policy implications of the proposed waste collection designs.

	Design (a)—Increased Walking Distance to Temporary Disposal Sites	Design (b)—The Use of Autonomous Pickup Mobility Technology
Challenges	- Resistance by those who are not willing to walk longer. - Determining a spacious area that can accommodate collective temporary disposal site.	- Increased demand of energy supply to run the technology. - The need for capital cost, operation, and reinvestment (such as battery replacement) of the technology.
Opportunities	- Encourages active lifestyle and sense of community. - Communal monitoring of waste separation compliance.	- The use of renewable energy for energy supply. - Introduction of automatization of the whole systems using the Internet of Things (IoT) and Artificial Intelligence (AI) for sensors and monitoring, resulting in a further reduction in human resources also at the management level.
Policy Implications	- The need for socialization of the new system to gain residents' cooperation and permission, especially by those living near the temporary disposal sites.	- The need for new policies regulating cyber security and liability issues.

6. Conclusions

Japan's cities are shrinking due to population aging, low birth rates, and the outflow of residents. The smart city strategy has recently been renamed Society 5.0 to highlight its ability to respond to social problems. Society 5.0 has been adopted to rejuvenate and accommodate the changing demography of the Japanese population and improve the maintainability of the city's services and infrastructures.

In the waste collection sector, the scarcity and aging of the physical workforce have put a strain on maintaining current practices. However, the proposed smart technology for replacing human workers in this area has raised questions about its acceptability by civil society. These concerns were raised because of (1) the characteristics of the new technology, which requires people to walk farther to carry their waste to the waste container for collective pickup; and (2) the aging general population, which is presumably less physically fit to carry their waste.

The present study collected surveys of 1000 Japanese people who were equally distributed in terms of location (Northern, Central, and Southern Japan), age (10s, 20s, 30s, 40s, 50s, 70s, 80s, and older), and gender. We analyzed the survey data using an exploratory statistical method, MCA, and nonparametric tests (Chi-square and Cramer's V) to identify the relationships, significance, and strength of the relationships. Four significant relationships were found: (1) older adults are willing to walk longer when carrying their waste to the temporary disposal site, (2) older adults are less concerned about disease infection from unintentional contact with other people's household waste, (3) single-person households prefer temporary waste disposal sites to be placed farther from their living place, and (4) plastic and packaging waste volume are positively related to household size. While these findings are significant, Cramer's V results indicated that they are only weakly significant. Interpreting such findings, the relationship may not be used to push for a certain strong policy or action. However, it might mediate a dispute about a solution or social-issue technological intervention. The present study uses this information to determine the appropriateness of a new smart waste collection technology that responds to the needs and challenges faced by waste collection workers.

The proposed new strategies in this study will reduce the number of workers and the contactless waste pickup desired by the workers. However, "design (a)", which is a more collective strategy, requires larger containers that must be placed farther than the currently practiced residents' house fronts and sidewalks. Since older respondents in this study are willing to walk farther to carry their waste bags, and those who live alone actually prefer a disposal site, replacing individual waste bags and small containers with larger waste containers placed between 1 and 5 min, the walking distance would be ideal. In this way, a more collective waste pickup by the smart contactless waste collection device can be realized. On the other hand, this strategy may be faced with challenges such as opposition by those unwilling to walk longer the disposal sites and by people who live near the temporary disposal sites. Meanwhile, "design (b)", which is the use of automatization technologies, may be faced with the challenge of increased energy demand to run the technologies. Ensuring that the power supply comes from renewable resources will improve the sustainability of this strategy.

For actual implementation, pilot studies in Japanese cities are required to test the acceptability of the new designs. This study's limitation is that it is specific to the Japanese population. However, Japan is leading globally in terms of the speed and magnitude of aging, population decline, city shrinkage, and technological advancement. Therefore, it can be used as a reference for other aging countries that face similar problems in terms of securing physical workers and maintaining waste collection services.

Future research in this field should be dedicated to the "soft" and "hard" needs of new waste collection designs to accommodate the challenges faced by aging and declining communities. The "soft" needs include ensuring active aging, community acceptance, and gaining residents' cooperation if the low-tech alternative is adopted. The "hard" needs include a more aggressive introduction of renewable energy to supply the expected increased energy needs and ensure enough financial resources for investment and reinvestment for the devices required and the energy storage replacement if the high-tech alternative is adopted.

Author Contributions: Conceptualization, A.H.P.; methodology, A.H.P.; software, A.H.P. and C.S.; validation, A.H.P. and C.S.; formal analysis, A.H.P. and C.S.; investigation, A.H.P. and R.T.; resources, A.H.P. and H.O.; data curation, A.H.P., C.S. and A.O.; writing—original draft preparation, A.H.P.; writing—review and editing, A.H.P.; visualization, A.H.P., C.S. and A.O.; supervision, A.H.P. and H.O.; project administration, A.H.P., R.T. and H.O.; funding acquisition, A.H.P. and H.O. All authors have read and agreed to the published version of the manuscript.

Funding: This study was funded by (1) The Environmental Restoration and Conservation Agency under project number JPMEERF20213G01, titled "Development of Contactless Garbage Collection Systems and Scenarios Construction for Social Implementation" project; and (2) The Japan Society for the Promotion of Science under project number 23K11548, titled "Improving the social acceptance of smart technologies: A people-centric approach for sustainable smart waste management and smart mobility technologies".

Data Availability Statement: Data will be made available upon request.

Acknowledgments: We thank the people at Rakuten Insight, who were commissioned to collect the survey on our behalf. We also thank the reviewers, who have tremendously helped us improve the quality of this manuscript.

Conflicts of Interest: Author Ryota Tsubouchi was employed by the company Daiei Kankyo Research Institute Co., Ltd. The remaining authors declare that the research was conducted in the absence of any commercial or financial relationships that could be construed as a potential conflict of interest.

Appendix A

Actual questions asked to the respondents (English translated from Japanese language):

In your best estimation, how many 45 L trash bags of plastic and packaging waste does your household dispose of in a week?	
a.	Less than 1 bag
b.	1 to 2 bags
c.	3 to 5 bags
d.	6 bags or more
e.	Unknown because my household does not separate the plastic and packaging waste
How long are you willing to walk carrying your trash from your residence or place of employment to the temporary disposal site?	
a.	Less than 1 min
b.	1 to 2 min
c.	2 to 5 min
d.	5 to 10 min
e.	More than 10 min would still be acceptable for me
f.	I have never experienced carrying trash to the temporary disposal site, so I do not know
g.	I do not know
What do you think about the necessity of a technology that can help you avoid the chance of touching other people's waste (to reduce the risk of diseases infection)?	
a.	Very necessary
b.	Necessary
c.	Neutral
d.	Unnecessary
e.	Not necessary at all

Appendix B

This study employed MCA and nonparametric test methods, namely Chi-square (Equation (A1)) and Cramer's V (Equation (A2)), to analyze the collected data, where X^2 is Chi-squared, O_i is observed value, E_i is expected value, V is Cramer's V coefficient, n is the total number of observations, r is the number of rows in the contingency table, and c is the number of columns in the contingency table.

$$X^2 = \sum (O_i - E_i)^2 / E_i \tag{A1}$$

$$V = \sqrt{X^2 / n \times \min[(r-1), (c-1)]} \tag{A2}$$

And then, we measured the count/expected count ratio (Equation (A3)) to test the direction of relationships between the observed and the expected data counts (positive relationship if more than 0, and negative if less than 0) [37].

$$Count\ ratio = O_i / E_i - 1 \tag{A3}$$

The detailed procedures of the methodologies were elaborated in [38,52]. MCA has several strength points, which is why it was selected for this study. Its first strength is its exploratory ability to provide insights out of a large set of data [53]. This is an important point, especially because this study deals with a new technology for which the acceptance in the society is still largely unknown. The second strength is its interpretation-friendliness. One can discover the relationships among respondents' attributes and responses simply by observing the distance proximity of variables presented in the MCA graphical output. Other aesthetical indicators, such as a change in colors, can also indicate the relationships' intensity. The drawback of MCA is its inability to indicate either the significance or high accuracy of strengths of the identified relationships among variables. To compensate for this drawback, we performed nonparametric tests following the MCA. The Chi-square test was used in this study to prove or disprove the significance, and Cramer's V was performed to tell the strength of the identified significant relationships [43].

Appendix C

Identified significant relationships based on Chi-square and Cramer's V.

Cross Table Parameters with Significant Chi-Square Test p-Value	Test Results		Largest Count/Expected Count Ratio	
	Chi-Square	Cramer's V	(RowVariable_Category/ ColVariable_Category: Count)	Expected Count
Age with Acceptable time to carry waste to temporary disposal site	<0.001	0.118	A_Lo/AT_Hi:67.3	41
Age with Need for a technology to reduce the possibility of touching other people's waste to reduce the risk of disease infection	<0.001	0.110	A_Hi/NC_Lo:18	30
Household size with Acceptable time to carry waste to temporary disposal site	<0.001	0.097	H_Lo/AT_Me:71.2	85
Household size with Weekly waste volume	<0.001	0.179	H_Lo/WV_Hi:13.9	5

References

1. Jarzebski, M.P.; Elmqvist, T.; Gasparatos, A.; Fukushi, K.; Eckersten, S.; Haase, D.; Goodness, J.; Khoshkar, S.; Saito, O.; Takeuchi, K.; et al. Ageing and Population Shrinking: Implications for Sustainability in the Urban Century. *Npj Urban Sustain.* **2021**, *1*, 17. [CrossRef]
2. OECD Demography—Elderly Population—OECD Data. Available online: https://data.oecd.org/pop/elderly-population.htm (accessed on 22 July 2022).
3. The World Bank. Population Ages 65 and above (% of Total Population)—Japan | Data. Available online: https://data.worldbank.org/indicator/SP.POP.65UP.TO.ZS?locations=JP (accessed on 16 June 2023).
4. United Health Foundation. Explore Population—Adults Ages 65+ in the United States | AHR. Available online: https://www.americashealthrankings.org/explore/measures/pct_65plus (accessed on 16 June 2023).
5. Hao, M.; Wang, X.; Hou, S.; Zhang, Q.; Xue, T. The Elderly Acceptance of Autonomous Vehicle Services in Beijing, China. *Int. Rev. Spat. Plan. Sustain. Dev.* **2023**, *11*, 64–84. [CrossRef]
6. McCurry, J. Japan Will Be Model for Future Super-Ageing Societies. *The Lancet* **2015**, *386*, 1523. [CrossRef]
7. UN DESA. World Population Prospects—Population Division—United Nations. Available online: https://population.un.org/wpp/ (accessed on 6 October 2022).
8. UNESCAP. *Asia-Pacific Report on Population Ageing 2022*; UNESCAP: Bangkok, Thailand, 2022.
9. Scott, A.J. The Longevity Society. *Lancet Healthy Longev.* **2021**, *2*, e820–e827. [CrossRef]
10. Prime Minister's Office Japan. *Japan's Smart Cities*; Prime Minister's Office Japan: Tokyo, Japan, 2019.
11. Japan Science and Technology Agency. *"Super Smart Society (Society 5.0)" Mission Area 2017*; Japan Science and Technology Agency: Saitama, Japan, 2017.
12. Trencher, G. Towards the Smart City 2.0: Empirical Evidence of Using Smartness as a Tool for Tackling Social Challenges. *Technol. Forecast. Soc. Chang.* **2019**, *142*, 117–128. [CrossRef]
13. Damianou, A.; Vayona, A.; Demetriou, G.; Katos, V. An Actionable Maturity Planning Model for Smart, Circular Cities. *Cities* **2023**, *140*, 104403. [CrossRef]
14. Miaoudakis, A. *Ideal Cities: Intelligence-Driven Urban Internet-of-Things Ecosystems for Circular, Safe and Inclusive Smart Cities*; European Comissions: Brussels, Belgium, 2018.
15. Hitachi-UTokyo Laboratory(H-UTokyo Lab.). *Society 5.0: A People-Centric Super-Smart Society*; Springer: Singapore, 2020; ISBN 9789811529894.
16. Shan, C.; Pandyaswargo, A.H.; Onoda, H. Environmental Impact of Plastic Recycling in Terms of Energy Consumption: A Comparison of Japan's Mechanical and Chemical Recycling Technologies. *Energies* **2023**, *16*, 2199. [CrossRef]
17. Peng, W.; Wu, Z.; Duan, J.; Gao, W.; Wang, R.; Fan, Z.; Liu, N. Identifying and Quantizing the Non-Linear Correlates of City Shrinkage in Japan. *Cities* **2023**, *137*, 104292. [CrossRef]
18. Peng, W.; Fan, Z.; Duan, J.; Gao, W.; Wang, R.; Liu, N.; Li, Y.; Hua, S. Assessment of Interactions between Influencing Factors on City Shrinkage Based on Geographical Detector: A Case Study in Kitakyushu, Japan. *Cities* **2022**, *131*, 103958. [CrossRef]

19. Uto, M.; Nakagawa, M.; Buhnik, S. Effects of Housing Asset Deflation on Shrinking Cities: A Case of the Tokyo Metropolitan Area. *Cities* **2023**, *132*, 104062. [CrossRef]
20. OECD Library. The Changing Shape of Cities: Density and Suburbanisation. 2023. Available online: https://www.oecd-ilibrary.org/docserver/252899a3-en.pdf?expires=1715652875&id=id&accname=guest&checksum=418EBD147E25DF13123D32F911CF993D (accessed on 24 January 2024).
21. Macomber, J. Smart Cities Are Complicated and Costly: Here's How to Build Them. 2018. Available online: https://hbswk.hbs.edu/item/smart-cities-are-complicated-and-expensive-here-s-how-to-build-them (accessed on 14 January 2024).
22. Hino, K.; Amemiya, M. Spatiotemporal Analysis of Burglary in Multifamily Housing in Fukuoka City, Japan. *Cities* **2019**, *90*, 15–23. [CrossRef]
23. MLIT. *Promotion of Urban Renovation and Compact Cities*; MLIT: Tokyo, Japan, 2012.
24. The Japan Times. Reimagining Japan's Growing Glut of Empty Homes. 2022. Available online: https://www.japantimes.co.jp/news/2022/04/04/business/japan-abandoned-houses-solutions/ (accessed on 23 January 2024).
25. Higashide, M. Japan's 27 Prefectures' Population, Area, and Density Ranking 2023. Available online: https://uub.jp/rnk/iwate/p_j.html (accessed on 10 January 2024).
26. Plastic Atlas Asia Edition: Facts and Figures about the World of Synthetic Polymers. Available online: https://www.iges.or.jp/en/pub/plastic-atlas-asia-edition-2021/en (accessed on 12 May 2023).
27. World Bank. *What a Waste 2.0*; The World Bank: Washington, DC, USA, 2018; ISBN 978-1-4648-1329-0.
28. Nishioka, S. Severe Labor Shortage Puts Upward Pressure on Wages in Japan. *JRI Res. J.* **2023**, *6*, 1–8.
29. Okamoto, A. Immigration Policy and Demographic Dynamics: Welfare Analysis of an Aging Japan. *J. Jpn. Int. Econ.* **2021**, *62*, 101168. [CrossRef]
30. Iijima, S.; Yokoyama, K. Socioeconomic Factors and Policies Regarding Declining Birth Rates in Japan. *Nihon Eiseigaku Zasshi. Jpn. J. Hyg.* **2018**, *73*, 305–312. [CrossRef]
31. Aronsson, A. Professional Women and Elder Care in Contemporary Japan: Anxiety and the Move toward Technocare. *Anthropol. Aging* **2022**, *43*, 17–34. [CrossRef]
32. Onoda, H. Smart Approaches to Waste Management for Post-COVID-19 Smart Cities in Japan. *IET Smart Cities* **2020**, *2*, 89–94. [CrossRef]
33. Ogawa, A.; Pandyaswargo, A.H.; Tsubouchi, R.; Onoda, H. Demonstration of a Contactless Waste Collection System: A Japanese Case Study. *IET Smart Cities* **2023**, *5*, 303–316. [CrossRef]
34. Rakuten Insight. *Rakuten Insight Rakuten Insight Panel Book*; Rakuten Insight: Tokyo, Japan, 2022.
35. Nakamura, H.; Abe, N.; Mizunoya, T. Factors Inhibiting the Use of Sharing Economy Services in Japan. *J. Clean. Prod.* **2021**, *278*, 123981. [CrossRef]
36. Singh, A.S.; Masuku, M.B. Sampling Techniques & Determination of Sample Size in Applied Statistics Research: An Overview. *Int. J. Commer. Manag.* **2014**, *2*, 1–22.
37. Shan, C.; Pandyaswargo, A.H.; Ogawa, A.; Tsubouchi, R.; Onoda, H. Japanese Public Perceptions on Smart Bin Potential to Support PAYT Systems. *Waste Manag.* **2024**, *177*, 278–288. [CrossRef]
38. Pandyaswargo, A.H.; Siregar, T.H.; Onoda, H. Exploring Japan's Older Adults Mobility Challenges and the Potential Role of Autonomous Vehicles. *Transp. Res. Part A Policy Pract.* **2023**, *176*, 103818. [CrossRef]
39. Abdi, H.; Valetin, D. Multiple Correspondence Analysis. In *Encyclopedia of Measurement and Statistics*; SAGE Publications: Thousand Oaks, CA, USA, 2007.
40. Roux, B.L.; Rouanet, H. *Multiple Correspondence Analysis*; SAGE Publications Inc.: Thousand Oaks, CA, USA, 2010.
41. Kassambara MCA—Multiple Correspondence Analysis in R: Essentials—Articles—STHDA. Available online: http://www.sthda.com/english/articles/31-principal-component-methods-in-r-practical-guide/114-mca-multiple-correspondence-analysis-in-r-essentials/ (accessed on 23 January 2023).
42. Sanchez, G. RPubs—Multiple Correspondence Analysis in R. Available online: https://rpubs.com/gaston/MCA (accessed on 20 January 2023).
43. IBM Cramér's V—IBM Documentation. Available online: https://www.ibm.com/docs/en/cognos-analytics/11.1.0?topic=terms-cramrs-v (accessed on 8 February 2022).
44. Koohsari, M.J.; Sugiyama, T.; Shibata, A.; Ishii, K.; Hanibuchi, T.; Liao, Y.; Owen, N.; Oka, K. Walk Score® and Japanese Adults' Physically-Active and Sedentary Behaviors. *Cities* **2018**, *74*, 151–155. [CrossRef]
45. Koohsari, M.J.; McCormack, G.R.; Nakaya, T.; Shibata, A.; Ishii, K.; Yasunaga, A.; Hanibuchi, T.; Oka, K. Urban Design and Japanese Older Adults' Depressive Symptoms. *Cities* **2019**, *87*, 166–173. [CrossRef]
46. Miura, K. Epidemiology and Prevention of Hypertension in Japanese: How Could Japan Get Longevity? *EPMA J.* **2011**, *2*, 59–64. [CrossRef]
47. Hirai, H.; Kondo, K.; Kawachi, I. Social Determinants of Active Aging: Differences in Mortality and the Loss of Healthy Life between Different Income Levels among Older Japanese in the AGES Cohort Study. *Curr. Gerontol. Geriatr. Res.* **2012**, *2012*, 701583. [CrossRef]
48. Haque, M.d.S.; Uddin, S.; Sayem, S.M.d.; Mohib, K.M. Coronavirus Disease 2019 (COVID-19) Induced Waste Scenario: A Short Overview. *J. Environ. Chem. Eng.* **2021**, *9*, 104660. [CrossRef]

49. Tripathi, A.; Tyagi, V.K.; Vivekanand, V.; Bose, P.; Suthar, S. Challenges, Opportunities and Progress in Solid Waste Management during COVID-19 Pandemic. *Case Stud. Chem. Environ. Eng.* **2020**, *2*, 100060. [CrossRef]
50. Kumagai, N. Care Preferences of Elderly People Living Alone in Japan. *Health Educ. Public Health* **2018**, *1*, 2.
51. Onoda, H. An Attempt to Implement Multi-Benefit Mobility in Society in Conjunction with Urban Development. *Inf. Process.* **2024**, *65*. Available online: https://www.ipsj.or.jp/dp/contents/publication/58/DP6505-S01.html (accessed on 13 May 2024).
52. Pandyaswargo, A.H.; Wibowo, A.D.; Onoda, H. Socio-Techno-Economic Assessment to Design an Appropriate Renewable Energy System for Remote Agricultural Communities in Developing Countries. *Sustain. Prod. Consum.* **2022**, *31*, 492–511. [CrossRef]
53. IBM. Multiple Correspondence Analysis—IBM Documentation. Available online: https://www.ibm.com/docs/en/spss-statistics/28.0.0?topic=categories-multiple-correspondence-analysis (accessed on 29 July 2022).

smart cities

MDPI

Article

AI-Driven Prediction and Mapping of Soil Liquefaction Risks for Enhancing Earthquake Resilience in Smart Cities

Arisa Katsuumi [1], Yuxin Cong [1] and Shinya Inazumi [2,*]

[1] Graduate School of Engineering and Science, Shibaura Institute of Technology, Tokyo 135-8548, Japan; ah20098@shibaura-it.ac.jp (A.K.); na23107@shibaura-it.ac.jp (Y.C.)

[2] College of Engineering, Shibaura Institute of Technology, Tokyo 135-8548, Japan

* Correspondence: inazumi@shibaura-it.ac.jp; Tel.: +81-3-5859-8360

Abstract: In response to increasing urbanization and the need for infrastructure resilient to natural hazards, this study introduces an AI-driven predictive model designed to assess the risk of soil liquefaction. Utilizing advanced ensemble machine learning techniques, the model integrates geotechnical and geographical data to accurately predict the potential for soil liquefaction in urban areas, with a specific focus on Yokohama, Japan. This methodology leverages comprehensive datasets from geological surveys and seismic activity to enhance urban planning and infrastructure development in smart cities. The primary outputs include detailed soil liquefaction risk maps that are essential for effective urban risk management. These maps support urban planners and engineers in making informed decisions, prioritizing safety, and promoting sustainability. The model employs a robust combination of artificial neural networks and gradient boosting decision trees to analyze and predict data points, assessing soil susceptibility to liquefaction during seismic events. Notably, the model achieves high accuracy in predicting soil classifications and N-values, which are critical for evaluating soil liquefaction risk. Validation against an extensive dataset from geotechnical surveys confirms the model's practical effectiveness. Moreover, the results highlight the transformative potential of AI in enhancing geotechnical risk assessments and improving the resilience of urban areas against natural hazards.

Keywords: AI-driven predictive model; ensemble machine learning; geotechnical data integration; soil liquefaction risk; urban resilience

1. Introduction

In an era of rapid urbanization and escalating environmental challenges, the concept of 'smart cities' as shown in Figure 1 has emerged as a pivotal framework for sustainable development [1–4]. These cities harness advanced technologies, such as IoT and big data analytics, to enhance quality of life, economic efficiency, and environmental sustainability [5]. A key element of this initiative is the integration of artificial intelligence (AI) into urban planning processes [6]. This integration not only revolutionizes resource management in urban areas but also enhances their resilience against natural hazards like earthquakes, torrential rain, and so on [7]. This study leads in applying smart technologies in geotechnical engineering, enhancing urban resilience through AI-driven risk prediction and zoning for soil liquefaction risk. By leveraging sophisticated AI algorithms to analyze extensive geotechnical data, this methodology facilitates the creation of detailed hazard maps. These maps are indispensable for effective urban planning and hazard mitigation, equipping urban planners and engineers with essential tools to make informed decisions that prioritize safety and sustainability.

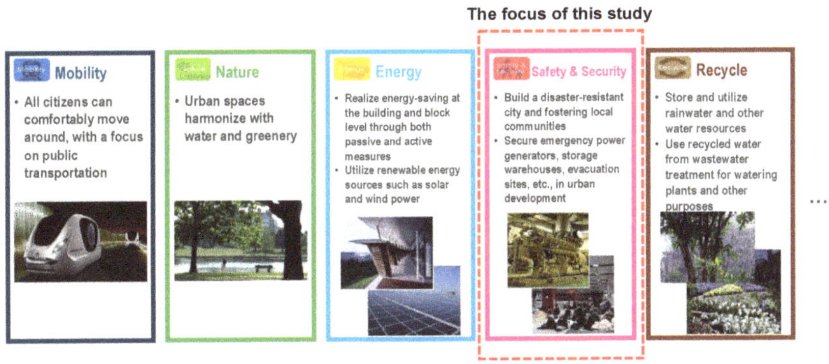

Figure 1. Key elements of a smart city.

The increasing frequency of urban hazards poses significant risks to densely populated areas, highlighting the critical importance of this study. By leveraging predictive analysis, smart cities can proactively address these challenges, ensuring preparedness and adaptability within a dynamic urban environment. This study significantly enhances conventional methods by delivering a nuanced, precise, and dynamic evaluation of soil liquefaction risks, particularly in regions sensitive to geological and topographical variations. Moreover, the smart city design strategy outlined in this study encompasses comprehensive urban development, incorporating strategic infrastructure placement, optimized resource management, and improved citizen engagement through transparent data practices. By integrating AI-driven tools into the urban fabric, we ensure that cities are not only functional but also resilient and responsive to the needs of their inhabitants. The introduction of such advanced technological solutions tackles long-standing urban challenges and aligns with the broader objectives of smart cities. The predictive capabilities developed in this study represent a significant advancement in protecting and enhancing urban livability, marking a critical contribution to the conceptualization and design of smart cities worldwide.

This study focuses on the significant natural hazard of soil liquefaction, as illustrated in Figures 2 and 3 [8–12]. Figure 3, taken from the 2011 Great Japan Earthquake, the largest recorded earthquake in Japan's history, shows particularly severe liquefaction. This figure shows the ground devastation caused by liquefaction, including toppled trees and the flooding of streets with swirling muddy water in the Urayasu residential area [13]. As a representation of liquefaction damage, Figure 3 highlights the frequent seismic activity in Japan, which often results in severe ground instability. The risk of liquefaction occurs when five specific conditions are met: the fine particle content of the soil is between 15% and 50%; water is present; the soil is loosely compacted; the particle sizes are uniform; and the area has experienced repeated earthquakes. Soil liquefaction primarily occurs due to the shaking of loosely packed, water-saturated soils during earthquakes. This shaking increases pore water pressure and reduces the effective stress between soil particles, leading to fluid-like soil behavior, as depicted in Figure 2. Notable events, such as the 2011 Great East Japan Earthquake and the 2024 Noto Peninsula Earthquake, highlight the devastating impacts of these phenomena and underscore the critical need for accurate prediction and effective mitigation strategies [7].

This study utilizes AI algorithms to predict risks associated with soil liquefaction by analyzing extensive datasets from geological surveys and seismic activity. The choice was made to use an ensemble method over a single machine learning approach. According to [14], in geological prediction studies, the ensemble method can yield a 20% improvement in model performance, demonstrating greater potential. Therefore, the goal is to develop an AI-driven predictive model that integrates various databases, including geotechnical and geographic information, using ensemble machine learning techniques. This model specifically aims to predict potential risks of soil liquefaction, with a particular focus on areas like Yokohama, Japan, which is known for its extensive reclaimed lands and high susceptibility to these hazards. By creating and validating hazard maps through case studies in such regions, the study aims to enhance urban planning and significantly contribute to the development of smart cities.

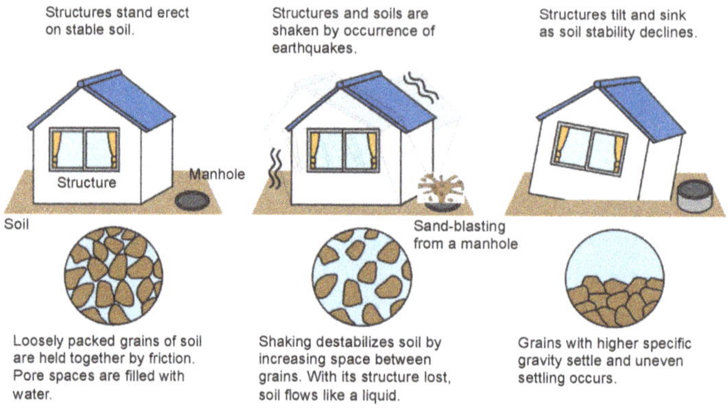

Figure 2. Soil liquefaction behavior during an earthquake.

Figure 3. Ground devastation in a residential area due to liquefaction following the 2011 Great East Japan Earthquake (Image courtesy of the Urayasu Office).

Yokohama is located in the Tokyo metropolitan area. Originally, it was planned to use construction debris from large-scale developments to fill the land, but in reality, most of the fill material consisted of sand from Mt. Asama. According to earthquake data for the past 30 days, Yokohama has experienced seven earthquakes of magnitude 1 or greater. Therefore, the liquefaction hazard in Yokohama is considered serious. Figure 4 summarizes past earthquake-induced liquefaction damage by topographic classification, which divides the country into grids approximately 250 m square. Each grid is classified into five levels to indicate the relative intensity of liquefaction susceptibility based on general ground characteristics as represented by the topography. However, this does not predict specific earthquake-induced liquefaction events.

In the integration and visualization of the smart city concept illustrated in Figure 5, an AI-driven predictive model that synthesizes data from diverse databases, including geotechnical and geographic information, enhances urban resilience and promotes the development of a safer and more sustainable society. This approach contributes to the sustainable growth of smart cities and ensures the safety of their inhabitants. In smart cities, prioritizing the reduction of natural hazard risks is crucial. The AI-driven predictive model leverages geo-information databases to assess liquefaction risks, minimizing potential damage to buildings and infrastructure during earthquakes. The model's high accuracy enables the identification of specific areas at risk of liquefaction, allowing for the proactive planning of effective countermeasures. Such capabilities expedite and streamline disaster responses, diminishing disruptions during emergencies. Additionally, this method provides a more detailed and accurate risk analysis than traditional approaches reliant on human judgment, thereby enhancing scientific decision-making in the planning and development within smart cities.

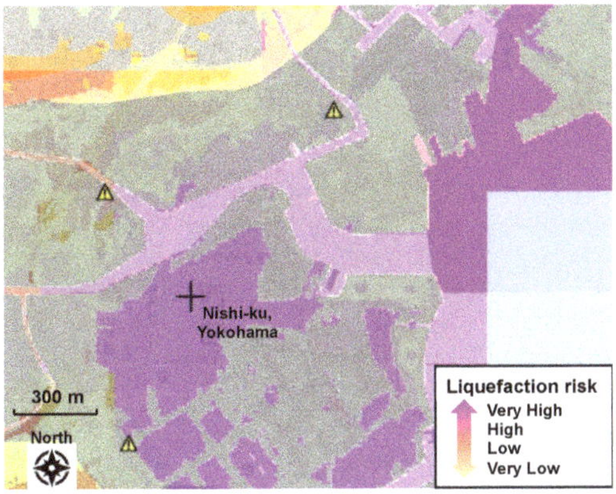

Figure 4. Summary of liquefaction damage in Yokohama for past earthquakes by topographic classification, indicating liquefaction tendency intensity in 250 m square grids.

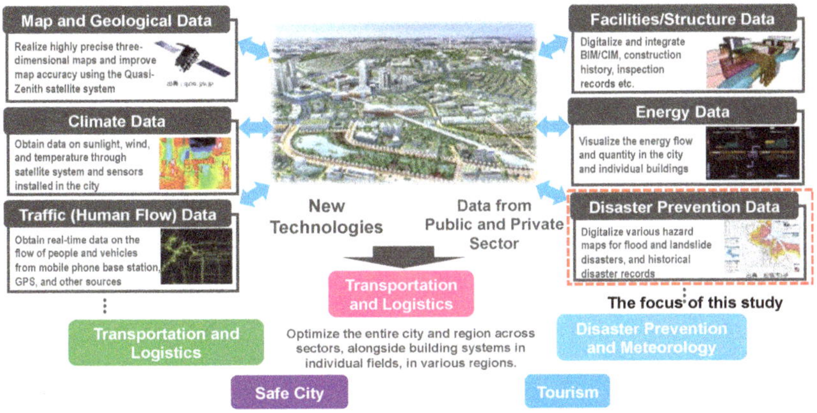

Figure 5. Data integration and visualization for smart city concept.

In this study, the authors develop a novel AI-driven predictive model that substantially advances the state of the art in assessing soil liquefaction risks, particularly in urban areas prone to seismic activity. Unlike previous studies [15–18], which primarily employed empirical or semi-empirical methods, this AI-driven model utilizes advanced ensemble machine learning techniques to integrate comprehensive geotechnical and geographic data. This integration yields a more precise, dynamic, and detailed assessment of soil liquefaction risks. The model is distinct in its ability to synthesize data from multiple sources, including real-time geological surveys and seismic records, thereby enhancing the accuracy of risk mapping and the efficacy of urban planning strategies. Moreover, the AI-driven predictive model addresses a critical gap in existing methodologies [15–18] by providing high-resolution hazard maps that are crucial for enhancing urban resilience in smart cities. By conducting a comparative analysis with models described in recent studies [15–18], this study not only fills a vital knowledge gap by improving the granularity and accuracy of predictions but also broadens the application of AI in geotechnical engineering, setting a new benchmark for future study and practical applications in earthquake-prone urban environments.

2. AI-Driven Predictive Models

In statistics and machine learning, ensemble machine learning trains multiple machine learning models (algorithms) and combines their predictions to obtain improved performance and better generalization ability than individual models. The fundamental idea behind ensemble machine learning is the recognition that each machine learning model has limitations and can make errors. Hence, ensemble machine learning aims to improve classification performance by harnessing the strengths of multiple models. Limitations of some machine learning models include high variance, high bias, and low accuracy [19,20]. However, several studies have shown that ensemble machine learning often achieves higher accuracy than a single machine learning model [21]. Ensemble machine learning can limit the variance and bias errors associated with a single model; for example, ensemble machine learning with a bagging algorithm reduces variance without increasing bias, while ensemble machine learning with a boosting algorithm reduces bias [22–24]. Overall, ensemble machine learning is more robust than an individual machine learning model [25]. Algorithms used in ensemble machine learning are mainly divided into three categories: bagging, boosting, and stacking. In this study, the boosting algorithm was selected for ensemble machine learning.

2.1. Ensemble Machine Learning

Boosting is an algorithm that sequentially fits "weak" classifiers to different weightings of the observations in a dataset. Those observations, that the previous classifier poorly predicts, receive greater weight in the next iteration. The final classifier is a weighted average of all the weak classifiers. In a wide variety of problems, the weighting scheme and final classifier of the boosting algorithm are merged, and the result has proven to be an effective method for reducing bias and variance.

The ensemble machine learning model is established using a boosting algorithm, and the procedure is illustrated in Figure 6, where sample data are randomly extracted from the training dataset. In this study, ensemble machine learning with the boosting algorithm was adopted, and an AI-driven predictive model was created by integrating a neural network with a gradient boosting decision trees model. By combining the strengths of different machine learning models, it is possible to achieve higher prediction accuracy than with a single model.

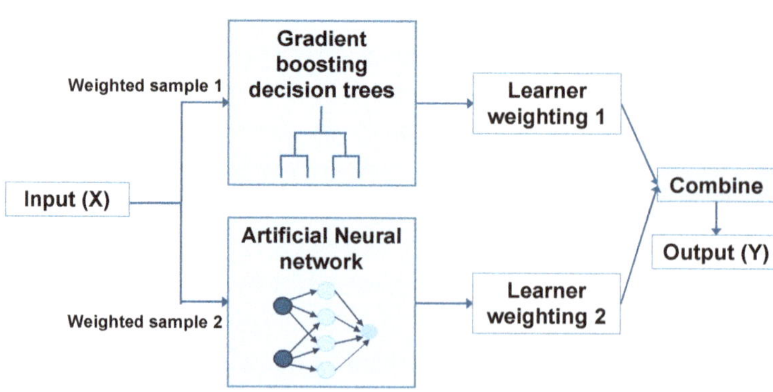

Figure 6. Schematic diagram of ensemble machine learning model with boosting algorithm.

An artificial neural network model as shown in Figure 7a is a soft computing technique used today by engineers around the world for a wide variety of nonlinear problems that can occur in civil engineering [26]. And it is an interconnected collection of simple processing elements, units or nodes, whose functionality is loosely based on the animal neuron. The processing capability of the network is stored in the inter-unit connection strengths or weights, which are obtained through a process of adaptation to, or learning from, a set of training patterns [27,28]. Recently, artificial neural networks achieved outstanding results in learning image representations for various vision tasks, including image classification [29] and object detection [30]. These artificial neural networks typically consist of several components, including convolutional, fully connected, and pooling layers. By stacking several of these layers, convolutional artificial neural networks are able to learn complex features that are highly invariant and discriminative [31]. Previous studies confirmed it as the most suitable and widely accepted statistical approach for soil liquefaction assessment [32,33].

A gradient boosting decision trees model shown in Figure 7b is a powerful machine learning tool that is widely used in many applications, including multi-class classification. This model uses decision trees as the base algorithm and sums the predictions of a series of trees. At each step, a new decision tree is trained to fit the residual errors between the ground truth and the current prediction. Gradient boosting decision trees are popular due to their accuracy, efficiency, and interpretability [34].

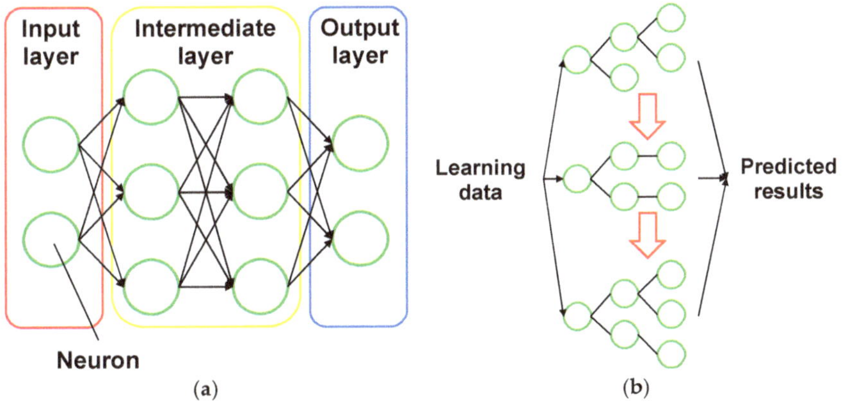

Figure 7. Architecture of artificial neural network and gradient boosting trees models. (**a**) Artificial neural network model. (**b**) Gradient boosting decision trees model.

In the ensemble machine learning model, the tuning of hyperparameters plays a critical role in optimizing model performance. Hyperparameters in ensemble algorithms, such as the number of base algorithms, the learning rate, and the depth of decision trees, require careful calibration to balance the tradeoff between bias and variance, thereby improving the generalizability of the model. The process of hyperparameter tuning involves a systematic search for the optimal settings that maximize the predictive accuracy of the ensemble. Techniques such as grid search, random search, and Bayesian optimization are commonly used for this purpose. These methods effectively explore the hyperparameter space and identify configurations that lead to the best performance on validation data sets. In this study, random search was employed to optimize model performance within the ensemble machine learning framework.

The main reason for choosing ensemble learning is the limited size and uneven distribution of the data set, which consists of only 214 data points, as shown in Figure 8. Ensemble learning is used to improve the accuracy and reliability of predictions. This method generates training sets by randomly drawing small samples from all categories and combining them with larger samples from the same categories. This process is repeated several times to create different training sets and models. Finally, a combination method, such as voting or weighted voting, is used to produce the classification predictions. Thus, when big data research methods are faced with data scarcity, the ensemble approach is chosen to increase accuracy.

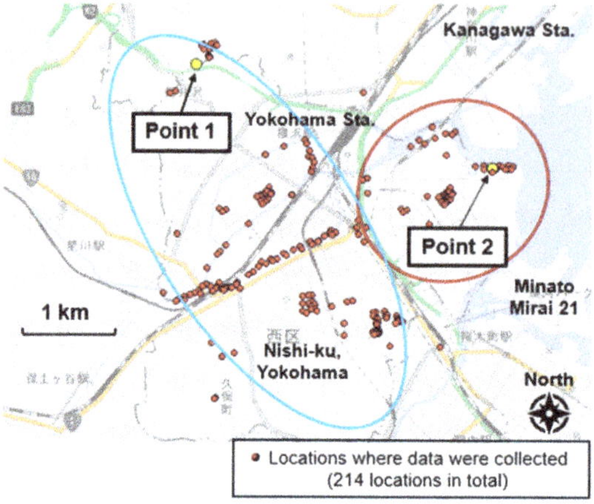

Figure 8. Geotechnical data collection points in Yokohama.

2.2. *Reliability of Predictive Results by AI-Driven Predictive Model*

In order to evaluate the performance of the AI-driven predictive model, "*Accuracy*", "Coefficient of determination (R^2)", "Average Precision (AP)", "Average Recall (AR)", and "F-measures" are selected as evaluation indicators.

"*Accuracy*" is the most basic index and is an evaluation index that indicates the percentage of correctly predicted predictions. This study calculated the "*Accuracy*" using Equation (1) [35,36] and the "*Accuracy*" was used as an evaluation index for predicting soil classification in this study:

$$Accuracy = \frac{TP + TN}{TP + FN + FP + TN} \quad (1)$$

where, TP is true positives, TN is true negatives, FP is false positives, and FN is false negatives.

"Coefficient of determination (R^2)" [37] is a key evaluation metric in predictive modeling. It quantifies the proportion of the variance in the dependent variable that is predictable from the independent variables. R^2 is bounded above by 1, which indicates a perfect fit. Although there is no theoretical lower bound, a value of 0 suggests that the model predicts as well as the mean of the dependent variable, represented by the horizontal line $y = K$, where K is the mean of the target values of all training points. Values of R^2 less than 0 indicate a fit worse than this trivial model. Consequently, R^2 values are typically considered within the range of 0 to 1. The coefficient of determination remains unchanged under linear transformations of the scale of measurement for the independent variables. A value close to 1 indicates a model with high predictive accuracy. R^2 was calculated using Equation (2):

$$R^2 = 1 - \frac{\sum_i (y_i - x_i)^2}{\sum_i \left(y_i - \overline{y}\right)^2} \quad (2)$$

where, y_i is the measured value, x_i is the predicted value, and $\overline{y}$ is the average of measured values.

"Average Precision (AP)" is a comprehensive metric used to evaluate model precision across various "Intersection over Union (IoU)" thresholds. Precision assesses whether all detected positive samples are indeed positive. The steps to calculate precision accuracy are as follows: 1. establish different IoU thresholds; 2. calculate the accuracy at each IoU threshold; 3. compute the average of these accuracies. Average precision was determined using Equation (3):

$$Precision = \frac{TP}{TP + FP} \tag{3}$$

"Average Recall" is a comprehensive metric that evaluates the recall rate of the model across different IoU thresholds. *Recall* measures the proportion of actual positive samples that the model correctly identifies. The steps to calculate average recall are as follows: 1. establish different IoU thresholds; 2. calculate the recall rate at each threshold; 3. compute the average of these recall rates. *Recall* was determined using Equations (4):

$$Recall = \frac{TP}{TP + FN} \tag{4}$$

The "F-measure" is the harmonic mean of *precision* and *recall*. The F-measure was calculated using Equation (5):

$$\text{F-measure} = 2 \times \frac{Precision \times Recall}{Precision + Recall} \tag{5}$$

3. Soil Liquefaction Risk Prediction with AI-Driven Predictive Model

3.1. Procedures on Soil Liquefaction Risk Prediction

In this study, data from the geological columnar section provided by the Ministry of Land, Infrastructure, Transport and Tourism, accessible through their online database, was utilized for the target location, Yokohama, Japan.

Data pertaining to "N-value", "soil classification", and "groundwater levels" were collected from the geological columnar section for a total of 214 locations, as depicted in Figure 8. Additionally, "elevation" data were obtained from the Geospatial Information Authority of Japan, using location information. An example of these collection records is shown in Figure 9. The criteria for determining soil liquefaction, according to the Highway Bridge Specifications and Commentary, are as follows:

(1) The groundwater level must be within 10 m of the ground surface, and saturation must occur within 20 m of the ground surface.
(2) The fine particle content (*FC*) must exceed 35%, with a plasticity index of 15 or less.
(3) The average grain size (D_{50}) must be 10 mm or less, and the 10% grain size (D_{10}) must be 1 mm or less.

The AI-driven predictive model identifies layers with an N-value of 50 or more extending over at least 2 m as indicative of a bearing layer. Generally, soil or rock with an N-value of 20 or more is considered suitable as a load-bearing layer. If the N-value is between 30 and 50, the layer is deemed appropriate for the load-bearing layer of civil engineering and building structures. An N-value of 50 or more suggests a very solid layer, suitable as a load-bearing layer capable of supporting large structures such as high-rise condominiums. Therefore, in this study, continuous strata with an N-value exceeding 50 over a length of more than 3 m are designated as load-bearing layers [38].

Major soil classification systems in Japan have evolved in conjunction with public works projects [39,40]. In Japan, soils with similar properties are grouped based on their particle size distribution. This study categorizes "soil type classifications," derived from geological columnar sections, into five categories: sand and gravel with fine content of 35% or less (prone to soil liquefaction), and silt, clay, and bedrock with fine content of 50% or more. Additionally, when multiple soil types, such as sandy silt or sandy clay mixed with gravel, are encountered, they are classified according to the component with the largest mass proportion, following the engineering classification method for soil materials.

(a)

Depth (m)	N-value	Soil classification
1	2	Clay
2	3	Clay
3	5	Clay
4	6	Clay
5	7	Clay
6	2	Clay
7	2	Clay
8	4	Sand
9	8	Sand
10	11	Gravel
11	12	Silt
12	15	Silt
13	17	Silt
14	15	Silt
15	16	Silt
16	18	Silt
17	19	Silt
18	14	Silt
19	18	Silt
20	46	Silt

(b)

Figure 9. An example of a data set containing geotechnical and geospatial information. (**a**) Longitude, latitude, and elevation data are obtained from the Geospatial Information Authority of Japan. (**b**) The data of N value, soil classification, and groundwater level were collected from the geological column section.

Concerning "groundwater levels," the geological columnar section contained sparse data, making it impractical to gather the necessary information for predictive analysis. Consequently, the borehole water level was assumed to represent the groundwater level, which was uniformly set at GL-1.0 m.

Soil liquefaction risk is assessed for layers that satisfy the previously mentioned criteria. The AI-driven predictive model was trained using the data on "N-value" and "soil classification" to predict these parameters at unknown locations. The predicted data was then incorporated into the soil liquefaction potential index (LPI) to assess the risk of soil liquefaction.

3.2. Assumed Earthquake Motions for Soil Liquefaction Risk Prediction

Level 1 earthquake motion refers to lower-intensity seismic events that structures are expected to withstand without significant damage, typically maintaining operational functionality post-event. In contrast, Level 2 earthquake motion pertains to higher-intensity events. Buildings are designed to endure these, possibly sustaining some structural damage, to ensure occupant safety.

Regulations require consideration of two specific types of Level 2 earthquake ground motions: Type I for large plate-boundary earthquakes and Type II for inland earthquakes, such as the 1995 Hyogoken–Nanbu Earthquake. Type I earthquake motion is characterized by shorter duration and higher frequency vibrations, commonly experienced in regions close to tectonic plate boundaries. Conversely, Type II earthquake motion involves longer duration and lower frequency vibrations, typically occurring in areas more distant from plate boundaries. This classification into Types I and II aids in designing structures tailored to the specific seismic risks associated with the frequency and duration of ground shaking typical of a region.

3.3. Soil Liquefaction Potential Index

The soil liquefaction potential index (LPI) was utilized for analysis of soil liquefaction risk in this study. LPI, introduced in 1982, serves as a metric to assess the potential for soil liquefaction at a site, with higher values indicating greater risk. This index aggregates the factor of safety against liquefaction (F_L value), which is weighted by depth, to evaluate the overall likelihood of soil liquefaction occurring at various depths within a site. A series of case studies applied the LPI to assess liquefaction risks at targeted locations [39–42].

The F_L value, essential for predicting soil liquefaction, quantifies the resistance to liquefaction at each depth. When the F_L value is 1.0 or less, the occurrence of soil liquefaction is deemed likely. In this study, the LPI was calculated using Equations (6)–(19) to comprehensively determine the risk of soil liquefaction at the study site:

$$LPI = \int_0^{20} F \times (1 - 0.5x) d_x \tag{6}$$

$$F = \begin{cases} 1 - F_L (F_L < 1.0) \\ 0 \quad (F_L \geq 1.0) \end{cases} \tag{7}$$

$$F_L = \frac{R}{L} \tag{8}$$

where, x represents the depth (m) from the ground surface; F is the soil liquefaction resistance factor, which is governed by the factor of safety against soil liquefaction (F_L), R is the dynamic shear strength ratio, and L represents the earthquake shear stress ratio. R is determined based on the standards for Level 1 and Level 2 earthquake motions:

$$R = C_W \times R_L \tag{9}$$

In the case of Level 1 and Type I of Level 2 earthquake motions:

$$C_W = 1.0 \tag{10}$$

In the case of Type II of Level 2 earthquake motion:

$$C_W = \begin{cases} 1.0 & (R_L \leq 0.1) \\ 3.3R_L + 0.67 & (0.1 < R_L \leq 0.4) \\ 2.0 & (0.4 < R_L) \end{cases} \tag{11}$$

$$R_L = \begin{cases} 0.0882\sqrt{(0.85N_a + 2.1)/1.7} & (N_a < 14) \\ 0.0882\sqrt{N_a/1.7} + 1.6 \times 10^{-6} \times (N_a - 14)^{4.5} & (N_a \geq 14) \end{cases} \tag{12}$$

In the case of sand:

$$N_a = cFC \times (N_1 + 2.47) - 2.47\ (D_{50} < 2\ mm) \tag{13}$$

$$N_1 = 170N/(\sigma_{vb}' + 70) \tag{14}$$

$$C_{FC} = \begin{cases} 1 & (0\% \leq FC < 10\%) \\ (FC + 20)/30 & (10\% \leq FC \leq 40\%) \\ (FC - 16)/12 & (40\% \leq FC) \end{cases} \tag{15}$$

In the case of gravel:

$$N_a = \left\{1 - 0.36\log_{10} \times \left(\frac{D_{50}}{2}\right)\right\} N_1 (D_{50} \geq 2\ mm) \tag{16}$$

$$L = r_d k_{hgL} \sigma_v / \sigma_v' \tag{17}$$

$$r_d = 1.0 - 0.015x \tag{18}$$

$$k_{hgL} = C_Z k_{hgL0} \tag{19}$$

where, the explanation and definition of each coefficient and variable symbol used in Equations (6)–(19) are summarized in Table 1.

Table 1. Explanation and definition of each coefficient and variable symbol used in Equations (6)–(19).

Coefficient and Variable Symbol	Explanation and Definition
C_W	Correction coefficient based on earthquake motion characteristics
R_L	Liquefaction strength ratio
N	N-value obtained from standard penetration test
N_1	N-value converted to effective overburden pressure equivalent to 100 (kN/m^2)
N_a	Corrected N-value considering the effect of grain size
$\sigma_{vb}{}'$	Effective overburden pressure at depth from ground surface when performing standard penetration tests (kN/m^2)
C_{FC}	Correction factor for N-value based on fine particle content
FC	Fine particle content (%) (Percentage of passing mass of soil particles with a particle size of 75 μm or less)
D_{50}	50% particle size (mm)
r_d	Depth reduction factor of seismic shear stress ratio
k_{hgL}	Design horizontal seismic intensity of the ground surface used to assess liquefaction (rounded to two decimal places)
C_Z	Regional correction factor (Yokohama is 1.0)
k_{hgL0}	Standard value of horizontal seismic intensity for design of ground surface used to judge liquefaction
σ_v	Total overburden pressure at depth x from ground surface (kN/m^2)
$\sigma_v{}'$	Effective overburden pressure at depth x from ground surface (kN/m^2)
X	Depth from ground surface (m)

Table 2 presents the correction coefficients based on the characteristics of earthquake motion. Table 3 details the physical properties of different soil classifications related to existing groundwater.

When the LPI equals 0, the risk or potential for soil liquefaction is very low. For $0 < LPI \leq 5$, the risk or potential is low; for $5 < LPI \leq 15$, it is high; and for $LPI > 15$, it is very high. Table 4 summarizes the correlation between LPI and liquefaction risk levels.

Table 2. Correction coefficients based on the characteristics of earthquake motion.

Type of Site	Level 1 Earthquake Motion	Level 2 Earthquake Motion	
		Type I	Type II
Site I	0.12	0.5	0.8
Site II	0.15	0.45	0.7
Site III	0.18	0.4	0.6

Table 3. Physical properties of different soil types related to existing groundwater.

Soil Classification	Unit Weight of Soil Below Groundwater Table (kN/m^3)	Unit Weight of Soil Above Groundwater Table (kN/m^3)	FC (%)
Clay	13	15	80
Silt	15.5	17.5	75
Sand	18	20	10
Gravel	19	21	0
Bedrock	20	20	5

Table 4. Correlation between liquefaction potential index (LPI) and liquefaction risk levels.

	$0 < LPI < 5$	$0 < LPI < 15$	$15 < LPI < 30$	$LPI > 30$
Liquefaction risk	Very low	Low	High	Very high

4. Results and Discussion

This study concentrates on geotechnical data derived from the geological columnar sections of Yokohama. Here, an AI-driven predictive model is employed to forecast the distribution of N-values and soil classifications. Case 1 involves predicting N-values, while Case 2 focuses on predicting soil classifications.

4.1. N-Values and Soil Classification Predictions

4.1.1. Case 1: Predictive Results for N-Value

Using geotechnical data collected from the geological columnar sections of Yokohama as training data, the AI-driven predictive model estimated the N-value at sites previously assumed to be unknown, though they were actually known.

To compare prediction accuracy based on data volume, the model's performance was evaluated at 214 sites and at half that number. Model 1 used data from 117 sites, while Model 2 used data from all 214 sites. The data set is divided into a training set, a test set, and a validation set in proportions of 60%, 20%, and 20%, respectively. In Model 1, the training set contains 70 values, the validation set contains 24 values, and the test set contains 23 values. In model 2, the training set contains 140 values, the validation set contains 48 values, and the test set contains 46 values.

Table 5 shows the coefficient of determination (R^2) for predicting the N-value in Model 1 and Model 2, respectively. From the evaluation results for the N-value, it was observed that this tends to increase with increasing depth. This may be attributed to the increasing number of points (load-bearing layer) where the N-value exhibits a consistent range of fluctuation at greater depths, and the range of N-values at these points becomes narrower. Therefore, it can be inferred that predictions in deeper ground can be made with relatively high accuracy. Comparing prediction accuracy based on the number of data points reveals that the accuracy of the N-value predictions improves as the number of points increases. By gathering more data points to create training data, more reliable and precise predictions can be expected.

Table 5. Coefficient of determination (R^2) for predicting N-value.

(a) Model 1.			(b) Model 2.		
Depth (m)	**Coefficient of Determination (R^2)**	**Root Mean Square Error (RMSE)**	**Depth (m)**	**Coefficient of Determination (R^2)**	**Root Mean Square Error (RMSE)**
1	0.0114	7.32	1	0.0720	6.38
2	0.0487	6.22	2	0.0695	7.67
3	0.0254	8.82	3	0.4005	8.65
4	0.2626	6.61	4	0.7248	6.18
5	0.2086	8.15	5	0.7445	7.00
6	0.4704	8.31	6	0.4879	11.13
7	0.2557	8.92	7	0.5324	12.09
8	0.6190	8.39	8	0.8231	8.23
9	0.7429	8.52	9	0.8096	9.21
10	0.6894	9.90	10	0.5211	15.14
11	0.6581	11.14	11	0.8628	8.29
12	0.9339	5.44	12	0.9112	6.80
13	0.9097	6.71	13	0.9518	5.72
14	0.9030	7.04	14	0.8981	7.46
15	0.9085	7.21	15	0.9314	6.18
16	0.8915	7.91	16	0.9763	3.63
17	0.9458	5.63	17	0.9379	5.90
18	0.8969	7.74	18	0.9130	6.87
19	0.8924	7.89	19	0.9168	6.73
20	0.8083	10.52	20	0.9063	7.11

Therefore, Model 2 was selected for prediction. The N-value was predicted using data from 214 sites in Yokohama. Given the differing geologies of reclaimed land and plateaus, Yokohama was categorized into two types: 'Landfill area (red)' and 'Plateau area (blue)'. Areas assumed to be unknown points were distinguished from known points in each category. Point 1 was selected from the landfill area and Point 2 from the plateau area, with their locations illustrated in Figure 8. Additionally, this study predicted the N-value, assumed to be an unknown point, using the following three prediction procedures. The method that achieved the highest accuracy was selected:

(1) A prediction procedure involving learning at every 1 m interval, without using the prediction results as data.
(2) A prediction procedure that makes predictions at every 1 m interval from the ground surface to the subsurface, incorporating prediction results shallower than the predicted depth into the learning process.
(3) A prediction procedure that makes predictions at every 1 m interval from 20 m below ground to the surface, incorporating prediction results deeper than the predicted depth into the learning process.

Figure 10 shows the results of predicting the N-value using each of the prediction procedures (1) to (3). Looking at the results for Point 1, while the results show that the N-value is low, prediction procedures (1) and (2) show that the N-value is 30 or more from points 10 m onwards. This is because the number of collected geotechnical data for a total of 214 points is more in the plateau area than in the landfill area, and there is much data that reaches the load-supporting layer from points around 10 m or higher and converges to an N-value of 30 or more. It is thought that this is having an influence. On the other hand, with prediction procedure (3), although the numerical value exceeds the survey results, it appears that similar behavior is predicted at points beyond 10 m while maintaining a relatively low N-value. As a result, at Point 1, prediction procedure (3) is more accurate than (1) and (2). Next, looking at the results for Point 2, all prediction procedures showed behavior similar to the survey results, and it can be said that predictions were made with relatively high accuracy. Among them, prediction procedure (3) can also predict the tendency of the survey N-value at a depth of around 5 m, and it is clear that the overall prediction accuracy is higher than (1) and (2). Based on the results of Points 1 and 2, it is thought that the prediction procedure (3) is better at making predictions at large depths because it learns and makes predictions from underground. Based on the above results, the following step was taken in this study: the prediction procedure (3) was chosen for prediction.

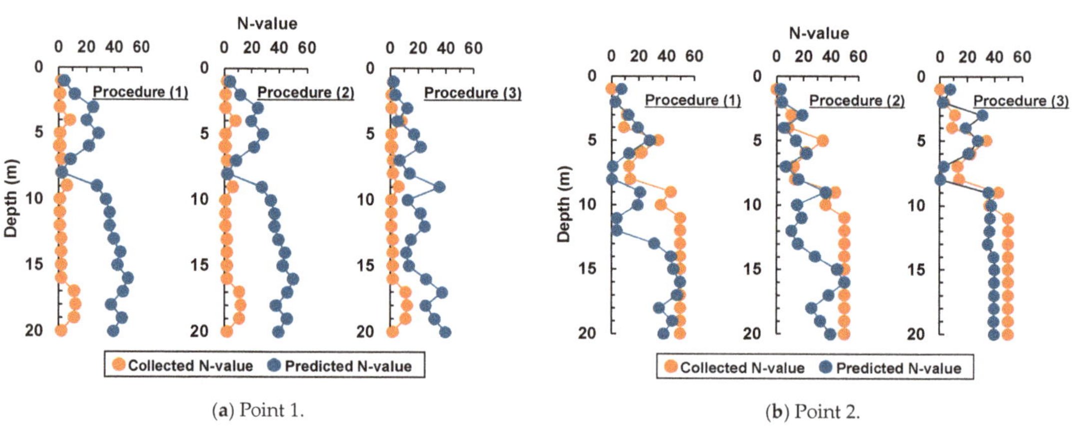

(**a**) Point 1. (**b**) Point 2.

Figure 10. N-value predictions using different prediction procedures (1) to (3).

4.1.2. Case 2: Predictive Results of Soil Classification

Similar to Case 1, geotechnical data collected from the geological columnar sections of Yokohama were utilized as training data to develop the AI-driven predictive model for soil classification. The results of the model's prediction accuracy are displayed in Table 6. Accuracy metrics were employed to assess the model's performance. The results indicate high accuracy starting from a depth of 1.2 m, suggesting that predictions are generally reliable. Observations show that surface-near points are predominantly sandy, and as depth increases, the load-bearing layer is encountered more frequently, and the presence of bedrock points increases, which correlates with improved accuracy. In addition, prediction accuracy improves as the number of data points increases. The accuracy of the model was evaluated using four metrics: accuracy, average precision, average recall, and the F-measure.

In addition, the three prediction procedures of (1) to (3) used for N-value estimation were also applied to predict soil classification at unknown points. The results are detailed in Figure 11. According to these tables, the results at Point 1 show that the soil type, as identified from geotechnical data gathered from geological columnar sections, is characterized as sand up to about 3 m deep and silt below this depth. Furthermore, the geotechnical data reveal intermittent clay layers from 8 m to 15 m deep, which prediction procedures (1) and (2) failed to detect. In contrast, the prediction procedure (3), which analyzes data from deeper layers upwards, successfully predicted the presence of clay with relative accuracy.

Table 6. Predicted accuracy for predicting soil classification.

(a) Model 1.					**(b) Model 2.**				
Depth (m)	***Accuracy* (%)**	**Average Precision (%)**	**Average Recall (%)**	**F-Measure (%)**	**Depth (m)**	***Accuracy***	**Average Precision (%)**	**Average Recall (%)**	**F-Measure (%)**
1	90.10	77.38	55.65	75.19	1	79.34	77.61	61.77	79.34
2	87.13	84.49	53.85	72.65	2	81.13	81.62	62.59	82.32
3	85.15	80.69	60.98	80.65	3	79.44	79.53	49.67	81.03
4	88.89	91.97	77.7	82.91	4	86.45	85.92	66.89	84.22
5	84.16	81.9	80.06	80.8	5	84.91	82.12	78.82	79.6
6	86.14	87.34	82.8	84.65	6	85.92	82.86	67.55	83.57
7	84.00	79.43	79.66	79.50	7	82.71	60.26	62.21	76.51
8	86.00	76.32	70.86	71.89	8	84.58	80.01	64.47	79.93
9	86.00	76.32	70.86	71.89	9	84.51	78.93	65.76	79.75
10	75.25	74.73	59.96	61.92	10	85.51	78.17	75.29	75.99
11	87.13	81.13	71.12	73.64	11	93.46	92.19	88.94	90.35
12	85.15	82.01	75.21	77.56	12	88.79	85.85	84.88	84.9
13	93.07	87.65	80.89	82.62	13	88.79	93.64	97.12	69.8
14	93.00	94.84	79.84	83.64	14	94.34	92.05	80.54	84.13
15	93.94	95.96	75	77.92	15	94.84	97.29	76.71	82.83
16	91.00	93.47	56.27	77.47	16	95.33	86.63	71.23	87.76
17	94.74	91.03	91.63	91.25	17	94.31	85.97	82.56	83.92
18	84.16	67.23	52.86	72.31	18	94.37	91.73	76.38	93.48
19	88.12	93.91	60.61	64.81	19	96.26	96.25	95.46	95.81
20	87.13	86.35	60.08	64.26	20	95.79	95.94	92.95	94.33

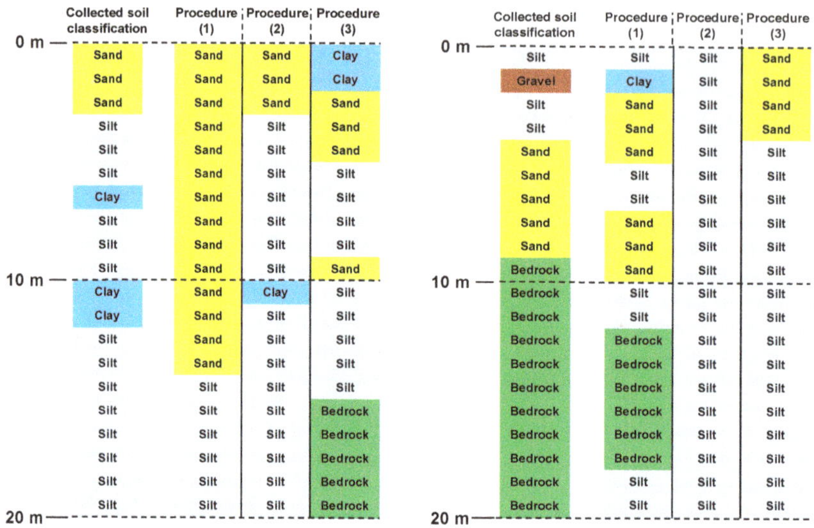

(**a**) Point 1. (**b**) Point 2.

Figure 11. Soil classification predictions using different prediction procedures (1) to (3).

For Point 2, mixed soil types are observed up to 10 m depth, beyond which the soil reaches the load-bearing layer and is identified as bedrock. The predictions accurately reflected the mixed soil types up to 10 m; however, they deviated from the geotechnical data at certain depths, resulting in less accurate classifications. At depths greater than 10 m, prediction procedures (1) and (2) predicted only silt and failed to identify the bedrock. This discrepancy is likely due to the dataset predominantly classifying the ground that reaches the load-bearing layer as silt rather than bedrock. However, akin to its performance at Point 1, the prediction procedure (3) successfully predicted the presence of bedrock layers and mixed soil types below 10 m with greater accuracy. Consequently, the prediction procedure (3) was chosen for further soil classification predictions in this study.

4.2. Creation of Soil Liquefaction Risk Mapping

4.2.1. Comparison of Self-Created and Existing Soil Liquefaction Risk Maps

During the site selection and planning stages for engineering structures and human settlements, the soil liquefaction potential of an area is a critical factor considered in geotechnical engineering. Accurately evaluating the soil liquefaction potential is the first step towards mitigating soil liquefaction hazards. Soil liquefaction risk maps have been prepared by many autonomous bodies in Japan to assess the impact of soil liquefaction in each target area [43–46]. These maps are essential tools in the development of smart cities, where resilient urban planning and infrastructure are paramount. By integrating these hazard maps into urban planning processes, smart cities can optimize their development strategies to enhance safety and sustainability in earthquake-prone areas.

The creation of these maps can generally be divided into three steps: The first step involves calculating the spatial distributions of the instrumental seismic intensity for ten recent earthquakes that resulted in soil liquefaction. In the second step, the results of these seismic intensity distributions are compared with liquefied sites from past earthquakes to estimate the ratio of soil liquefaction occurrence. The third step involves merging geomorphological classes with similar soil liquefaction susceptibilities and applying a regression method to estimate the final probability of soil liquefaction occurrence based on instrumental seismic intensity.

Unlike previous approaches [15–18], this study first makes predictions for unknown locations and then uses these results to create hazard maps. Soil liquefaction risk maps were developed for three types of earthquake motions. Level 1 earthquake motion refers to seismic activity that is likely to occur once or twice during the lifespan of a structure; however, the intensity of such shaking is relatively low and is not expected to damage the structure. Figure 12a displays a self-created soil liquefaction risk map for Level 1 earthquake motion, while Figure 12b depicts the official soil liquefaction risk map for the same level published by Yokohama. These maps feature a 100 m × 100 m grid and are color-coded according to the soil liquefaction risk. Conversely, Level 2 earthquake motion represents a seismic event that is less likely to occur but could be extremely severe, potentially causing significant damage. This level is subdivided into Type I, which anticipates a major earthquake at the plate boundary, and Type II, which considers an inland earthquake. Figure 13a shows a self-created soil liquefaction risk map for Level 2 earthquake motion with Type I, while Figure 13b shows the official soil liquefaction risk map for the same level and type published by Yokohama. Figure 14a shows a self-created soil liquefaction risk map for Level 2 earthquake motion with Type II, while Figure 14b shows the official soil liquefaction risk map for the same level and type published by Yokohama. The two prediction models differ primarily in their variables. While the existing official map of soil liquefaction risk is based solely on terrain differentiation, our self-generated map incorporates additional variables beyond terrain differentiation.

The soil liquefaction risk maps created in this study reveal a relatively larger area at high risk of soil liquefaction compared to the maps published by Yokohama. Another major advantage of using AI models is the ability to quickly update maps as new geotechnical information and seismic data become available. This is a time-consuming task using standard methods, but can be done efficiently using AI. The maps created in this study were divided into detailed risk levels, making it easier to develop countermeasures for each level of risk. The highly accurate maps created by AI will provide important information to local residents and governments, helping them to formulate specific action plans to improve the safety of local communities. This is expected to minimize damage in the event of a liquefaction disaster. It can be confirmed that the liquefaction risk maps created in this study have many advantages over the existing maps provided by Yokohama. This will enable safer and more accurate disaster planning and further improve preparedness for future disasters.

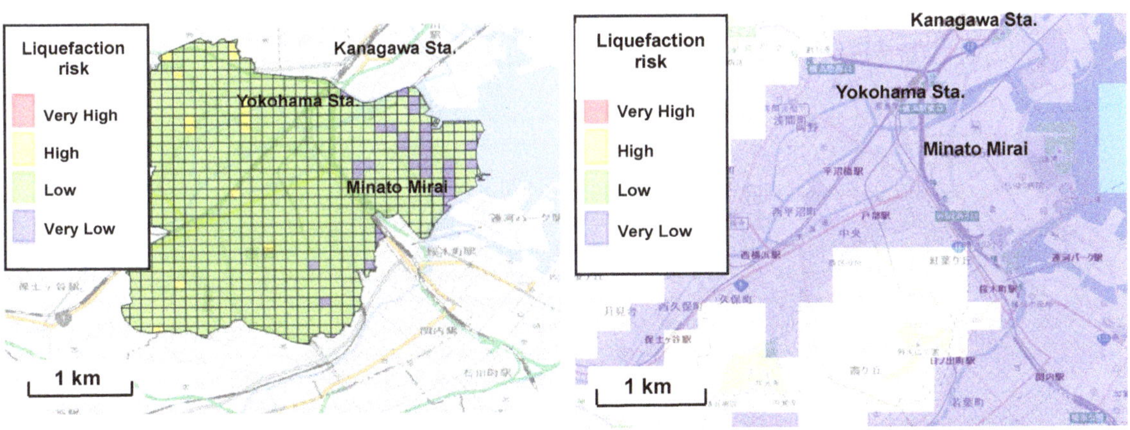

(**a**) Self-created soil liquefaction risk map.

(**b**) Officially published soil liquefaction risk map.

Figure 12. Soil liquefaction risk map for Level 1 earthquake motion.

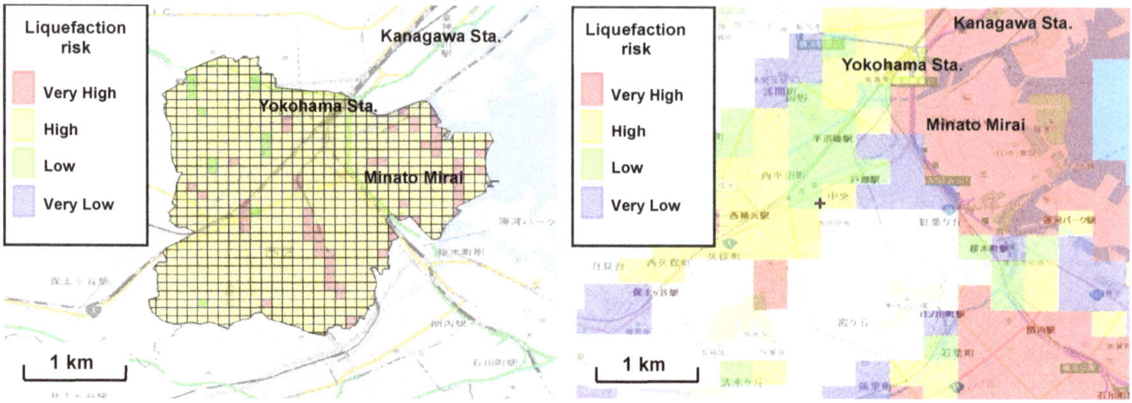

(**a**) Self-created soil liquefaction risk map. (**b**) Officially published soil liquefaction risk map.

Figure 13. Soil liquefaction risk map for Level 2 earthquake motion with Type I.

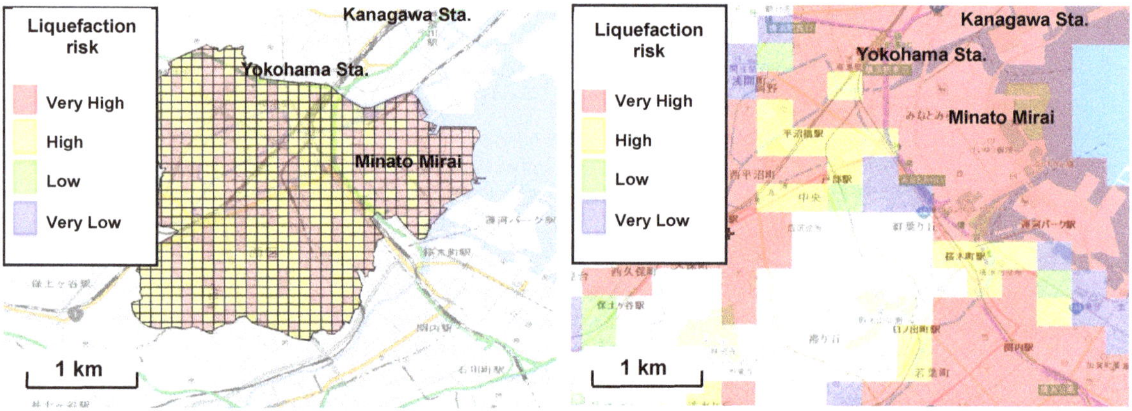

(**a**) Self-created soil liquefaction risk map. (**b**) Officially published soil liquefaction risk map.

Figure 14. Soil liquefaction risk map for Level 2 earthquake motion with Type II.

However, the model developed in this study exhibits significant limitations. First, the training dataset is inadequate, comprising just over 200 data points, which is insufficient for most machine learning algorithms that require large datasets to function effectively. This constitutes a major limitation. Second, the representativeness of the training data is questionable. As illustrated in Figure 7, the data distribution is quite dispersed, potentially compromising the model's generalization capability. Therefore, a critical step for future development is to acquire more comprehensive data, ideally exceeding 500 data points. A literature review [7] identified a dataset that includes latitude, longitude, elevation, and 'n' values for 433 data points in Tokyo, which could provide a basis for enhancing the data pool and refining the model.

4.2.2. Advantage of Soil Liquefaction Risk Maps Created by AI-Driven Predictive Model

The integration of AI into soil liquefaction risk prediction and mapping represents a transformative development in geotechnical engineering, particularly within the framework of smart cities. This paper explores the advantages of using AI-driven predictive models to produce detailed and dynamic soil liquefaction risk maps, enhancing urban resilience to earthquakes and other seismic activities. AI-driven predictive models represent a significant advancement over traditional liquefaction risk assessment methods, which typically rely on historical data and empirical formulas. While useful, traditional methods often fail to capture the complex, dynamic interactions within urban environments or adapt to rapid changes in urban landscapes. In contrast, AI models can continuously learn and adapt by integrating multiple data sources, including real-time geotechnical, geographic, and seismic data. This integration allows for a more nuanced and accurate assessment of liquefaction risk, providing urban planners and engineers with powerful tools for proactive urban development and disaster mitigation.

As shown in Table 7, one of the key advantages of AI-driven liquefaction risk maps is their ability to dynamically update and refine predictions as new data become available. This capability is critical in urban areas where conditions can change rapidly due to construction, land use changes, and environmental factors. AI models can adapt to these changes, maintaining the accuracy and relevance of risk maps. Additionally, AI ability to process large datasets enables the analysis of complex variables and interactions that would be too time-consuming or impossible for traditional methods, resulting in more detailed and comprehensive risk assessments. Another significant advantage of AI-driven hazard maps is their contribution to emergency preparedness and response strategies. Detailed hazard maps created from AI predictions can guide emergency planning and response efforts, helping to allocate resources more efficiently and improve response times during crises. For example, during an earthquake, AI-driven maps can quickly identify areas most at risk of soil liquefaction, allowing emergency services to prioritize these areas for immediate attention. In addition, these AI-driven maps facilitate better urban planning and infrastructure development. By predicting potential liquefaction zones, city planners can avoid building critical infrastructure in high-risk areas or implement special construction techniques to mitigate the risks. This proactive approach not only saves costs associated with repairing damage after a disaster but also protects lives by reducing the risk of catastrophic failure during seismic events.

Table 7. Advantages of AI-driven liquefaction risk maps to improve urban planning, safety, and sustainability in smart city.

Advantage	Description
Dynamic Updates	The AI-driven predictive model can continuously update and refine predictions as new data become available, which is critical in rapidly changing urban areas.
Adaptability	The AI-driven predictive model adapts to changes in the urban landscape, such as construction and land use changes, maintaining the accuracy and relevance of risk maps.
Complex Data Analysis	The ability of the AI-driven predictive model to process large data sets enables the analysis of complex variables and interactions, providing more detailed and comprehensive risk assessments.
Emergency Preparedness	The AI-driven predictive model guides emergency planning and response efforts, improving resource allocation and response times during seismic events.
Proactive City Planning	Prediction of potential liquefaction zones helps to avoid building in high-risk areas or to implement special construction techniques to mitigate risk.
Support for Smart Cities	Integrates with smart city goals to improve sustainability, safety, and quality of life by applying advanced technology to urban planning.
Enhanced Urban Resilience	Improves resilience to earthquakes and seismic activity, contributing to safer and more sustainable urban environments.

The advantages of soil liquefaction risk maps created by AI-driven predictive models are numerous and significant. They provide a more accurate, detailed, and adaptive tool for managing geotechnical risk in urban environments. These maps enhance urban planning, improve emergency preparedness, and contribute to the overall resilience and sustainability of cities. As urban areas continue to grow and face complex challenges, the role of AI in geotechnical engineering will become increasingly important, marking a pivotal shift toward more intelligent, responsive, and resilient urban environments.

5. Conclusions

In this study, a self-developed AI-driven predictive model was trained to analyze geotechnical data collected from geological columnar sections at known points to predict the 'N-value' and 'soil classification' at unknown points. The predicted data were then used in the liquefaction potential index (LPI) to assess soil liquefaction risk and to create maps for various levels of seismic motion. The findings include the following:

(1) It was confirmed that the larger the training dataset used in the AI-driven predictive model, the higher the accuracy of the predictions.
(2) The prediction procedure, which estimates the N-value and soil classification from 20 meters below ground to 1 meter above ground and incorporates learning from results deeper than the predicted depth, was found to be the most accurate.
(3) The AI-driven predictive model provided more detailed soil liquefaction risk mapping for the seismic motion level compared to existing mappings.

Future challenges include enhancing prediction accuracy. Despite utilizing the most accurate procedure available in this study, the predictions were not entirely consistent, with some demonstrating low accuracy due to biases in the training data. Further data collection, exploration of data processing methods, combinations of explanatory variables, and the development of new prediction techniques are necessary. In terms of assessing trigger risks, further improvements in prediction accuracy and the exploration of new evaluation methods may facilitate the creation of more precise hazard maps. Beyond assessing risks associated with soil liquefaction, this methodology could feasibly be applied to comprehensively evaluate other hazard risks and to generate maps predicting potential damage from various hazards.

Author Contributions: Conceptualization, S.I.; methodology, A.K.; software, A.K. and Y.C.; validation, A.K. and S.I.; formal analysis, A.K. and Y.C.; investigation, A.K. and Y.C.; resources, S.I.; data curation, A.K. and Y.C.; writing—original draft preparation, Y.C.; writing—review and editing, S.I.; visualization, A.K. and Y.C.; supervision, S.I.; project administration, S.I.; funding acquisition, S.I. All authors have read and agreed to the published version of the manuscript.

Funding: This research received no external funding.

Data Availability Statement: The data that support the findings of this study are available from the corresponding author, Shinya Inazumi, upon reasonable request.

Conflicts of Interest: The authors declare no conflicts of interest.

References

1. Trindade, E.P.; Hinnig, M.P.F.; Costa, E.M.D.; Marques, J.S.; Bastos, R.C.; Yigitcanlar, T. Sustainable development of smart cities: A systematic review of the literature. *J. Open Innov. Technol. Mark. Complex.* **2017**, *3*, 1–14. [CrossRef]
2. Sharifi, A.; Allam, Z.; Bibri, S.E.; Garmsir, A.R.K. Smart cities and sustainable development goals (SDGs): A systematic literature review of co-benefits and trade-offs. *Cities* **2024**, *146*, 104659. [CrossRef]
3. Su, Y.; Fan, D. Smart cities and sustainable development. *Reg. Stud.* **2023**, *57*, 722–738. [CrossRef]
4. Mishra, R.K.; Kumari, C.L.; Krishna, P.S.J.; Dubey, A. Smart cities for sustainable development: An overview. In *Smart Cities for Sustainable Development*; Springer: Singapore, 2022; pp. 1–12. [CrossRef]
5. Ismagilova, E.; Hughes, L.; Dwivedi, Y.K.; Raman, K.R. Smart cities: Advances in research—An information systems perspective. *Int. J. Inf. Manag.* **2019**, *47*, 88–100. [CrossRef]
6. Cugurullo, F.; Caprotti, F.; Cook, M.; Karvonen, A.; Guirk, P.M.; Marvin, S. The rise of AI urbanism in post-smart cities: A critical commentary on urban artificial intelligence. *Urban Stud.* **2024**, *61*, 1168–1182. [CrossRef]

7. Cong, Y.; Inazumi, S. Integration of smart city technologies with advanced predictive analytics for geotechnical investigations. *Smart Cities* **2024**, *7*, 1089–1108. [CrossRef]
8. Cong, Y.; Motohashi, T.; Nakao, K.; Inazumi, S. Machine learning predictive analysis of liquefaction resistance for sandy soils enhanced by chemical injection. *Mach. Learn. Knowl. Extr.* **2024**, *6*, 402–419. [CrossRef]
9. Hazout, L.; Zitouni, Z.E.A.; Belkhatir, M.; Schanz, T. Evaluation of static liquefaction characteristics of saturated loose sand through the mean grain size and extreme grain sizes. *Geotech. Geol. Eng.* **2017**, *35*, 2079–2105. [CrossRef]
10. Bao, X.; Ye, B.; Ye, G.; Zhang, F. Co-seismic and post-seismic behavior of a wall type breakwater on a natural ground composed of liquefiable layer. *Nat. Hazards* **2016**, *83*, 1799–1819. [CrossRef]
11. Bao, X.; Jin, Z.; Cui, H.; Chen, X.; Xie, X. Soil liquefaction mitigation in geotechnical engineering: An overview of recently developed methods. *Soil Dyn. Earthq. Eng.* **2019**, *120*, 273–291. [CrossRef]
12. Nakao, K.; Inazumi, S.; Takahashi, T.; Nontananandh, S. Numerical simulation of the liquefaction phenomenon by MPSM-DEM coupled CAEs. *Sustainability* **2022**, *14*, 7517. [CrossRef]
13. Ishihara, K.; Araki, K.; Toshiyuki, K. Liquefaction in Tyoko Bay and Kano Regions in the 2011 Great East Japan Earthquake. *Earthq. Geotech. Eng. Des.* **2013**, *28*, 93–140.
14. Wang, Z.Z.; Hu, Y.; Guo, X.F.; He, X.G.; Kek, H.; Ku, T.; Goh, S.H.; Leung, C.F. Predicting Geological Interfaces using Stacking Ensemble Learning with Multi-scale Features. *Can. Geotech. J.* **2022**, *60*, 1036–1054. [CrossRef]
15. Kajihara, K.; Okuda, H.; Kiyota, T.; Konagai, K. Mapping of liquefaction risk on road network based on relationship between liquefaction potential and liquefaction-induced road subsidence. *Soils Found.* **2020**, *60*, 1202–1214. [CrossRef]
16. Honda, K.; Takeyama, T.; Tachibana, S.; Iizuka, A. Liquefaction risk assessment in the 23 wards of Tyoko using elastoplastic analysis. *Int. J. Geomate* **2021**, *21*, 48–54. [CrossRef]
17. Matsuoka, M.; Wakamatsu, K.; Hashimoto, M.; Midorikawa, S. Evaluation of liquefaction potential for large areas based on geomorphologic classification. *Earthq. Spectra* **2015**, *31*, 2375–2395. [CrossRef]
18. Karimzadeh, S.; Matsuoka, M. A weighted overlay method for liquefaction-related urban damage detection: A case study of the 6 September 2018 Hokkaido eastern Iburi earthquake, Japan. *Geosciences* **2018**, *8*, 487. [CrossRef]
19. Mishra, S.; Shaw, K.; Mishra, D.; Patil, S.; Kotecha, K.; Kumar, S.; Bajaj, S. Improving the accuracy of ensemble machine learning classification models using a novel bit-fusion algorithm for healthcare AI systems. *Front. Public Health* **2022**, *10*, 858282. [CrossRef]
20. Sun, Y.; Li, Z.L.; Li, X.W.; Zhang, J. Classifier selection and ensemble model for multi-class imbalance learning in education grants prediction. *Appl. Artif. Intell.* **2021**, *35*, 290–303. [CrossRef]
21. Wu, H.; Levinson, D. The ensemble approach to forecasting: A review and synthesis. *Transp. Res. Part C Emerg. Technol.* **2022**, *132*, 103357. [CrossRef]
22. Doroudi, S. The bias-variance tradeoff: How data science can inform educational debates. *AERA Open* **2020**, *6*, 4. [CrossRef]
23. Ghosal, I.; Hooker, G. Boosting random forests to reduce bias; One-step boosted forest and its variance estimate. *J. Comput. Graph. Stat.* **2020**, *30*, 493–502. [CrossRef]
24. Alelyani, S. Stable bagging feature selection on medical data. *J. Big Data* **2021**, *8*, 11. [CrossRef]
25. Miemye, I.D.; Sun, Y. A Survey of ensemble learning: Concepts, algorithms, applications, and prospects. *IEEE Access* **2022**, *10*, 99129–99149. [CrossRef]
26. Ghnai, S.; Kumari, S.; Jaiswal, S.; Sawant, V.A. Comparative and parametric study of AI based models for risk assessment against soil liquefaction for high intensity earthquakes. *Arab. J. Geosci.* **2022**, *15*, 1262. [CrossRef]
27. Zhong, L.; Guo, X.; Xu, Z.; Ding, M. Soil properties: Their prediction and feature extraction from the LUCAS spectral library using deep convolutional neural networks. *Geoderma* **2021**, *402*, 115366. [CrossRef]
28. Pham, B.T.; Nguyen, M.D.; Ly, H.B.; Pham, T.A.; Hoang, V.; Le, V.H.; Le, T.T.; Nguyen, H.Q.; Bui, G.L. Development of artificial neural networks for prediction of compression coefficient of soft soil. In Proceedings of the 5th International Conference on Geotechnics 2020, Civil Engineering Works and Structures, Hanoi, Vietnam, 31 October–1 November 2019; Springer: Singapore, 2020; pp. 1167–1172.
29. He, K.; Zhang, X.; Ren, S.; Sun, J. Deep residual learning for image recognition. In Proceedings of the IEEE Conference on Computer Vision and Pattern Recognition (CVPR), Las Vegas, NV, USA, 27–30 June 2016; pp. 770–778.
30. Ren, S.; He, K.; Girshick, R.; Sun, J. Faster R-CNN: Towards real-time object detection with region proposal networks. *arXiv* **2015**, arXiv:1506.01497. [CrossRef]
31. Ji, S.; Yu, D.; Shen, C.; Li, W.; Xu, Q. Landslide detection from an open satellite imagery and digital elevation model dataset using attention boosted convolutional neural networks. *Landslides* **2020**, *17*, 1337–1352. [CrossRef]
32. Ghani, S.; Kumari, S. Liquefaction study of fine-grained soil using computational model. *Innov. Infrastruct. Solut.* **2021**, *6*, 58. [CrossRef]
33. Ghani, S.; Kumari, S. Prediction of liquefaction using reliability-based regression analysis. In *Advances in Geo-Science and Geo-Structures, Lecture Notes in Civil Engineering*; Springer: Singapore, 2022; p. 154. [CrossRef]
34. Zhang, Z.D.; Jung, C. GBDT-MO: Gradient-boosted decision trees for multiple outputs. *IEEE Trans. Neural Netw. Learn. Syst.* **2020**, *32*, 3156–3167. [CrossRef]
35. Chekhaba, C.; Rebatchi, H.; ElBoussaidi, G.; Moha, N.; Kpodjedo, S. Coach: Classification-based architectural patterns detection in Android apps. In Proceedings of the 36th Annual ACM Symposium on Applied Computing, Virtual Event, 22–26 March 2021; pp. 1429–1438. [CrossRef]

36. Komolov, S.; Dlamini, G.; Megha, S.; Mazzara, M. Towards predicting architectural design patterns: A machine learning approach. *Computers* **2022**, *11*, 151. [CrossRef]
37. Chicco, D.; Warrens, M.J.; Jurman, G. The coefficient of determination R-squared is more informative than SMAPE, MAE, MAPE, MSE and RMSE in regression analysis evaluation. *PeerJ-Comput. Sci.* **2021**, *5*, e623. [CrossRef] [PubMed]
38. Shan, S.; Pei, X.; Zhan, W. Estimating deformation modulus and bearing capacity of deep soils from dynamic penetration test. *Adv. Civ. Eng.* **2021**, *2021*, 1082050. [CrossRef]
39. Obara, H.; Maejima, Y.; Kohyaha, K.; Ohkura, T.; Takata, Y. Outline of the comprehensive soil classification system of Japan-first approximation. *Jpn. Agric. Res. Quartely JARQ* **2015**, *49*, 217–226. [CrossRef]
40. Inazumi, S.; Intui, S.; Jotisankasa, A.; Chaiprakaikeow, S.; Kojima, K. Artificial intelligence system for supporting soil classification. *Results Eng.* **2020**, *8*, 100188. [CrossRef]
41. Rahman, M.Z.; Siddiqua, S.; Kamal, A.S.M.M. Liquefaction hazard mapping by liquefaction potential index for Dhaka city, Bangladesh. *Eng. Geol.* **2015**, *188*, 137–147. [CrossRef]
42. Wu, M.H.; Wang, J.P.; Wu, Y.J.; Chen, Z.B. Relationship between liquefaction potential index and liquefaction probability. *J. GeoEngineering* **2020**, *15*, 135–144. [CrossRef]
43. Kajihara, K.; Pokhrel, R.M.; Kiyota, T.; Konagai, K. Liquefaction-induced ground subsidence extracted from digital surface models and its application to hazard map of Urayasu city, Japan. *Soil Mech. Geotech. Eng.* **2016**, *2*, 829–834. [CrossRef]
44. Kiyota, T.; Ikeda, T.; Yokoyama, Y.; Kyokawa, H. Effect of in-situ sample quality on undrained cyclic strength and liquefaction assessment. *Soils Found.* **2016**, *56*, 691–703. [CrossRef]
45. Imaide, K.; Nishimura, S.; Shibata, T.; Shuku, T.; Murakami, A.; Fujisawa, K. Evaluation of liquefaction probability of earth-fill dam over next 50 years using geostatistical method based on CPT. *Soils Found.* **2019**, *59*, 1758–1771. [CrossRef]
46. Nakao, K.; Yamaguchi, H.; Hoshino, S.; Inazumi, S. Applicability of weighting method as measure for existing manholes against uplifting during liquefaction. *Appl. Sci.* **2022**, *12*, 3818. [CrossRef]

smart cities

MDPI

Article

Transitioning to a Low-Carbon Lifestyle? An Exploration of Millennials' Low-Carbon Behavior—A Case Study in China

Yan Wu [1,*], Pim Martens [2] and Thomas Krafft [1]

[1] Department of Health, Ethics & Society, CAPHRI Care and Public Health Research Institute, Faculty of Health, Medicine and Life Sciences, Maastricht University, P.O. Box 616, 6200 MD Maastricht, The Netherlands; thomas.krafft@maastrichtuniversity.nl

[2] System Earth Science, University College Venlo, Faculty of Science and Engineering, Maastricht University, P.O. Box 616, 6200 MD Maastricht, The Netherlands; p.martens@maastrichtuniversity.nl

* Correspondence: yan.wu@maastrichtuniversity.nl

Abstract: The Sustainable Development Goals (SDGs) have set the agenda for 2030, calling for collective global efforts to deal with climate change while seeking a balance between economic development and environmental protection. Although many countries are exploring emission reduction paths, mainly from government and corporate perspectives, addressing climate change is also an individual responsibility and requires public participation in collective action. The millennial generation constitutes the current workforce and will be the leaders in climate action for the next 30 years. Therefore, our study focuses on the Chinese millennial generation, conducting in-depth semi-structured interviews with 50 participants in qualitative research to explore their low-carbon lifestyles, the barriers, and enablers in switching to a wider range of low-carbon lifestyles. There are three main results: (1) Based on our study samples, there is an indication that Chinese millennials have a positive attitude towards transitioning to a low-carbon lifestyle. Women demonstrate a stronger willingness to adopt low-carbon behaviors in their daily household activities compared to men. However, their involvement in governance in the context of transitioning to a low-carbon society is limited, with most women assuming execution roles in climate action rather than decision-making positions. (2) Millennial's low-carbon life transition is accompanied by technological innovation and progress. However, this progress brings some new forms of resource waste, and reasonable policy-making is essential. (3) Personal economic interests and the satisfaction of their consumption needs will drive millennials to reduce carbon emissions in their daily lives, but it requires the guidance of reasonable policy-making and synergies among various stakeholders. This research will help policymakers better understand the current status and potential issues related to people's low-carbon actions, enabling the formulation of more rational guiding policies. It can also help other stakeholders learn about millennials' demands and take more effective collective action toward carbon reduction.

Keywords: climate change; low-carbon behavior; millennials' lifestyle; city transformation; China; qualitative research

1. Introduction

July 2023 is considered the hottest month on record for the Earth's temperature [1]. Some locations in the USA, China, Europe, and Africa were hit by a record-breaking heatwave. For example, Phoenix, Arizona in America recorded 31 days, as of 30 July, of daytime temperatures above 43.3 °C [2]; on 16 July, Turpan City in China's Xinjiang province, had a temperature of 52.2 °C [3,4], which would have been about a 1 in 250-year event [2]. Extreme heat poses serious threats to people's health and the environment [3]. This alarming trend underscores the urgent need for concerted action toward mitigating climate change.

Low-carbon transition goals were integrated into the global development agenda [5]. In the past decades, governments have announced their reducing carbon emission goals, which are in line with the Paris Agreement [6]. For example, the European Union aimed to realize climate neutrality by 2050 and achieve an economy with net-zero greenhouse gases [7]; Germany set the goal of becoming net zero emissions by 2045 [8]; China aimed to reach peak CO_2 emissions by 2030 and achieve carbon neutrality before 2060. However, the current climate change positions of various countries are often based on their economic interests. We are halfway to 2030, and despite the passage of seven years, we have made little progress toward achieving the Sustainable Development Goals (SDGs); thus, the next seven years will be the key [9]. Therefore, this article believes that coping with climate change is not only the responsibility of governments but also of individuals. It requires civil society, businesses, and others to work collectively [10]. This paper focuses on exploring the attitudes and behaviors of millennial stakeholders towards low-carbon initiatives and the potential for collaborative efforts in reducing carbon emissions.

Urban areas contribute to over 70% of CO_2 emissions and are major contributors to climate change [11]. China's urbanization rate has exceeded 65%. Reducing urban greenhouse gas emissions is crucial for China to achieve its peak carbon and carbon neutrality goals [12]. Achieving carbon neutrality requires companies, organizations, or individuals to reduce their total greenhouse gas (GHG) emissions over a certain period. Energy conservation and emission reduction will be helpful in ultimately reaching "zero emissions" [13]. Previous studies on low-carbon cities in China, from the research content perspective, primarily focused on the development or impact of low-carbon policy from conceptual theory [14–19], the discussion of pilot low-carbon city development patterns through case studies [20–24], and carbon emission calculation and energy consumption in the industry [25–28]. Citizens are the most important part of cities; they are energy users and consumers in cities and have the responsibility to take climate action and join low-carbon activities. However, there is limited research focusing on the public's low-carbon behavior and attitude toward low-carbon lifestyles in China.

Previous studies on low-carbon cities in China, from the research method perspective, most utilized quantitative research based on statistical analysis. For example, Zhang and Zheng [29] used a penal data analysis to show that pilot low-carbon city policy can promote urban residents to transfer to a green lifestyle. Some researchers calculated residents' household carbon emissions through questionnaire surveys [30,31], while others employed the difference-in-differences model based on experimental evidence to explore the impact of pilot low-carbon city policies on urban carbon intensity [32–34]. While quantitative analysis excels in measuring and analyzing large datasets to provide clear numerical evidence, it may overlook crucial qualitative aspects such as personal experiences, emotions, and psychology that influence human behavior [35,36]. Some complex social phenomena are not easy to quantify. Low-carbon city transformations encompass various aspects, including policy-making, environmental education, public participation, and social equity. However, there is inadequate literature utilizing qualitative research to explore the potential synergistic and incentive policies in low-carbon city transformation in China. This research will address this gap, using a semi-structured interview method within a qualitative framework to examine individuals' attitudes toward a low-carbon society transformation as well as their willingness to reduce carbon emissions and participate in climate action.

Climate change is a challenge of collective action, not only within generations but also across generations [37]. Previous research indicates that age and education could affect people's environmental perception [38]. Millennials, the generation born between 1981 and 1996 (they were between 27–42 years old in 2023) [39], have already entered the job market, and they are poised to become the future workforce and shake things up across various fields [40]. Additionally, they bear the responsibility of raising and nurturing the next generation. Therefore, they play an important role in dealing with climate change and realizing carbon neutrality in the next 30 years. Petrescu-Mag et al. [41] used qualitative research to interview 41 Romanians, comparing millennials and Gen Z's attitudes toward health concerns regarding climate change in Romania. Some researchers used the questionnaire to explore millennials' attitudes towards climate change in America [42] or social media's impact on individual behavior in South Korea [43]. There is limited research focusing on millennials' attitudes towards climate change or low-carbon transition in China.

Chinese millennials grew up with the policy of reform and opening up in 1978, and most of them are the only child in their families due to the one-child policy from the 1980s, which was loosened in 2015. Unlike previous generations in China, millennials have more opportunities to receive higher education in universities and colleges. The gross tertiary education rate in China increased from 2.7% in 1978, when the national college entrance examination (known as Gaokao) was resumed, to 59.6% in 2022 [44]. In the past 40 years, the growth of this generation was accompanied by China's rapid economic development. China's GDP growth has averaged over 9 percent per year since 1978, and over 700 million Chinese have been lifted out of poverty [45,46]. Due to disparities in education, economic conditions, and social culture, the younger generation's lifestyle diverges from that of their parents or grandparents [47]. They exhibit higher levels of indulgence, sophistication, and a propensity for consumption [48]. Although they are the beneficiaries of economic development, they are the bearers of environmental risks, and they will also be responsible for the mitigation of climate change in the future. There is significant potential for public participation in climate action in China. Existing research on public environmental awareness and green behaviors has primarily focused on the behaviors of teenagers in China [49–51], with a lack of studies targeting millennials. Lifestyle is regarded as individual behavioral patterns within the sociomedical discourse [52]. Embracing lifestyle change forms an essential element of transformative paths towards achieving net-zero emissions [53]. Given the unique context of millennials, this article seeks to explore low-carbon behaviors and lifestyles from their perspective in China. By understanding their motivations and challenges, we can foster meaningful engagement and accelerate progress toward a sustainable future. Three questions will be addressed in this research:

(1) What are the low-carbon behaviors of millennials in their daily lives?
(2) What are the barriers to a wider low-carbon lifestyle for millennials?
(3) What factors enable millennials to adopt a low-carbon lifestyle?

Using in-depth interviews within qualitative research, this research investigates millennials' low-carbon awareness and behaviors in low-carbon city transformation in China. It explores barriers to adopting low-carbon lifestyles and identifies motivating factors for engaging in climate action. This study contributes to the literature on low-carbon lifestyles and urban transitions in the following three innovative ways.

Firstly, we focus on the Chinese millennial generation, a demographic with 2–20 years of work experience transitioning from formal education to broader social learning. They often contribute significantly to carbon emissions as the main body of the consumption market and are more likely poised to assume managerial positions in various industries. Their adoption of low-carbon practices could influence organizational strategies, thereby impacting a wider range of people. This focus is different from previous studies, which primarily explore the low-carbon awareness and behavior of teenagers or college students in schools [54].

Secondly, we employ in-depth interviews to investigate respondents' low-carbon awareness and behaviors, as well as their perceptions of peer behavior. The sample extends previous studies' analysis quality by engaging stakeholders from diverse industry sectors. The flexible interview format enables us to delve beyond surface-level insights, exploring respondents' initial reactions, perceptions, and emotions. More importantly, we discussed in depth the underlying factors shaping their low-carbon choices. Therefore, our research overcomes the limitations of traditional survey methods within quantitative research prevalent in the existing literature [55,56].

Thirdly, we summarize and analyze the respondents' low-carbon behaviors in their lives, work, and social interactions, offering a holistic view beyond the focus on household-level carbon emissions often found in current research [31,57]. Our study not only advances the understanding of millennial attitudes towards low-carbon lifestyles but also enriches the methodology for studying such complex societal shifts.

The findings of this study can help policymakers improve their strategies for low-carbon urban transitions. Furthermore, our research will benefit stakeholders across various fields by improving their low-carbon products and services, thus contributing to building a more sustainable future through cooperation.

2. Method

2.1. Semi-Structured Interviews

In this study, we used semi-structured interviews within qualitative research to explore (1) millennials' low-carbon attitudes and behavior, (2) the barriers to switching to a low-carbon lifestyle, and (3) the enablers in turning to a low-carbon lifestyle for millennials. The semi-structured interview is widely used in qualitative research for its ability to elicit comprehensive insights into a topic of the daily world or social phenomenon [58–60]. This kind of interview collects information through conversation under the guidance of open-ended questions, and the researchers can deepen the questions according to the interviewee's answers [61,62]. We chose a semi-structured interview because it provides the flexibility needed to deeply understand the interviewees' first impressions and reactions to a low-carbon lifestyle from their unique perspectives. This method facilitates one-to-one conversations, which is better suited to learning about participants' mental processes [63].

A semi-structured interview is suitable for this research because, unlike questionnaires, it is more flexible in content and is not constrained by written language. This allows us to conduct in-depth investigations and explore various aspects of our research questions. Chinese millennials, as products of the one-child policy, grew up alongside China's rapid economic growth during the reform and opening-up period, with increased opportunities to receive higher levels of education and have distinctive growth experiences. They are oriented towards experiences, seeking individuality, and pursuing lifestyle upgrades [64]. However, they are also the main body of the consumption market, which causes huge amounts of carbon emissions [65]. This approach enables us to capture the complexities of participants' attitudes and behaviors, providing deep insights into their willingness to adopt and switch to a low-carbon lifestyle.

Since qualitative semi-structured interviews typically involve small sample studies, the purpose is to gain an in-depth understanding of participants' individual experiences rather than calculating the number of occurrences of an event or assessing the extent of an effect [66–69]. The dialog between researchers and participants is flexible rather than following pre-set questionnaires or scales [70]. Therefore, statistical analysis such as standard deviation and confidence interval analyses are not applicable in this context [71–75], as they are more aligned with probability inferences and error estimation in large-sample quantitative research [76]. Following the guidance of qualitative research, providing detailed method descriptions could enhance the reliability and effectiveness of the research results [77]. Therefore, this study provides detailed method descriptions about Sampling and Participant Recruiting in Section 2.2, Data Collection Processes in Section 2.3, and Data Analysis in Section 2.4. This study also introduces the potential limitations of this method in Section 5 to maintain the rigor of research.

2.2. Sampling and Participant Recruiting

We combined purposive sampling and snowballing methods to recruit participants. Purposive sampling was used to select different cases according to our research questions and aim. In qualitative research, scholars widely use purposive sampling techniques to identify participants who can potentially provide the most effective information based on their related knowledge or experience about the discussion phenomenon [78,79]. This method requires researchers to select cases with variation in order to collect diverse and rich information [80] to achieve a depth of understanding. Therefore, we first created a potential participant list based on our personal networks. Our research targets millennials who were aged 27–42 in 2023 (born between 1981 and 1996) who work in various organizations, including government agencies, academic institutions, media institutions, NGOs, residents, and business companies, with a focus on different fields. Practitioners in these fields are stakeholders in the low-carbon city transformation [20], and they are also knowledgeable about millennials' lifestyles. This age group is 20–35 years away from retirement, and most of them have some work experience. Many people have played or will play managerial roles in their fields, and they will be the key figures over the next 30 years in China's pursuit of carbon neutrality by 2060. Therefore, millennials' awareness and behavior regarding low-carbon lifestyles will not only affect the education of the next generation but will also play a significant role in the low-carbon transition and decision-making within their organizations or industries.

In addition, the snowball sampling approach was also used to recruit participants because this method is effective in accessing difficult-to-reach or 'hidden' populations by leveraging the existing social connections of individuals with relevant characteristics [81–83]. Based on the first stage of conducting the preliminary interviews, we sought recommendations for additional interviewees who could also be suitable participants for this study. Moreover, we utilized social platforms such as LinkedIn, WeChat, and Weibo to identify and contact relevant interviewees from various sectors and fields. There were three selection criteria: (1) they were millennials aged between 27 and 42 in 2023 (born between 1981 and 1996); (2) they were familiar with low-carbon-related terms such as climate change, reuse, recycling, garbage sorting, etc.; (3) they know about low-carbon development within their fields.

Based on previous studies on sample size in qualitative research, many researchers adhere to the notion that data saturation, as defined by Glaser and Strauss, depends on the presence of new relevant information in subsequent interviews [84–86]. Sample size guidelines suggest a range of 20 to 30 interviews as sufficient [87–92]. Interview data collection is deemed complete when no further discoveries surface [91,93]. This should be distinguished from the larger sample size of written surveys [94] because the advantage of qualitative interviews lies in their interactive nature, their ability to capture the complexity of experience, and their allowance for unexpected topics to emerge, enriching the data due to their depth and duration [95]. In line with the research advice of Loraine, B. et al. [96], this study was first carried out through online interviews (through video calling and voice calling) with 21 participants between 1 August 2022 and 16 September 2022 due to the COVID-19 travel restrictions. Following the initial data collection, the subsequent step involved data analysis. To meet our research needs, we then carried out face-to-face interviews with nine participants from 1 February 2023 to 29 February 2023. Through this further data collection, we were able to identify new information during this phase of analysis. This iterative process persisted until no new information emerged. Redundancy of information occurred after we conducted an additional 20 online in-depth interviews in January and February of 2024 to enrich our data, reaching a point termed saturation. Each interview duration ranged from 30 min to 2 h, depending on whether the interviewees fully answered our questions or if there were any additional in-depth explorations related to our research topics and aims.

In total, we reached out to 74 interviewees; 50 individuals met our research criteria and were willing to participate in interviews, and the other 24 potential participants declined due to time constraints or a lack of related knowledge.

2.3. Data Collection

Semi-structured interviews can be conducted in person or online, and researchers can flexibly arrange the interview questions [60]. Before the interviews, we sent interview invitation letters to all potential participants, introducing details, including research aims, interview questions, privacy protection, data confidentially, and the information collection and data usage process. When participants agreed to participate, we made an appointment for an interview time slot with them. Before each interview, we also introduced the background of our research and reconfirmed the process and anonymity. The interview questions began with general questions about the participants' personal information, like age, education, and role in their institution. Then, the full interview was guided by four sections with 10 questions (Sheet S5 and Appendix A). The order of interview questions was arranged flexibly based on respondents' knowledge and background to facilitate deeper communication. All participants were free to decide whether to participate and were informed they could abort the interview at any time. Interviews were recorded using the cell phone's recording function, and the interviewer made notes with important information, with the participants' agreement. To ensure the protection of participants' privacy and anonymity, we implemented a coding structure that used numerical identifiers instead of names in the transcripts. Table 1 shows the basic demographic information of the interviewees.

Table 1. Basic demographic information of interviewees.

Characteristic	Demographic	Frequency	Percentage
Age	27–29	14	28%
	30–34	25	50%
	35–39	8	16%
	40–42	3	6%
Gender	Male	24	48%
	Female	26	52%
Education	Junior college	2	4%
	Bachelor's degrees	18	36%
	Master's degree	25	50%
	Doctor's degree	5	10%
Organizations	Government	4	8%
	Traditional medias	16	32%
	Corporates	16	32%
	Educational institutions	3	6%
	Residents	9	18%
	NGOs	2	4%
Areas	First-tier cities in China	24	48%
	New first-tier cities in China	16	32%
	Second or third-tier cities	10	20%

While interviews provide in-depth insights and researchers can capture participants' emotions, perspectives, and experiences, it is also important for researchers to observe the environment and context [97]. To gain a better understanding of the current urban lifestyle in China and examine the interviewees' responses, we conducted fieldwork as an additional approach during the in-person interview phase in Beijing, Hohhot, Nanjing, and Suzhou from 1 February 2023 to 29 February 2023. Our fieldwork involved observing waste-sorting practices in 20 communities, the utilization of shared bicycles and new energy vehicles in these cities, and a low-carbon company.

2.4. Data Analysis

The interviews were digitally recorded and fully transcribed. Our study follows an inductive approach based on grounded theory in qualitative research to analyze the data, focusing on obtaining synergistic knowledge and showing the empirical evaluations [98]. Data collection and analysis were conducted simultaneously, in line with the iterative approach of qualitative research [99,100], until no new information appeared to reach the data saturation [101]. After each interview, we promptly organized and analyzed the data. This research involved several analysis steps. Figure 1 shows the data analysis process of this research. We began by transcribing the interviews in Atlas.ti 22.2.0 qualitative software, and then reviewed all transcriptions twice to ensure the accuracy of the original interview data.

Next, we proceeded with the coding process. According to grounded theory, the coding process includes three phases. First is open coding. We used an inductive approach based on the text of the interviews to analyze data [102,103]. The transcripts were reviewed line-by-line, word-by-word, or phrase-by-phrase. Preliminary codes were generated based on the 'in vivo' terms used by the participants themselves after comparing for similarities and differences [104]. This process involved selecting paragraphs, fragments, and significant quotes from the transcripts [105]. Subsequently, after the initial coding, we checked and added the coding, if necessary, and removed duplicates.

The second phase is axial coding. This involves drawing connections between the initial open codes to form categories [106]. The acquired codes were compared with previous ones, and conceptually comparable codes were grouped together to form categories. In addition, categories were compared and, if necessary, combined. In some cases, one category was divided into two or more [107]. We organized the initial codes into fourteen categories based on their commonality and the textual content of the interviews [108].

Thirdly, selective coding, the conclusive step in grounded theory analysis, facilitates forming a larger theoretical scheme by integrating categories [109]. It aims to unveil the interrelations among categories to establish the ultimate thematic frameworks [110]. We organized the previously established categories into three overarching themes aligned with our research questions: millennials' low-carbon behavior, the barriers to a low-carbon lifestyle, and the possible enablers for millennials to live a low-carbon life.

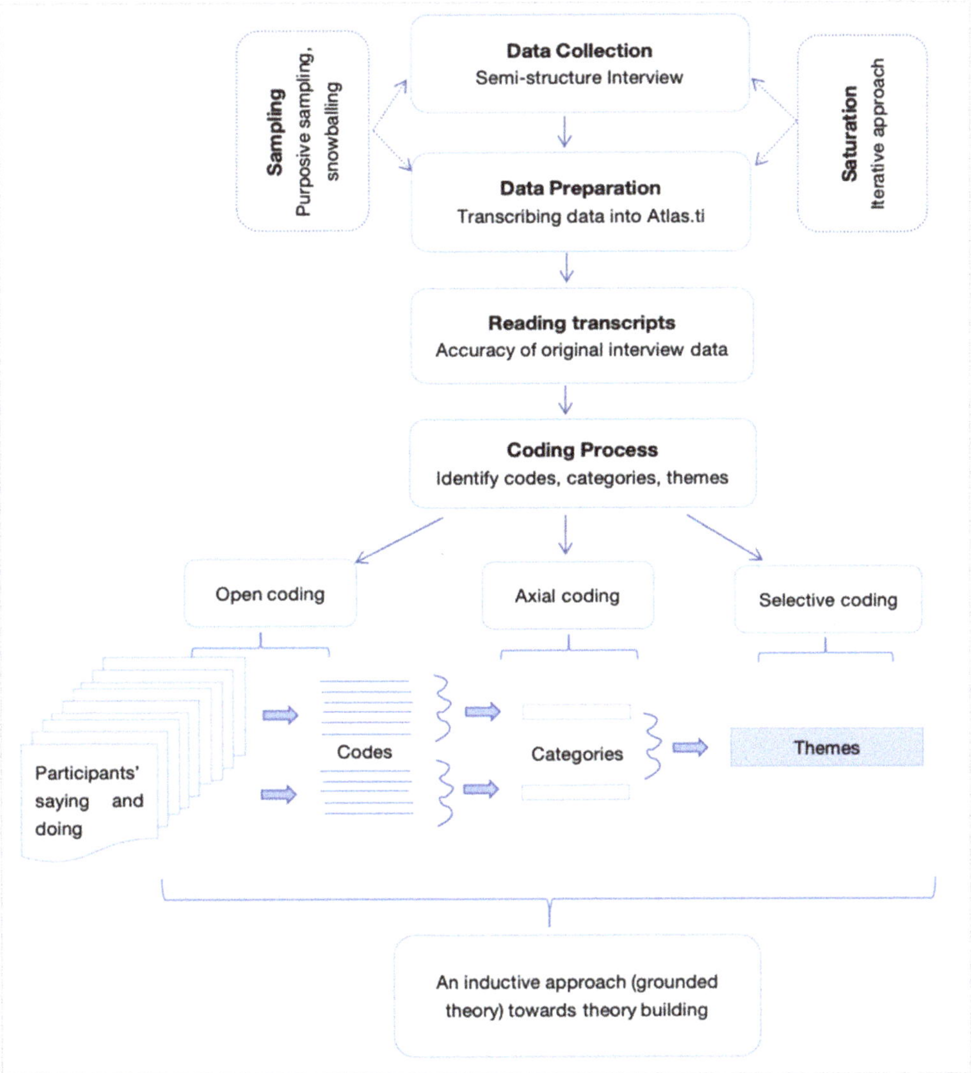

Figure 1. The data analysis process of this research.

To ensure the accuracy of our final themes and categories in reflecting the phenomena under investigation, we applied respondent validation by inviting participants to provide feedback for our results [111]. Additionally, to further enhance the validity and strengthen the credibility of our research outcomes, we invited a colleague to comment on our data analysis [112]. Finally, we read through and checked all documents, coding, and categories, extracted important information, and translated it into English. The last iteration of coding resulted in a final set of 94 codes. Figure 2 shows the structure of categories and codes; more examples of code details and analyses can be seen in Sheets S2–S4.

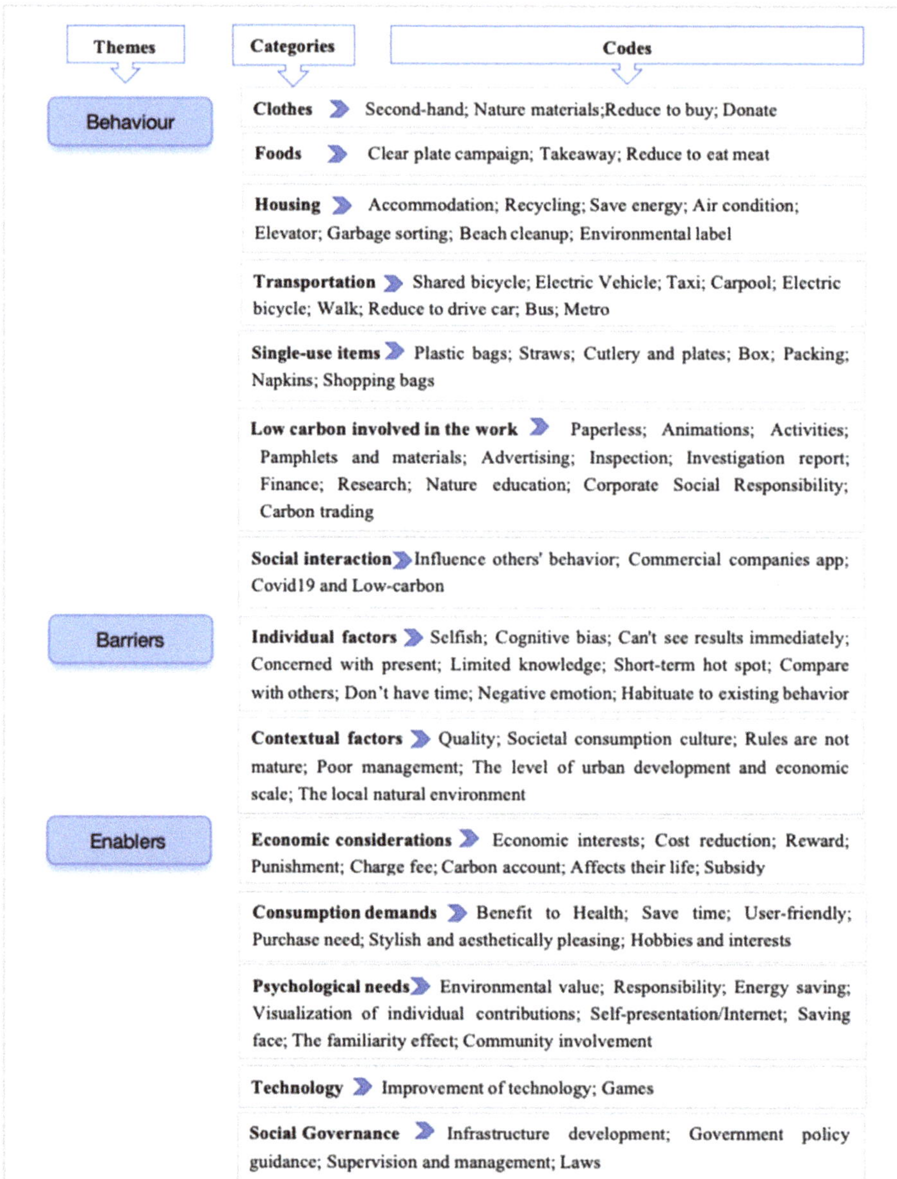

Figure 2. The structure of categories and codes.

3. Results

3.1. General Results

In terms of age, all of our participants were 27–42, belonging to the millennial generation. Among the interviewees, 18 held bachelor's degrees, 25 held master's degrees, 5 had Ph.D. degrees, and only 2 graduated from junior colleges. This represents that our article's results will show the daily life and consumption behavior of the well-educated millennial generation.

In our study of 50 participants, 24 were male, and 26 were female. We found that both male and female respondents had positive attitudes towards a low-carbon lifestyle. However, when it came to daily actions, women were more inclined to adopt low-carbon practices in household energy conservation because most of them are responsible for energy bill payments within their households. Additionally, married women played a key role in environmental education within their families, influencing their husbands and children through their low-carbon behaviors. One participant stated the following:

> In our new neighborhood, property management hasn't started, so we're not paying for water right now. We only pay the monthly electricity bill because it's managed by the energy company directly. I try to save water when I wash vegetables, do laundry, and take showers, and I also turn off lights promptly when leaving. However, my husband doesn't do so, I always remind him. P5

Our participants come from various backgrounds, including government (n = 4), traditional media (n = 16), corporate (n = 16), educational institutions (n = 3), residents (n = 9), and NGOs (n = 2). Figure 3 shows the specific work fields of participants. We found that some participants knew about a low-carbon lifestyle due to their work. Participants with prior experience in environmental protection are more willing to embrace low-carbon lifestyles because of their deeper understanding of environmental degradation and climate change impacts. In addition, female participants with low-carbon work experience or involvement in climate change action at work are fewer than male participants. Among the females who have related experience, most occupy execution roles rather than decision-making positions within their organizations.

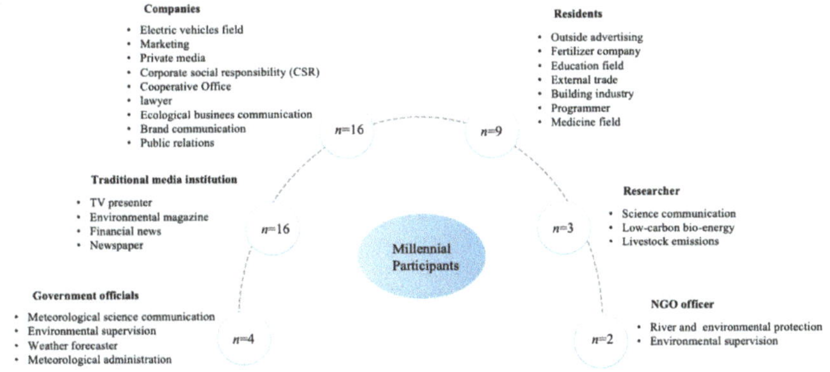

Figure 3. Work fields of participants.

In terms of residential areas, shown in Sheet S1 and Figure 4, 24 respondents live in China's first-tier cities. Among them, 18 participants live in Beijing, 4 participants are from Shanghai, and 2 of them are living in Guangzhou. Sixteen participants live in China's new first-tier cities (Nanjing, Suzhou, Hangzhou, Chengdu, Xi'an, Changsha, and Fuzhou), while ten respondents live in second or third-tier cities (Guiyang, Nanchang, Hohhot, Rizhao, Zhanjiang, and Changzhou). Our research indicates that city size impacts the millennials' lifestyle. For example, participants living in first-tier cities often rely on subway commutes, and they also have a preference for ordering takeout. On the other hand, respondents living in second-tier or third-tier cities are more inclined to take buses or drive cars for commuting.

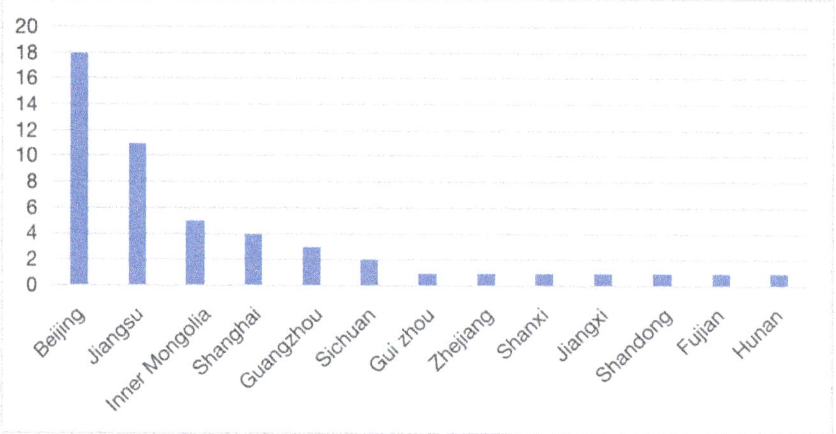

Figure 4. Areas of participants.

3.2. *Low-Carbon Behaviors of Millennials*

3.2.1. Low-Carbon Behaviors in Their Daily Life

Clothes. Participants practice low-carbon behaviors related to clothing by buying clothes with natural materials, reducing the buying of clothes, purchasing second-hand clothes, and donating or selling old clothes. Most of them prefer online shopping for new clothes, utilizing platforms such as Taobao or buying second-hand clothes through apps like Xianyu. They believe that buying less is key to reducing carbon emissions. Regarding recycling, the majority of respondents choose to either donate their old clothes at nearby second-hand clothing collection bins within their community or arrange for doorstep pickups with second-hand clothing sorting businesses. These businesses often offer compensation based on the weight of the clothes. Most respondents' motivation for dealing with second-hand clothing is not only to free up closet space but also to avoid wasting.

Foods. When it comes to low-carbon diets, respondents' initial reactions include reducing takeout orders, clean-plate campaigns, and consuming less meat. They believe that ordering takeout generates excessive single-use packaging, disposable utensils, and plastic bags. However, some respondents still would choose to takeout when they do not want to cook or need to save some time. Most respondents actively avoid wasting food and believe that the Clean-Plate Campaign is effective in China. This is because preventing food waste is a traditional value deeply ingrained in Chinese culture, and people have a higher awareness of this. During the Clean-Plate Campaign, some restaurants and cafeterias implemented rules to minimize food waste, and ordinary citizens participated in online monitoring. Furthermore, there is a growing trend among young adults to reduce meat consumption or transition to vegetarian diets, driven by health awareness and social environments.

Housing. Regarding housing, participants' low-carbon behaviors primarily include household energy conservation, waste recycling and utilization, and the purchase of eco-friendly home products. Firstly, the majority of participants exhibit household energy conservation habits, such as saving water and electricity and reducing the use of air conditioning and elevators. Among these practices, respondents demonstrate a stronger awareness of electricity conservation, as their monthly electricity bills exceed their water bills. Additionally, some participants like to reuse or renovate old items out of personal interest, and others actively participate in beach clean-up activities due to the proximity of pollution sources to their homes. While many participants mentioned garbage sorting, most admitted to not practicing it in their daily lives; a lack of enforcement and management is the main reason. Thirdly, some participants expressed consideration for insulation effectiveness when purchasing a house and exhibited a preference for eco-friendly paints and other decorative products with eco-friendly labels. These choices are primarily motivated by concerns for energy efficiency and health.

Transportation. Compared to other low-carbon behaviors, participants show a higher level of awareness regarding low-carbon transportation. Many of them mentioned frequent use of public transportation, such as subways and buses, for their daily commute. Additionally, many participants regularly choose to use shared bicycles, particularly after disembarking from the subway or bus. Participants are conscious of the emissions associated with private cars and have opted for alternatives, such as taking taxis, carpooling, or purchasing electric vehicles. However, there are still some participants who continue to rely on gasoline cars for the commute because they purchased such vehicles early and do not want to buy new ones. Furthermore, some participants mentioned commuting via electric bicycles or walking to work regularly. Regardless of the chosen mode of transportation, most participants prioritize time and cost savings over environmental factors when making their commuting decisions.

Single-use items. Most participants mentioned that they try their best to reduce the use of plastic bags. Some participants stated that they use eco-friendly bags for shopping and avoid excessive packaging. Additionally, some participants mentioned that they actively reduce their use of disposable utensils, straws, and napkins while reusing disposable boxes. We found that the primary reason for respondents to reduce their use of plastic bags is due to plastic bag regulations, with many supermarkets now charging for plastic bags. Therefore, many people bring their own shopping bags.

Example quotes of participants can be seen in Table 2, which shows the above personal choices in terms of clothing, food, housing, transportation, and daily life about low-carbon behavior.

Table 2. Example quotes of participants' low-carbon behaviors.

	Categories	Example of Quotes
Section 3.2.1	Clothes	Many communities now have individuals who collect clothes and offer doorstep pickups. It was convenient to gave them my clothes without the need for sorting, and they gave me a little money, usually ranging from 0.5 to 2 RMB per kilogram. P30 If there is a stylish coat with a fur collar, I would choose faux fur rather than fur sourced from animals. P34
	Foods	The Clean Plate Campaign in China is quite effective. Even among us adults, you see most people packing up their leftovers. I think this concept is not just about propaganda; it aligns with a traditional Chinese way of life. It's also about saving face. P29 I try to avoid ordering takeout as much as possible because it generates a lot of waste. P15
	Housing	The situation of waste sorting is vary from city to city. In Suzhou, my community has a specific schedule for garbage collection: from 6:00 AM to 8:00 AM, and from 6:00 PM to 8:00 PM. During these times, two garbage bins are placed outside, and residents can dispose of their garbage. Dedicated individuals supervise the process, and they also help people for further sorting. We mainly separate the kitchen waste and other types of garbage. P19 In our community, we do have separate garbage bins for classification, but few people do so, including me. Most people are simply throwing the garbage into the bins without separation. P5

Table 2. *Cont.*

	Categories	Example of Quotes
	Transportation	Shared bicycles have perfectly solved the last-mile problem because it takes 10 min to walk from the subway to my home, but I can get there in three minutes by riding a bicycle. I don't even have to buy a bicycle, and there are many shared bicycles at the subway station. However, a terrible thing is that the bicycles at the subway station are often placed in a disorderly manner. P23
Section 3.2.2	Low carbon behaviour involved in the work	I'm involved in green financing, which supports the financing of industries related to renewable energy, such as carbon-neutral initiatives and the electric vehicle sector. Currently, green financing also benefits from policy and financial support. We introduce our financial products to certain businesses, and if a company reduces its carbon emissions and meets the standards, their financing costs can be lowered. P26 I participated in environmental monitoring reports previously. The provincial environmental protection department visits to some cities and companies then compiled the results, which were handed over to the local governments for follow-up. We then produced special reports based on these information. We visit the same places where they had been, and we assess how well these companies had implemented corrective measures. We then broadcast these findings on television. P7 Last year, I cooperated with a research institution to Mount Everest to collect samples for a month and filmed a documentary about glaciers. Through this experience, I gained a deeper understanding of how climate change is closely related to our lives. P36
Section 3.2.3	Low-carbon behavior in Social interaction	I like to use Ant Forest because their concept is innovative, allowing users to convert their daily low-carbon lifestyle actions recorded through Alipay mobile payments into energy. Once a certain amount is accumulated, we can virtually plant a tree on the Alipay platform, and Ant Financial will also plant a real tree in the severely western desert areas of China. I believe participating in this activity is meaningful. My earliest awareness of low-carbon life came through Ant Forest. This approach, combining internet technology, finance, and low-carbon principles, makes it relatively easy for everyone to participate without any costs. P17 Waste sorting in Shanghai was progressing well before the pandemic. However, the pandemic had a significant impact on waste sorting efforts, and very few people in my community continued to do so. People's priorities shifted during the pandemic as they had to address various immediate challenges in their lives. P10

(Note: More quotes of this part can be seen in Sheet S2).

3.2.2. Low Carbon Behavior Involved in the Work

Interviewees' low-carbon behaviors in the work include conserving energy in the office, engaging in low-carbon education, environmental monitoring and communication, carbon finance, etc. Some respondents mentioned that they make an effort to reduce paper usage and opt for double-sided printing, primarily due to their organization's requirements and to minimize waste.

In addition, some interviewees have work experience in environmental monitoring. Some of them work for the government, where they carry out environmental inspections. Others work in the media, investigating environmental pollution, producing reports, and making climate change-related promotional videos. There are also interviewees from NGOs who collaborate with the government and media to supervise polluting companies. These interviewees believe that their strong environmental awareness in daily life is due to their exposure to environmental degradation through their work. They also believed that low-carbon knowledge and personal behavior changes are mainly from their work.

Example quotes of interviewees' low-carbon behaviors in their work can be seen in Table 2.

3.2.3. Low-Carbon Behavior in Social Interaction

The participants' low-carbon behaviors also extend to social interactions, including influencing others' low-carbon actions, using Ant Forest, and behavior changes due to the pandemic (COVID-19). Some participants mentioned that in their daily lives, they remind their family members to conserve energy, recommend environmentally friendly products, and purchase links to friends. Given their reliance on online shopping, entertainment, and social interactions, online product reviews and recommendations from popular social media influencers also play a significant role in shaping their purchasing choices. Additionally, participants reported that the environmentally friendly and low-carbon behaviors their children learn at school influence their own low-carbon behaviors and awareness. Furthermore, some participants mentioned Ant Forest, and most of them have been participating in the game for over three years because of its simple design, interactive rankings with friends, and easy integration into daily life. Ant Forest is an energy accumulation charity program initiated by Alipay, and as of April 2019, it had over 500 million users and had planted over 100 million trees in desert areas [113].

Furthermore, some participants reported significant changes in their lifestyles during the pandemic (COVID-19). Participants believe that during the pandemic, people were more focused on their personal lives and less inclined to consider low-carbon issues. While some individuals proactively reduced their carbon footprint, such as reducing shopping due to decreased personal income, others reported that they were in passive low-carbon behaviors, such as working from home, reducing takeout orders, and using public transportation less to maintain social distancing.

The above quotes from participants' low-carbon behaviors reflect the shift from the high-carbon lifestyle of the millennial generation, which requires not only an increase in individual low-carbon awareness and behavioral changes but is also influenced by personal interests, external pressures, social contexts, and policy support. The development of the internet has a significant impact on their shopping, dining, and travel behaviors. In addition, the quotes also demonstrate that low-carbon actions in the work fields are often connected to other individuals or organizations and are primarily achieved through collaboration with others. Our participants' work experiences related to low-carbon efforts range from raising environmental awareness among a broader audience to facilitating corporate transitions toward low-carbon practices, as well as supervising and rectifying high-pollution and high-emission enterprises.

3.3. Barriers to a Wider Range of Low-Carbon Behavior

Although many respondents maintain a positive attitude toward low-carbon living, their current low-carbon behaviors primarily stay in saving electricity, adopting low-carbon transportation, and reducing the use of plastic bags. Transitioning to a wider low-carbon lifestyle still has a long way to go. Therefore, we explore the barriers that influence the millennials' low-carbon behaviors.

3.3.1. Individual Factors

From a personal perspective, respondents believe that many people are primarily concerned with short-term personal satisfaction and lack deeper knowledge of low-carbon concepts. Most respondents pointed out that people tend to focus on the visible aspects of their lives or prioritize convenience, habits, saving time, and personal interests. They may not see their contributions to the environment or the impact of their actions in the short term. Furthermore, some participants reported limited knowledge of low-carbon products, and cognitive biases are also the reasons why people are reluctant to take action. For instance, participants mentioned that they were unclear about how to properly classify their waste despite being aware of the concept of waste sorting. This indicates the importance of climate change education, raising low-carbon awareness, and promoting more sustainable and low-carbon lifestyles.

> Some products are made of recycled materials, and some consumers may have cognitive bias, they think that these recyclable materials are extracted from trash. Therefore, some commercial advertisements may not specifically emphasize that

> the products are made from recyclable materials, which can make consumers doubt the quality and safety of the products. P3

> I believe that many people may feel that their small daily wasteful habits do not have a significant impact on the broader environment and climate change. Consequently, they may not consider changing their behavior. P5

> The government and the media have been talking about carbon neutrality and climate change, but for ordinary people, it's just a concept. There is no clear understanding of what we need to do to support or participate in climate change mitigation. P41

3.3.2. Contextual Factors

As for external factors, participants reported that when engaging in low-carbon consumption, they consider the quality and safety of low-carbon products. Participants also mentioned that there is currently no established culture of low-carbon consumption, and the availability of low-carbon products on the market is limited, providing them with fewer choices. They also believe that limited media coverage about low-carbon lifestyles affects people's behavior. Furthermore, some participants indicated that their modes of transportation and the natural environment of their cities play a significant role. For example, in cities with hilly terrain, local residents are less likely to commute by bicycle. Moreover, most participants believe that bringing about change solely through individual voluntary actions, without rules and reasonable management, is not easy. However, some participants also believe that excessive constraints and mandates could lead to negative emotions and reluctance to act. This shows that guiding people toward low-carbon lifestyle changes requires balanced rules while considering individuals' freedom and well-being.

> Some low-carbon products may have quality issues. For instance, certain degradable plastic bags or straws may not be very durable. P2

> Due to space constraints, most media outlets do not cover climate change or low-carbon lifestyles on a daily basis. To some extent, this affects the scope of information that people receive. The level of professionalism in media coverage also influences people's interest in reading. P33

These barriers to adopting a more comprehensive low-carbon lifestyle indicate that achieving a broader transition toward a low-carbon lifestyle among millennials requires the cooperation of different stakeholders. For example, improving education and raising awareness about climate change through educational institutions and the media, encouraging businesses to transition toward low-carbon production, and having governments strengthen supervision and improve regulations. Sheet S3 shows additional example quotes of the barriers influencing participants' transition toward a low-carbon lifestyle.

3.4. *Enablers to Switch to Low-Carbon Lifestyle*

3.4.1. Satisfy Personal Needs

Economic considerations. Participants mentioned that people's interest in and attempts at low-carbon transformation are driven by their own economic interests. Some participants believe that when climate change disasters start affecting people's lives and causing more economic losses, there will be a stronger awareness of changing high-carbon lifestyles. However, by that time, it might be too late. Participants also indicated that if low-carbon products were more affordable and helped them save money, they would consider purchasing them. Conversely, if the prices of non-environmentally friendly products increased or if previously free services started charging fees, they would reduce their consumption and use. In addition, participants mentioned rewards and punishment mechanisms. Some participants believed that both material and psychological rewards, such as cash subsidies, reward points accumulated in carbon accounts for vacation benefits, or certificates of honor, could encourage low-carbon behaviors. However, others felt that rewards might be optional for some individuals, but punishment could be more effective.

Consumption demand. Participants reported that they would consider changing their consumption habits if low-carbon products could meet their daily consumption needs. They mentioned that their primary consideration would be whether they actually need the product rather than consuming it solely because it is low carbon. Some participants indicated that factors such as the product's potential health benefits, time-saving attributes, and ease of use would motivate their low-carbon consumption. Others mentioned that they are driven by the desire for personalized design and the fun aspect of products, as these help them showcase their individuality and aesthetics.

Psychological needs. Participants believe that certain personal values and social needs can drive them to change lifestyles to satisfy their inner needs, gain self-fulfillment, and obtain social acceptance. In terms of personal values, some participants mentioned that environmental values influence their behavior, such as reducing meat consumption for animal protection and organizing environmental protection projects in their work. Participants also reported that a sense of responsibility for the next generation motivates them to save energy and educate their children. Additionally, participants mentioned that if they could visualize their individual contributions to reducing carbon footprints, it would improve their confidence and motivation to change their lifestyles.

In addition, from a social perspective, some participants mentioned that they sometimes would consider the 'familiarity effect' or 'saving face' in their low-carbon actions. For example, when they see their colleagues turning off lights in the workplace or avoiding food wastage, they also follow. This helps them better integrate into their social circles. Furthermore, participants also mentioned that their self-presentation needs, particularly on social media, drive their low-carbon actions because they want to shape their social identity online.

3.4.2. Technology

Participants believe that technological advancements can drive people to change their lifestyles. In terms of consumption, they believe that technological progress in low-carbon products can lower production costs while improving product quality, meeting people's consumption needs, and increasing trust in these products. For example, many participants consider purchasing electric vehicles because of the maturity of the technology. In addition, participants also believe that the development of internet technology has transformed their lives, providing convenience and more choices in various aspects such as food delivery services, shared bicycles, online shopping, social media, virtual studio of TV programs, and live-streaming platforms. Therefore, they believe that future transformations toward a broader range of low-carbon lifestyles will also require technological innovation to guide low-carbon actions. However, some participants expressed concerns that while they enjoy the faster and more convenient services facilitated by technology, it could also lead to the emergence of new forms of waste, such as excessive packaging in takeout and online shopping. They believe it is necessary for the government to enhance relevant policies and further advance technology.

3.4.3. Social Governance

Participants believe that the government plays a crucial role in social governance, which includes low-carbon policy making, strengthening supervision and management, improving relevant laws, and enhancing urban low-carbon infrastructure. Participants reported that a broader transition to low-carbon lifestyles requires the government to establish reasonable policies and guidance, such as price controls and subsidies for new energy vehicles. Participants also mentioned that their awareness of low-carbon lifestyles improved due to the dual carbon goals impacting various industries, consequently affecting their lives.

In addition, participants mentioned that the government should strengthen supervision and management of waste and environmental pollution. They believe that if the government enhances management and improves relevant regulations, more people will participate in waste sorting. Furthermore, participants also mentioned that the government needs to provide infrastructure support, which can influence their choice of transportation. For example, building more green pedestrian paths or bike lanes can encourage them to choose low-carbon transportation methods. Example quotes could be seen in Table 3.

Table 3. Example quotes of enablers to switch to low-carbon lifestyle.

	Categories	Example of Quotes
Section 3.4.1	Economic considerations	Economic incentives are quite practical, and the simplest form is economic subsidies. People are willing to participate; for instance, many people purchase new energy vehicles because of substantial government subsidies, which make the purchase price cheaper. P4 I have an electric vehicle, since I frequently business travel, using the electric vehicle instead of my previous gasoline car saves me a significant amount of money. There are numerous charging stations along the roads, and I can also avail of free battery swaps for recharging. P45
	Consumption demand	If it's related to health, I tend to prioritize it. For example, when renovating a house, using eco-friendly adhesive that emits fewer harmful gases than regular adhesive or using water-based paint instead of solvent-based paint that emits fewer harmful fumes is not only environmentally friendly but also chosen primarily for the sake of one's own health. P3
	Psychological needs	We need a specific data to gain confidence and hope, understanding that everyone's individual efforts, when accumulated, can contribute to mitigating climate change. This way, we will be more motivated to participate, as otherwise, many people may feel that our individual impact is insignificant. P2 Sometimes, I have a need for self-presentation. For example, I want to portray myself as an environmental enthusiast and refine my image on social networks. Building this personal brand may help me get more followers and more recognition from others. P28 We engage in environmental protection, such as organizing volunteer-led clean-up activities in Tibet or arranging charity events to assist pneumoconiosis patients. Our primary motivation stems from our values, even though the impact of these activities were limited. P48
Section 3.4.2	Technology	Compared to traditional cars, electric vehicles now have some advantages. Smart features of an electric vehicles can provide users with a better experience. In addition, as the quality of electric cars continues to improve and charging infrastructure expands, users' criticism towards electric vehicles has decreased. These improvements are mostly attributed to technology. P16 The improvement of technology in low-carbon communication, such as utilizing AI technology to simulate climate change scenarios and future potential risks, can enhance people's engagement and leave a more profound impression. P34
Section 3.4.3	Social governance	Policy making: I believe policy influence is significant. Influenced by national policies, now some big cities have implemented restrictions on driving, license plate issuance, and car purchase. For example, in Beijing, the government controls the ratio of issuing license plates for gasoline cars and new energy vehicles each year, making it easier for new energy vehicles to obtain licenses. Moreover, the government provides subsidies for new energy vehicles, which is a very important reason for people to purchase them. P8 Supervision and management: The reason why waste sorting is difficult to implement is because of the lack of supervision and reasonable rules. Unless the community is very strict, like in Shanghai, where each community has someone responsible, otherwise it is difficult for people to take action. P21 Infrastructure support. Some cities have bike lanes that are not very cyclist-friendly and are often encroached upon by private cars. Encouraging people to choose low-carbon transportation requires a better road infrastructure support. P5

(Note: more quotes on this part can be seen in Sheet S4).

The above quotes show the perspectives of the participants on the enablers of low-carbon behavioral change. The transition to a low-carbon society requires synergy among stakeholders, considering individuals' economic interests, consumption needs, and psychological needs. Reasonable policies and technological developments play a crucial role in encouraging millennials to participate in low-carbon transition. Figure 5 shows the theoretical model of the relationship between behavior, barriers, and enablers of millennials, as identified in our research.

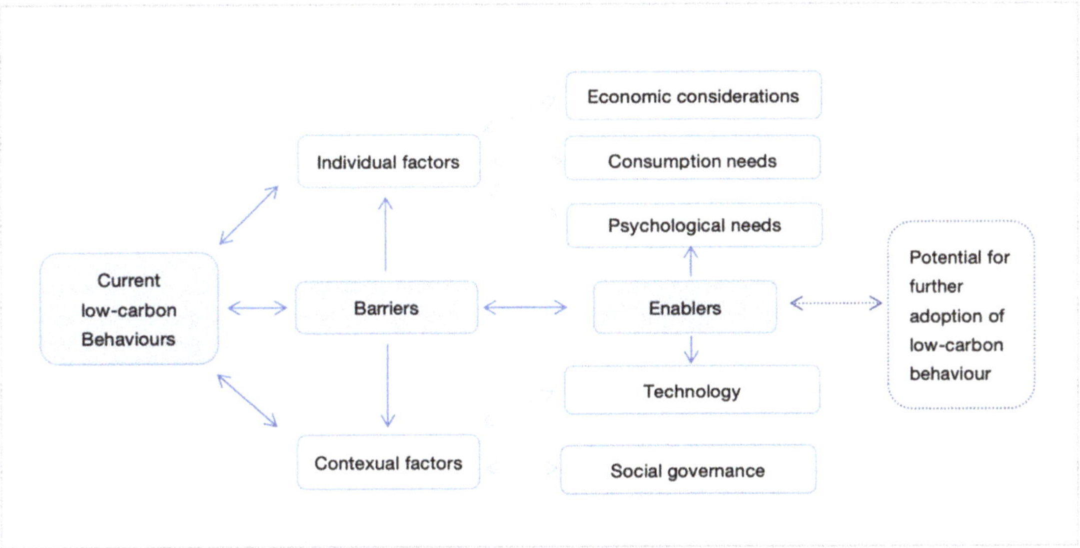

Figure 5. Theoretical model of the relationship between key themes in this research.

4. Discussion

This research provides a comprehensive overview of the low-carbon behaviors adopted by Chinese millennials in their daily lives based on our study samples. It explores the barriers and enablers that influence young adults in transitioning to a wider range of low-carbon lifestyles through in-depth interviews. Our research indicates that many Chinese millennials have positive attitudes towards a low-carbon lifestyle. Currently, their low-carbon behaviors mainly focus on saving electricity, using shared bicycles, reducing plastic bag usage, minimizing paper usage in the workplace, and taking public transportation for commuting. The main barriers for millennials in transitioning to a wider low-carbon lifestyle include the difficulty of seeing immediate contributions and impacts on the environment, more concern for the current situation rather than climate change in the future, and a lack of deeper knowledge about low-carbon behaviors. Millennials would change their lifestyle mainly for economic benefits, consumption needs, and social interaction. Reasonable policies, technical innovations, and cooperation from stakeholders in different fields will encourage more people to join in low-carbon action. Our findings provide the following three core insights.

Firstly, our study samples indicate that Chinese millennials have a positive attitude towards transitioning to a low-carbon lifestyle. Women demonstrate a stronger willingness to adopt low-carbon behaviors in their daily household activities compared to men. However, their involvement in governance in the context of transitioning to a low-carbon society is limited. Our research indicates that married women play a crucial role in environmental education within their families, and their low-carbon behaviors influence their husbands and children. This is primarily because modern urban women in the Chinese millennial generation have received a good education, leading to a higher environmental awareness. Over the past 40 years of China's reform and opening up, the proportion of millennial women receiving higher education has gradually increased. The percentage of female graduate students in China reached 50.9% in 2020, surpassing that of their male students [114]. However, our case study shows that, in the context of transitioning to a low-carbon society, female participants with experience in climate change or low-carbon activities are fewer than male participants. Within organizational management, women often assume execution roles in climate action rather than decision-making positions. Based on previous research, climate change and related policies will involve issues of gender equality, particularly in developing countries [115–119]. Encouraging more women to participate in the social innovation process will effectively contribute to mitigating climate change [120]. Some millennials are already or will soon become part of organizational management; they will play a key role in climate action over the next 30 years. Jiang et al. researched Chinese companies in manufacturing industries with serious carbon emissions and found that executives' low-carbon attitudes indirectly affect the performance of carbon emission reduction in their companies [121]. Martínez et al. [122] analyze the relationship between gender on company boards and CO_2 emissions in European companies, and they find that a higher proportion of female members on corporate boards correlates with lower carbon emissions. Altunbas et al. [123] observed that an increase of 1% in female managers is associated with a 0.5% decrease in CO_2 emissions in the organizations. Therefore, our research suggests that the transformation towards a low-carbon society needs to consider gender equality. It is essential to amplify the voices of women and provide them with more opportunities in the decision-making process. Additionally, this requires the improvement of more female managers' low-carbon transition awareness and active participation in more low-carbon activities at work.

Secondly, technological innovation and progress play a crucial role in the transition to a low-carbon lifestyle among the millennials. However, they also come with new forms of resource waste, so making sensible policy formulation is the key. In the millennials' current life, whether it is food delivery services, bike-sharing, online shopping, social media platforms, or payment platforms, all rely on the development of mobile internet technology. Technological innovation could also enhance the quality of low-carbon products, improve household energy efficiency, and reduce people's living costs. However, the millennials, while enjoying the convenience of rapid online consumption and services, also leading to new forms of resource waste and environmental issues. For instance, excessive packaging generated by takeout and online shopping, the widespread use of disposable utensils, and the damage and idle occupancy of public spaces by shared bicycles. A study conducted by Echegaray et al. [124] in Brazil indicates that while most people have a positive awareness of recycling electronic appliances, only a few of them actually implemented e-waste recycling practices. The tension between economic development and environmental protection is a common challenge faced by contemporary society [125]. Ecomodernism believes that technological development can protect nature and improve human well-being, but currently, there is limited explanation of how to protect the substantive freedom of vulnerable populations during the climate change crisis [126]. The "Jevons Paradox", proposed by the British economist William Jevons, suggests that although modern technology can enhance resource efficiency, the decrease in usage costs leads to an increase in demand [127]. Consequently, this increase in demand results in higher, rather than lower, resource consumption [128]. In other words, relying solely on technological advancement cannot help humanity reduce the overall consumption of resources and energy [129,130]. Our research believes that in the transition to a low-carbon society, technological innovation and application need reasonable policy guidance and effective supervision to prevent new forms of waste [131]. People's low-carbon awareness should be improved through education and media by disseminating more details of low-carbon knowledge and guiding individuals on how to take action in their daily lives. Collective action will contribute to ensuring that technological progress genuinely serves the achievement of sustainable development in terms of resources, environment, and society.

Thirdly, personal economic interests are the main driver for millennials to switch to a low-carbon lifestyle. Reasonable policies can guide them toward a low-carbon lifestyle, but this requires synergies among stakeholders. Our research has found that millennials, whether in the purchase of electric vehicles, reducing the use of plastic bags, or increasing low-carbon activities at work, have been influenced by relevant policies. These policies include incentives, price adjustments, supervision, and management. The low-carbon transformation not only relates to the economy and the environment but also involves social interactions and policy-making, which requires multidimensional synergy [132–134]. Improving public low-carbon awareness and behavior is crucial to low-carbon city transformation [135]. Most millennials have transitioned from formal school education to broader social learning. We found that many of them learn about low-carbon practices through their work, as the dual carbon goals and related policies in China have impacted various industries, further affecting their lives, so the improvement of low-carbon knowledge training is necessary in related industry. Our research suggests that universities, enterprises, and the government should strengthen cooperation to enhance the quality of low-carbon products and services, align with the life needs of millennials, and reduce the prices of low-carbon products. Moreover, efforts should be made to offer more easily understandable and actionable low-carbon guidance. These will encourage more millennials to switch to a low-carbon lifestyle. In addition, due to the impact of COVID-19, an increasing number of millennials are focusing on their health, leading to a shift in their low-carbon lifestyle. Our research believes that a wider range of low-carbon society transitions requires the establishment of a coordinated and comprehensive social governance that is credible, flexible, and reflective. This should incorporate different backgrounds, cultures, values, and anticipate future changes to continuously revise policies [136]. Additionally, it should consider people's well-being and happiness.

5. Contributions, Limitations, and Future Research

Previous research about China's low-carbon transition has mainly focused on exploring the concepts and policies related to low-carbon cities, analyzing the models of low-carbon pilot cities, and calculating household carbon emissions. Our research will have several contributions. In terms of theoretical contributions, first, we use in-depth interviews to focus on the attitudes, behaviors, and past low-carbon activities of the Chinese millennials. Although we have 50 sample interviewees, compared to questionnaire methods with larger sample sizes that only provide approximate written information, interview methods prioritize depth and understanding of the experiences and sentiments of the respondents. This emphasis enhances the reliability of the research's authenticity. Through interviews and in-depth conversations with participants, we also learned about their attitudes toward their peers' lifestyles. By deeply communicating and discussing with participants according to our theme, the discussion of low-carbon lifestyles extends beyond the 50 participants. All of our participants are from different industry fields in China; many of them are related experts and journalists who are knowledgeable about related social issues and low-carbon ideas. Our research will help scholars get a better understanding of people's daily low-carbon choices and practices, which provides a foundation for future scholars to further explore individual low-carbon behavioral preferences, personal carbon footprint calculations, public participation in low-carbon societal transitions, etc. Secondly, by showing real-life cases from the respondents, we analyzed the barriers and driving factors of low-carbon actions from psychological, sociological, and economic perspectives. This multidisciplinary approach will assist scholars in exploring the influencing factors of low-carbon consumption behaviors. Thirdly, our study highlights the significant role played by women in addressing climate change and the potential problems related to technology participation in climate action. This offers new perspectives for future scholars to explore climate action, helping deepen the understanding of the roles and challenges faced by different social groups in low-carbon transitions. Furthermore, in terms of practical contributions, understanding the lifestyles, driving enablers, and barriers faced by the millennial generation will help stakeholders consider the actual needs of this generation when formulating low-carbon policies and providing related services. This understanding will guide millennials to adopt low-carbon lifestyles and actively participate in collective climate action. During our in-depth interviews, our research also helped our participants rethink their low-carbon behaviors and related work. This encouraged them to contribute more to the transformation of low-carbon society in their respective fields.

This study has some limitations. Firstly, our research focuses solely on China's millennial generation. Therefore, the findings might not be directly applicable to low-carbon societal transitions in other countries with different policies, cultural backgrounds, and levels of economic development, although our case studies could serve as a reference. Secondly, this study may face criticism for the limitation of sample size in qualitative research. We chose a semi-structured interview due to its depth and flexibility in conversation, as well as the lack of research focusing on the millennials' low-carbon behavior in China. While our study provides in-depth insights and rich data based on our sample, addressing gaps in the literature, our research results may not capture the entirety of millennials' lifestyles in China. For example, our study is based on participants from different cities, most of whom reside in first-tier or capital cities in China. Due to differences and imbalances in economic development, education, natural conditions, and cultural diversity among cities in China's eastern, middle, and western regions [137], our research results might not be universally applicable to other cities in China. Thirdly, with over half of our participants holding master's degrees or higher, our results only reflect the highly educated Chinese millennials' daily life and consumption behavior.

For future research, we suggest conducting studies focused on specific cities. These studies could integrate local customs and urban development characteristics, explore the low-carbon lifestyles of local residents, and compare consumption behaviors across different genders and generations. Additionally, we recommend future research to further explore women's role in climate change action in developing countries and investigate the relationship between female leadership and carbon emissions in China. We also encourage scholars to explore other research methods to discuss people's low-carbon lifestyle and behavior, such as combining interviews with panel data analysis or creating workshops as a qualitative research tool to observe the dialogs between participants from different fields and generations.

6. Conclusions

Climate change issues exhibit temporal lag, the misalignment between perpetrators and victims, and the non-renewability of climate resources. Different regions and populations have varying demands for greenhouse gas emissions, mitigation, and adaptive capacities, which not only involve environmental concerns but also intersect with social development, fairness, justice, human rights, and sustainable development issues. This study focuses on the millennials' low-carbon behaviors in China, who are the mainstay of the current labor market and will be the decision-makers and managers in the future for climate change action. They have significant potential to participate in collective climate action. This study uses in-depth interviews to explore the millennials' low-carbon behaviors, the barriers, and enablers for a wider range of low-carbon actions, which enhances the understanding of low-carbon lifestyles in China. We summarize and analyze the respondents' low-carbon behaviors in their lives, work, and social interactions, offering a holistic view beyond the focus on household-level carbon emissions often found in previous research. Our research provides a foundation for future studies on individual preferences, carbon footprint calculations, and public engagement in low-carbon transitions. From practical contributions, this research will help policymakers in formulating more rational incentive policies. It can also help business managers better comprehend consumer demands and find a balance between economic interests and emissions reduction, facilitating their own low-carbon transformations.

Based on our study samples, there is an indication that Chinese millennials have a positive attitude towards transitioning to a low-carbon lifestyle. Current low-carbon behaviors mainly focus on saving electricity, using bike-sharing services, reducing plastic bag usage, and reducing paper in the workplace. Women in households show a higher willingness for low-carbon actions compared to men. However, women who have experience in climate change or low-carbon activities in the workplace are fewer than males. Most women play an execution role in climate change actions rather than a decision-making position. Further, millennials' low-carbon life transition is accompanied by technological innovation and progress. However, this progress would bring some new forms of resource waste, and reasonable policy-making and management will be the key. The difficulty in immediately perceiving their contribution and impact on the environment, coupled with a lack of deeper knowledge about low-carbon lifestyles, are the main barriers for millennials to switch to low-carbon action. Personal economic interests and the satisfaction of their consumption needs will drive millennials to reduce carbon emissions in their daily lives, but it requires reasonable policy-making and synergy among various stakeholders. As most millennials have transitioned from formal school education to broader social learning, many of them acquire low-carbon knowledge through their work. Therefore, enhancing low-carbon knowledge training in related industries is necessary. Technological advancements could also drive millennials to join in low-carbon actions by reducing the cost of low-carbon products and providing more easily understandable and actionable guidance.

Supplementary Materials: The following supporting information can be downloaded at: https://www.mdpi.com/article/10.3390/smartcities7040080/s1, Sheet S1: Area of participates living in; Sheet S2: Example quotes of participants' low-carbon behaviors; Sheet S3: The barriers influencing participants' transition toward a low-carbon lifestyle; Sheet S4: Enablers to switch to low-carbon lifestyle; Sheet S5: Interview Questions.

Author Contributions: Research design: Y.W., P.M. and T.K.; interview and data analysis: Y.W.; writing—original draft: Y.W.; writing—review: P.M. and T.K.; supervision: P.M. and T.K. All authors have read and agreed to the published version of the manuscript.

Funding: Yan Wu is funded by the China Scholarship Council (CSC) with grant No. 202008320395. The funder had no role in research design, data collection and analysis, decision to publish, or preparation of the manuscript.

Institutional Review Board Statement: This project has been reviewed and approved by the Faculty Niet-WMO Verplicht Research Ethics Committee, Maastricht University. Reference Number: FHML-REC/2023/090. Approval date: 23 October 2023.

Informed Consent Statement: Presented at the end of the article.

Data Availability Statement: The original contributions presented in the study are included in the article/Supplementary Materials, further inquiries can be directed to the corresponding authors.

Acknowledgments: We want to thank all participants for joining the interviews.

Conflicts of Interest: The authors declare no conflicts of interest.

Appendix A

Interview Invitation Letter

Dear Sir/Madam,

My name is Yan Wu, I am a PhD researcher at Maastricht University, supervised by Prof. Dr. Pim Martens and Prof. Dr. Thomas Krafft, currently doing my research project, 'Public awareness on low-carbon city transformation in China'. I'm funded by China Scholarship Council (CSC) with grant No. 202008320395.

Responding to global climate change is not only a national responsibility but also an individual responsibility. This interview is part of the research project "public awareness on low-carbon city transformation". In order to realize a low-carbon future, we need to understand the public's low-carbon awareness and behavior. We aim to learn about factors affecting public awareness towards low-carbon city transformation.

This interview is being used for research purposes only, not for any commercial purposes. All the answers will just represent individual opinions instead of your institution. To protect the privacy of the interviewees, this study will establish a numbering system for each interviewee, using a number instead of the name in the transcript. Thank you for accepting our interview.

Research Questions

1. Age, gender, location, highest level of education, job occupation.
2. Current role in your organization. Describe your past environmental work experience.
3. What do you know about low-carbon lifestyle and low-carbon economy?
4. How do you know these?
5. How do you feel about the shift to a low-carbon economy and low-carbon life?
6. What type of transportation do you prefer to choose for short trips and long trips, and why?
7. What factors do you think affect people's awareness about low-carbon (why or for example)?
8. Do you buy low-carbon products in daily life? Why?
9. Do you think many people know that low-carbon will contribute to climate change and our environment, but sometimes it is difficult for them to change behavior?
10. What do you think can interest people to join in low-carbon activities and turn to low-carbon life and low-carbon consumption?

Informed Consent Statement

1. I have read the interview invitation letter and understand that my role is interviewee in this interview.
2. I voluntarily agree to participate in the interview and know that I can stop the interview at any time.
3. I know I have the right not to answer or skip questions if I don't want to answer.
4. I agree that this study could use my responses anonymously as research data and be published in journals.
5. I allow the interviewers to take notes and record during the interview.
6. I agree that this research team may use my interview for scientific research on the project "Public Awareness of Low-Carbon City Transformation".
7. I have read and understood the points and statements of this consent form.
8. I have had the opportunity to ask questions about this interview.
9. I confirm and accept the interview.

____________________ ____________________

Participants Date

References

1. World Meteorological Organization. July 2023 is Set to Be the Hottest Month on Record. 2023. Available online: https://wmo.int/media/news/july-2023-set-be-hottest-month-record (accessed on 2 August 2023).
2. World Meteorological Organization. Heatwaves, Wildfires Mark Summer of Extremes. 2023. Available online: https://wmo.int/media/news/exceptional-heat-and-rain-wildfires-and-floods-mark-summer-of-extremes (accessed on 2 August 2023).
3. United Nations. Northern Hemisphere Summer Marked by Heatwaves and Wildfires: WMO. 2023. Available online: https://news.un.org/en/story/2023/07/1139307 (accessed on 2 August 2023).
4. China Meteorological Administration. Turpan: Extreme Record Broken with 52.2 °C High Temperature, Meteorological Services Followed throughout. 2023. Available online: https://www.cma.gov.cn/2011xwzx/2011xqxxw/2011xjctz/202307/t20230717_5652897.html (accessed on 2 August 2023).
5. Gomez-Echeverri, L. Climate and development: Enhancing impact through stronger linkages in the implementation of the Paris Agreement and the Sustainable Development Goals (SDGs). *Philos. Trans. R. Soc. A Math. Phys. Eng. Sci.* **2018**, *376*, 20160444. [CrossRef] [PubMed]
6. Bauer, A.; Menrad, K. Standing up for the Paris Agreement: Do global climate targets influence individuals' greenhouse gas emissions? *Environ. Sci. Policy* **2019**, *99*, 72–79. [CrossRef]
7. European Union. 2050 Long-Term Strategy. Available online: https://climate.ec.europa.eu/eu-action/climate-strategies-targets/2050-long-term-strategy_en (accessed on 2 August 2023).
8. Özdemir, Ö.; Hartmann, C.; Hafner, A.; König, H.; Lützkendorf, T. Next generation of life cycle related benchmarks for low carbon residential buildings in Germany. *IOP Conf. Ser. Earth Environ. Sci.* **2022**, *1078*, 012053. [CrossRef]
9. United Nations. Halfway to 2030, World 'Nowhere Near' Reaching Global Goals, UN Warns. 2023. Available online: https://news.un.org/en/story/2023/07/1138777 (accessed on 3 August 2023).

10. Boehnke, R.F.; Hoppe, T.; Brezet, H.; Blok, K. Good practices in local climate mitigation action by small and medium-sized cities; exploring meaning, implementation and linkage to actual lowering of carbon emissions in thirteen municipalities in The Netherlands. *J. Clean. Prod.* **2019**, *207*, 630–644. [CrossRef]
11. United Nations. Progress towards SDG 11. Available online: https://unhabitat.org/sites/default/files/2023/07/sdg11_booklet_final_full_5_2.pdf (accessed on 3 August 2023).
12. Su, N. Urban Energy Low-Carbon Transformation Faces Three Key Challenges. China Energy News. 2023. Available online: http://paper.people.com.cn/zgnyb/html/2023-05/29/content_25992909.htm (accessed on 4 August 2023).
13. Yang, S.; Yang, D.; Shi, W.; Deng, C.; Chen, C.; Feng, S. Global evaluation of carbon neutrality and peak carbon dioxide emissions: Current challenges and future outlook. *Environ. Sci. Pollut. Res.* **2023**, *30*, 81725–81744. [CrossRef] [PubMed]
14. Liu, W.; Qin, B. Low-carbon city initiatives in China: A review from the policy paradigm perspective. *Cities* **2016**, *51*, 131–138. [CrossRef]
15. Jiang, J.; Ye, B.; Sun, Z.; Zeng, Z.; Yang, X. Low-carbon energy policies benefit climate change mitigation and air pollutant reduction in megacities: An empirical examination of Shenzhen, China. *Sci. Total Environ.* **2023**, *892*, 164644. [CrossRef] [PubMed]
16. Wang, Z.; Liang, F.; Li, C.; Xiong, W.; Chen, Y.; Xie, F. Does China's low-carbon city pilot policy promote green development? Evidence from the digital industry. *J. Innov. Knowl.* **2023**, *8*, 100339. [CrossRef]
17. Han, J.; Li, T.; Philbin, S.P. Does low-carbon pilot policy in China improve corporate profitability? The role of innovation and subsidy. *Innov. Green Dev.* **2023**, *2*, 100050. [CrossRef]
18. Yang, G. Can the low-carbon city pilot policy promote firms' low-carbon innovation: Evidence from China. *PLoS ONE* **2023**, *18*, e0277879. [CrossRef]
19. Yang, L.; Li, Y. Low-carbon city in China. *Sustain. Cities Soc.* **2013**, *9*, 62–66. [CrossRef]
20. Zhou, N.; He, G.; Williams, C. *China's Development of Low-Carbon Eco-Cities and Associated Indicator Systems*; Lawrence Berkeley National Lab (LBNL): Berkeley, CA, USA, 2012. [CrossRef]
21. Su, M.; Liang, C.; Chen, B.; Chen, S.; Yang, Z. Low-carbon development patterns: Observations of typical Chinese cities. *Energies* **2012**, *5*, 291–304. [CrossRef]
22. Zeng, S.; Jin, G.; Tan, K.; Liu, X. Can low-carbon city construction reduce carbon intensity? Empirical evidence from low-carbon city pilot policy in China. *J. Environ. Manag.* **2023**, *332*, 117363. [CrossRef]
23. Cheng, J.; Yi, J.; Dai, S.; Xiong, Y. Can low-carbon city construction facilitate green growth? Evidence from China's pilot low-carbon city initiative. *J. Clean. Prod.* **2019**, *231*, 1158–1170. [CrossRef]
24. Shen, L.; Wu, Y.; Lou, Y.; Zeng, D.; Shuai, C.; Song, X. What drives the carbon emission in the Chinese cities?—A case of pilot low carbon city of Beijing. *J. Clean. Prod.* **2018**, *174*, 343–354. [CrossRef]
25. Ma, X.; Wang, C.; Dong, B.; Gu, G.; Chen, R.; Li, Y.; Zou, H.; Zhang, W.; Li, Q. Carbon emissions from energy consumption in China: Its measurement and driving factors. *Sci. Total Environ.* **2019**, *648*, 1411–1420. [CrossRef]
26. Zhang, Y.; Yan, D.; Hu, S.; Guo, S. Modelling of energy consumption and carbon emission from the building construction sector in China, a process-based LCA approach. *Energy Policy* **2019**, *134*, 110949. [CrossRef]
27. Lyu, K.; Yang, S.; Zheng, K.; Zhang, Y. How does the digital economy affect carbon emission efficiency? Evidence from energy consumption and industrial value chain. *Energies* **2023**, *16*, 761. [CrossRef]
28. Xu, L.; Du, H.; Zhang, X. Driving forces of carbon dioxide emissions in China's cities: An empirical analysis based on the geodetector method. *J. Clean. Prod.* **2021**, *287*, 125169. [CrossRef]
29. Zhang, J.; Zheng, T. Can dual pilot policy of innovative city and low carbon city promote green lifestyle transformation of residents? *J. Clean. Prod.* **2023**, *405*, 136711. [CrossRef]
30. Xu, X.; Tan, Y.; Chen, S.; Yang, G.; Su, W. Urban household carbon emission and contributing factors in the Yangtze River Delta, China. *PLoS ONE* **2015**, *10*, e0121604. [CrossRef]
31. Ye, H.; Ren, Q.; Hu, X.; Lin, T.; Xu, L.; Li, X.; Zhang, G.; Shi, L.; Pan, B. Low-carbon behavior approaches for reducing direct carbon emissions: Household energy use in a coastal city. *J. Clean. Prod.* **2017**, *141*, 128–136. [CrossRef]
32. Liu, X.; Li, Y.; Chen, X.; Liu, J. Evaluation of low carbon city pilot policy effect on carbon abatement in China: An empirical evidence based on time-varying DID model. *Cities* **2022**, *123*, 103582. [CrossRef]
33. Pan, A.; Zhang, W.; Shi, X.; Dai, L. Climate policy and low-carbon innovation: Evidence from low-carbon city pilots in China. *Energy Econ.* **2022**, *112*, 106129. [CrossRef]
34. Yu, Y.; Zhang, N. Low-carbon city pilot and carbon emission efficiency: Quasi-experimental evidence from China. *Energy Econ.* **2021**, *96*, 105125. [CrossRef]
35. Howitt, D.; Cramer, D. *Introduction to Qualitative Methods in Psychology*; Ashford Colour Press Ltd.: Gosport, UK, 2010.
36. Delle Fave, A.; Brdar, I.; Freire, T.; Vella-Brodrick, D.; Wissing, M.P. The eudaimonic and hedonic components of happiness: Qualitative and quantitative findings. *Soc. Indic. Res.* **2011**, *100*, 185–207. [CrossRef]
37. Andina, T. Climate issue: The principle of transgenerational responsibility. *Riv. Estet.* **2020**, *75*, 17–32. [CrossRef]
38. Ben-Elia, N. Environmental Values and the Subjective Assessment of Residential Quality. Ph.D. Thesis, University of California, Berkeley, CA, USA, 1979.
39. Dimock, M. Defining generations: Where Millennials end and Generation Z begins. *Pew Res. Cent.* **2019**, *17*, 1–7.

40. Kroth, A.; Young, S.J. New professionals, new desires: What millennials want in their work. *Recreat. Sports J.* **2014**, *38*, 23–32. [CrossRef]
41. Petrescu-Mag, R.M.; Petrescu, D.C.; Ivan, A.; Tenter, A. An intergenerational reading of climate change-health concern nexus: A qualitative study of the Millennials' and Gen Z participants' perceptions. *BMC Public Health* **2023**, *23*, 484. [CrossRef]
42. Ross, A.D.; Rouse, S.M. (Young) generations as social identities: The role of Latino* Millennial/Generation Z in shaping attitudes about climate change. *Political Behav.* **2022**, *44*, 1105–1124. [CrossRef]
43. Son, Y.A.; Jun, J.W. Effects of risk perception, SNS uses, personal characteristics on climate change participation behaviors of millennials in South Korea. *Cogent Bus. Manag.* **2024**, *11*, 2318793. [CrossRef]
44. Zha, Q. Reimagining China–US university relations: A global 'ecosystem' perspective. *Stud. High. Educ.* **2023**, 1–21. [CrossRef]
45. Yang, Y. China's Economic Growth in Retrospect. China 2049: Economic Challenges of a Rising Global Power, 2020; pp. 3–28. Available online: https://www.brookings.edu/wp-content/uploads/2019/08/9780815738053_ch1.pdf (accessed on 20 November 2023).
46. The World Bank. The World Bank in China. Available online: https://www.worldbank.org/en/country/china/overview (accessed on 17 February 2024).
47. Kim, J.H.; Hsu, M.M.; Yuen, C.A. Individual and social factors impacting Chinese millennials' luxury consumption. *Int. J. Costume Fash.* **2020**, *20*, 27–43. [CrossRef]
48. Shin, H.; Eastman, J.K.; Mothersbaugh, D. The effect of a limited-edition offer following brand dilution on consumer attitudes toward a luxury brand. *J. Retail. Consum. Serv.* **2017**, *38*, 59–70. [CrossRef]
49. Wang, X.-Q.; Chen, J. Factors affecting climate change mitigation behavior intention and behavior in Chinese teenagers. *Adv. Clim. Change Res.* **2021**, *17*, 212.
50. Han, P.; Tong, Z.; Sun, Y.; Chen, X. Impact of climate change beliefs on youths' engagement in energy-conservation behavior: The mediating mechanism of environmental concerns. *Int. J. Environ. Res. Public Health* **2022**, *19*, 7222. [CrossRef]
51. Wang, X.; Chen, J. Fear emotion reduces reported mitigation behavior in adolescents subject to climate change education. *Clim. Chang* **2022**, *174*, 1. [CrossRef]
52. Brivio, F.; Viganò, A.; Paterna, A.; Palena, N.; Greco, A. Narrative Review and Analysis of the Use of "Lifestyle" in Health Psychology. *Int. J. Environ. Res. Public Health* **2023**, *20*, 4427. [CrossRef]
53. Agnew, M.D.; Pettifor, H.; Wilson, C. Lifestyle, an integrative concept: Cross-disciplinary insights for low-carbon research. *Wiley Interdiscip. Rev. Energy Environ.* **2023**, *12*, e490. [CrossRef]
54. Zong, X.; Tian, J.; Zheng, X.; Tong, Z.; She, S.; Sun, Y. Influence of climate change beliefs on adolescent food saving behavior: Mechanisms mediating environmental concerns. *Environ. Dev. Sustain.* **2024**, 1–19. [CrossRef]
55. Ge, J.; Lin, B. Impact of public support and government's policy on climate change in China. *J. Environ. Manag.* **2021**, *294*, 112983. [CrossRef]
56. Wang, B.; Shen, Y.; Jin, Y. Measurement of public awareness of climate change in China: Based on a national survey with 4025 samples. *Chin. J. Popul. Resour. Environ.* **2017**, *15*, 285–291. [CrossRef]
57. Jiang, X.; Ding, Z.; Liu, R. Can Chinese residential low-carbon consumption behavior intention be better explained? The role of cultural values. *Nat. Hazards* **2019**, *95*, 155–171. [CrossRef]
58. Al Balushi, K. The use of online semi-structured interviews in interpretive research. *Int. J. Sci. Res.* **2016**, *57*, 2319–7064.
59. Adeoye-Olatunde, O.A.; Olenik, N.L. Research and scholarly methods: Semi-structured interviews. *J. Am. Coll. Clin. Pharm.* **2021**, *4*, 1358–1367. [CrossRef]
60. Irvine, A.; Drew, P.; Sainsbury, R. 'Am I not answering your questions properly?' Clarification, adequacy and responsiveness in semi-structured telephone and face-to-face interviews. *Qual. Res.* **2013**, *13*, 87–106. [CrossRef]
61. Harrell, M.C.; Bradley, M. Data Collection Methods: Semi-Structured Interviews and Focus Groups. 2009. Available online: https://ntrl.ntis.gov/NTRL/dashboard/searchResults/titleDetail/ADA512853.xhtml (accessed on 10 December 2023).
62. Adams, W.C. Conducting semi-structured interviews. In *Handbook of Practical Program Evaluation*; John Wiley & Sons: Hoboken, NJ, USA, 2015; pp. 492–505. [CrossRef]
63. Dolczewski, M. Semi-structured interview for self-esteem regulation research. *Acta Psychol.* **2022**, *228*, 103642. [CrossRef]
64. Luo, J.; Dey, B.L.; Yalkin, C.; Sivarajah, U.; Punjaisri, K.; Huang, Y.A.; Yen, D.A. Millennial Chinese consumers' perceived destination brand value. *J. Bus. Res.* **2020**, *116*, 655–665. [CrossRef]
65. Axsen, J.; Kurani, K.S. Social influence, consumer behavior, and low-carbon energy transitions. *Annu. Rev. Environ. Resour.* **2012**, *37*, 311–340. [CrossRef]
66. Shelton, C.L.; Smith, A.F.; Mort, M. Opening up the black box: An introduction to qualitative research methods in anaesthesia. *Anaesthesia* **2014**, *69*, 270–280. [CrossRef] [PubMed]
67. Alam, M.K. A systematic qualitative case study: Questions, data collection, NVivo analysis and saturation. *Qual. Res. Organ. Manag. Int. J.* **2021**, *16*, 1–31. [CrossRef]
68. DeJonckheere, M.; Vaughn, L.M. Semistructured interviewing in primary care research: A balance of relationship and rigour. *Fam. Med. Community Health* **2019**, *7*, e000057. [CrossRef] [PubMed]
69. Draper, A.; Swift, J.A. Qualitative research in nutrition and dietetics: Data collection issues. *J. Hum. Nutr. Diet.* **2011**, *24*, 3–12. [CrossRef] [PubMed]

70. Opie, C. Research Procedures. In *Getting Started in Your Educational Research: Design, Data Production and Analysis*; Sage Publications Ltd.: London, UK, 2019; Volume 159, p. 176.
71. Hammarberg, K.; Kirkman, M.; De Lacey, S. Qualitative research methods: When to use them and how to judge them. *Hum. Reprod.* **2016**, *31*, 498–501. [CrossRef] [PubMed]
72. Mays, N.; Pope, C. Assessing quality in qualitative research. *BMJ* **2000**, *320*, 50–52. [CrossRef]
73. Welch, C.; Piekkari, R. How should we (not) judge the 'quality' of qualitative research? A re-assessment of current evaluative criteria in International Business. *J. World Bus.* **2017**, *52*, 714–725. [CrossRef]
74. Davies, D.; Dodd, J. Qualitative research and the question of rigor. *Qual. Health Res.* **2002**, *12*, 279–289. [CrossRef]
75. Stenbacka, C. Qualitative research requires quality concepts of its own. *Manag. Decis.* **2001**, *39*, 551–556. [CrossRef]
76. Yilmaz, K. Comparison of quantitative and qualitative research traditions: Epistemological, theoretical, and methodological differences. *Eur. J. Educ.* **2013**, *48*, 311–325. [CrossRef]
77. Vasileiou, K.; Barnett, J.; Thorpe, S.; Young, T. Characterising and justifying sample size sufficiency in interview-based studies: Systematic analysis of qualitative health research over a 15-year period. *BMC Med. Res. Methodol.* **2018**, *18*, 1–18. [CrossRef] [PubMed]
78. Palinkas, L.A.; Horwitz, S.M.; Green, C.A.; Wisdom, J.P.; Duan, N.; Hoagwood, K. Purposeful sampling for qualitative data collection and analysis in mixed method implementation research. *Adm. Policy Ment. Health Ment. Health Serv. Res.* **2015**, *42*, 533–544. [CrossRef] [PubMed]
79. Campbell, S.; Greenwood, M.; Prior, S.; Shearer, T.; Walkem, K.; Young, S.; Bywaters, D.; Walker, K. Purposive sampling: Complex or simple? Research case examples. *J. Res. Nurs.* **2020**, *25*, 652–661. [CrossRef] [PubMed]
80. Tongco, M.D.C. Purposive sampling as a tool for informant selection. *Ethnobot. Res. Appl.* **2007**, *5*, 147–158. [CrossRef]
81. Heckathorn, D.D. Comment: Snowball versus respondent-driven sampling. *Sociol. Methodol.* **2011**, *41*, 355–366. [CrossRef] [PubMed]
82. Valerio, M.A.; Rodriguez, N.; Winkler, P.; Lopez, J.; Dennison, M.; Liang, Y.; Turner, B.J. Comparing two sampling methods to engage hard-to-reach communities in research priority setting. *BMC Med. Res. Methodol.* **2016**, *16*, 146. [CrossRef] [PubMed]
83. Atkinson, R.; Flint, J. Accessing hidden and hard-to-reach populations: Snowball research strategies. *Soc. Res. Update* **2001**, *33*, 1–4.
84. Glaser, B.; Strauss, A. *Discovery of Grounded Theory: Strategies for Qualitative Research*; Routledge: New York, NY, USA, 2017.
85. Constantinou, C.S.; Georgiou, M.; Perdikogianni, M. A comparative method for themes saturation (CoMeTS) in qualitative interviews. *Qual. Res.* **2017**, *17*, 571–588. [CrossRef]
86. Braun, V.; Clarke, V. To saturate or not to saturate? Questioning data saturation as a useful concept for thematic analysis and sample-size rationales. *Qual. Res. Sport Exerc. Health* **2021**, *13*, 201–216. [CrossRef]
87. Creswell, J.W.; Poth, C.N. *Qualitative Inquiry and Research Design: Choosing among Five Approaches*; Sage Publications: London, UK, 2016.
88. Mason, M. Sample size and saturation in PhD studies using qualitative interviews. *Forum Qual. Sozialforschung/Forum Qual. Soc. Res.* **2010**, *11*, 3. [CrossRef]
89. Bekele, W.B.; Ago, F.Y. Sample size for interview in qualitative research in social sciences: A guide to novice researchers. *Res. Educ. Policy Manag.* **2022**, *4*, 42–50. [CrossRef]
90. Hennink, M.M.; Kaiser, B.N.; Marconi, V.C. Code saturation versus meaning saturation: How many interviews are enough? *Qual. Health Res.* **2017**, *27*, 591–608. [CrossRef] [PubMed]
91. Baker, S.E.; Edwards, R. How Many Qualitative Interviews Is Enough. 2012. Available online: https://eprints.ncrm.ac.uk/id/eprint/2273/4/how_many_interviews.pdf (accessed on 26 March 2024).
92. White, D.R.; Palmieri, P.A. Women caring for husbands living with Parkinson's Disease: A phenomenological study protocol. *J. Pers. Med.* **2022**, *12*, 659. [CrossRef] [PubMed]
93. Saunders, B.; Sim, J.; Kingstone, T.; Baker, S.; Waterfield, J.; Bartlam, B.; Jinks, C. Saturation in qualitative research: Exploring its conceptualization and operationalization. *Qual. Quant.* **2018**, *52*, 1893–1907. [CrossRef] [PubMed]
94. Hyde, K.F. Recognising deductive processes in qualitative research. *Qual. Mark. Res. Int. J.* **2000**, *3*, 82–90. [CrossRef]
95. Morgan, D.L.; Eliot, S.; Lowe, R.A.; Gorman, P. Dyadic interviews as a tool for qualitative evaluation. *Am. J. Eval.* **2016**, *37*, 109–117. [CrossRef]
96. Loraine, B.; Wick, W.; Christoph, G. How to use and assess qualitative research methods. *Neurol. Res. Pract.* **2020**, *2*, 14. [CrossRef]
97. Mikėnė, S.; Gaižauskaitė, I.; Valavičienė, N. Qualitative Interviewing: Field-Work Realities. *Soc. Darb.* **2013**, *12*, 49–62.
98. Corbin, J.; Strauss, A. *Basics of Qualitative Research: Techniques and Procedures for Developing Grounded Theory*; Sage Publications: London, UK, 2014.
99. Lock-Wah-Hoon, J.; Zheng, Y.; Braks, M.; van Asten, L.; Liu, Q.; Sushama, P.; Doreleijers, S.; Krafft, T.; van der Hoek, W.; Fanoy, E.; et al. Exploring vector-borne disease surveillance and response systems in Beijing, China: A qualitative study from the health system perspective. *Int. J. Environ. Res. Public Health* **2020**, *17*, 8512. [CrossRef]
100. Srivastava, P.; Hopwood, N. A practical iterative framework for qualitative data analysis. *Int. J. Qual. Methods* **2009**, *8*, 76–84. [CrossRef]

101. Kerr, C.; Nixon, A.; Wild, D. Assessing and demonstrating data saturation in qualitative inquiry supporting patient-reported outcomes research. *Expert Rev. Pharmacoeconomics Outcomes Res.* **2010**, *10*, 269–281. [CrossRef] [PubMed]
102. Braun, V.; Clarke, V. Using thematic analysis in psychology. *Qual. Res. Psychol.* **2006**, *3*, 77–101. [CrossRef]
103. Bengtsson, M. How to plan and perform a qualitative study using content analysis. *NursingPlus Open* **2016**, *2*, 8–14. [CrossRef]
104. Lacey, A.; Luff, D. *Qualitative Data Analysis*; Trent Focus Group: Nottingham, UK, 2001; pp. 320–357.
105. Wetherell, M.; Taylor, S.; Yates, S.J. (Eds.) *Discourse as Data: A Guide for Analysis*; Sage: London, UK, 2001.
106. Williams, M.; Moser, T. The art of coding and thematic exploration in qualitative research. *Int. Manag. Rev.* **2019**, *15*, 45–55.
107. Holt, N.L.; Dunn, J.G. Toward a grounded theory of the psychosocial competencies and environmental conditions associated with soccer success. *J. Appl. Sport Psychol.* **2004**, *16*, 199–219. [CrossRef]
108. Graneheim, U.H.; Lundman, B. Qualitative content analysis in nursing research: Concepts, procedures and measures to achieve trustworthiness. *Nurse Educ. Today* **2004**, *24*, 105–112. [CrossRef] [PubMed]
109. Wolfswinkel, J.F.; Furtmueller, E.; Wilderom, C.P. Using grounded theory as a method for rigorously reviewing literature. *Eur. J. Inf. Syst.* **2013**, *22*, 45–55. [CrossRef]
110. Baskerville, R.; Pries-Heje, J. Grounded action research: A method for understanding IT in practice. *Account. Manag. Inf. Technol.* **1999**, *9*, 1–23. [CrossRef]
111. Milne, J.; Oberle, K. Enhancing rigor in qualitative description. *J. Wound Ostomy Cont. Nurs.* **2005**, *32*, 413–420. [CrossRef]
112. Torrance, H. Triangulation, respondent validation, and democratic participation in mixed methods research. *J. Mix. Methods Res.* **2012**, *6*, 111–123. [CrossRef]
113. Zhang, Y.; Chen, J.; Han, Y.; Qian, M.; Guo, X.; Chen, R.; Xu, D.; Chen, Y. The contribution of Fintech to sustainable development in the digital age: Ant forest and land restoration in China. *Land Use Policy* **2021**, *103*, 105306. [CrossRef]
114. Lu, Y.; Du, W. Women's Education in China: Past and Present//SHS Web of Conferences. *EDP Sci.* **2023**, *152*, 02001. [CrossRef]
115. Terry, G. No climate justice without gender justice: An overview of the issues. *Gend. Dev.* **2009**, *17*, 5–18. [CrossRef]
116. Aliani, K. Does board diversity improve carbon emissions score of best citizen companies? *J. Clean. Prod.* **2023**, *405*, 136854. [CrossRef]
117. Alston, M. Women and adaptation. *Wiley Interdiscip. Rev. Clim. Change* **2013**, *4*, 351–358. [CrossRef]
118. Eagly, A.H. *Sex Differences in Social Behavior: A Social-Role Interpretation*; Psychology Press: New York, NY, USA, 2013.
119. Pearse, R. Gender and climate change. *Wiley Interdiscip. Rev. Clim. Change* **2017**, *8*, e451. [CrossRef]
120. Le Loarne-Lemaire, S.; Bertrand, G.; Razgallah, M.; Maalaoui, A.; Kallmuenzer, A. Women in innovation processes as a solution to climate change: A systematic literature review and an agenda for future research. *Technol. Forecast. Soc. Change* **2021**, *164*, 120440. [CrossRef]
121. Jiang, Y.; Hu, Y.; Asante, D.; Ampaw, E.M.; Asante, B. The Effects of Executives' low-carbon cognition on corporate low-carbon performance: A study of managerial discretion in China. *J. Clean. Prod.* **2022**, *357*, 132015. [CrossRef]
122. Martínez, M.C.V.; Santos-Jaén, J.M.; Román, R.S.; Martín-Cervantes, P.A. Are gender and cultural diversities on board related to corporate CO_2 emissions? *J. Clean. Prod.* **2022**, *363*, 132638. [CrossRef]
123. Altunbas, Y.; Gambacorta, L.; Reghezza, A.; Velliscig, G. Does gender diversity in the workplace mitigate climate change? *J. Corp. Financ.* **2022**, *77*, 102303. [CrossRef]
124. Echegaray, F.; Hansstein, F.V. Assessing the intention-behavior gap in electronic waste recycling: The case of Brazil. *J. Clean. Prod.* **2017**, *142*, 180–190. [CrossRef]
125. Purvis, B.; Mao, Y.; Robinson, D. Three pillars of sustainability: In search of conceptual origins. *Sustain. Sci.* **2019**, *14*, 681–695. [CrossRef]
126. Symons, J.; Karlsson, R. Ecomodernist citizenship: Rethinking political obligations in a climate-changed world. *Citizsh. Stud.* **2018**, *22*, 685–704. [CrossRef]
127. Jevons, W.S. *The Coal Question: Can Britain Survive*; Yale University Press: New Haven, CT, USA, 1865; p. 719.
128. Polimeni, J.M.; Polimeni, R.I. Jevons' Paradox and the myth of technological liberation. *Ecol. Complex.* **2006**, *3*, 344–353. [CrossRef]
129. Magee, C.L.; Devezas, T.C. A simple extension of dematerialization theory: Incorporation of technical progress and the rebound effect. *Technol. Forecast. Soc. Change* **2017**, *117*, 196–205. [CrossRef]
130. Korinek, M.A.; Schindler, M.M.; Stiglitz, J. *Technological Progress, Artificial Intelligence, and Inclusive Growth*; International Monetary Fund: Washington, DC, USA, 2021.
131. Dong, F.; Yu, B.; Hadachin, T.; Dai, Y.; Wang, Y.; Zhang, S.; Long, R. Drivers of carbon emission intensity change in China. *Resour. Conserv. Recycl.* **2018**, *129*, 187–201. [CrossRef]
132. Sovacool, B.K.; Martiskainen, M.; Hook, A.; Baker, L. Beyond cost and carbon: The multidimensional co-benefits of low carbon transitions in Europe. *Ecol. Econ.* **2020**, *169*, 106529. [CrossRef]
133. Büchs, M.; Bahaj AB, S.; Blunden, L.; Bourikas, L.; Falkingham, J.; James, P.; Kamanda, M.; Wu, Y. Promoting low carbon behaviours through personalised information? Long-term evaluation of a carbon calculator interview. *Energy Policy* **2018**, *120*, 284–293. [CrossRef]
134. Cai, B.; Guo, H.; Cao, L.; Guan, D.; Bai, H. Local strategies for China's carbon mitigation: An investigation of Chinese city-level CO_2 emissions. *J. Clean. Prod.* **2018**, *178*, 890–902. [CrossRef]
135. Li, X.; Xing, H. Better cities better lives: How low-carbon city pilots can lower residents' carbon emissions. *J. Environ. Manag.* **2024**, *351*, 119889. [CrossRef]

136. Hildingsson, R.; Johansson, B. Governing low-carbon energy transitions in sustainable ways: Potential synergies and conflicts between climate and environmental policy objectives. *Energy Policy* **2016**, *88*, 245–252. [CrossRef]
137. Zhang, Y.; Zhang, J.; Yang, Z.; Li, S. Regional differences in the factors that influence China's energy-related carbon emissions, and potential mitigation strategies. *Energy Policy* **2011**, *39*, 7712–7718. [CrossRef]

smart cities

MDPI

Article

Probabilistic Causal Modeling of Barriers to Accessibility for Persons with Disabilities in Canada

Mouri Zakir [1,*,†], Gregor Wolbring [2,†] and Svetlana Yanushkevich [1,*,†]

1 Biometric Technologies Laboratory, Department of Electrical and Software Engineering, Schulich School of Engineering, University of Calgary, Calgary, AB T2N 1N4, Canada
2 Community Rehabilitation and Disability Studies, Cumming School of Medicine, University of Calgary, Calgary, AB T2N 1N4, Canada; gwolbrin@ucalgary.ca
* Correspondence: mouri.zakir@ucalgary.ca (M.Z.); syanshk@ucalgary.ca (S.Y.)
† Current address: Department of Electrical and Software Engineering, Schulich School of Engineering, University of Calgary, 2500 University Drive NW, Calgary, AB T2N 1N4, Canada.

Highlights:

This research investigates the causal relationship between accessibility barriers and demographic and other traits of people with disabilities based on the Canadian Disability Survey 2022. We apply a unique computational intelligence tool—probabilistic causal models such as structural equation modeling and Bayesian networks—to discover those relationships and model scenarios to assess the risks and barriers in accessing services, transportation and communications in the context of inclusive smart cities. In addition, we apply both generative and causal models to produce synthetic data in order to test the accuracy of the causal models' structure learning.

What are the main findings?

- Probabilistic models are a feasible tool to derive reasoning from data about vulnerable groups.
- A causal relation between accessibility experience, demographics and disabilities is confirmed.
- The synthetic data generated using probabilistic models preserve the structure of real data.

What is the implication of the main finding?

- Insights into accessibility for diverse groups will be essential in designing inclusive smart cities.
- Synthetic data can help overcome the insufficiency of real data while maintaining confidentiality.

Abstract: This paper utilizes a methodological two-step process incorporating statistical and causal probabilistic modeling techniques to investigate factors affecting the accessibility experiences of persons with disabilities in Canada. We deploy a network-based approach using empirical data to perform a holistic assessment of the relations between various demographic features (e.g., age, gender and type of disability) and accessibility barriers. A statistical measurement method is applied that utilizes structural equation modeling supported by exploratory factor analysis. For causal probabilistic modeling, Bayesian networks are employed as a straightforward and compact way to interpret knowledge representation. This causal reasoning approach analyzes the nature and frequency of encountering barriers based on data to understand the risk factors contributing to pressing accessibility issues. Furthermore, to evaluate network performance and overcome any data limitations, synthetic data generation techniques are applied to create and validate artificial data built on real-world knowledge. The proposed framework strives to provide reasoning to understand the prevalence of physical, social, communication or technological barriers encountered by persons with disabilities in their daily lives. This study contributes to the identification of areas for prioritization in facilitating accessibility regulation and practices to realize an inclusive society.

Keywords: accessibility barriers; persons with disabilities; inclusive smart city; causality; exploratory factor analysis; structural equation modeling; regression; machine reasoning; Bayesian network; synthetic data generation; statistical hypotheses testing

1. Introduction

A growing emphasis has been placed on providing equal opportunities for all people, particularly people with disabilities, to support their participation in society without discrimination [1]. In recent years, the world has witnessed increased smart city initiatives that employ emerging data and technologies to tackle complex urban challenges [2]. This life-enhancing process is promoted by inclusion and social sustainability to build a community for people to thrive with equal opportunities [3,4].

The global "smart city" movement ideates the application of Information and Communication Technologies (ICT) as a crucial means of addressing the negative consequences of urban development and improving the quality of life of residents [2]. In this context, inclusion is understood in many ways, such as social cohesion and sense of belonging, equity, participation, quality of life, and human rights [2]. As a tool to discover whether equity and inclusion have been achieved, accessibility is investigated here using data on accessibility barriers for the case of Canada. An overview of the motivations for this research is presented in Figure 1 and discussed in this section.

Cities are complex systems that face a growing number of societal challenges that city governments are struggling to address [5]. An individual's ability to integrate into a city, navigate the environment and comprehend information affects their well-being and overall quality of life [5,6]. Vulnerable groups are particularly exposed to exclusion from social activities and full enjoyment of their rights and duties as citizens in urban environments [5].

Impacted by the rapidly increasing global population, cities face an increase in the number of people with various disabilities, whether sensory or physical. Recent findings from the 2022 Canadian Survey on Disability (CSD) [7] show that 27% of Canadians aged 15 years and older, or 8.0 million people, have one or more disabilities that limit them in their daily activities. This indicates a 5% increase since 2017 [8], which can be partially attributed to factors such as demographics and the type of disability. Despite representing a significant portion of the population and the diversity of conditions, people with disabilities face common barriers that prevent their full participation in society [9].

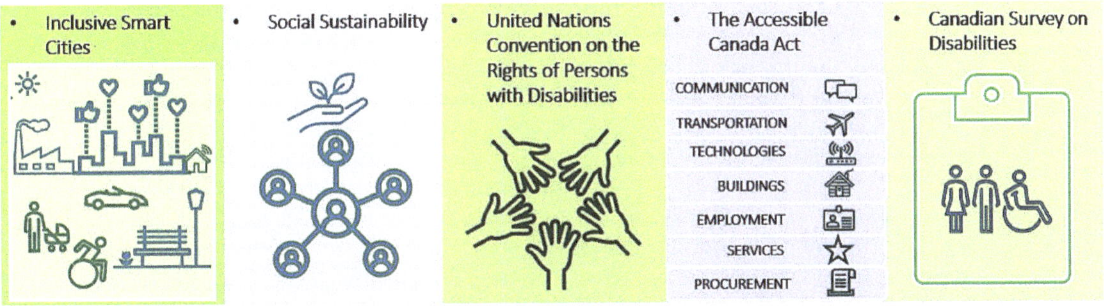

Figure 1. Motivations for this investigation on accessibility in Canada: Inclusive smart cities, social sustainability, the United Nations Convention on the Rights of Persons with Disabilities, the Accessible Canada Act and the Canadian Survey on Disabilities [2,3,6,7,9].

Smart cities are promising communities that leverage intelligent technologies to connect citizens, which is especially crucial for citizens with disabilities, who face substantial challenges in urban living [10]. But this falls short of a comprehensive urban development paradigm, which needs the critical analysis of the extent to which smart cities also promote social sustainability of urban areas [11]. Social sustainability is a dynamic concept that promotes infrastructure to support social needs and concerns equally for all. While smart cities primarily aim to enhance performance through the innovative use of data and technology, a social sustainability perspective stresses critical interconnections between people and places [11].

An inclusive city should accommodate everyone, regardless of their economic means, gender, race, ethnicity, and religion, enabling and empowering people to fully participate in the social, economic and political opportunities that cities have to offer [5]. For instance, in the city of Toronto, a barrier-free, well-run, modern, and connected community enabling technology is envisioned to improve the quality of life of its residents as a long term vision of the city administration's [12].

However, research has shown that the idea of an inclusive city is mostly perceived in general terms of poverty and social inclusion, while "disability-inclusion" is neglected [5]. Despite some disability-specific political, legislative and standardization frameworks in place, such as the UN Convention on the Rights of Persons with Disabilities [13], there is a lack of applicable approaches and tools that can help relevant stakeholders with translating these frameworks into more pragmatic terms to better understand the overall complexity of their requirements [5].

The UN Convention on the Rights of Persons with Disabilities (CRPD) aims to recognize the inherent dignity and equal rights of all members of the human family [13]. The CRPD was ratified and signed by the Government of Canada (GC) in 2010 [9] and is fulfilled for federal organizations by the Accessible Canada Act (ACA) [9]. The ACA is landmark federal legislation that aims to realize a barrier-free Canada in the area of communications, Information and Communication Technologies (ICT) and physical/transportation aspects by 2040 [9].

The findings discussed above align with the goals of the Accessible Canada Act [9] and serve as the motivation for our study on the application of smart tools such as causal models to investigate the challenges faced by persons with disabilities when accessing services, physical spaces, communication, transportation and other areas.

1.1. Objective

Smart cities are increasingly targeted to create inclusive environments that cater to the needs of all residents, including those facing limitations and hindrances in the use of "everyday" technologies [6]. This objective can be achieved by ensuring accessibility for vulnerable groups such as persons with disabilities in any environment so that all individuals can be independent and safe enough to reach their full potential [14].

The technological solutions driving smart cities do not counter the fundamental barriers to exclusion, especially for vulnerable populations, as there are no one-size-fits-all solutions for marginalization [2]. To improve the living conditions of persons with disabilities within smart city initiatives, it is essential to improve the accessibility of federal services in terms of physical spaces, communications, transportation, and other areas for population with disabilities. This requires the investigation of the nature of any barriers to accessibility and the influence of demographic traits on the extent the experience of such barriers utilizing advanced frameworks beyond simple statistical/correlation analysis.

In addition, data collection is largely dependent on self-identification, which people with disabilities might be reluctant to participate in. This may result in insufficient data, which can be addressed by generating synthetic data using Artificial Intelligence (AI) techniques, which would also aid in protecting sensitive information.

The objective of our paper is to confirm that probabilistic causal reasoning models are feasible tools for understanding the effects of barriers encountered by persons with disabilities depending on their demographic attributes when accessing federal services and facilities. We deploy a modeling tool to discover the cause-and-effect relationships between variables to investigate the nature of accessibility barriers and the influence of demographic traits on the extent of those barriers. This approach presents the discovery of the underlying factors that drive changes in accessibility experiences.

1.2. Prior Work, Novelty and Outline

This investigation lays the groundwork for the prediction of trends in social development research using causal reasoning algorithms and first-hand accounts of user experience from surveys.

1.2.1. Prior work

Our preliminary research on causal analysis reported in [15] involved accessibility experiences of persons with disabilities based on the nature of their reported long-term conditions based on the CSD 2017 survey [16]. The prevalence of accessibility barriers related to communications, information and communication technology (ICT) and transportation was assessed based on the type of disabilities reported.

1.2.2. Novelty

We utilize probabilistic causal models to provide a comprehensive assessment of the relationships between various demographic features (e.g., age, gender and type of disability) and accessibility barriers in the public and private sectors. Additionally, synthetic data generation and statistical testing are applied using the created Bayesian probabilistic model to handle any limitation that may occur due to occasional and, in many cases, unbalanced data, especially concerning persons with disabilities.

Based on our research and to the best of our knowledge, that application of causal modeling is a new approach to the qualitative assessment of accessibility experience. This study aims to help stakeholders and policy makers in strategic planning and execution with respect to accessible environments for persons with disabilities.

1.2.3. Outline

This paper is structured as follow: Section 2 outlines the problem formulation for this investigation and literature review of related work. Section 3 elaborates on the proposed approach for the methodology, including a brief description of data pre-processing. Sections 4–7 comprehensively discuss the definitions, hypothesis formulation and implementation of Exploratory Factor Analysis (EFA), Structural Equation Modeling (SEM) and Bayesian Networks (BNs), respectively, followed by the generation of Synthetic Data (SD), along with analyses of the experimental results for each method. The discussion and conclusion are presented in Sections 8 and 9, respectively.

Collectively, these sections provide a comprehensive narrative of the application of AI modeling techniques for the analysis of accessibility for persons with disabilities, unraveling the intricacies of their effects on decision making, engaging the smart city discourse and ultimately envisioning a future where cities are not just digitally enabled but dynamically understood and optimized through the lens of inclusion.

2. Background and Literature Review

This section includes the background of the problem addressed by this study and a discussion of the various techniques adopted by other investigators in the analysis of underlying factors in various social development studies.

2.1. Problem Formulation

Cities increasingly rely on digital analytics and smart solutions to address key challenges associated with the recent proliferation of big data driving smart city transformations [17]. Big data refer to large and complex sets of data that generally represent digital traces of human activities characterized by the scale or volume, analysis method or effect on organizations [18]. Cities around the world collect large quantities of data related to urban living to produce useful content for various stakeholders, including citizens, patrons, local governments and companies [19]. For instance, in San Francisco, crime records were analyzed to improve public security services; in Seoul, patterns and demands for the usage of the city buses were identified to improve public bus services; and in Rio de Janeiro, data from cameras and sensors were used to address safety concerns [19].

Despite the emergence of such cases, the understanding of accessibility in the public and private sectors for smart cities remains limited in the literature [19]. Smart cities' potential to improve the lives of people with disabilities also remains underexplored [20]. By leveraging data-driven insights, smart cities can create more inclusive environments that enhance the overall quality of life of people with disabilities [18,20]. This paper discusses the multitude of challenges encountered by individuals with disabilities reflected in the CSD data and proposes techniques for utilizing AI-driven smart tools to effectively identify the patterns in these challenges.

The CSD is premised on the social model of disability, which perceives disability as the result of the interaction between a person's functional limitations and barriers in the surrounding environment [7,21]. This includes social and physical barriers that make it difficult to function day to day. Thus, disability is a social disadvantage that an unfavorable environment imposes on top of an individual's functional limitation [7]. Disabilities can encompass significant activity limitations that result from the association of a long-term condition with contextual factors [22], directly affecting more than one billion people worldwide [23]. Hence, the experience is unique for each individual living with a disability, shaped by their physiological condition, along with social and physical environments [22].

Disabling barriers stem from inadequate policies or standards, negative attitudes and prejudices, service deficiencies, inaccessible built environments and a lack of evidence and data to inform effective policies and programs [23]. Moreover, people affected by disabilities are at an increased risk of poor health outcomes, lower education attainment, reduced employment and earning potential, living in poverty and higher dependency on others [23].

Accessibility is about removing barriers and ensuring equal participation in everyday activities. A variety of societal barriers still prevalent in Canada contribute to inadequate accessibility faced by people with disabilities and often lead to undesirable consequences for people with disabilities, such as social exclusion, negative stereotypes and perceptions and challenges that often lead to disability burnout [24].

Although disabilities affect people of all ages, genders and geographical regions, some groups may be more likely to develop disabilities, while individuals and groups may differ in their ability to manage adverse health conditions [22]. Hence, it is crucial to identify the obstacles and risk factors to improve accessibility in federal services and private facilities to ensure equal participation in society for persons with disabilities.

Our study seeks to leverage probabilistic causal models to determine the likelihood of experiencing specific barriers given a particular type of disability, age or gender. These projections have the potential to infer implications for entities under the federal government in providing barrier-free services and further predict trends in accessibility experiences a decade or more ahead of time to help with evaluation and preparedness. Subsequently, we strive to provide a comprehensive understanding of the correlation between data on multiple biological traits to decipher the risk factors and pathways contributing to varying accessibility experiences, ultimately providing recommendations for policy-making decisions and resource utilization.

2.2. Related Work

This section discusses published work regarding a range of accessibility issues experienced by individuals with disabilities, as well as related methodologies addressing real-world user-experience scenarios for quality assessment.

- General Considerations for Accessibility

The inherent problems associated with measuring accessibility are discussed in [25] for people with disabilities in terms of surfaces, barriers and travel modes. The proposed method includes the absolute measure currently used in ADA compliance [26] to map accessibility, as well as to help select the mitigation or renovation projects that yield the greatest increases in accessibility. The author of [27] discussed limitations of AI personalization algorithms driven by data in creating inclusive technology such as device accessibility settings or driving guides for autonomous vehicles. The study concluded with suggestions for assistive technologies developed with each unique disability in mind rather than for a broad category of disability for optimum accessibility experience.

- Probabilistic Modeling

The authors of [28] combined knowledge with available demographic and health survey data to construct a BN and investigate the factors associated with childhood health conditions. They investigated the effects of missing values, sample size and knowledge-based constraints and assessed their accuracy with multiple scoring functions.

- Statistical Modeling

Statistical modeling in the context of causal analysis is explored in a tribute to researchers who have significantly contributed to the advancement of SEM [29]. An overview of SEM including its beginnings and usefulness in social sciences, is presented here. Another paper [30] focused on SEM and a test confirmatory factor analysis system to better explain how social networking systems could be utilized by students for educational purposes, outlining the relations among different types of independent and dependent variables.

- Factor Analysis with Statistical Modeling

The research described in [31] developed a self-aware computational model to assess student satisfaction in a university engineering faculty depending on different aspects of academic experience using FA and SEM. This model provides the basis for probabilistic inference using a reasoning mechanism.

- Probabilistic Analysis and Statistical Modeling

In [32], the techniques of BN and SEM were combined to model service quality in the rail transit service of Seville, Spain. The proposed approach initiates with a BN to find certain hypotheses about the relationships between the dimensions that define service quality, followed by the use of SEM to validate the overall model. Moreover, SEM helped with validation of the relationships of the service quality framework extracted from the BN. Hence, this approach allowed researchers to examine more than one relationship and hypotheses with a large amount of information at the same time.

An approach combining deep learning with BN techniques was employed in [33] for the analysis of mass migration patterns. Migration crises were addressed in the context of crisis management by using a combination of modeling and imbalance assessment tools. By using deep learning for forecasting and causal reasoning via Bayesian networks, this methodology enabled the evaluation of risks and imbalances in the socio-technological landscape. The causal analysis revealed significant impacts of economic conditions, crises and government policies on migration patterns.

The study reported in [34] proposed linking a BN to SEM, providing an advantage in testing causal relationships in order to overcome the limitation of BN in differentiating causal from spurious relationships among decision factors. This study applied the integrated approach to decision support for customer retention in a virtual community, showing a pathway for Bayesian modeling at the latent variable level and extending SEM for managerial prediction and diagnosis.

Another study [35] assessed the risk of a stroke given stroke patient data through the application and validation of causal graphical models. The risk factors identified by a stroke expert were modeled with a BN, while SEM was later applied to provide a quantitative assessment of causality among the variables in the BN. The authors of [36] explore assistive technologies for people with disabilities in smart cities and emergency assistance provided by first responders. Technologies designed for interactions between responders and the assisted persons were discussed based on AI formalism.

- Generative AI

This literature reported in [37] examined the interplay between urban form and social behavior by identifying the key indicators of urban morphology and presenting state-of-the-art methodologies for data collection. The computational capability of foundation models and generative AI models to simulate interactions between individuals and urban built environments in a diverse group across the globe was also discussed. Through this exploration, the potential of advanced AI systems to assist urban scientists in understanding cities was demonstrated for the optimal application of generative AI in urban development research and policy making.

The literature review revealed that while quantitative data have been analyzed using probabilistic modeling in many references, causal models have not been applied, as far as we know, to investigate accessibility barriers faced by people with disabilities.

3. Proposed Approach and Materials

This section describes the proposed approach for developing a computational tool for causal modeling, along with an overview of the primary dataset [38] used in this study.

3.1. Methodology

Smart cities, which involve citizens leveraging advanced data analytics, predictive models and digital twin techniques, can offer a transformative model for socially sustainable and inclusive urban development [17]. The probabilistic framework describes how to represent and manipulate uncertainty about models and predictions, which has a central role in scientific data analysis and AI [39]. On the other hand, predictive analytics is critical to proactive planning, enabling cities to cultivate urban environments that are centered around the needs of citizens [17].

This paper proposes an approach constituting the above-mentioned methods, promoting the development of such urban public spaces catering to a diverse community, with a focus on people with disabilities. Our proposed approach comprises a multivariate statistical analysis framework and probabilistic reasoning techniques, as illustrated in Figure 2, to identify causal relations between different combinations of demographic traits and barriers to accessibility for persons with disabilities. Statistical analysis is applied to empirical data to provide a quantitative assessment of causal relationships among the variables used in the Bayesian network, thus statistically validating expert knowledge.

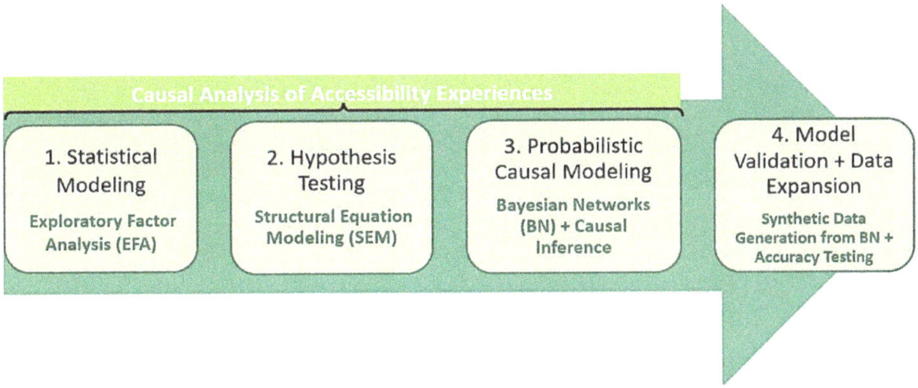

Figure 2. Proposed methodology: A probabilistic causal reasoning framework is employed to identify the underlying causes of barriers for varying demographics with disabilities. Here, EFA, SEM and BN techniques are combined for causal analysis, followed by SD generation using the probabilistic model to overcome limitations of the real data.

We use the causal probabilistic method to identify challenges and considerations of the use of synthetic data for smart cities. In this direction, diagnostic analysis and predictive analysis are keystones for identifying reasons behind specific outcomes and preparing for future accessibility concerns. Leveraging historical data, these tools identify potential challenges and predict trends for hypothetical scenarios, empowering cities to adopt a proactive stance in addressing issues [17]. The integration of predictive analytics into smart city frameworks allows decision makers to make informed choices regarding infrastructure development, resource distribution and policy implementation to thrive in an ever-evolving urban landscape.

For statistical modeling, we use EFA, an exploratory data analysis method, to search factors or latent variables from a set of observed variables. It helps with data interpretations by reducing the number of variables. Next, we implement a linear model framework called SEM to represent the various aspects of accessibility thought to be causally connected.

To support the confirmatory analysis of SEM, we then apply probabilistic reasoning based on a causal model such as a BN. Several scenarios are demonstrated for risk assessment using the inference mechanism of the network. Finally, the created BN is used for data generation to explore the possibility of using Synthetic Data (SD) for future investigations in case of inconclusive or limited real-world data.

Additionally, the BN is used as a template to replicate the structure and variables in the dataset to generate a new set of data by applying parameter learning techniques. This synthetic set of data is then compared with the original data through statistical hypothesis testing to evaluate the accuracy of the former. Furthermore, using the structure learning method, the BN is recreated in an automated manner to explore the precision of this learning algorithm with the synthetic data. Adjustment is applied to observe improvement in the results in the form of conditional probability.

3.2. Dataset

The primary data (PD) used for this paper were obtained from the Statistics Canada data hub and published by the Government of Canada in May 2024. This section contains provides an overview, in addition to addressing accuracy, limitations and data processing steps leading to statistical and causal probabilistic modeling later.

The dataset [38] is based on the Canadian Survey on Disability (CSD) 2022, a national survey that collects information about Canadians aged 15 and over whose everyday activities are limited because of a long-term condition or health-related problem [7]. As demonstrated in (Figure 3), a range of data is provided on 10 different disability types leading to activity limitations related to *seeing, hearing, vision, mobility, flexibility, dexterity, pain, learning, memory, developmental and mental health*. The total sample size for the CSD is approximately 54,000 individuals, with an overall collection response rate of 61.1% [7]. The data includes barrier scenarios categorized under four groups: (1) communications, (2) attitude, (3) information and communication technologies (ICT) and (4) physical/transportation estimated by both number and percentage.

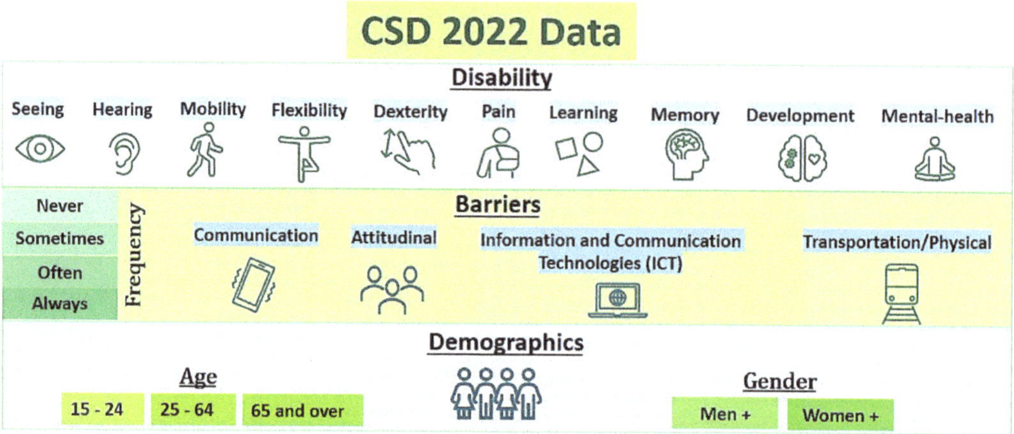

Figure 3. Primary dataset (PD)—Categorical variables and their subcategories presented in the CSD 2022 accessibility barrier data. The "+" symbol next to each class under **Gender** denotes "non-binary persons" [38].

In the dataset, there are two types of variables: numerical and categorical. The categorical variables and their subcategories are presented in Figure 3. Observing the categorical features, there two kinds of categorical data: ordinal and nominal. An ordinal variable has values defined by an order relation between the different categories, whereas the categories of nominal data are purely descriptive and do not possess any quantitative or numeric value [40].

For the disability types, the dataset focuses on activity limitations related to *hearing, vision, mobility, flexibility, dexterity, pain, learning, mental health, memory* and *developmental* disabilities. Brief descriptions of these categories, along with the corresponding recommended necessary accessibility measures, are included in Table A2. The data for demographic with disabilities are categorized in terms of barrier scenarios related to *Communications, attitude, information and communication technologies (ICT)* and *physical/transportation*. The collected data are estimated by both number and percentage in the CSD dataset.

To prepare the data to implement the modeling algorithms, we performed some data normalization steps. The data cleaning steps were followed by splitting the 'number' and 'percentage' data. One-hot encoding was performed on nominal features; however, ordinal data were mapped into numerical data so that the inner order could be preserved.

Since the source is a sample survey, all estimates are subject to both sampling and non-sampling errors. Estimates and confidence interval that do not meet Statistics Canada's quality standards are suppressed [7].

4. Exploratory Factor Analysis (EFA) for Identification of Data Structure

With the accessibility barrier data, the goal was to investigate the causality between demographic attributes and accessibility experiences. To establish a causal connection between variables, it was assumed that there are underlying latent constructs in the data. However, with no prior knowledge, dimension reduction and structure identification were needed before proceeding with statistical modeling.

4.1. EFA Overview

Factor analysis is a widely used unsupervised machine learning algorithm in dimension reduction. It creates factors from the observed variables to represent the common variance, i.e., variance due to correlation among the observed variables [31,41]. In experimental design, the term "factor" refers to a categorical explanatory variable and is the reason the observable variables have the correlations they do [42]. Factors are also known as latent or hypothetical variables.

EFA aims to explain the covariances or correlations among the variables and to understand what constructs underlie the data [43]. Hence, EFA was selected as a statistical framework to identify latent constructs or factors that explain the patterns of correlations within the observed variables. It is often used as a first analytic step to determine the most appropriate method of conceptualizing scores in a given dataset without any modeling limitations [44].

With EFA, the goal is to describe and summarize a dataset by explaining a set of observed variables in terms of a smaller number of latent variables or factors [42]. The equation below represents features in terms of factors (β) independently for i = 1, ..., n. Given a variable (x), factor F, factor loading (l) and error term (ϵ) (weight of the factor for the corresponding variable), the number of factors is equal to the number of variables [41]:

$$x_i - \mu_i = l_1 F_1 + \ldots + l_n F_n + \epsilon_i$$

To determine the most effective structure with an appropriate number of factors for statistical modeling with SEM later, factor validity is assessed here using EFA.

4.2. Application of the EFA Method

The pre-processed dataset (Section 3.2) was used for EFA as illustrated in Figure 4, employing the following steps. To check the suitability of data for EFA, correlation tests were performed. The co-linearity variable obtained from these tests, as included in Table 1, indicates how strongly a single variable is correlated with other variables.

4.2.1. Bartlett's Test of Sphericity

Bartlett's test checks whether correlation is present in the given data. It tests the null hypothesis that the correlation matrix is an identical matrix [41]. The identical matrix consists of all diagonal elements, represented as 1, so, the null hypothesis assumes that no correlation is present among the variables.

We needed to reject this null hypothesis because factor analysis aims to explain the common variance, i.e., the variation due to correlation among the variables. If the p-test statistic value is less than 0.05 [32,34], we can decide that the correlation is not an identical matrix, i.e., correlation is present among the variables with a 95% confidence level.

Table 1. Bartlett's sphericity and KMO test parameters to check variable correlations for EFA.

	Bartlett Sphericity Test		**KMO Test**
Component	Chi-squared value: 5890.8469	p value: 0.0	KMO test statistic: 0.6489

Since the p-test statistic is less than 0.05, as shown in Table 1, we can conclude that correlation is present among the variables, which allows us to apply factor analysis.

4.2.2. Kaiser–Meyer–Olkin (KMO) Test

The KMO test measures the proportion of variance that might be common among the variables [41]. Larger proportions are expected to show several correlations present among the variables, thereby ensuring the application of dimension reduction techniques such as factor analysis [42]. The KMO score is always between 0 and 1, and values more than 0.6 are highly desirable as a measure of how well-suited our data are for factor analysis [45].

Table 1 includes the co-linearity variable obtained from these tests, which indicates how strongly a single variable is correlated with the other variables. There are eight observed variables in the pre-processed dataset. It can be observed that age, disability, barriers and the four levels of the frequency of encountering barriers can all influence the accessibility experiences of persons with disability. Since these variables are correlated, we can form factors to explain the common variance, i.e., variance due to correlation among these features [41].

4.3. *Interpretation of the EFA Results*

In the EFA method, every factor explains a certain variance in observed variables. The factors with the lowest variance were dropped to obtain optimal number of factors, i.e., four in our case.

4.3.1. Description of the EFA Components

Here, we interpret the factors influencing variable correlation by making use of components such as loadings, variance, eigenvalues and communalities, as included in Table 2.

- **Factor loading:** is a matrix that shows the relationship of each variable to the underlying factor. It shows the correlation coefficient for an observed variable and factor, along with variance explained by the observed variables [41].
- **Eigenvalues:** represent variance explained by each factor against the total variance , also known as characteristic roots. We selected a number of factors whose eigenvalues are greater than 1 [41]. In a standard normal distribution with a mean of 0 and standard deviation of 1, the variance is 1 [42]. Since we standard-scaled the data, the variance of a feature is 1. This is the reason for selecting factors whose eigenvalues (variance) are greater than 1, i.e., the factors that explain more variance than a single observed variable.
- **Communalities:** are the sum of the squared loadings for each variable. They represents the common variance [41,42], ranging from 0 to 1, with a value close to 1 representing more variance.

Table 2. EFA parameters for the frequency of encountering barriers.

	Factor Loadings				
Variable	**Factor 1**	**Factor 2**	**Factor 3**	**Factor 4**	**Communalities**
Age	0.8270	0.0228	0.1227	0.0258	0.7002
Gender	−0.0067	−0.0018	0.0423	−0.2223	0.0512
Disability	0.0022	0.0746	−0.0438	0.0277	0.0083
Barriers	0.0094	0.0052	0.0423	−0.0024	0.0019
Never	0.4332	0.5239	0.0901	0.6612	0.9074
Sometimes	0.1653	0.9587	0.2186	−0.0425	0.9961
Often	−0.5055	−0.1514	0.7928	−0.1716	0.9365
Always	−0.5543	−0.3019	0.2974	−0.1847	0.5209

4.3.2. Determining the Optimum Number of Factors

The factors and their eigenvalues, which represent the amount of variance the factor explains, are plotted in Figure 4. We selected the number of factors whose eigenvalues are greater than 1, since in a standard normal distribution with a mean of 0 and standard deviation of 1, the variance is 1.

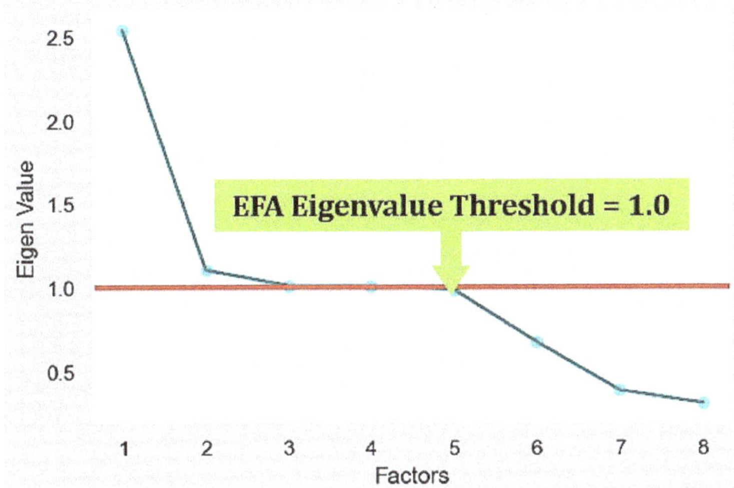

Figure 4. Exploratory factor analysis: The scree plot shows the effect of eigenvalues with an increase in the number of factors.

The number of factors is equal to the number of variables in our dataset. Not all of the factors provide a significant amount of useful information about the common variance among the variables, so we decided the number of factors on the basis of the common variance the factors explain [41].

For our data, the eigenvalues are greater than 1 for four factors (or unobserved variables), given the eight observed variables. This matches the number of factors that we have in our model, as shown in Figure 5. For example, factor 4 has an eigenvalue of 2.5k = m whereas the eigenvalue for factor 5 is 0.99, which is less than 1.

Considering only the original eigenvalues from the graph in Figure 4, we can see that the eigenvalues drop below 1 starting for the fifth factor for dimension reduction. Therefore, four latent variables were selected for SEM in Figure 5 based on these EFA results.

The EFA method examined the data to find possible underlying factors given a set of observed variables. This helped with data interpretations by reducing the number of unobserved variables or factors. After extracting the maximum common variance from all variables, there were four optimum latent factors discovered. This simplification made the data more manageable and easier to interpret to define the structure of our SEM model.

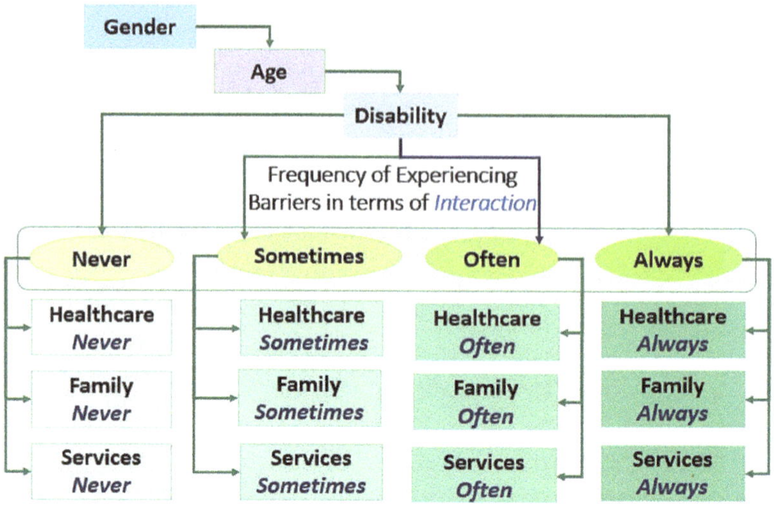

Figure 5. SEM structure 1: SEM application for measuring the correlation between accessibility experience and overall user attributes for *interaction*-related barriers. Each rectangle is an observed variables with multiple subcategories, whereas the oval shapes represent the latent variable of *accessibility experience* attributed to *gender, age, disability, barriers* and all four responses for *frequency*.

5. SEM Architecture for Statistical Analysis

SEM is a multivariate analysis technique widely applied in the field of social sciences [32,46]. Its factor analysis element allows for assessment of the relationships among different observed variables (available data) through causal networks. These relationships allow for understanding of the unobserved or latent variables. The multiple regression element estimates the hypothesized causal relationships among variables to be tested [47].

5.1. Application of the SEM Method

In order to define the structure in Figure 5 to model the dataset, we used the Semopy tool in Python [48], which facilitates parameter estimation through a restricted Maximum Likelihood Estimation (MLE) approach while simultaneously obtaining unbiased estimates to analyze structural relationships among different variables in the data.

5.1.1. Measurement Model

The definition of the SEM structure typically consists of left-sided values and right-sided values separated by an operator character that constitutes the nature of the relationship between them [30,48].

The model was outlined by determining the number and relationships of measured and latent variables. Only the variables that are valid indicators of the constructs of disability and accessibility were selected carefully for SEM. Measured or observed variables are presented in the dataset with values, whereas latent variables are usually inherent in the data. The latent variables were chosen to quantify factors that can only be understood as a concept but cannot be measured. Next, a path diagram depicting the structural and measurement models was generated for SEM analysis.

5.1.2. Defining the Variables

To define the structure, the data used for factor analysis were considered for SEM. The purpose of the SEM analysis was to assess the holistic causal connections among the given personal attributes. Each barrier scenario includes four separate data points based on the frequency of encountering them in the PD.

The extent of the barriers lies in the frequency with which they were experienced, i.e., if 80% of a barrier had been reported as **never** being experienced, then it would be less significant than one that had been reported as **always** being experiences in a majority of responses. Hence, our goal is to assess the barriers in terms of the frequency of encountering them, and as shown in Figure 5, we used frequency to center the SEM. The latent variables for this model are *never, sometimes, often* and *always*, measured by the observed variables for each barrier category as described in Figure 3.

The goal was to explore all possible structures in order to determine the most efficient one reflecting the true nature of the variable dynamics while minimizing the number of insignificant relationships. Therefore, another structure that we regarded is shown in Figure A1 with latent variables of *adjustment, understanding* and *assistance* to reflect issues people with disabilities can experience while interacting in person with *healthcare, family* and *services*, respectively.

5.2. Analysis of the SEM Results

SEM models determine the relationship between the measured and latent variables [30,46–48]. In the SEM definition, we attempted statistical modeling with two structures (Figures 5 and A1). Since the results from of EFA suggest four possible latent variables, we selected the first structure for our analysis.

5.2.1. Description of the SEM Parameters

P-values and regression estimates collectively determine which relationships in the model are statistically significant and their nature [29,30,47]. To investigate the effects of the independent variable on the dependent ones, parameter analysis is presented in Table 3 for all barriers based on the model in Figure 5.

- **Regression Estimate (RE):** The primary result of an SEM analysis is a set of estimates of the regression coefficients. The sign of the estimate tells us whether there is a positive or negative correlation between variables [29,30,46]. Equation (1) expresses the formula for calculating the estimate (β), given an independent variable (X), a dependent variable (Y) and their means ($\bar{X}$ and $\bar{Y}$, respectively) [49].

$$\beta = \frac{\sum(X - \bar{X})(Y - \bar{Y})}{\sum(X - \bar{X})^2} \quad (1)$$

- **Standardized Regression Estimates (SRE):** Since the regular estimates use different scales for different variables in a model, they cannot be compared to identify their level of importance. However, standardized estimates all use the same scale and can be compared to determine which independent variables are crucial in the model [46–48]. The standardized estimate (β) is calculated by multiplying the estimate (β) by the standard deviation of X and the standard deviation of Y [29,49] as shown in Equation (2).

$$\beta^* = \beta \times \frac{\sigma_X}{\sigma_Y} \quad (2)$$

- ***P*-value:** The p-value indicates whether relationships between variables are statistically significant or not. It measures how likely it is that any observed association between variables is due to chance [29,47]. A common threshold for the p-value is 0.05, stating that 5% of the time, variable correlation might have been falsely concluded.

The first variable in the measurement model for each barrier category does not produce any p-value, as seen in Table 3. A p-value > 0.05 indicates an insignificant relationship [50].

Table 3. SEM Parameter analysis for the frequency of encountering barriers.

Barrier Scenario	Regression Analysis for Frequency of Encounter							
	Never		Sometimes		Often		Always	
	SRE	*p*-Value	*SRE*	*p*-Value	*SRE*	*p*-Value	*SRE*	*p*-Value
Attitude (Family)	0.9923	-	0.9355	-	0.7295	-	0.7086	-
Attitude (Healthcare)	0.9987	0.0001	0.9605	0.0025	0.7630	0.0002	0.7530	0.0005
Attitude (Business)	0.9993	0.0002	0.9443	0.0012	0.7065	0.0003	0.4539	0.0066
Alarms	0.9996	0.0011	0.7371	0.0019	0.6466	0.0004	0.8754	0.0017
Lighting/ Sound	0.9997	0.0003	0.9395	0.0013	0.8613	0.0007	0.9030	0.0004
Videoconferencing	0.9996	0.0021	0.9230	0.0003	0.7604	0.0005	0.5729	0.0048
Automated message	0.9973	0.0023	0.9175	0.0001	0.8475	0.0006	0.5274	0.0344
Interaction (Family)	0.9938	0.0014	0.9547	0.0009	0.7543	0.0013	0.9178	0.0045
Interaction (Services)	0.9967	0.0001	0.9823	0.0011	0.8688	0.0001	0.8806	0.0002
Interaction (Healthcare)	0.9966	0.0001	0.9708	0.0002	0.8663	0.0001	0.9124	0.0003
Signs	0.9988	0.0001	0.8289	0.0009	0.5889	0.0106	0.7729	0.0001
Traffic signals	0.9977	0.0010	0.2630	0.0017	0.2268	0.0071	0.6647	0.0244
Online (Business)	0.9986	0.0023	0.4603	0.0120	0.1098	0.1933	0.6428	0.0198
Online (Booking)	0.9986	0.0021	0.5110	0.0383	0.2354	0.0051	0.8007	0.0001
Online (Services)	0.9986	0.0002	0.6653	0.0010	0.4839	0.0062	0.9017	0.0029
Phone (health care)	0.9989	0.0011	0.9876	0.0001	0.8989	0.0007	0.8217	0.0007
Phone (Family)	0.9974	0.0003	0.9774	0.0013	0.9036	0.0001	0.8701	0.0002
Phone (Services)	0.9980	0.0004	0.9842	0.0001	0.9005	0.0012	0.7396	0.0002
Doorway	0.9898	0.0001	0.7763	0.0002	0.7069	0.0001	0.6429	0.0019
Sidewalks	0.9882	0.0003	0.8821	0.0002	0.7557	0.0001	0.5785	0.0057
Sidewalks (snow)	0.9790	0.0001	0.9194	0.0005	0.6211	0.0051	0.1154	0.1702
Washrooms	0.9972	0.0013	0.7480	0.0001	0.6924	0.0002	0.8274	0.0002
Queues	0.9904	0.0001	0.9352	0.0006	0.7060	0.0001	0.5674	0.0143
Self-serve	0.9970	0.0019	0.8059	0.0103	0.7447	0.0001	0.7377	0.0002
Social media	0.9996	0.0210	0.7963	0.0001	0.4823	0.0069	0.8881	0.0011

SEM parameter values below the threshold are highlighted in red in the table.

5.2.2. SEM Parameter Analysis

The SEM parameters include the standardized regression coefficients or simply *coefficient* and *p*-values derived from the analysis of data on experiencing barriers: ***never, sometimes, often*** and ***always*** for all observed barrier scenarios. The threshold for significance is greater than 0.50 for *SRE* and 0.05 for the *p*-value. The values that are not significant are highlighted in red.

Communication Barriers: Accessibility in communication via *interaction, phone, automated message and videoconferencing*, as shown in Table 3, implies that all relationships are significant, with *p*-values being less than 0.05. All barrier scenarios with SRE values close to 1 imply strong connections between the disability type and barriers while accessing services for persons with disabilities .

ICT Barriers: Most *p*-values for ICT-related barriers encountered through online services using are below 0.0200 (Table 3). However, the frequency category of ***sometimes*** for *online (business)* has a *p*-value of 0.1933, which is above the threshold, rendering this connection insignificant.

Transportation Barriers: The parameters tabulated in Table 3 show that disability type significantly relates to transportation, where *physical, navigation* and *announcement*-related barriers can be experienced. The majority of *p*-values are below 0.03, with a *p*-value of only 0.1702 for *sidewalks (snow):always*.

Observing the statistical modeling analysis, the majority of *p*-values are less than 0.05, indicating that the associations between disabilities and barriers are rationally well-grounded. The SEM confirms the assumed strong correlation between the frequency of experiencing accessibility barriers and the type of disability. The analysis confirms the hypotheses about the valid connection between specific accessibility needs that comes with the varying nature of disabilities and the nature of barriers experienced due to a lack of accommodation in federal sector facilities and services.

6. Probabilistic Modeling with Causal Bayesian Network

This section presents the application of Bayesian networks (BNs) and experimental results of causal Bayesian inference to evaluate the probabilities of various accessibility scenarios. The objective of this application is to examine how a BN can serve as an effective system for decision making in social development, including with respect to accessibility in federal sector organizations.

6.1. BN Overview

Although SEM techniques have been utilized to learn causal relationships from data, they do not entail the predictive and diagnostic capabilities of BNs when handling non-linear relationships [34,46]. Thus, we apply BNs for reasoning under uncertainty conditions, which, in our dataset, may exist due to limited knowledge about systems, an ambiguous understanding of disabilities or randomness in the demographic attributes driving accessibility experiences [28,34,35].

The network was modeled after the SEM structure shown in Figure 5. For demonstration, the BN for barriers related to in-person *Interactions* is illustrated in (Figure 6). Here, the graph represents connections between each node, both causal and non-causal. The causality between nodes such as *Disability* and the nature/ frequency of *Barriers* would be assumed during inference later in this process.

6.1.1. Fundamental Concepts

A BN represents conditional probability and causality relationships between variables expressed by two key concepts in probability: conditional probability and Bayes' theorem.

- **Conditional probability** is a measure of the change in likelihood of an event in relation to information about another event, as expressed by the following equation [51].

$$P(A|B) = \frac{P(A \cap B)}{P(B)} \quad (3)$$

This states that the probability of event A occurring given that event B has occurred is equal to the probability of both events divided by the probability of event B [51]. The calculation for conditional probability involves the joint probability of both events and the probability of the given event.

- **Bayes' theorem** propagates probability through the model, which relates the conditional and marginal probabilities of random events. It is an equation that allows for the manipulation of conditional probabilities in a BN [52]. The theorem is expressed below and is used to update the probabilities of hypotheses based on new evidence:

$$P(A|B) = \frac{P(B|A) * P(A)}{P(B)} \quad (4)$$

Here, A and B are events and $P(B) \neq 0$. P(A | B) and P(B | A) are conditional probabilities that indicate the probability of an event occurring given another event, also called posterior probability. P(A) and P(B) are the probabilities of observing A and B, respectively, without any given conditions, also known as the prior and marginal probability respectively [53].

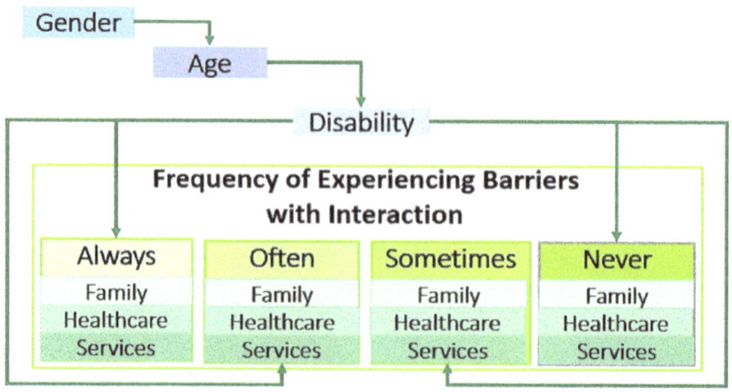

	Often		
Disability	**family**	**services**	**healthcare**
Development	0.3333	0.3333	0.3333
Dexterity	0.3334	0.3333	0.3333
Flexibility	0.4060	0.2997	0.2943
Hearing	0.2811	0.3582	0.3607
Learning	0.3448	0.3249	0.3303
Memory	0.2838	0.3553	0.3608
Mental	0.3537	0.3219	0.3243
Mobility	0.4059	0.2981	0.2960
Pain	0.3841	0.2402	0.3757
Seeing	0.3342	0.3315	0.3343
Unknown	0.3333	0.3333	0.3333

	Always		
Disability	**family**	**services**	**healthcare**
Development	0.3333	0.3333	0.3333
Dexterity	0.3333	0.3333	0.3333
Flexibility	0.3333	0.3333	0.3333
Hearing	0.3333	0.3333	0.3333
Learning	0.3333	0.3333	0.3333
Memory	0.3333	0.3333	0.3333
Mental	0.3151	0.3141	0.3708
Mobility	0.3333	0.3333	0.3333
Pain	0.3151	0.3142	0.3706
Seeing	0.3333	0.3333	0.3333
Unknown	0.3333	0.3333	0.3333

	Sometimes		
Disability	**family**	**services**	**healthcare**
Development	0.3381	0.3286	0.3332
Dexterity	0.3255	0.3370	0.3375
Flexibility	0.3210	0.3379	0.3411
Hearing	0.3391	0.3294	0.3315
Learning	0.2850	0.2749	0.4401
Memory	0.3443	0.3422	0.3135
Mental	0.3631	0.3292	0.3077
Mobility	0.2582	0.3752	0.3667
Pain	0.2929	0.3149	0.3922
Seeing	0.3372	0.3373	0.3254
Unknown	0.3333	0.3333	0.3333

	Never		
Disability	**family**	**services**	**healthcare**
Development	0.3301	0.3406	0.3294
Dexterity	0.3385	0.3301	0.3314
Flexibility	0.3403	0.3297	0.3299
Hearing	0.3290	0.3379	0.3331
Learning	0.3166	0.3406	0.3428
Memory	0.3152	0.3330	0.3517
Mental	0.3080	0.3414	0.3505
Mobility	0.3408	0.3270	0.3321
Pain	0.3341	0.3294	0.3365
Seeing	0.3317	0.3315	0.3368
Unknown	0.3336	0.3335	0.3329

Figure 6. BN probabilistic graph model designed according to the SEM structure in Figure 5, showing dependencies among the data variables. The Conditional Probability tables (CPT) contain probabilities (color-coded values between 0 and 1) of barriers under 4 nodes associated with the frequency of experiencing barriers (*never, sometimes, often* and *always*) under the parent node of *disability*. Here, shades closer to red is the lower and closer to green represent higher probabilities.

6.1.2. Graphical Model

The BN model consists of a qualitative part to represent the dependencies and a quantitative part to specify the probabilistic relationships.

- **Graph Structure:** A BN is a type of causal probabilistic model that is described qualitatively by a directed acyclic graph as illustrated in Figure 6 for in-person *Interaction*. The vertices of the graph representing variables are called nodes. The nodes are represented as circles containing the variable name. The connections between the nodes are called arcs or edges, which are drawn as arrows between the nodes representing dependence between the variables. Therefore, for any pair of nodes indicating that one node is the parent of the other, there is no independence assumptions, which are implied in a BN by the absence of a link. Moreover, a conditional probability table (CPT) is associated with each node containing the probability distribution of the node given its parent nodes [54].
- **Conditional Probability Tables (CPTs):** Once the network topology is constructed, the CPTs are initialized. The values in a CPT represents prior probabilities used in the inference, which is the calculation of posterior probabilities using Bayes' formula [28]. In Bayesian statistics, a posterior probability is the revised or updated probability of an event occurring after factoring in new information [55]. To draw conclusions about the causal dependencies, the posterior probabilities are deduced using prior knowledge with Bayesian inference [54].

6.2. BN Experimental Results from Causal Inference

By representing conditional dependence, probabilistic graphical models such as BNs allow for the identification of causality from data via inference for the probability computation [56] of frequency of experiencing barriers depending on disability type. Here, the BN application is analyzed in two primary ways for prediction and diagnosis [57]:

1. **Diagnostic reasoning** involves inspecting causality by reasoning backwards from the observed effects of an event.
2. **Predictive reasoning** involves obtaining the expected probabilities of certain outcomes given some observed evidence.

6.2.1. Diagnostic Reasoning with Evidence of *Barriers*

The diagnosis of accessibility experiences is carried out with BN inference to learn about the likelihood of of a demographic trait being present in individuals experiencing specific barriers. The probability of the cause (*disability*) is continuously shifting either upward or downward, depending on the information gathered about the effect (*barriers*) during the process [58].

The purpose of diagnostic testing is to move from the probability of disability before considering the causality with barriers (prior probability) to the probability after considering the causality with barriers based on the report/data (posterior probability) [58] using Bayes' theorem (Equation (4)). This aims to determine the most and least considerable effects of the selected barriers on disabilities.

- **Demographic traits correlated with *interaction barriers*:**

The inference of *disability* is illustrated in Figure 7, assuming the evidence of *interaction* barriers to estimate posterior probabilities for all 11 types of ***disability*** (*development, dexterity, flexibility, hearing, learning, memory, mental health, mobility, pain, seeing and unknown*).

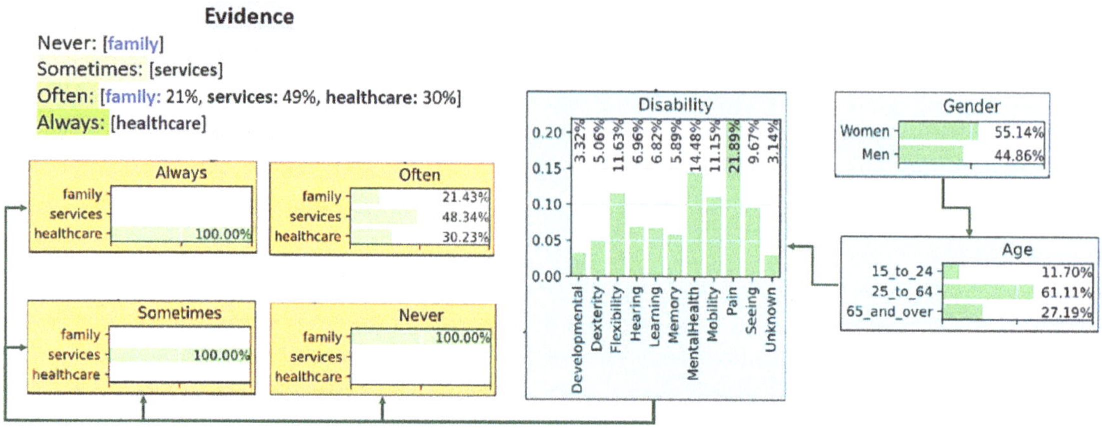

Figure 7. BN inference:likelihood of disability for persons encountering communication barriers during *interaction* with family, healthcare and other services. Evidence of the nature and frequency of barriers (rectangles highlighted in orange) was randomly selected to demonstrate the diagnostic reasoning.

To update the probabilities, ***interaction barriers*** were considered to deduce their effects on demographics with disabilities. For example, $P(Disability|Never = family)$ is the expression for the evidence of ***never*** in the BN definition to predict the chances of a person having various disability types when no barrier is experienced interacting with family.

In Figure 7, although the partial evidence is defined initially as whole numbers for ***often***, such as 20% for *family*, the algorithm adjusts it to a decimal value, such as 21.43%, to match real-world conditions.

While individuals with *pain-related disability* (21.89%) are most likely to have this experience, *mental health* (14.48%) and *mobility* (11.63%) also seem to have a substantial likelihood. Also *Women* between the ages of 25 and 64 years also appear to be most vulnerable when it comes to facing challenges with in-person interactions.

- **Likelihood of Disability types for each barrier scenario:**

The holistic process of updating the posterior probabilities of the BN ***disability*** parent node is executed by providing evidence of barriers experienced in specific scenarios. The overall inference results are included in Table 4 for each type of disability when barriers are encountered ***always***.

The observations from these updated posterior probabilities indicate that for attitudinal barriers (*attitude (business), attitude (family)* and *attitude (healthcare)*), an individual is most likely to have *pain-related disabilities*. Communications scenarios seem to mostly affect persons with the following disabilities: *videoconferencing—learning (*20.80%*); interaction* with *family—mental (*33.82%*), services—mobility (*20.59%*) and healthcare—pain (*22.73%*); automated message: mental (*25.62%*).* All issues related to *phone* with *family, services and healthcare* show the greatest likelihood for *pain-related disabilities* .

Additionally, ICT barriers appear to impact persons with *pain-related disability* (*online (booking):* 34.49%; *social media:* 32.69%;*online (services):* 18.49%). For *self-serve technology, flexibility disability* has the highest probability (26.11%).

Furthermore, Physical and transportation-related issues yield the greatest probability for individuals with *pain-related disability* (*traffic signals*: 37.15%; wait lines or *queues*: 28.67%; level of *lighting/sound* indoors: 24.94%). Substantial effects can be observed for *flexibility dis-*

ability (*doorways*: 23.50%; *sidewalks*: 19.35%). *Mental health* issues seem to be greatly affected by barriers pertinent to announcements or *alarms* (28.59%) and *signs* or direction (21.94%).

Table 4. Probability of an individual having a disability given ***always*** encountering a specific barrier.

Barrier	Development	Dexterity	Flexibility	Hearing	Learning	Memory	Mental	Mobility	Pain	Seeing	Unknown
Attitude (business)	0.0371	0.0539	0.0005	0.2261	0.0587	0.1584	0.0139	0.1946	0.2399	0.0106	0.0063
Attitude (family)	0.0553	0.0404	0.0168	0.0988	0.0081	0.0129	0.0655	0.2552	0.2900	0.1567	0.0002
Attitude (healthcare)	0.0250	0.0200	0.1071	0.0408	0.1593	0.0718	0.1484	0.0657	0.2341	0.1047	0.0232
Alarms	0.0566	0.0641	0.1485	0.0822	0.0134	0.1113	0.2859	0.0918	0.0261	0.0560	0.0640
Lighting/ sound	0.0069	0.0760	0.2066	0.0341	0.1154	0.0768	0.0858	0.1305	0.2494	0.0033	0.0153
Video conferencing	0.0370	0.0964	0.0772	0.0949	0.2080	0.0888	0.0738	0.0829	0.1740	0.1443	0.0228
Automated message	0.0428	0.0734	0.1634	0.1259	0.0109	0.0900	0.2562	0.0192	0.0582	0.1298	0.0301
Interaction (family)	0.0064	0.0461	0.1663	0.0062	0.0976	0.0069	0.3382	0.1067	0.0074	0.1450	0.0732
Interaction (services)	0.0573	0.0891	0.0183	0.0142	0.0229	0.0913	0.1627	0.2059	0.1458	0.0801	0.1123
Interaction (healthcare)	0.0282	0.0477	0.1208	0.0073	0.1228	0.0563	0.2230	0.0529	0.2273	0.0996	0.0142
Signs	0.0297	0.0652	0.1701	0.1246	0.1000	0.0226	0.2194	0.0460	0.0884	0.1276	0.0064
Traffic signals	0.0574	0.0506	0.0173	0.0036	0.1042	0.0689	0.0777	0.0666	0.3715	0.1198	0.0624
Online (booking)	0.0261	0.0170	0.0750	0.0694	0.0982	0.0695	0.1188	0.0440	0.3449	0.1011	0.0359
Online (services)	0.0341	0.0347	0.0669	0.1295	0.0412	0.0932	0.1599	0.0955	0.1849	0.0925	0.0675
Doorways	0.0712	0.0117	0.2350	0.0392	0.0394	0.0568	0.0985	0.0874	0.1802	0.0939	0.0868
Sidewalks	0.0813	0.0070	0.1935	0.0796	0.1017	0.0145	0.0752	0.1587	0.1216	0.1538	0.0131
Washrooms	0.0083	0.0718	0.1580	0.0785	0.1133	0.0953	0.0779	0.1356	0.1704	0.0324	0.0584
Queues	0.0371	0.0155	0.0782	0.0359	0.0477	0.0260	0.2211	0.1527	0.2867	0.0908	0.0083
Phone (healthcare)	0.0602	0.0368	0.1450	0.0596	0.0676	0.0034	0.1032	0.1774	0.3101	0.0368	0.0001
Phone (family)	0.0360	0.0382	0.0627	0.0678	0.0869	0.0853	0.1584	0.1026	0.2438	0.1021	0.0162
Phone (services)	0.0558	0.0551	0.0435	0.0539	0.0673	0.0761	0.1407	0.1181	0.2515	0.1106	0.0274
Self-serve tech	0.0819	0.1038	0.2611	0.1701	0.0771	0. 0534	0.0443	0.0965	0.0420	0.0374	0.0324
Social media	0.0150	0.0270	0.1427	0.0382	0.0643	0.0083	0.2294	0.0253	0.3269	0.1095	0.0134

For each barrier, the sum of the probabilities across all disabilities is 1. The highest probability for each barrier scenario is highlighted in green.

6.2.2. Predictive Reasoning with the Evidence of *Disability*

This section describes BN inference for the nature and frequency of experiencing barriers given the presence of an attribute such as *disability* (Figure 8 and Table 5. To measure the likelihood of an event given that another event has occurred, conditional probability (Equation (3) and Table 5) is employed. The evidence used here is one or a combination of multiple attributes, including hard (100% certainty) and soft (with some level of uncertainty) evidence.

- ***Interaction* barriers for particular demographics:** For the inference in Figure 8, we use hard evidences for each attribute node, using the BN for *interaction* barriers in Figure 6. The algorithm estimates probabilities of encountering various barriers based on specified attributes i.e., *women* with *mobility* disabilities aged *15 to 24* years. The results show that this group is most likely to face barriers from interaction with *family* (**sometimes** (48%); **often** (72%); **always** (57%)) and *healthcare* (**never** (39%)).
- ***Interaction* barriers for persons with *mobility* disability:** Based on the Bayesian analysis through inference, the probability of an individual with *mobility* disabilities experiencing each type of barrier is illustrated in Table 5. In the table, each column represents a barrier scenario except for 'Frequency'. The evidence is randomly assumed here, pertaining to the parent *disability* parent node with subtypes to predict the

probabilities for child nodes (frequency) ***never, sometimes, often*** and ***always*** subject to those types.

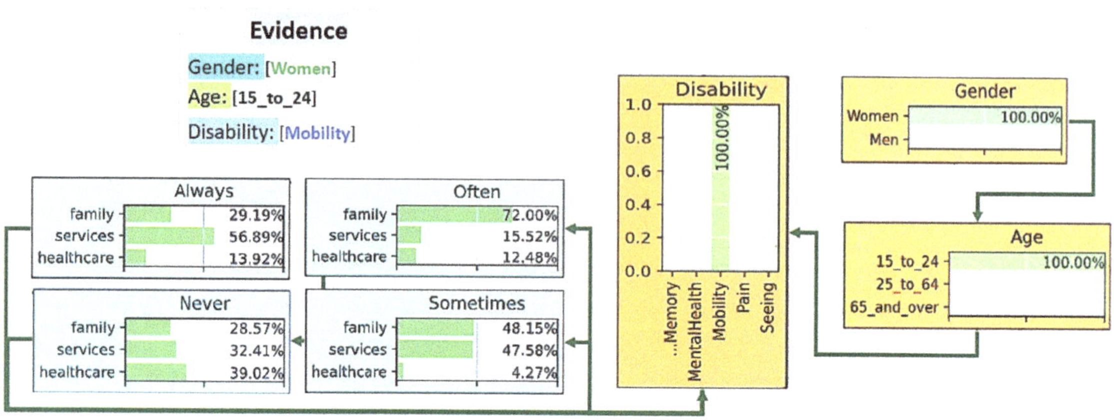

Figure 8. BN inference: probability of encountering barriers to communication via *interaction* with *family, healthcare* and other *services*. Only hard evidence (100% probability of the presence of an attribute) is randomly selected here (rectangles highlighted in orange). The inference of experiencing barriers by Women between the ages of 15 to 24 with *Mobility* disability is illustrated here.

Table 5. Causal inference: overall probability of encountering barriers for persons with *mobility* disabilities.

cy	Attitude (Family)	Attitude (Healthcare)	Attitude (Business)	Alarms	Lighting /Sound	Video Conference	Automated Message	Interaction (Family)	Interaction (Services)	Interaction (Healthcare)
	0.0310	0.0714	0.0432	0.0760	0.0326	0.0503	0.0601	0.0659	0.0668	0.0314
es	0.0334	0.0133	0.0354	0.0036	0.0169	0.0596	0.0182	0.0339	0.0777	0.0787
	0.0409	0.0595	0.0759	0.0105	0.0765	0.0113	0.0564	0.0188	0.0209	0.0630
	0.0950	0.0252	0.0534	0.0260	0.0533	0.0291	0.0088	0.0400	0.0490	0.0278

cy	Signs	Traffic Signals	Online (Booking)	Online (Service)	Door Ways	Side Walks	Wash Rooms	Queue	Phone (Healthcare)	Phone (Family)	Phone (Service)	Self-serve Technology	Social Media
	0.0660	0.0732	0.0462	0.0065	0.0623	0.0238	0.0771	0.0095	0.0358	0.0202	0.0442	0.0042	0.0023
es	0.0777	0.0308	0.0688	0.0733	0.0010	0.0299	0.0087	0.0788	0.0746	0.0188	0.0276	0.0690	0.0705
	0.0646	0.0768	0.0297	0.0120	0.0374	0.0073	0.0603	0.0166	0.0622	0.0376	0.0142	0.0770	0.0709
	0.0218	0.0316	0.0228	0.0445	0.0279	0.0561	0.0625	0.0794	0.0783	0.0650	0.0575	0.0321	0.0130

Results from this inference indicate that attitudinal barriers when dealing with family and friends (*attitude (family)*; Table 5) have the greatest likelihood of occurring among all other barriers for a person with mobility issues. *Queue* and accessing healthcare via phone (*phone (healthcare)*) also show significant probabilities for *always* encountering a barrier.

The overall analysis shows that individuals with conditions related to *mental health* and *pain* are more susceptible to barriers, among other conditions. Moreover, persons with disabilities between the ages of *15 and 24 years* are most likely to have conditions related to *mental health*, while those aged between *25 and 64 years* are most likely to report *pain* and *mobility* disabilities. For persons with disabilities aged over 65 years, BN analysis infers that *men* are 3% more prone to overall barriers compared to *women*.

7. Synthetic Data Generation from BN

The methodology completes with the description of approaches to generating the SD and evaluation of the performance of the BN recreated using the SD. This paper aims to lay the groundwork for future research by introducing a framework to generate SD that mimics the statistical properties of the PD [38]. These data can be not only used for more precise inference and to overcome limited data but also for the protection of any sensitive information about persons with disabilities.

7.1. SD Generation: Purpose and Overview

Synthetic data (SD) generation is the process of creating artificial data that mimic the statistical patterns and properties of real-life data using algorithms, models and other techniques [59]. The goal is to have realistic data, similar to the original data, for a greater sample size. Such data are needed for a more comprehensive analysis and to overcome the scarcity or limitations of real-life data in areas that are usually dependent on survey-based collection.

Moreover, the SD generation method aids in effectively testing the performance of the BN created earlier in Section 6. The data generation process directly in the network allows for observation of the precision of the parameter learning capabilities of the BN.

Since the PD is collected and released occasionally [7], some limitation in analysis or prediction may occur. Nevertheless, merging SD with the PD can help to obtain more accurate CPTs in the BN. This also leads to better inference and a more robust network. Hence, this avenue was explored for future use in a more comprehensive data analysis and for time-series forecasting of accessibility trends. The purpose, framework for data generation and validation are illustrated in Figure 9 and described in the following sections.

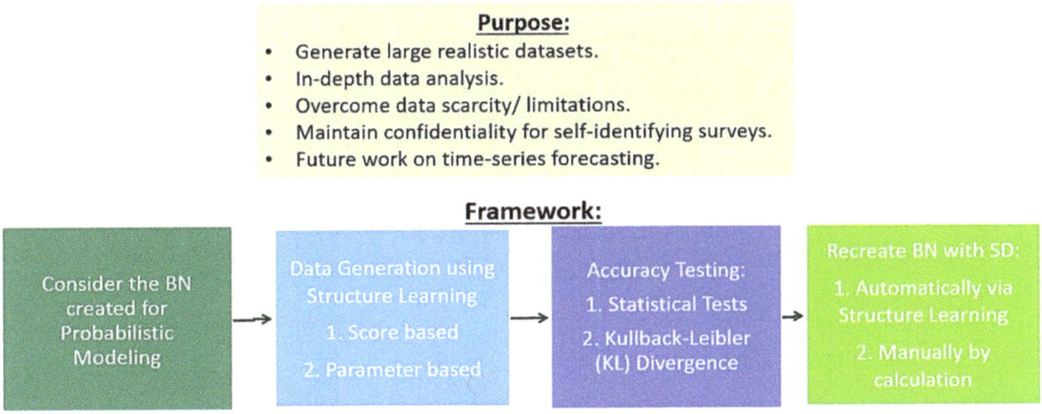

Figure 9. Synthetic data (SD) generation: purpose and methodological process for generative model application and validation.

7.2. Application of the SD Generation Method

The application of the SD generation technique, including the process for data generation and rebuilding of the BN model using the data, is detailed in this section.

7.2.1. Data Generation Process

To create the data, a subset of the PD is considered that only concerns barriers to *interaction* with *family, healthcare* and *services*. For the network in Figure 6, conditional probability were previously calculated and stored using the PD. Hence, these data were utilized to create the SD using the PyAgrum library in Python [54]. The BN was provided as a parameter for the learner in order to obtain the variables and the order of the labels for each variable.

With the network in place, the aim was to generate SD that maintain the statistical characteristics of the PD to achieve greater precision in the analyses [59,60]. The process is outlined in Figure 10. First, parameter and structure learning techniques are applied to the BN to learn the variables and structure of the actual dataset. Parameter learning is the process of using data to learn the BN distributions. Structure learning picks up both the structure and parameters of the provided data [28].

Then, data are generated using the BN by specifying the desired data sample size and variable order. These SD are rearranged, with minor changes to the structure of the dataset to ensure that it completely aligns with the overall structural appearance of the PD. This also helps with the data validation process later.

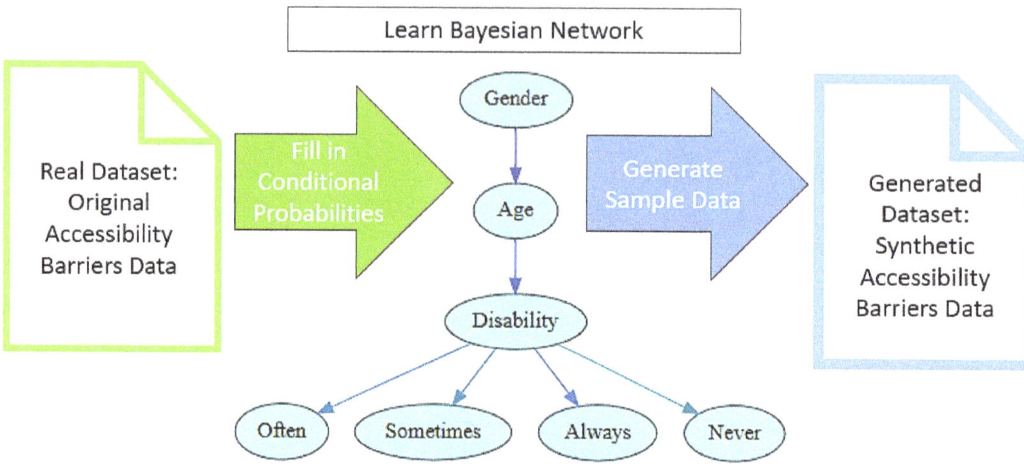

Figure 10. Synthetic data (SD) Generation: purpose and methodological process for the proposed framework.

7.2.2. Recreating the Network

The BN learning process gives access directly to the complete learning algorithm and its parameters, such as prior, scores and constraints [54]. It also facilitates the development of new learning algorithms through statistical tests and the creation of a database from the BN based on the conditional likelihood [54].

A. Automatic recreation via parameter learning:

To recreate the BN from SD, the type of learning algorithms (both parametric and score-based), priors (such as smoothing), constraints, order of the variables and conditional likelihood were specified. This way, the algorithm was able to learn the structure and CPTs of the BN from the provided data.

Structure learning refers to learning of the structure of the DAG from data. We employ this method to estimate the graph by determining how the variables depend on each other based on the SD. Parameter learning is applied when the graph structure is known in order to estimate the factors. A combination of the score-based and constraint-based approach was utilized for this.

- **Score-based learning:** To score and selecting a model, a score-based approach, namely the Bayesian Information Criterion (BIC), was employed. It is a statistical technique to estimate how well a given model fits a dataset and the model's complexity. It is derived from Bayesian probability and is appropriate for models under the MLE framework. The BIC statistic is expressed as follows [61]:

$$BIC = -2 * LL + log(N) * k \quad (5)$$

Here log() has a base-e called the natural logarithm, LL is the log likelihood of the model, N is the number of examples in the training dataset and k is the number of parameters in the model. The score defined above is minimized, e.g., the model with the lowest BIC is selected [61].

- **Constraint-based learning:** For the constraint-based approach, the Multivariate Information-based Inductive Causation (MIIC) is used in recreating the BN. Starting from a complete graph, the method iteratively removes dispensable edges by uncovering significant information from indirect paths. This also assesses edge-specific confidences resulting from the randomization of provided data. Remaining edges are oriented based on the signature of causality in the data [54,62,63].

B. Manual recreation via direct probability calculation:

The direct way to recreate the BN from the SD is to calculate the probabilities for different variable conditions and fill in the CPTs with them for each BN node. The process is similar to how the probabilities are calculated for the CPTs in Figure 6. Once the SD are generated, it is essential to evaluate how well they preserves the PD's characteristics. Hence, the synthetic data generation process is evaluated by comparing specific attributes between these two datasets.

7.3. Accuracy Evaluation of SD Generated by the BN

The experiments and statistical tests performed to validate the accuracy of the SD created from the BN learned from the PD are described in this section.

7.3.1. Using Structure and Parameter Learning Algorithm

The SD generated using the BN in Figure 6 is compared with the PD in Figure 11. Here, adjustment techniques are applied on the automatically recreated BN using the SD introduced earlier in this chapter. The BN was recreated from the SD through structure learning techniques using conditional probabilities. Then, Laplace smoothing was added as a prior to prevent the multiplication of probabilities from resulting in zero, ensuring that no feature was entirely disregarded [54].

SD adjustment: Laplace adjustment or Laplacian smoothing refers to an interactive mesh smoothing process that adjusts parameters such as a the weighting factor and number of iterations to achieve real-time adjustment. It involves utilizing suitable data structures to allow for efficient implementation of the smoothing algorithm [60].

$$P_{i,empirical} = x_i / N \tag{6}$$

$$P_{(i,\alpha-smoothed)} = x_i + \alpha / (N + \alpha d) \tag{7}$$

A notional value (α) weighs into the posterior distribution. If the frequency of each item (i) is x_i out of N samples, the empirical probability of event i is Equation (6). The process of the posterior probability being additively Laplace-smoothed, as if to increase each count (x_i) by α [60], is expressed by Equation (7).

In Figure 11, the red line graph shows the PD, while the dark-green and light-green bar plots illustrate probabilities for all 10 ***disability*** types. The values corresponding to the BN learned from SD both before and after smoothing (Equations (6) and (7), respectively) seem to be closest to the values when the PD was used to create the BN. The prior to smoothing, the bar plots show considerable fluctuations for some disability types, such as *pain*. The value greatly diverges from conditional probabilities in the PD, with a substantial decline from 0.36 to 0.09. In particular, *dexterity* and *flexibility* show probabilities that are evidently higher (4.2% and 6.7%, respectively) than those for the PD.

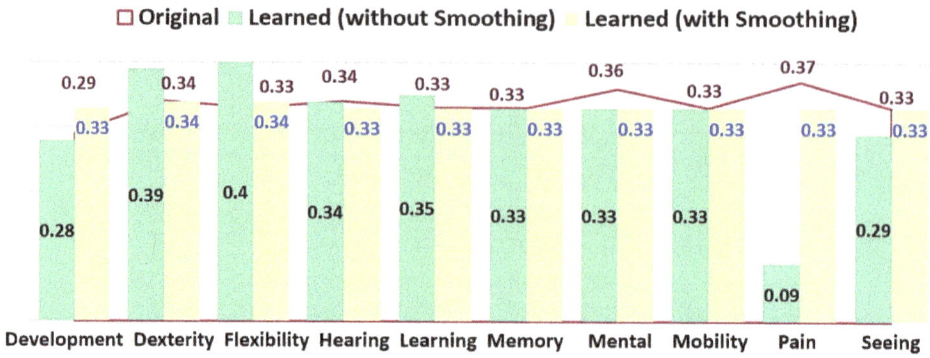

Figure 11. BN recreated from SD: comparison of conditional probabilities for ***always***: *interaction (healthcare)* (Figure 6). Representation: red line plot—PD; dark- and light-green bar plots—SD prior to and post smoothing, respectively.

7.3.2. Hypothesis Testing

Statistical tests are performed to determine if the SD and PD are significantly different from each other. The assumptions that can be made [64] are whether:

- The two samples' data groups are independent;
- Respective data elements follow a normal distribution;
- Samples have equal variance (homogeneity assumption).

The original data include both categorical and numerical variables [40]. In reality, the SD include categorical variables with the same set of values or categories as the PD, but the numerical variable are different, since they are decimal values for probability, which, upon simulation, do not follow the same counts as the PD. Hence, the statistical tests differ based on the variable type. The numerical variables are tested using an independent T-test whereas the categorical data are tested with the Chi-square test. The results from these tests involve two hypotheses, H0 and H1 [64].

Hypothesis 1. *The variables have no relationship (independent).*

Hypothesis 2. *There is a relationship (dependent) between variables.*

All tests performed below use the *statistic* and *p*-value parameters to assess their results, similar to the SEM analysis earlier, where the concept of *statistic* is equivalent to RE in SEM. These tests how the SD conform to the hypothesis of no similarity relative to the original distribution in the PD.

1. Test for Numerical Data: To check the accuracy of the newly generated SD from the BN, we conduct a parametric test called a two-sample T-test in Python with the PD and the SD.

- **Two-sample independent T-test:**

The independent two-sample t-test (referred to as T-test here) allows for confirmation that the means of two datasets are equal [65] using the following formula:

$$t = \frac{M_1 - M_2}{\sqrt{\frac{{SD_1}^2}{n_1} + \frac{{SD_2}^2}{n_2}}} \tag{8}$$

The T-test quantifies the difference between the arithmetic means of the two samples. Equation (8) includes the sample means (M_1 and M_2), variance or squared standard deviations (SD_1 and SD_2) and the sample sizes in the two data groups ($n1$ and $n2$) [66].

Before conducting the T-test, it needed to be checked if the given data groups had the same variance [67]. In Table 6, the quotient of variances is approximately 1, which qualifies for the homogeneity assumption [64]. The *statistic* and p-value parameters were ready to be calculated for numerical data with the condition met for the T-test.

Table 6. Accuracy evaluation for SD—numerical variable.

	Variance (V)			Two-Sample T-Test	
Variable	**o = V (PD)**	**s = V (SD)**	**o/s**	***Statistic***	***p*-Value**
Probability	0.0013	0.0014	1.1228	0.5954	0.0468

The test results in a p-value of 0.0468 in Table 6, which is be;ow the threshold of 0.05. This indicates that the respective distributions of the *Probability* numerical variable in the PD and SD are not equal (first hypothesis (H_0) in Section 7.3.2).

2. Test for Categorical Data: The distribution of a categorical variable in a sample often needs to be compared with the distribution of a categorical variable in another sample to examine their differences [65]. In this study, they are computed to contrast the PD and SD.

- **Pearson's chi-squared test:**

This is a hypothesis test that is used to determine whether there is a significant association between two categorical variables in the data [68]. The equation below expresses the test parameters, where x^2 is chi-square or the *statistic* and O_i and E_i are observed and expected values, respectively.

$$x^2 = \sum(O_i - E_i)^2/E_i \tag{9}$$

All categorical variables from both the PD and SD were accounted for in implementing the chi-square test to check whether there was a relation between the pair of variables with the same name from each dataset. This tests the null hypothesis (assumption) that the variables are independent.

The chi-square statistic is a non-parametric (distribution-free) tool that helps analyze differences within data distributions [69]. Like all non-parametric statistics, the chi-square is robust with respect to the distribution of the data and does not require equality of variance [70].

The results are included in Table 7; the p-values are insignificant for all variables (over the threshold of a p-value < 0.05, the chi-square test hypotheses do not hold, which implies that the values and distributions for the categorical variables are the same for both datasets).

Table 7. Accuracy evaluation for SD—categorical variables.

	Chi-Square Test Parameters				
Parameter	***Gender***	***Age***	***Disability***	***Frequency***	***Barriers***
Statistic	13.3469	18.7759	37.1769	19.6190	33.3070
p-value	0.9942	0.9269	0.1418	0.9044	0.2655

Hypothesis Testing Results: The results of the statistical tests, as included in Tables 6 and 7, indicate that there is significant difference between the two datasets for numerical data, whereas the categorical data are evidently similar. The p-values obtained with these statistical tests also confirm that. This aligns with the rationale that, evidently, the subcategories for each of these variables are exactly the same in both datasets. For instance, the **age** variable only takes on any of the three subcategories from the list of *15_to_24, 25_to_64 and 65_and_over* throughout the dataset in the SD. This is due to this data being generated from the BN that was initially learned from the PD.

7.3.3. Accuracy Estimation with the KL Divergence

To estimate the difference between the SD and PD, Kullback–Leibler (KL) divergence was utilized. It is used in AI as a loss function to compare predicted data with true values [71]. It can also be utilized to compare two BNs' probability distributions when one is a true distribution and the other one is an approximation of it [72].

The KL divergence can be expressed as the below equation to measure the dissimilarity of two probability distributions, P and Q [72].

$$D_{K_L}(P\|Q) = \sum_{x \in X} P(x) log(\frac{P(x)}{Q(x)}) \quad (10)$$

Here, X is a possible event in the probability space; $P(x)$ and $Q(x)$ are probabilities of x in P and Q, respectively; and $P(x)/Q(x)$ represents the relative likelihood of event x according to P in comparison with Q.

Exact and (Gibbs) approximated KL divergence

To compare the initial and previously recreated BN, the exact and Gibbs-approximated KL divergence are considered. Networks for both distributions have the same structure, but the values are different for every CPT.

- **Exact KL divergence:** Here, the brute-force KL divergence is calculated using Equation (10), which yields 0.0712, representing the exact distance between the true and randomly generated probability distribution [54].
- **Gibbs KL divergence:** For the approximated distance, the calculation uses pyAgrum's built-in function for KL divergence using Gibbs inequality [54], which states that the KL divergence is always positive ($KL(F|G) >= 0$) unless $F(x) = G(x)$ [72]. Gibbs-approximated computation of the difference between the BN is depicted in Figure 12. In the computation of the distance for the KL divergence, the equations to convert the absolute value to log and vice versa are:

$$\begin{aligned} y &= log_b x \\ x &= log_b^{-1}(y) = b^y \end{aligned} \quad (11)$$

The divergence or distance between the original BN and the one recreated with the SD is plotted in Figure 12. The time period for each oscillation is 500 ms. Epsilon is a small value to address cases in the KL computation where division by zero occurs to avoid infinite distances.

In the plot, the divergence is calculated for each iteration and expressed on a logarithmic scale. As presented in Equation (11), y is the plotted divergence; b is the base of the logarithm, which is considered to be 10 here; and x is the actual divergence before converting to the log scale. Then, the anti-logarithm (inverse) can be calculated by raising the base (b) to the logarithm (y). In the figure, the longest distance can be seen at around -4.0, which is equivalent to a value of 0.0010. As the distance approaches 0, the iteration comes to a halt in the approximation process in Figure 12 so that the logarithmic value does not become undefined.

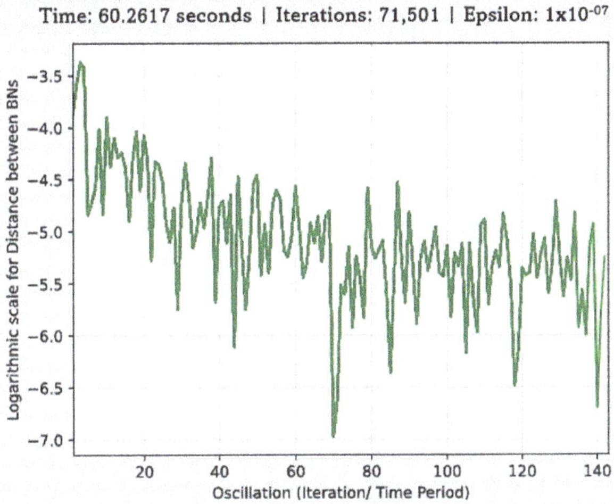

Figure 12. Gibbs -approximated KL divergence: the distance between respective BNs learned from the PD and SD is calculated and plotted here. The number of oscillations is represented on the horizontal axis with a time period of 500 millisecond. The logarithmic scale ($log10$) for the distance is represented by the vertical axis. The approximation stops when the actual distance approaches zero.

The KL divergence quantifies the 'distance' between probability distributions in the two BNs developed from the PD and SD. This is useful to see how much information is lost when one is used to approximate the other, as expressed by Equation 10. This also shows how the quality of parameter learning of the algorithm gets more precise and improved with every iteration.

7.4. Discussion on SD Generation

Applying parameter and structure learning techniques to create the SD helped in exploring the aspect of employing artificial data to explore different scenarios, validate models and test the hypotheses without using real-world data.

The accuracy with which these methods capture the underlying patterns of the real data greatly depends on the data generation method and adjustment techniques. Assessing the quality of the SD based on the selected generation method revealed the areas of improvement for continuous optimization of these BN learning algorithms.

This new dataset was then compared with the original PD through statistical testing to evaluate the accuracy of the former. The conclusion from the tests aligns with the real-life observation that categorical variables are the same in both datasets, while the numerical variables have different values. Moreover, KL divergence was approximated to visualize the distance between the two datasets, which eventually approaches zero after a few iterations, indicating equal distribution.

The synthetic data (SD) were then used to recreate the Bayesian network, and Laplace smoothing was applied to normalize the values. The conditional probabilities before and after the adjustment show that the probabilities learned with synthetic data are closer to the real data after adjustment.

Despite uncertainty around the necessary number of iterations for the algorithm to produce data with 100% similarity to the real data pattern, this method holds significant potential in the expansion and enhancement of insufficient and unbalanced data in any field of research. Overall, the methodology proposed in this paper covers the complete process of implementation and testing for data modeling and analysis.

8. Summary and Discussion

This investigation contributes to the comprehensive understanding of the nature and extent of the demographic–accessibility dynamic. This may be essential for legislators to plan and evaluate policies and programs to achieve social sustainability, a key objective of smart cities, by enabling full participation in society for persons with disabilities.

Each individual has their unique needs that can only be fulfilled by creating barrier-free communities, environments, services, technology and opportunities. Ensuring the removal of existing accessibility barriers and the prevention of new barriers will help realize smart cities and an overall society that are also inclusive. This way, all members of society can be empowered and equipped to reach their full potential in an environment that accommodates everyone, regardless of their diverse needs.

To facilitate the development of such inclusive smart cities, it is imperative that factors causing disadvantage be recognized. Identifying theses crucial areas will accelerate the prioritization and allocation of resources to support those experiencing stress and pressure that come with these disadvantages. Moreover, to efficiently overcome discrimination and inequality, it is necessary that persons with disabilities be integral to the policy- and decision-making process. This study incorporates first-hand accounts from persons with disabilities themselves in the analysis, ensuring a realistic view of accessibility barriers. For decision making via data utilization and analytics, unique computational intelligence tools such as causal modeling are envisioned to be a part of smart cities. In particular, information on probable areas where issues are encountered by each group of individuals with diverse attributes is essential for the effective planning and development of specialized programs.

This study aims to determine the demographic and disability profiles of groups that are significantly impacted by barriers to further assist in risk mitigation. Based on the lived experiences of persons with disabilities, the proposed methodology can help design and model other aspects of the accessibility goals of cities and communities, such as employment disparities, lack of accommodation, social isolation and food security.

8.1. Difference Between Preliminary Investigation and Extended Study

This study utilizes the Canadian Disability Survey from 2022. The last survey cycle was in 2017, which was incorporated in the preliminary investigation for this research [15] on the nature of accessibility barriers linked to an individual's disability type.

While the majority of questions remained the same between the 2017 and 2022 Disability Survey, the content of the 2022 survey was expanded to add several new indicators. Some subject areas were revised and updated, while others were removed, with new survey modules added for 2022 in order to address data gaps and emerging data needs. The CSD 2022 includes information on gender and a total of 25 additional, more specified barrier scenarios. This is reflected in the accessibility barrier dataset in terms of frequency used in this study. The data used for the preliminary investigation based on the 2017 survey only included 11 scenarios. Similarly, the 6 types of disabilities represented in the 2017 survey were more narrowly defined in 2022 to include 10 disabilities.

The 2022 survey also classifies data for each barrier in terms of the frequency of encountering. Confidence intervals in data estimation in both whole numbers and percentage were introduced in the 2022 survey. Due to updating of the sorting criteria and method, this research was able to perform a more rigorous analysis of accessibility experiences compared to the preliminary study.

Unlike a complete framework with interconnected methods and the latest 2024 accessibility data used in this research, the preliminary study developed a computational model where each method is a standalone tool for confirmation of causality using an older dataset from 2021.

8.2. Summary of Experiments and Results

The application of EFA aimed to identify the factors that define the underlying relationships between observed variables. This dimension reduction method confirmed the correlation of data variables and resulted in more manageable data. The number of factors with eigenvalues greater than 1.0 (factors that explain more variance than a single variable) were selected for further analysis with SEM. Here, variance is a measure of how far a set of numbers is spread out its average value. The results were used to define the structure to categorize each observed variable under one of these factors.

Parameter analysis of the SEM tool confirmed that all barriers experienced at different frequencies have significant connections with the type of disability, except for the *online (business)* and *sidewalk (snow)* scenarios. The scenarios 'always' encountered in each barrier category that yielded the highest standard estimates are **communication**—*phone (family)*, **ICT**—*online (booking)*, **transportation**—*Washrooms* and **attitudinal**—*attitude (healthcare)*.

Causal Bayesian inference was performed on the model that was determined by the SEM process, excluding the abovementioned insignificant scenarios. Barriers with the greatest probability under each category turned out to be the same those determined for the SEM above. Overall diagnosis of the demographic attributes depending on reported barriers showed that individuals with *mental health* and *pain-related* disability are the most susceptible to experiencing barriers in general. The intersectionality of these attributes was also considered for inference.

To explore the possibility of using synthetic data for future investigations in cases of inconclusive or limited real-world data, the BN learned from real data served as a template for artificial data generation. By checking the similarities between the real and synthetic distribution samples through statistical tests, it was verified that the latter were different for numerical values but similar for categorical ones, attesting to the expected outcome based on observation of the two datasets. Furthermore, recreating the Bayesian model from synthetic data allowed for the application of adjustment techniques, which visibly improved the resulting conditional probabilities.

The experimental results confirm the initial hypothesis that probabilistic causal reasoning tools such as structural equation modeling and Bayesian networks are feasible methods for understanding varying accessibility experiences attributed to demographics and disability types. This study also explored synthetic data generation techniques with structures identical to real data for more comprehensive analysis, which helped in testing the precision of the probabilistic model.

9. Conclusions and Future Work

It is a human right for all individuals to have access to basic needs so that they can thrive and do more than just survive within a community. Learning about the aspects and areas associated with higher probabilities of facing obstacles can help to relieve anxiety around accessibility issues for persons with disabilities. Hence, the work presented here can significantly benefit individuals with accessibility anxiety by identifying significant scenarios through the modeling of real-world data to enhance preparedness and the predictability of barriers.

For persons with disabilities, it is crucial to be recognized as a vulnerable population, and disability should be regarded in research as another demographic attribute, such as gender or age. Incorporating first-hand accounts such as through surveys to exercise the inclusion of people with disabilities can enhance the scientific integrity of research findings to develop real-world policy applications.

As a stepping stone to address disparities through research, this work proposes a scientific approach for systematic evaluation of accessibility experiences of the millions of people living with disabilities across Canada. The findings can provide a closer view of the degree of accommodation in federal services and transportation that can serve as a guide for prospective studies and detailed identification of the priority areas through causal reasoning. This work also provides insights into accessibility barriers, which are essential in designing inclusive smart cities.

For subsequent work in the future, synthetic data generation techniques can be adopted to expand real-world data to predict accessibility trends and to enable in-depth analysis in any social development research. This research lays the groundwork for projection of the level of accommodation available at any facility using advanced machine learning techniques to help with accessibility anxiety for people with disabilities. The synthetic data generation techniques for the expansion and replacement of real data used here can enable the production of large datasets for time-series forecasting.

This methodological approach incorporating causal modeling of data can be applied in any investigative research regarding vulnerable groups in the equity, diversity and inclusion paradigm. Inference derived from this framework paired with the relevant data can be used to develop and redesign improved assistive technology that would fit the population with special needs. Furthermore, future work could facilitate the development of means and technology for loved ones and caregivers to help and look after individuals with functional limitations.

Through this research, we hope that governments adopting smart city solutions for urban transformation incorporate smart tools and available applications to understand the needs and experiences of patrons in order to make federal sector organizations and services fully accessible. Moreover, frequent and efficient initiatives to share such information with the public would improve not only the quality of government services but also citizen well-being. This avenue of research could serve to streamline inter-organizational collaboration, supporting accessibility initiatives and humanitarian efforts in communities around the world to ensure equal participation through having meaningful options and the freedom to make their own choices for persons with disabilities.

Author Contributions: Conceptualization, M.Z., G.W. and S.Y.; methodology, S.Y.; software implementation, M.Z.; formal analysis, G.W., S.Y. and M.Z.; writing—original draft preparation, M.Z.; writing—review and editing, G.W., S.Y. and M.Z.; visualization, M.Z.; supervision and funding acquisition, S.Y. All authors have read and agreed to the published version of the manuscript.

Funding: This work was supported in part by the Social Sciences and Humanities Research Council of Canada (SSHRC) through the project "Emergency Management Cycle-Centric R&D: From National Prototyping to Global Implementation" with Funding opportunity "New Frontiers in Research Fund" under Grant NFRF-2021-00277.

Institutional Review Board Statement: Not applicable. Data collection was performed as part of the Canadian Disability Survey by the Government of Canada. The authors declare no involvement in the process.

Informed Consent Statement: Not Applicable. The data used in this study were collected and published by Statistics Canada for the 2022 Canadian Survey on Disability. The authors of this paper confirm no involvement in the collection or publication process.

Data Availability Statement: The original dataset used for the case study is publicly available in the Statistics Canada Data hub at https://www150.statcan.gc.ca/t1/tbl1/en/tv.action?pid=1310089901 (accessed on 14 August 2024) and cited here [7].

Conflicts of Interest: The authors declare no conflicts of interest. The funders had no role in the design of the study; in the collection, analyses or interpretation of data; in the writing of the manuscript; or in the decision to publish the results.

Abbreviations

The following abbreviations are used in this manuscript:

AI	Artificial Intelligence
BIC	Bayesian Information Criterion
BN	Bayesian Network
CDF	Cumulative Distribution Function
CPT	Conditional Probability Table
CRPD	Convention on the Rights of Persons with Disabilities
CSD	Canadian Survey on Disability
EFA	Exploratory Factor Analysis
ICT	Information and Communication Technology
KL	Kullback–Leibler
KMO	Kaiser–Meyer–Olkin
ML	Machine Learning
MIIC	Multivariate Information-based Inductive Causation
MLE	Maximum Likelihood Estimation
NA	Not Applicable
PD	Primary Dataset
RE	Regression Estimate
SD	Synthetic Data
SEM	Structural Equation Modeling
SRE	Standardized Regression Estimate

Appendix A

Figure A1 show the alternative structure that was considered for the SEM. The results from the EFA suggest four latent variables for the statistical modeling, which were was not possible with this structure. Therefore, we proceeded with the structure shown in Figure 5 for all modeling and analysis purposes in this article.

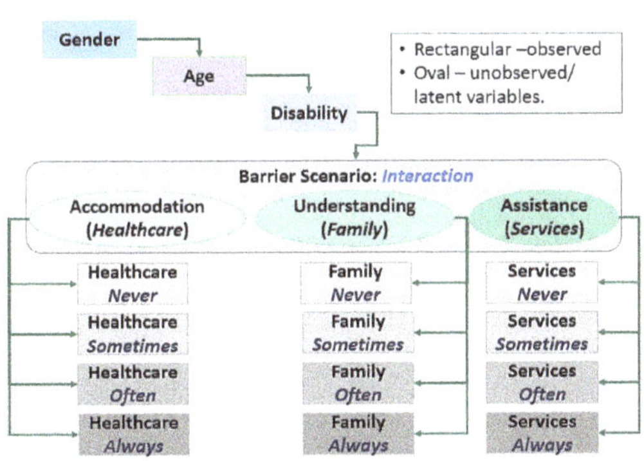

Figure A1. SEM structure 2: Each rectangle is an observed variable with multiple subcategories, whereas the oval shapes represent the latent variables. The latent variables of *accommodation (healthcare), understanding (family) and assistance (services)* represent issues persons with disabilities may experience at frequencies of *never, sometimes, often* or *always*.

Appendix B

The survey questionnaire prepared and provided by the Government of Canada for the CSD 2022 is included Table A1 and a brief description of the different disability types are presented in Table A2 in alignment with the CSD data [73] released by Statistics Canada.

Table A1. CSD 2022 Questionnaire for Barriers encountered by Persons with Disabilities.

Introduction: "We now want you to think about accessibility barriers that might limit your access to different areas and activities in your daily life due to your condition. By barriers we mean something that could be removed, modified, or done differently" [74].		
i. In the past 12 months, have you experienced barriers with any of the following features inside or outside public spaces because of your condition(s)?	A. Entrances or exits to buildings	1 2 3 4 5
	B. Floorplans inside buildings	1 2 3 4 5
	C. Lighting or sound levels inside buildings	1 2 3 4 5
	D. Public washrooms	1 2 3 4 5
	E. Wait lines	1 2 3 4 5
	F. Self-serve technology	1 2 3 4 5
	G. Announcements or alarms	1 2 3 4 5
	H. Signs or directions	1 2 3 4 5
	I. Pedestrian signals at intersections or crosswalks	1 2 3 4 5
	J. Sidewalks when covered in ice or snow	1 2 3 4 5
	K. Sidewalks in general	1 2 3 4 5
ii. In the past 12 months, have you experienced any barriers in terms of behaviours, misconceptions or assumptions made about you from any of the following because of your condition(s)?	A. Family or close friends	1 2 3 4 5
	B. Staff of a business	1 2 3 4 5
	C. Medical or health care professionals	1 2 3 4 5
	D. Staff of government services or programs	1 2 3 4 5
iii. In the past 12 months, have you experienced barriers communicating in any of the following situations because of your condition(s)?	A. In person with family or close friends	1 2 3 4 5
	B. In person with medical or health care professionals	1 2 3 4 5
	C. In person with others such as the general public, customer service or government representatives	1 2 3 4 5
	D. By voice over the phone with family or close friends	1 2 3 4 5
	E. By voice over the phone with medical or health care professionals	1 2 3 4 5
	F. By voice over the phone with others such as the general public, customer service or government representatives	1 2 3 4 5
	G. With an automated phone message system	1 2 3 4 5
	H. Using video conferencing	1 2 3 4 5
	I. Using social media or online chat forums to interact with others	1 2 3 4 5
iv. In the past 12 months, have you experienced barriers using the Internet for any of the following activities because of your condition(s)?	A. Online banking or online shopping	1 2 3 4 5
	B. Online booking for appointments, services or reservations	1 2 3 4 5
	C. Online access to government information, services or supports	1 2 3 4 5

1 : not applicable; 2 : never; 3 : sometimes; 4 : often; 5 : always.

Table A2. Overview and Accessibility Recommendation for various types of Disabilities.

Disability	**Description of Associated Difficulties**	**Recommended Accessibility Practices to be Incorporated**
Developmental	Communicating or understanding written or spoken language, overlap with learning disability.	Reduce communication barriers by speaking at a steady pace, using clear, short sentences; avoid words or phrases that have multiple meanings.
Hearing	Hearing and how well a person hears or whether they hear at all.	Information in sign language; assistive listening devices; proper sound and lighting; closed captioning for videos and slide shows.
Mobility	Ability to move around, walk or use stairs.	Option to stand or sit at events and extra time to complete documents.
Flexibility	Ability to stand, pick things up, reach out/ upwards.	Accommodation at events; assistive Web features; speaking-based participation on videoconferencing platforms; Accessible parking
Dexterity	Affects way of using fingers, wrist, any extremities.	Assistive devices to perform tasks; Accessible sidewalks and restrooms.
Learning	Affects ability to learn, read, write, process information, organize or manage time.	Reduce auditory and visual distractions; providing information in simple, clear and concise language and in alternate formats.
Memory	Difficulty processing or retaining information.	Similar to learning and developmental disabilities
Mental health	Difficulty managing emotions, thoughts and sensations in different ways (usually not visible).	Avoid sudden noises, changes in lighting and other things that could trigger anxiety; Set aside a quiet, private room to sit or take medication.
Pain-related	Chronic pain that may fluctuate unpredictably over time, disrupting daily activities.	Adjust light, sound and seating arrangements; prohibit strong-smelling products; be aware of specific needs and conditions.
Seeing	Difficulty in seeing light, color, distance or texture or seeing at all.	Provide audio versions of Web pages and electronic documents; ensure color contrast between text and backgrounds.

References

1. Gossett, A.; Mirza, M.; Barnds, A.K.; Feidt, D. Beyond access: A case study on the intersection between accessibility, sustainability, and universal design. *Disabil. Rehabil. Assist. Technol.* **2009**, *4*, 439–450. [CrossRef] [PubMed]
2. Chen, T.; Ramon Gil-Garcia, J.; Gasco-Hernandez, M. Understanding Social Sustainability for Smart Cities: The Importance of Inclusion, Equity, and Citizen Participation as Both Inputs and Long-term Outcomes. *J. Smart Cities Soc.* **2022**, *1*, 135–148. [CrossRef]
3. Wolbring, G.; Rybchinski, T. Social Sustainability and Its Indicators through a Disability Studies and an Ability Studies Lens. *Sustainability* **2013**, *5*, 4889. [CrossRef]
4. Vallance, S.; Perkins, H.C.; Dixon, J.E. What is social sustainability? A clarification of concepts. *Geoforum* **2011**, *42*, 342–348. [CrossRef]
5. Rebernik, N. Smart and Sustainable Cities Inclusive of People with Disabilities: An Exploratory Study Using Digital Tools and Participatory Practices. Ph.D. Thesis, University of Deusto, Bilbo, Spain, 2020.
6. Makkonen, T.; Inkinen, T. Inclusive smart cities? Technology-driven urban development and disabilities. *Cities* **2024**, *154*, 105334. [CrossRef]
7. Pianosi, R.; Presley, L.; Buchanan, J.; Lévesque, A.; Savard, S.A.; Lam, J. *Canadian Survey on Disability, 2022: Concepts and Methods Guide*; Technical report; Government of Canada: Ottawa, ON, Canada, 2023.
8. Statistics Canada. *Canadian Survey on Disability, 2017 to 2022*; Technical report; Government of Canada: Ottawa, ON, Canada, 2023.
9. Statistics Canada. *About an Accessible Canada*; Technical Report; Government of Canada: Ottawa, ON, Canada, 2023.
10. Zhou, S.; Loiacono, E.T.; Kordzadeh, N. Smart Cities for People with Disabilities: A Systematic Literature Review and Future Research Directions. *Eur. J. Inf. Syst.* **2023**, *33*, 845–862. [CrossRef]
11. Sugandha.; Freestone, R.; Favaro, P. The social sustainability of smart cities: A conceptual framework. *City Cult. Soc.* **2022**, *29*, 100460. [CrossRef]
12. City of Toronto. 2019. Available online: https://www.toronto.ca/city-government/accountability-operations-customer-service/long-term-vision-plans-and-strategies/connected-community/ (accessed on 24 November 2024).
13. United Nations. *Article 1—Purpose*; Technical Report; Department of Economic and Social Affairs: New York, NY, USA, 2016.
14. Grecoa, A. An introduction to causal inference. *EGE Rev. Expr. Graf. Edif.* **2020**, *6*, 18–27. [CrossRef]
15. Zakir, M.; Wolbring, G.; Yanushkevich, S. A Causal Approach to Investigating Accessibility Experiences of Persons with Disabilities. In Proceedings of the 2024 Sixth International Conference on Intelligent Computing in Data Sciences (ICDS), Marrakech, Morocco, 23–24 October 2024; pp. 1–8. [CrossRef]
16. Statistics Canada. *Canadian Survey on Disability, 2017*; Technical report; Government of Canada: Ottawa, ON, Canada, 2018.
17. Gkontzis, A.F.; Kotsiantis, S.; Feretzakis, G.; Verykios, V.S. Enhancing Urban Resilience: Smart City Data Analyses, Forecasts, and Digital Twin Techniques at the Neighborhood Level. *Future Internet* **2024**, *16*, 47. [CrossRef]
18. Gkontzis, A.F.; Kalles, D.; Paxinou, E.; Tsoni, R.; Verykios, V.S., A Big Data Analytics Conceptual Framework for a Smart City: A Case Study. In *Building on Smart Cities Skills and Competences*; chapter Internet of Things; Springer: Cham, Switzerland, 2022; pp. 259–279. [CrossRef]
19. Lim, C.; Kim, K.J.; Maglio, P.P. Smart cities with big data: Reference models, challenges, and considerations. *Cities* **2018**, *82*, 86–99. [CrossRef]
20. Bringolf, J. The smart city and disability. In *Research Papers for Built Environment*; Academic Paper; Centre for Universal Design Australia: Moorebank, NSW, Australia, 2024. Available online: https://universaldesignaustralia.net.au/the-smart-city-and-disability/ (Accessed on 2 December 2024).
21. People with Disability Australia. Social Model of Disability. Available online: https://pwd.org.au/resources/models-of-disability/ (accessed on 4 October 2024).
22. Hosseinpoor, A.R.; Bergen, N.; Kostanjsek, N.; Kowal, P.; Officer, A.; Chatterji, S. Socio-demographic patterns of disability among older adult populations of low-income and middle-income countries: Results from World Health Survey. *Int. J. Public Health* **2015**, *61*, 337–345. [CrossRef] [PubMed]
23. Statistics Canada. *World Report on Disability*; Technical Report; World Health Organization: Geneva, Switzerland, 2011.
24. Wolbring, G.; Lillywhite, A. Burnout through the Lenses of Equity/Equality, Diversity and Inclusion and Disabled People: A Scoping Review. *Societies* **2023**, *13*, 131. [CrossRef]
25. Church, R.L.; Marston, J.R. Measuring Accessibility for People with a Disability. *Geogr. Anal.* **2010**, *35*, 83–96. [CrossRef]
26. Americans with Disabilities Act (ADA). *ADA Standards for Accessible Design*; Technical Report; U.S. Department of Justice: New York, NY, USA, 2010.
27. Wald, M. AI Data-Driven Personalisation and Disability Inclusion. *Front. Artif. Intell.* **2021**, *3*, 571955. [CrossRef] [PubMed]
28. Kitson, N.K.; Constantinou, A.C. Learning Bayesian networks from demographic and health survey data. *J. Biomed. Inform.* **2021**, *113*, 103588. [CrossRef]

29. Tarka, P. An overview of structural equation modeling: Its beginnings, historical development, usefulness and controversies in the social sciences. *Qual. Quant.* **2018**, *52*, 313–354. [CrossRef]
30. Abrahim, S.; Mir, B.A.; Suhara, H.; Mohamed, F.A.; Sato, M. Structural equation modeling and confirmatory factor analysis of social media use and education. *Int. J. Educ. Technol. High. Educ.* **2019**, *16*, 32. [CrossRef]
31. Abid, N.; Pond, L.; Yanushkevich, S. Causality Exploration in Modeling Engineering Student Satisfaction. In Proceedings of the 2024 IEEE Global Engineering Education Conference (EDUCON), Kos Island, Greece, 8–11 May 2024; pp. 1–10. [CrossRef]
32. Diez-Mesa, F.; de Oña, R.; de Oña, J. Bayesian Networks and Structural Equation Modelling to develop service quality models: Metro of Seville case study. *Transp. Res. Part Policy Pract.* **2018**, *118*, 1–13. [CrossRef]
33. Lai, K.; Yanushkevich, S. Causal Models Applied to the Patterns of Human Migration due to Climate Change. In Proceedings of the 2023 IEEE Symposium Series on Computational Intelligence, Mexico City, Mexico, 5–8 December 2023; pp. 1644–1649.
34. Gupta, S.; Kim, H.W. Linking structural equation modeling to Bayesian networks: Decision support for customer retention in virtual communities. *Eur. J. Oper. Res.* **2008**, *190*, 818–833. [CrossRef]
35. Oliveira, H.C.R.; Yanushkevich, S.; Almekhlafi, M. Sensitivity Analysis of Stroke Predictors Using Structural Equation Modeling and Bayesian Networks. In Proceedings of the 2022 IEEE Conference on Computational Intelligence in Bioinformatics and Computational Biology (CIBCB), Ottawa, ON, Canada, 15–17 August 2022; pp. 1–8. [CrossRef]
36. Shaposhnyk, O.; Lai, K.; Wolbring, G.; Shmerko, V.; Yanushkevich, S. Next Generation Computing and Communication Hub for First Responders in Smart Cities. *Sensors* **2024**, *24*, 2366. [CrossRef]
37. Balsa-Barreiro, J.; Cebrián, M.; Menéndez, M.; Axhausen, K. Leveraging Generative AI Models in Urban Science. *Curr. Top. Behav. Neurosci.* **2024**, *23*, 143–149. [CrossRef]
38. Statistics Canada. *Barriers to Accessibility for Persons with Disabilities Aged 15 Years and Over, by Disability Type, Age Group and Gender*; Technical Report; Government of Canada: Ottawa, ON, Canada, 2024.
39. Ghahramani, Z. Probabilistic machine learning and artificial intelligence. *Nature* **2015**, *521*, 452–459. [CrossRef] [PubMed]
40. Statistics Canada. Statistics: Power from Data! Available online: https://www150.statcan.gc.ca/n1/edu/power-pouvoir/toc-tdm/5214718-eng.htm (accessed on 15 September 2024).
41. Babu, D. Machine Learning Exercises. Available online: https://github.com/Dhamodaran-Babu/Machine-Learning-Exercises/tree/master/12.Factor%20Analysis (accessed on 15 July 2024).
42. University of Toronto. Exploratory Factor Analysis. Available online: https://www.utstat.toronto.edu/~brunner/oldclass/431s23/textbook/Ch2.pdf (accessed on 18 August 2024).
43. Crossman, A. Principal Components and Factor Analysis. 2023. Available online: https://www.thoughtco.com/principal-factor-analysis-3026699 (accessed on 5 November 2024).
44. Swami, V.; Maïano, C.; Morin, A.J. A guide to exploratory structural equation modeling (ESEM) and bifactor-ESEM in body image research. *Body Image* **2023**, *47*, 101641. [CrossRef] [PubMed]
45. Kaiser, H.F.; Rice, J. Little jiffy, mark IV. *Educ. Psychol. Meas.* **1974**, *34*, 111–117. [CrossRef]
46. Yoo, C.; Oh, S. Combining Structure Equation Model with Bayesian Networks for predicting with high accuracy of recommending surgery for better survival in Benign prostatic hyperplasia patients. In Proceedings of the 20th International Congress on Modelling and Simulation, MODSIM, Adelaide, Australia, 1–6 December 2013.
47. Violato, C.; Beran, T.N. Structural Equation Modeling in Medical Research: A primer. *BMC Res. Notes* **2010**, *3*, 267. [CrossRef]
48. Igolkina, A.A.; Meshcheryakov, G. Semopy: A Python Package for Structural Equation Modeling. *Struct. Equ. Model. Multidiscip. J.* **2020**, *27*, 952–963. [CrossRef]
49. Siegel, A.F.; Wagner, M.R. Chapter 12—Multiple Regression: Predicting One Variable From Several Others. In *Practical Business Statistics*, 8th ed.; Academic Press: Cambridge, MA, USA, 2022; pp. 371–431. [CrossRef]
50. Dahiru, T. P-value, a true test of statistical significance? A cautionary note. *Ann. Ib. Postgrad. Med.* **2008**, *6*, 21–26. [CrossRef]
51. Hebert, S.; Lee, V.; Morabito, M.; Polan, J. 13.5: Bayesian Network Theory. 2023. Available online: https://eng.libretexts.org/Bookshelves/Industrial_and_Systems_Engineering/Chemical_Process_Dynamics_and_Controls_(Woolf)/13%3A_Statistics_and_Probability_Background/13.05%3A_Bayesian_network_theory (accessed on 28 August 2024).
52. Albright, E.A. Probability: Joint, Marginal and Conditional Probabilities. 2024. Available online: https://sites.nicholas.duke.edu/statsreview/jmc/ (accessed on 15 October 2024).
53. Kendall's Advanced Theory of Statistics, D.T. *Stuart, Alan and Ord, Keith*; John Wiley & Sons: Hoboken, NJ, USA, 2010.
54. Ducamp, G.; Gonzales, C.; Wuillemin, P.H. aGrUM/pyAgrum: A Toolbox to build models and algorithms for Probabilistic Graphical Models in Python. In Proceedings of the 10th International Conference on Probabilistic Graphical Models, PMLR, Online, 23–25 September 2020; Jaeger, M., Nielsen, T.D., Eds.; Volume 138, pp. 609–612.
55. Hayes, A. Posterior Probability: Definition, Formula for Calculation. 2023. Available online: https://www.investopedia.com/terms/p/posterior-probability.asp (accessed on 28 September 2024).
56. Pearl, J. An introduction to causal inference. *Int. J. Biostat.* **2010**, *6*, 7. [CrossRef]

57. Fenton, N.; Neil, M. *Risk Assessment and Decision Analysis with Bayesian Networks*; Chapman and Hall/CRC: Boca Raton, FL, USA, 2018. [CrossRef]
58. Bours, M.J. Bayes' rule in diagnosis. *J. Clin. Epidemiol.* **2021**, *131*, 158–160. [CrossRef]
59. Ali, M. Synthetic Data is the Future of Artificial Intelligence. 2023. Available online: https://moez-62905.medium.com/synthetic-data-is-the-future-of-artificial-intelligence-6fcfd2ce1a14 (accessed on 2 September 2024).
60. Preim, B.; Botha, C. *Visual Computing for Medicine*, 2nd ed.; Morgan Kaufmann: Boston, MA, USA, 2014; p. iii. [CrossRef]
61. Brownlee, J. Probabilistic Model Selection with AIC, BIC, and MDL. 2020. Available online: https://machinelearningmastery.com/probabilistic-model-selection-measures/ (accessed on 15 September 2024).
62. Tocci, T.; Cabeli, V; Li, H; Simon, F. GitHub—miicTeam/miic_R_package: Learning Causal or Non-Causal Graphical Models Using Information Theory. Available online: https://github.com/miicTeam/miic_R_package (accessed on 25 September 2024).
63. Verny, L.; Sella, N.; Affeldt, S.; Singh, P.; Isambert, H. Learning causal networks with latent variables from multivariate information in genomic data. *PLoS Comput. Biol.* **2017**, *13*, e1005662. [CrossRef]
64. Fein, E.C.; Gilmour, J.; Machin, T.; Hendry, L. *Statistics for Research Students*; University of Southern Queensland: Toowoomba, Australia, 2022.
65. Campbell, M.J. *Statistics at Square One*; BMJ Publication Group: London, UK, 1998.
66. Price, P.C.; Chiang, I.C.A.; Jhangiani, R.S. *Research Methods of Psychology—2nd Canadian Edition*; BCcampus: Victoria, BC, Canada, 2015.
67. Reddy, T. Statistical Functions (scipy.stats). 2024. Available online: https://docs.scipy.org/doc/scipy/reference/generated/scipy.stats.ttest_ind.html (accessed on 15 July 2024).
68. Pearson, K.X. On the criterion that a given system of deviations from the probable in the case of a correlated system of variables is such that it can be reasonably supposed to have arisen from random sampling. *Lond. Edinb. Dublin Philos. Mag. J. Sci.* **1900**, *50*, 157–175. [CrossRef]
69. Light, C. Tutorial: Pearson's Chi-Square Test for Independence. 2008. Available online: https://www.ling.upenn.edu/~clight/chisquared.htm (accessed on 15 September 2024).
70. McHugh, M.L. The chi-square test of independence. *Biochem. Med.* **2023**, *23*, 143–149. [CrossRef] [PubMed]
71. Murphy, K.P. *Machine Learning: A Probabilistic Perspective*; The MIT Press: Cambridge, MA, USA, 2012.
72. Moral, S.; Cano, A.; Gómez-Olmedo, M. Computation of Kullback-Leibler Divergence in Bayesian Networks. *Entropy* **2021**, *23*, 1122. [CrossRef] [PubMed]
73. Employment and Social Development Canada. *Annex: Understanding Disabilities*; Technical Report; Goverment of Canada: Ottawa, ON, Canada, 2022.
74. Statistics Canada. Canadian Survey on Disability—2022: Questionnaire. Available online: https://www23.statcan.gc.ca/imdb/p3Instr.pl?Function=assembleInstr&lang=en&Item_Id=1400719#qb1403532 (accessed on 15 August 2024).

smart cities

MDPI

Article

Game-Theoretic Analysis of Policy Impacts in Competition Between Reverse Supply Chains Involving Traditional and E-Channels

Asra Aghaei [1,2], Fulin Cai [1,2] and Teresa Wu [1,2,*]

1 School of Computing and Augmented Intelligence, Arizona State University, Tempe, AZ 85281, USA; aaghaei@asu.edu (A.A.); fulin.cai@asu.edu (F.C.)
2 ASU-Mayo Center for Innovative Imaging, Arizona State University, Tempe, AZ 85287, USA
* Correspondence: teresa.wu@asu.edu

Highlights:

What are the main findings?

- This study explores the competition among reverse logistic-based supply chains to assess the impact of policies such as incentive pricing, return quality standards, advertising and transportation investments, and processing times on economic and environmental objectives.
- A hybrid supply chain, utilizing both traditional and e-channels, attains higher return rates and profitability, and both hybrid and traditional supply chains reach optimal profitability with traditional SC leadership, underscoring its effectiveness in achieving economic objectives.

What is the implication of the main finding?

- The strategic advantage of integrating traditional and e-channels in a supply chain can enhance economic and environmental outcomes in smart cities.
- Through adjusting policies such as incentive pricing, return quality level, advertising investments, transportation investments, and processing time, a balanced economic and environmental outcome can be achieved in sustainable practices.

Abstract: Smart cities aim to enhance the quality of life by advancing infrastructure, leveraging technology, and promoting sustainability, balancing environmental, societal, and economic needs for long-term efficiency. Given resource scarcity and environmental regulations, advanced supply chains play a crucial role in developing smart cities by adopting the circular economy concept, which emphasizes maximizing resource efficiency through recycling and remanufacturing. This study delves into the competition between two types of supply chains in the context of reverse logistics: the hybrid supply chain, which utilizes a dual channel including traditional and e-channels for collecting used products, and the traditional supply chain, which relies solely on a traditional channel. Both supply chain models are actively involved in remanufacturing and recycling used products, and each considers varied policies, including incentive-based policies, advertising investments, the acceptance return quality level, the return processing time, and transportation investments, to enhance their performance. Specifically, this research has two primary objectives: (1) evaluating the economic and environmental outcomes across three competitive scenarios, and (2) analyzing the impact of varied policy settings on these outcomes. These objectives frame the analysis of optimal incentive values, return rates, and profitability across the Nash equilibrium and Nash–Stackelberg structures, providing insights into both the strategic and policy dimensions of supply chain operations. The findings indicate that a hybrid supply chain in this case achieves higher return rates and profitability, highlighting the success of its dual-channel strategy and associated policies. Regarding economic goals, both supply chains achieve the highest profits under the Nash–Stackelberg traditional supply chain leadership structure. However, for environmental goals, the traditional supply chain favors Nash equilibrium for higher return rates, while the hybrid supply chain prefers Nash–Stackelberg with traditional leadership. These scenario-specific results emphasize the importance of aligning economic and environmental goals through tailored strategies. A sensitivity analysis, supported by Pareto prioritization, identifies the return quality level and processing time as critical for the hybrid supply chain, and advertisement investments and the return processing time as key for the traditional supply chain. These insights suggest that H-SCs should prioritize stricter quality standards, efficient inspection protocols, and automation (e.g., AI or optical scanning) to improve the quality and processing time efficiency. Meanwhile, T-SCs should focus on advertising traditional channels by emphasizing faster processing time and less restrictive quality standards, while adopting automated time management strategies similar to H-SCs to enhance engagement and profitability. These findings show that by integrating smart city internet-based initiatives and managing related policies, supply chains can enhance circular economy objectives by optimizing both the economic and environmental performance, ultimately fostering more resilient and sustainable supply chains.

Keywords: circular economy; reverse supply chain; game theory; collecting channels

1. Introduction

Smart cities balance competitiveness and sustainability by leveraging technology, smart growth, and knowledge-driven approaches to enhance resource efficiency and quality of life [1]. Two key dimensions of these encompass smart economy (productivity), and smart environment (resource management) [2]. Central to this vision is the circular economy (CE), which aims to optimize resource usage, minimize waste, and extend product life cycles, thereby supporting sustainable development and enhancing economic, social, and environmental value [3,4]. The concept of CE has been integrated into the supply chain (SC), enabling a shift from a linear economy to a circular model [5]. This transition impacts SC by promoting sustainability through product design focused on longevity and recyclability [6]. In contrast to the traditional linear economy, CE emphasizes the reuse, repair, refurbishment, and recycling of existing materials, which help reduce operational costs, generate revenue, and lower environmental footprints [7,8]. At the core of this transition is reverse logistics (RL), a strategic process that manages the return journey of goods from the point of consumption back to the point of origin [9]. This process aims to recapture value and ensure the proper disposal of used products, thereby supporting sustainability and minimizing environmental impact [10]. By optimizing material flows, reducing waste, and enhancing resource efficiency, reverse logistics strengthens economic resilience while contributing to the development of smart, sustainable cities. Across industries, leading companies such as Microsoft, Panasonic, Apple, and Huawei in electronics; Best Buy and Amazon in retail; and Mac, Eileen Fisher, and H&M in fashion have implemented RL systems to enhance their operations. The critical role of well-designed RL systems has been highlighted in improving efficiency and aligning with CE objectives, further emphasizing the importance of this process in driving both environmental and economic benefits [11,12]. Within the RL process, products that do not meet remanufacturing standards are either recycled or disposed of, while those that do are remanufactured. Recycling involves extracting critical and valuable components from used products for production while remanufacturing refurnishes products and restores the original raw materials [13]. Much of the added value from the initial production process is thus preserved in remanufacturing, resulting in significant cost savings [14] and energy conservation [15]. While many existing studies primarily focus on recycling processes within RL, the dual consideration of both recycling and remanufacturing remains underexplored, despite their combined potential to enhance SC sustainability and operational effectiveness, as demonstrated by Dell's partnership with Goodwill [16]. An essential process in RL is the collection system, where an SC is responsible for collecting used products. This sub-process includes acquiring them from consumers, inspecting products, and transporting them to remanufacturing systems [17]. There are three main channels for collecting used products from consumers: traditional channels (Figure 1a), e-channels (Figure 1b), and third-party collecting channels [18,19]. In the traditional channel, retailers inspect the used devices in retail stores. Immediate incentives are paid for eligible devices, while others, which are ineligible for remanufacturing, are offered recycling without incentives [20]. The traditional channel provides direct interaction and immediate feedback for consumers who prefer in-person engagement. On the other hand, the e-channel offers an online questionnaire to estimate initial incentives on the manufacturer's website. The products are then shipped to and inspected at the inspection center. After a thorough inspection, consumers receive a final incentive offer. If they accept the offer, the process continues; if they decline, the used product is shipped back to them by the carrier. Despite the rise of e-channels, the integration of traditional and e-channels in SCs from environmental and economic perspectives remains underexplored, highlighting the need for further analysis to optimize RL and efficiency. The third-party collection channel relies on external expertise and infrastructure to collect used products. However, employing a third-party company may increase overall costs and introduce potential inefficiencies in the SC compared to e-channels and traditional channels owned by the manufacturer [21]. Therefore, this study focuses on the traditional channel and the e-channel structures as key components of its competitive SCs.

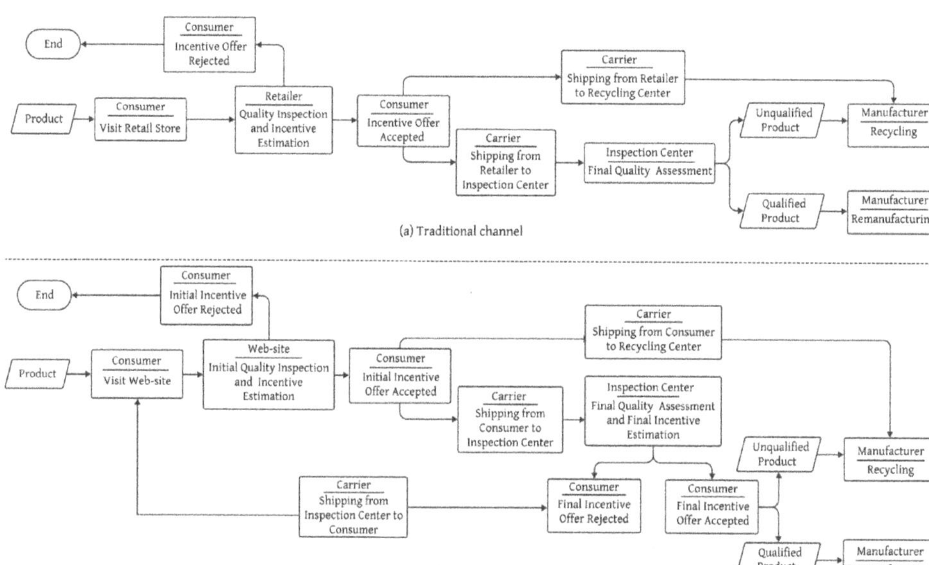

Figure 1. The illustration of reverse logistics processes in (**a**) traditional and (**b**) e-channels.

For an efficient RL-based channel, five critical policies are identified as pivotal facilitators: (1) Price: incentive-based policies that offer discounts, credits, or other forms of compensation for returned items, effectively motivating consumers to participate in the return process [19,22–25]; (2) Quality: acceptance return quality level for remanufacturing, an essential policy associated with RL's economic and environmental values and customer experience [26,27]; (3) Advertisement: advertisement investments that raise awareness of return programs, encouraging consumer participation and building loyalty [23,24,28]; (4) Service: transportation services, such as free shipping, which enhance participation motivation and profitability through reduced consumer costs [23,29]; (5) Time: efficient return processing times linked to shipment, inspection, and refunding, which significantly impact customer satisfaction. Existing studies acknowledge the impacts of these policies, by examining various subsets of them [23,24,27]; however, to the best of our knowledge, no research has comprehensively examined the combined interactions among all five policies. This study presents a model that incorporates all five policies. This model consists of two distinct two-level SCs, each comprising a manufacturer and a retailer. One SC is implemented based on a dual channel (both the traditional and e-channels) for collecting used items, referred to as a hybrid supply chain (H-SC). This structure has been proven as a superior approach compared with the single-channel approach [12,30–32]. In contrast, the other relies solely on the traditional channel, which is called a traditional supply chain (T-SC). Solely online channels, while relevant in specific contexts, are less common, reduce model generalizability, and, also, they introduce complexities beyond the scope of this study, such as limited data on consumer preferences [33]. The two SCs compete to improve the profit and return rate of used products for recycling and remanufacturing by offering different policies and utilizing distinct structures.

To facilitate this competition, each SC has established a centralized decision-making structure that prioritizes the overall profit of the entire SC rather than individual members. Game-theoretic scenarios are used to analyze the economic and environmental performance of these SCs under varying power dynamics. Specifically, we consider three scenarios: Nash equilibrium, representing a peer-to-peer relationship where both SCs have equal power, and Nash–Stackelberg models, which examine leader–follower dynamics when the H-SC acts as the leader and the T-SC as the follower, and vice versa.

As smart cities embrace digital transformation, this study explores how leveraging its initiatives such as internet-based channels (e-channel), alongside key policies including incentive pricing, return quality standards, advertising investments, transportation investments, and return processing times can enhance the SC efficiency in achieving both economic (profitability) and environmental (return rate) objectives within a competitive landscape. This analysis aims to assess how smart city implementation can further strengthen CE goals by optimizing resource utilization, reducing waste, and improving sustainability.

The key contributions of this study are threefold:

- Game-theoretic insights and competitive dynamics in SCs: This study provides a detailed examination of the competitive interactions and performances of H-SCs and T-SCs, comparing the advantages of dual-channel and single-channel structures in the context of RL. By evaluating these dynamics through both Nash equilibrium and Nash–Stackelberg frameworks, it offers a deeper understanding of how strategic decision-making and policy adjustments influence both economic (profitability) and environmental (return rate) performance under different power structures.
- Integrated remanufacturing and recycling with policy-driven incentive models: This research integrates remanufacturing and recycling, moving beyond conventional studies that have primarily examined these processes separately, to demonstrate their combined impact on advancing CE goals. Additionally, it introduces a policy-driven

approach by incorporating acceptance return quality levels, advertising investments, transportation investments, and return processing times into the incentive pricing model. This novel contribution extends to traditional models and highlights how policy variations influence the SC performance.

- Real-world validation and managerial implications: This study employs real-world data from Apple, which closely aligns with the proposed models, thereby enhancing their validity and practical relevance. This strong alignment underscores the applicability of our insights, providing valuable strategic guidance for managers in large-scale SCs.

This study is structured with Section 2 covering the relevant literature. Section 3 introduces the proposed models and outlines the underlying assumptions. Section 4 presents the computational results, while Section 5 provides a detailed sensitivity analysis. Finally, Section 6 concludes the findings and discusses implications for future research.

2. Literature Review

The CE concept aims to minimize material consumption by recovering and reusing materials, enhancing resource efficiency, and promoting sustainable consumption [6,34]. The concept of CE has been extensively applied in multiple SC research studies [35]. The influential factors considered in these studies can be categorized into RL processes, channels, and policies. Table 1 summarizes a literature review focusing on channels, policies, and RL processes. Channels are categorized into traditional channels, e-channels, and third-party channels, alongside policies such as price, quality, service, and advertisement. Notably, to the best of our knowledge, processing time has not been thoroughly investigated in existing literature. This gap highlights the novelty of this study, as it explicitly incorporates processing time as a critical decision-making factor. Thus, four policies, including price, quality, advertisement, and services, are included. In addition to channels and policies, Table 1 includes RL processes such as recycling and remanufacturing.

Table 1. Summary of literature review. Channels: traditional channel (TC)/e-channel (EC)/third-party channel (ThC); policies: price (P)/quality (Q)/advertisement (A)/service (S); RL processes: recycling (RC)/remanufacturing (RM); —: not included in the study.

Ref.	Channels	Policies	RL Processes
[36]	TC/—/—	P/Q/—/—	—/RM
[37]	TC/EC/—	P/—/—/—	—/RM
[23]	TC/—/—	P/—/S/A	—/RM
[38]	TC/—/—	P/—/S/A	—/RM
[39]	TC/—/—	P/—/—/—	—/RM
[40]	TC/—/—	P/—/—/—	—/RM
[32]	TC/—/ThC	P/—/—/—	—/RM
[41]	TC/—/—	P/—/—/—	—/RM
[42]	TC/—/—	P/—/—/—	—/RM
[24]	TC/—/ThC	P/—/—/A	—/RM
[43]	TC/—/—	P/—/—/—	RC/—
[44]	TC/—/—	P/Q/—/—	—/RM
[45]	TC/—/—	P/—/—/—	—/RM
[46]	TC/—/—	P/—/S/A	—/RM

Table 1. *Cont.*

Ref.	Channels	Policies	RL Processes
[47]	TC/—/—	P/—/S/—	—/RM
[11]	TC/—/ThC	P/—/—/—	—/RM
[28]	TC/—/—	P/—/—/A	—/RM
[48]	TC/EC/—	P/—/—/—	—/—
[22]	TC/EC/—	P/—/—/—	—/RM
[30]	TC/EC/—	P/—/—/—	RC/—
[49]	TC/—/ThC	P/—/—/—	RC/—
[25]	TC/EC/ThC	P/Q/—/—	RC/—
[50]	TC/EC/—	P/—/—/—	—/RM
[26]	TC/—/—	P/Q/—/—	—/RM
[51]	TC/—/ThC	P/—/—/—	—/RM
[12]	TC/EC/—	P/—/—/—	—/RM
[27]	TC/—/—	P/Q/—/—	—/RM
[52]	TC/EC/—	P/—/—/—	—/RM
[53]	TC/EC/—	P/—/—/—	—/RM
[54]	TC/EC/—	P/—/S/—	—/RM
[55]	TC/EC/—	P/—/—/—	RC/RM
[19]	—/EC/ThC	P/—/—/—	RC/—
Our study	TC/EC/—	P/Q/S/A	RC/RM

One essential process in RL is the collection system, including traditional, e-, or dual channels, which may be influenced by factors such as the SC members' preferences [56], consumer preferences [30,54], and the preferences of both consumers and SC members [50,53,57]. The effectiveness of single or multi-channel structures on the operational efficiency and profitability of SCs has been intensively investigated in studies [18,46,58]. Studies have indicated that dual-recycling channels, involving an e-channel or third-party channel alongside the traditional channel, outperform single-channel approaches in terms of profitability [12,30–32]. Specifically, a dual-channel strategy provides flexibility and improvements in profitability by effectively handling and reselling returned items, compared to a single-channel strategy [22].

Employing different policies in channels impacts the effectiveness and environmental sustainability of SCs, influencing consumer behavior regarding the return of used products [17]. The policies that have been considered from existing studies include price, quality, service, and advertisement (see Table 1). Incentive-based policies seek to convince consumers to return used products by offering prices (rewards or benefits). This policy has been extensively studied, and its impact on enhancing consumer engagement, increasing return rates, and the overall SC performance has been noted. For instance, [59] demonstrates that optimal incentive pricing strategy significantly influences product returns, driving sustainable operations. Returns are encouraged by higher incentive values, ensuring a steady supply of remanufacturable materials and improving profitability. In the context of competition and channel differentiation in reverse SC, the significance of optimal incentive pricing decisions for maximizing profits has been investigated [19,55]. Together, these studies provide a framework for optimizing incentive pricing in both traditional and e-channels. In the context of competition between SCs, the positive influence of offered incentives by each chain and its competitors has been illustrated by [50]. Furthermore, the effectiveness of contract-based optimal incentive pricing decisions in a reverse SC with dual recycling channels has been suggested by [11,52] to maximize profits. Inspecting the acceptance return quality level for remanufacturing can influence the SC's economic and environmental outcomes [36]. The implementation of optimal return quality standards lead to improved economic and environmental outcomes by optimizing remanufactured product quality and the cost of remanufacturing processes. The low-quality returned products could increase the SC's operational costs [44,60]. In the context of the channel, it has been found that the dual-channel model generally benefits manufacturers by yielding a higher product quality than the single-channel model [25]. Furthermore, an investigation of the influence of product quality on incentive values [26] suggests that coordinated models increase incentive values for high-quality products and decrease prices for low-quality items. This approach aims to maximize profits by prioritizing higher-quality returns. Encouraging product returns through advertising is an efficient strategy for increasing consumer awareness and improving return rates [17]. While previously mentioned research has mainly focused on incentive pricing and the product return quality level, advertising investment must also be considered a critical policy for increasing return rates [61]. Studies by [24,28,62] highlight the positive influence of advertising investments on profit and return rates in RL. Additionally, the effectiveness of coordinated advertising investments in enhancing recycling rates and maximizing SC profits is highlighted in [48]. Also, the positive effects of optimal advertising investments on product return rates and profit maximization have been emphasized [23]. Another essential policy in promoting consumer participation in return programs is the quality of transportation services provided, with transportation convenience highlighted in this study. Offering free shipping in the H-SC's e-channel increases consumer motivation for returns by reducing financial burdens, despite raising operational costs. This strategy ultimately offsets costs and enhances profitability [23,28]. By critically analyzing these policies alongside the processing time policy within both single- and dual-channel frameworks, our study provides a detailed comparison, highlighting the unique advantages and challenges of each approach. This contributes to a more nuanced understanding of policy-driven SC performance and sustainability.

Existing studies highlight the transition from traditional linear models to achieve more sustainable practices utilizing RL [9,63]. RL begins with collecting used products from consumers and managing end-of-life items through recycling and remanufacturing [64]. Recycling and remanufacturing are distinct processes that aim to reduce waste and conserve resources by extracting valuable components and reusing and repairing them [56]. Incorporating both remanufacturing and recycling within an SC can optimize resource utilization and enhance environmental performance, a focus explicitly addressed in this study. However, as summarized in Table 1, the simultaneous consideration of both processes has been overlooked.

3. Model Description

Game-theoretic models are practical tools for analyzing and solving optimization problems in competitive environments [62]. Numerous research studies have delved into the importance of strategic game-theoretical models in competition among SCs [19,51,52,54,65–70]. Two well-studied game models are Nash equilibrium, where the two competitors have the same power (e.g., two local restaurants), and Nash–Stackelberg, where the competing companies have different powers (e.g., local and global markets) [50]. This study examines the economic and environmental outcomes of T-SCs and H-SCs under the Nash equilibrium and Nash–Stackelberg scenarios. By capturing different competitive dynamics that mirror real-world market conditions, these scenarios provide insights into the influence of varying competitive power levels on SC's performance. The competition and decision-making processes are illustrated in Figure 2.

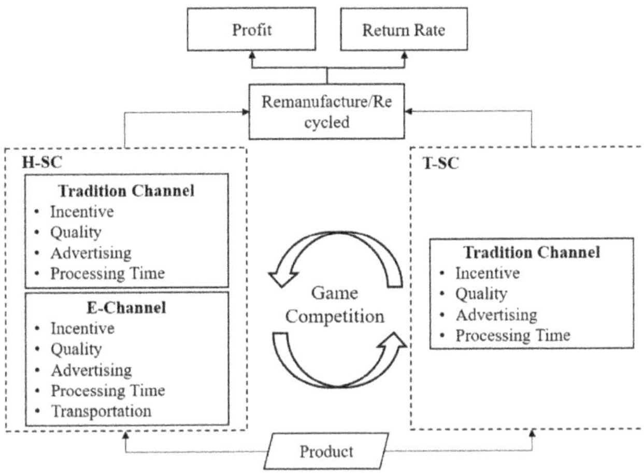

Figure 2. Illustration of the competitive dynamics and decision-making processes between two SC models.

We consider various policies, including price, quality, advertisement, service, and time. The following notations are used to formulate the competition model. The table of parameters is provided below (See Table 2). The superscript notation respectively denotes the type of channel (t or e) and the SC structure (H or T). For instance, $AiC^{t,\,T}$ denotes the cost of advertising investment for the traditional channel in a T-SC. Additionally, the coefficients are marked with subscripts to denote the impact of the applied policies on either the current channel (1) or other channels (2). For instance, $\tau_1^{t,\,T}$ represents the advertising coefficient of the traditional channel in the H-SC, with the number 1 indicating the related effect on the current channel.

Table 2. Description of parameters and decision variables.

Parameters	Description
Q	Return quality level
AQ	Acceptance return quality level for remanufacturing
AiC	Advertising investment cost
SCc	Shipping cost to the consumer
SCi	Shipping cost to the inspection center
$ProcT$	Processing time for each returned item
$ProbF$	Probability that a consumer agrees to recycle for free
$ProbE$	Probability of inspection error in assessing product quality
IC	Inspection cost per returned item at the inspection center
ICr	Inspection cost at the retail store
RC	Remanufacturing cost per item
SP	Selling price per remanufactured item
PR	Profit of recycling per item
N	Number of returned items

Table 2. *Cont.*

Parameters	Description
$PercA$	Percentage of returned items accepted for remanufacturing
Max_{IncPr}	Maximum incentive price for a return item
Coefficients	
α	Coefficient related to incentive price
λ	Coefficient related to acceptance return quality level
τ	Coefficient related to advertising investment cost
β	Coefficient related to transportation investments
θ	Coefficient related to processing time for each returned item
Independent Decision Variable	
$IncPr$	Incentive price per returned item
Dependent Variables and Functions	
P	Total profit
R	Return rate

The coefficients in the model quantify the proportional influence of various policies on return rates and profitability, providing a foundational framework for evaluating policy effectiveness within the SC. Key assumptions regarding sensitivity coefficients are established to measure their impact on the channel performance, supported by prior research [51,57,71–73]. Specifically, we assume that changes in a policy's sensitivity coefficient primarily affect the return rate of the current channel more significantly than either its influence on other rival channels or the influence from the other channels. This approach reflects real-world SC behavior, where the immediate advantages of a channel policy are often more prominently observed within the specific channel implementing it. For example, for the incentive price coefficient of the traditional channel in the H-SC $\left(\alpha_1^{t,\,H}\right)$, we posit that $\alpha_1^{t,\,H} \geq \alpha_2^{e,\,H}$, $\alpha_2^{t,\,T}$, and $\alpha_2^{t,\,H}$. This inequality indicates that adjustments to $\alpha_1^{t,\,H}$ impact its own channel's return rate more substantially than either its effect on other channels or the effects exerted by other channels within the H-SC or in the competing T-SC. This combined assumption allows for a more precise representation of channel-specific responses to policy adjustments, essential for constructing a nuanced competitive model that closely aligns with real-world SC interactions.

3.1. Competition Models and Solution Methods

To calculate the return rates for each channel, we utilize a linear model that considers the positive impacts of policies within the channel deducted from the influences arising from competing channels. The analysis focuses on the return rate functions for three channels: (1) the traditional channel in the H-SC, denoted as $R^{t,\,H}$; (2) the e-channel in the H-SC, denoted as $R^{e,\,H}$; and (3) the traditional channel in the T-SC, denoted as $R^{t,\,T}$. For instance, we can depict the return rate function of the traditional channel in the T-SC as follows:

$$R^{t,\,T} = Imp^{t,\,T} - \left(Imp^{e,\,H} + Imp^{t,\,H}\right), \tag{1}$$

where the impact of policies within the t channel in the T-SC (denoted as Imp_t^T) encompasses a set of positive coefficients, each reflecting the impact of traditional operational policies specifically tailored for the incentive price, return quality level, advertisement investment, transportation investments, and processing time within T-SC, respectively:

$$Imp^{t,\,T} = \left(\alpha_1^{t,\,T} \times IncPr^{t,\,T}\right) + \left(\tau_1^{t,\,T} \times AiC^{t,\,T}\right) + \left(\theta_1^{t,\,T} \times ProcT^{t,\,T}\right) + \left(\lambda_1^{t,\,T} \times AQ^{t,\,T}\right) + \left(\beta_1^{t,\,T} \times SCi^{t,\,T}\right), \quad (2)$$

in which it represents the respective impact of policies within competing channels in the H-SC, encompassing negative coefficients accounting for similar operational policies in two t and e-channels that adversely affect the t channel's performance in the H-SC:

$$Imp^{e,\,H} + Imp^{t,\,H} = \left(\alpha_2^{e,\,H} \times IncPr^{e,\,H} + \tau_2^{e,\,H} \times AiC^{e,\,H} + \theta_2^{e,\,H} \times ProcT^{e,\,H} + \lambda_2^{e,\,H} \times AQ^{e,\,H} + \beta_2^{e,\,H} \times SCi^{e,\,H}\right) + \left(\alpha_2^{t,\,H} \times IncPr^{t,\,H} + \tau_2^{t,\,H} \times AiC^{t,\,H} + \theta_2^{t,\,H} \times ProcT^{t,\,H} + \lambda_2^{t,\,H} \times AQ^{t,\,H} + \beta_2^{t,\,H} \times SCi^{t,\,H}\right). \quad (3)$$

In Equation (3), the coefficients $\alpha_i,\ \tau_i,\ \theta_i,\ \lambda_i,\ \beta_i$ indicate the influence of each policy within the channels on itself ($i = 1$) and its competitors ($i = 2$). Thus, in Equation (2), all of the indices are considered as $i = 1$, while in Equation (3), they are $i = 2$, capturing the impact of competitor actions on the T-SC. By quantifying these influences, the model explains how internal strategies and external competitive dynamics collectively impact return rates. The following section outlines profit models for competitive SCs, categorizing costs associated with each member under different conditions to simplify the analysis.

$$MC_qTC = RC + IncPr^t + IC + SCi + \frac{IC^t}{PercA} \quad (4)$$

$$MC_eTC = IncPr^t + \frac{ICr}{PercA} + SCi + IC \quad (5)$$

$$MC_qEC = RC + IncPr^e + SCi + \frac{IC^e}{PercA} \quad (6)$$

$$MC_{q'}TC = ProbF^t \times SCi + \frac{IC^t}{PercA} \quad (7)$$

$$MC_{q'}EC = SCi + (1 - ProbF^e) \times SCc + \frac{IC^e}{PercA} \quad (8)$$

Equation (4) outlines the costs incurred by the manufacturer from the traditional channel in the absence of errors which happen only from the retailer's side. When there are no errors, these costs include remanufacturing expenses, traditional channel's incentive prices, inspection costs, and shipping fees to the inspection center. Equation (5) details the costs for the manufacturer in the traditional channel when errors occur. This includes the traditional channel's incentive prices, inspection costs per accepted item, shipping fees to the inspection center, and the overall inspection costs related to the traditional channel. Equation (6) describes the costs for the manufacturer utilizing the e-channel, which encompasses remanufacturing expenses, e-channel incentive prices, shipping costs, and inspection fees per accepted item. Equation (7) pertains to the costs linked to non-qualified items in the traditional channel that will be recycled. This is calculated based on the percentage of individuals willing to recycle their devices for free through the traditional channel, multiplied by the associated shipping costs. Equation (8) reflects the costs for the manufacturer using the e-channel for non-qualified items that will not be recycled. This includes shipping costs as well as the percentage of individuals who opt not to recycle their devices for free, multiplied by the remanufacturing expenses.

The flowchart of the formation of the total profit function is presented in Figure 3, which includes four main processes, each representing a distinct operational pathway. These processes are defined by the following equations, which outline their contribution to P^H and P^T, taking into account various factors such as the number of products (N), return quality level (Q), acceptance rates ($PercA$), consumer acceptance rate for recycling for free ($PercF$), channels' return rate (R), incentive price ($IncPr$), inspection error rate ($ProbE$), and cost and price parameters (PR, AiC, and RC). The profit functions are defined as follows:

$$\begin{aligned} Profit_1 = N \times Q \times PercA^t \times R^t \times (1 - ProbE) \times (SP - MC_q TC) \\ + ProbF^t \times PercA^t \times R^t \times \left(PR - MC_{q'} TC\right) \end{aligned} \tag{9}$$

$$Profit_2 = N \times (1 - Q) \times PercA^t \times R^t \times \left(ProbF^t \times PR - MC_{q'} TC\right) - AiC^t \tag{10}$$

$$Profit_3 = N \times Q \times PercA^e \times R^e \times (SP - MC_q EC) \tag{11}$$

$$Profit_4 = N \times (1 - Q) \times PercA^e \times R^e \times \left(ProbF^e \times PR - MC_{q'} EC\right) - AiC^e \tag{12}$$

$$P^H = Profit_1 + Profit_2 + Profit_3 + Profit_4 \tag{13}$$

$$P^T = Profit_1 + Profit_2 \tag{14}$$

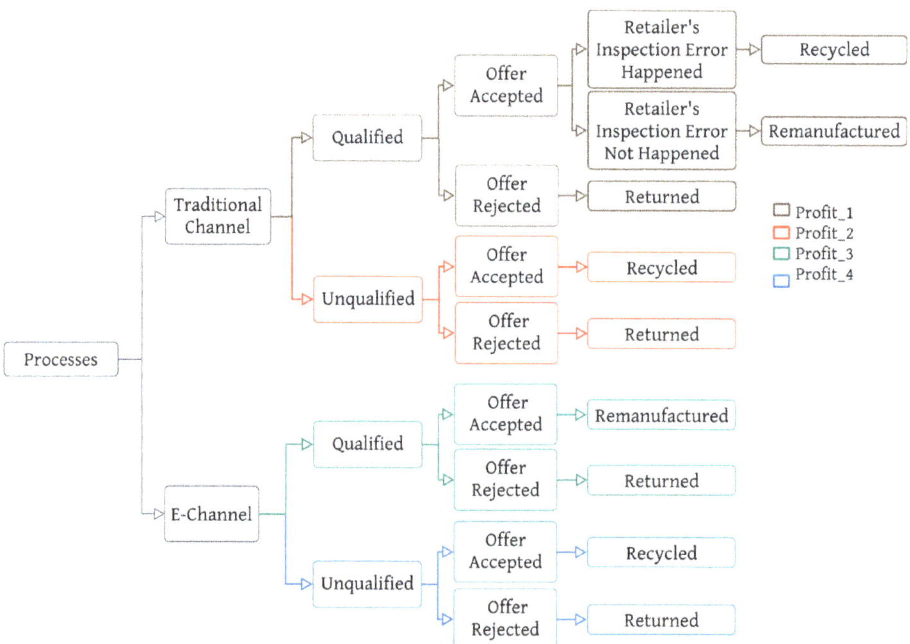

Figure 3. Operational procedures and their related profits within each channel.

Before going through the computational results and competition scenarios, it is crucial to examine how key policies interact within real-world contexts. Consumer-friendly strategies effectively increase return rates and consumer participation but also impose substantial costs. Incentive pricing enhances return rates while elevating operational costs, whereas free shipping improves convenience and consumer engagement but significantly increases logistical costs. Similarly, advertising investments boost awareness and participation but require careful cost optimization to maintain financial viability. Stricter quality standards for returns enhance the value of remanufactured products and reduce remanufacturing costs but may deter consumers due to longer processing times or the risk of rejection. These interconnected dynamics underscore the delicate balance SCs must achieve between maximizing consumer satisfaction and ensuring operational efficiency, forming the foundation for the analysis that follows.

3.2. Competition Scenarios

To model SC competition, three scenarios are examined: Nash equilibrium and Nash–Stackelberg leadership (with either H-SC or T-SC leadership). The Nash equilibrium represents a competitive scenario where equally powerful players simultaneously set their policies without influencing one another. In contrast, the Nash–Stackelberg leadership framework introduces a hierarchical structure, where a leader sets its policies first, subsequently shaping the decisions of the follower. This approach mirrors real-world dynamics, where market leaders shape the strategies of smaller competitors.

For each scenario, the decision-making structure is analyzed. Proving the concavity of the profit function is essential to ensure a global maximum, as failing to do so may result in suboptimal or incorrect solutions. Concavity is established by verifying that the second derivatives of the profit functions with respect to the decision variables are negative. All detailed structures and proposition proofs are provided in the corresponding Appendix A section.

3.2.1. Nash Equilibrium

In this scenario, SCs independently and simultaneously determine their policies while considering market conditions and competitor behavior. Both competitors are assumed to have equal power, like well-known companies such as Apple and Dell [21].

Proposition 1. *The profit function of the H-SC, P^H, is concave with respect to $IncPr^{t,\,H}$ and $IncPr^{e,\,H}$(see Appendix A.1 for details).*

As P^H is concave, the best decisions regarding $IncPr^{t,\,H}$ and $IncPr^{e,\,H}$ to maximize H-SC's profit can be accurately determined by setting the first derivatives of P^H with respect to $IncPr^{t,\,H}$ and $IncPr^{e,\,H}$ to zero.

$$\left(IncPr^{t,\,H}\right)' = \frac{\partial P^H}{\partial IncPr^{t,\,H}} = 0 \tag{15}$$

$$\left(IncPr^{e,\,H}\right)' = \frac{\partial P^H}{\partial IncPr^{e,\,H}} = 0 \tag{16}$$

Proposition 2. *The profit function of the T-SC, P^T, is concave with respect to $IncPr^{t,T}$ (see Appendix A.1: Proof of Proposition 2 for details).*

With the same logic of Proposition 1, given the concavity of the T-SC profit function, a unique optimal value of $IncPr^{t,\,T}$ could be obtained by setting the first-order derivative of the equation P^T with respect to $IncPr^{t,\,T}$ to zero.

$$\left(IncPr^{t,\,T}\right)' = \frac{\partial P^T}{\partial IncPr^{t,\,T}} = 0 \tag{17}$$

The decision-making structure and detailed derivation can be found in Appendix A.1.

3.2.2. Nash–Stackelberg H-SC Leadership Scenario

Nash–Stackelberg leadership occurs when two SCs have unequal power, with one acting as the leader. For instance, this dynamic can be seen in the competition between a global brand like Apple and a local brand such as Micromax. In the Nash–Stackelberg H-SC leadership, the H-SC assumes the role of the leader, while the T-SC acts as a follower. The decision-making process starts by analyzing the follower's behavior, and then the leader determines its optimal strategies based on this analysis. So, the T-SC first calculates its optimal incentive, denoted as $\left(IncPr^{t,T}\right)'$, in response to the H-SC's decisions under the Nash game framework. Using this, the H-SC recalculates its return $R^{t,\,H}$ and $R^{e,\,H}$, labeled as $\left(R^{t,\,H}\right)''$ and $\left(R^{e,\,H}\right)''$, respectively. These updated return rates are then used to derive the H-SC's updated profit function, $\left(P^{H}\right)''$. To optimize the leader's profit function, $\left(P^{H}\right)''$, its concavity with respect to $IncPr^{t,\,H}$ and $IncPr^{e,\,H}$ must be proven. This enables calculation of the leader's optimal incentives, after which the follower (T-SC) adjusts its own incentives $\left(IncPr^{t,T}\right)''$.

Proposition 3. *The profit function of the H-SC under the Nash–Stackelberg H-SC leadership scenario, denoted as* $\left(P^{H}\right)''$*, is concave over* $IncPr^{t,\,H}$ *and* $IncPr^{e,\,H}$ *(see Appendix A.2 for detail).*

The decision-making structure and detailed derivation can be found in Appendix A.2.

3.2.3. Nash–Stackelberg T-SC Leadership Scenario

In this scenario, the leadership role shifts to the T-SC. The same Nash–Stackelberg integrated game approach is employed as in the second scenario.

Proposition 4. *The profit function of the T-SC under the Nash–Stackelberg T-SC leadership scenario,* $\left(P^{T}\right)'''$*, is concave with respect to* $IncPr^{t,\,T}$ *(see Appendix A.3 for detail).*

The decision-making structure and detailed derivation can be found in Appendix A.3.

4. Computational Results

The competition models, designed to gain economic and environmental insights, are applied to the competitive H-SC and T-SC frameworks using data from a real-world case study of Apple Inc., headquartered in Cupertino, California, USA, and its Give Back trade-in program. These data were sourced from Apple reports, informative websites, third-party retail platforms, and prior studies [41,50]. This program, which encourages consumers to return used devices for remanufacturing or recycling through traditional channels (retail Apple stores) or e-channels (Apple's online website), aligns closely with the study's objectives and provides valuable insights into the RL operations. The dataset encompasses one year (2017–2018) of detailed information, including sales volume, repair, shipping, and advertising costs, incentive pricing, return processing time, and quality criteria for phones eligible for remanufacturing (as summarized in Table 3).

Table 3. Qualified properties for per-item data selection criteria and sources for SC analysis.

Parameter	Value Range/Description	Unit	Data Source
Quality Criteria	Low (0.55) to High (0.90) (1 = highest)	Score	Apple Trade-In Program
Processing Time	4 to 370	Hours	Apple Trade-In Program

Table 3. *Cont.*

Parameter	Value Range/Description	Unit	Data Source
Repair Costs	63.6 (average)	USD	Apple Website
Recycling Profits	90.5 (average)	USD	Environmental Responsibility Report
Remanufactured Product Price	690 (average)	USD	Apple Website/Best Buy
Transportation Costs	12.5 (average)	USD	Courier Service Providers

4.1. Incentive Price

As seen in Figure 4, the H-SC offers higher incentive values than the T-SC under all scenarios, with the highest value obtained under the Nash equilibrium scenario. Table 4 summarizes the optimal incentive values of each channel under different scenarios. The findings demonstrate that, the e-channel in the H-SC consistently offers more incentives than traditional channels. The T-SC's traditional channel offers more modest incentives even when assuming a leadership role. Also, the H-SC's traditional channel has shown the lowest incentive under all scenarios. The maximum incentive values of all three channels have been offered under the Nash scenario. To understand the rationale behind these decisions and their potential outcomes, it is essential to conduct a further analysis of the return rates and profit outcomes, and a sensitivity analysis of the policies employed.

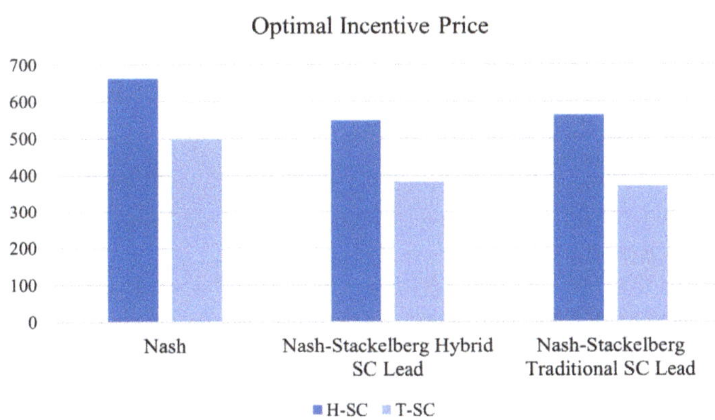

Figure 4. Optimal incentive price in H-SC and T-SC under different scenarios (USD per unit).

Table 4. Optimal average incentive values for each product within different channels (USD per unit).

Scenario	Incentive	Product 1	Product 2	Product 3	Product 4	Product 5
Nash Equilibrium	$IncPr^{e,H}$	**156.74**	**183.60**	**195.37**	**192.82**	**244.88**
	$IncPr^{t,H}$	82.46	72.90	67.48	62.12	66.70
	$IncPr^{t,T}$	102.81	97.32	93.11	86.03	100.70
Nash–Stackelberg H-SC Leadership	$IncPr^{e,H}$	**150.45**	**167.73**	**169.65**	**160.67**	**214.65**
	$IncPr^{t,H}$	75.99	55.53	39.08	26.96	33.54
	$IncPr^{t,T}$	98.78	83.72	69.53	56.30	73.56
Nash–Stackelberg T-SC Leadership	$IncPr^{e,H}$	**152.33**	**170.15**	**172.25**	**163.82**	**218.73**
	$IncPr^{t,H}$	78.14	58.29	42.08	30.35	37.93
	$IncPr^{t,T}$	97.14	81.53	67.20	53.48	69.85

4.2. Return Rates

As shown in Table 5, the total return rate in the H-SC exceeds that of the T-SC under all scenarios. Upon closer examination of individual channels, it becomes apparent that the e-channel experiences a lower return rate than traditional channels despite offering higher incentive values. This paradox can be attributed to the channel's stringent return quality-checking process, which may discourage some consumers due to the increased likelihood of rejection and longer return processing times, despite its other consumer-friendly policies, such as higher incentives, free shipments, and greater ease of access. Conversely, the higher return rate in the traditional channels can be attributed to faster inspection, shorter processing times, and less stringent return quality standards, which appeal to consumers who prioritize these policies over the consumer-friendly policies provided by e-channels. In our analysis, the highest return rate for the entire H-SC was observed under the Nash–Stackelberg scenario with T-SC leadership. Conversely, for the T-SC, the highest return rate occurred under the Nash equilibrium scenario.

Table 5. Results of return rate under different scenarios.

Scenario	$R^H(R^{t,H}, R^{e,H})$	$R^T(R^{t,T})$
Nash Equilibrium	**59.80% (37.00%, 22.80%)**	40.20%
Nash–Stackelberg H-SC Leadership	**62.63% (34.28%, 28.35%)**	37.37%
Nash–Stackelberg T-SC Lead	**62.69% (34.46%, 28.50%)**	37.04%

4.3. Profit

The profit analysis in Table 6 reveals that the H-SC generates significantly higher profits than the T-SC, indicating the superiority of the H-SC dual-channel structures and offered policies. The highest profit for both the H-SC and T-SC arises under the Nash–Stackelberg T-SC leadership. Although the T-SC experiences a significantly increased return rate with Nash equilibrium, its total profit is the lowest under this scenario, which might imply that despite the high return rates, the associated costs or operational inefficiencies reduce the profitability. In contrast, the highest return rate for the H-SC is also obtained under Nash–Stackelberg T-SC leadership, suggesting an increasing relationship between the H-SC's return rate and overall profit.

Table 6. Profits across different scenarios (billion USD).

Scenario	Profit	Product 1	Product 2	Product 3	Product 4	Product 5
Nash Equilibrium	p^H	**101.323**	**80.2032**	**56.6943**	**75.4965**	**111.239**
	p^T	54.8832	44.986	30.6644	44.6155	66.7623
Nash–Stackelberg H-SC Leadership	p^H	**100.091**	**77.7913**	**54.0021**	**71.6727**	**107.275**
	p^T	55.0541	42.26	30.97	45.08	67.24
Nash–Stackelberg T-SC Leadership	p^H	**100.844**	**80.1214**	**56.8909**	**75.8301**	**111.278**
	p^T	55.4401	45.6827	31.3361	45.6277	67.9116

In conclusion, the dual-channel structure of the H-SC outperforms the T-SC by combining traditional channels' faster processing and less restrictive quality standards with e-channels' higher incentives and convenience. This synergy leads to higher return rates and as shown by the positive correlation, greater profitability. These results show how strategic policy adjustments can enhance profitability in SCs. Next, we explore the specific role of sensitivity analysis in identifying optimal policies.

5. Sensitivity Analysis

Five policies (incentive, quality, advertisement, service, and time) play significant roles in analyzing both economic and environmental objectives. The influences of the policy changes on both SCs are investigated using the one-at-a-time (OAT) method [72]. This sensitivity analysis highlights the variations in profitability, while also providing insights into the corresponding changes in return rates, as demonstrated by their correlation discussed in Sections 4.2 and 4.3. Therefore, both economic and environmental objectives are considered and evaluated. Each policy is adjusted within a feasible range, defined by the mean value of the parameter and its specific deviation in an increasing direction, and the resulting effects are observed. The analysis was conducted for both SCs. In the following sections, each policy is analyzed separately (see Table 7). Additionally, a Pareto-based analysis is conducted to identify the most impactful policies for each SC, highlighting those with the greatest influence on profit outcomes.

Table 7. Summary of analysis of profit variation resulting from changes in policies (+/−: positively/negatively correlated).

Policy	Targeted SCs	Profit Change	
		H-SC	T-SC
Incentive price	H-SC	−	−
	T-SC	−	−
Return quality level	H-SC	+	−
	T-SC	−	+
Advertisement investments	H-SC	+	−
	T-SC	−	+
Transportation investments	H-SC	−	+
Return processing time	H-SC	−	+
	T-SC	+	−

5.1. Analysis of Profit Variations Resulting from Changes in Incentive Price

Starting with the incentive policy, we note that incentive values are optimized to maximize profitability, and any deviation from these values could negatively impact the SC's profit; however, to understand the impact of variations, a sensitivity analysis on the incentives has been conducted.

By analyzing the optimal incentive prices ($IncPr^{t,H}$, $IncPr^{e,H}$, and $IncPr^{t,T}$), it can be observed that variations in these policies reveal distinct patterns (see Figure 5). Traditional channels demonstrate stability by sustaining profitability across a range of incentive values before experiencing a decline, which is more pronounced in the e-channel and less so in the traditional channels. This stability allows traditional channels to increase incentive prices up to a certain threshold when needed to attract higher returns or effectively address competitive pressures, without resulting in profit reduction. In contrast, e-channels demonstrate a higher sensitivity to such increases, requiring precise calibration to balance customer satisfaction and operational efficiency.

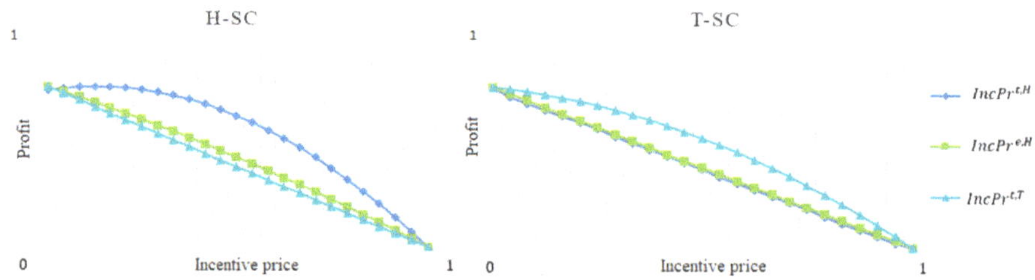

Figure 5. Analysis of profit variations (in billions of USD) resulting from changes in incentive price (ranging between 0 and 1) for (**left**) H-SC and (**right**) T-SC.

5.1.1. Analysis of Profit Variations Resulting from Changes in Return Quality Level

Quality levels (Q) play a critical role in driving profitability across both SCs. As illustrated in Figure 6, higher quality standards are strongly associated with improved profitability, as they enhance the value of remanufactured products and reduce remanufacturing costs. However, stricter inspections, while ensuring higher quality, may deter some returns. In traditional channels, relaxed quality standards can increase return rates but fail to significantly boost profitability, as the higher remanufacturing costs associated with lower-quality returns offset potential gains.

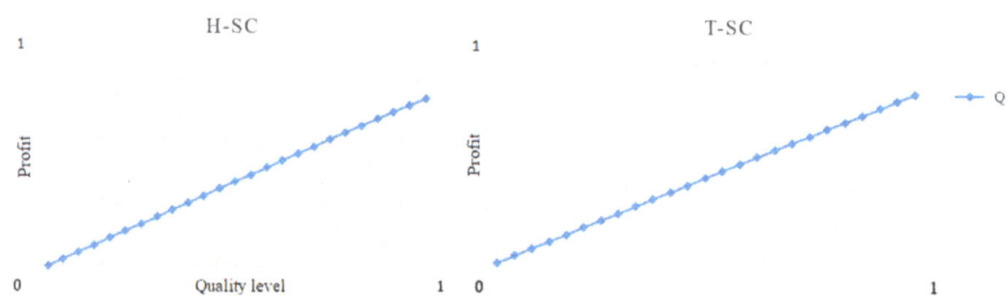

Figure 6. Analysis of profit variations (in billions of USD) resulting from changes in return quality level (ranging between 0 and 1) for (**left**) H-SC and (**right**) T-SC.

5.1.2. Analysis of Profit Variations Resulting from Changes in Advertisement Investments

Increasing advertising investment ($AiC^{t,H}$, $AiC^{e,H}$, and $AiC^{t,T}$) drives profitability by enhancing customer awareness and engagement, fostering stronger participation in return programs (see Figure 7). However, these profitability gains are accompanied by increased operational costs, underscoring the need for a well-balanced and strategic approach to investment.

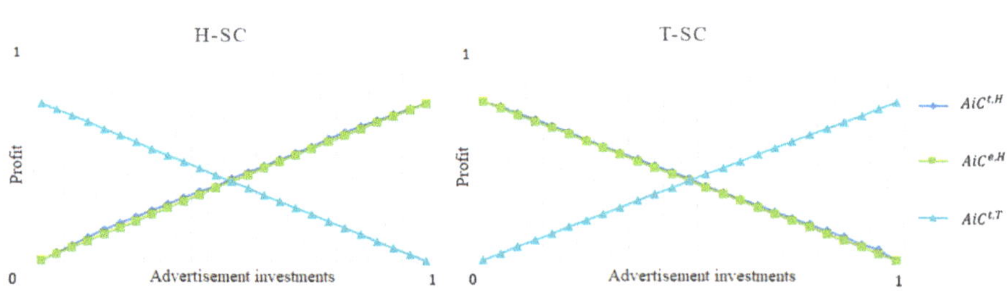

Figure 7. Analysis of profit variations (in billions of USD) resulting from changes in advertisement investments (ranging between 0 and 1) for (**left**) H-SC and (**right**) T-SC.

5.1.3. Analysis of Profit Variations Resulting from Changes in Transportation Investments

Based on Figure 8, higher transportation investments ($SCi^{e,\ H}$) in H-SC's e-channel initially reduce profits as operating costs outweigh the benefits. However, beyond a certain threshold, additional increases in transportation investments do not significantly impact H-SC's profit, suggesting that the e-channel reaches an efficiency plateau where further investments no longer result in significant financial losses.

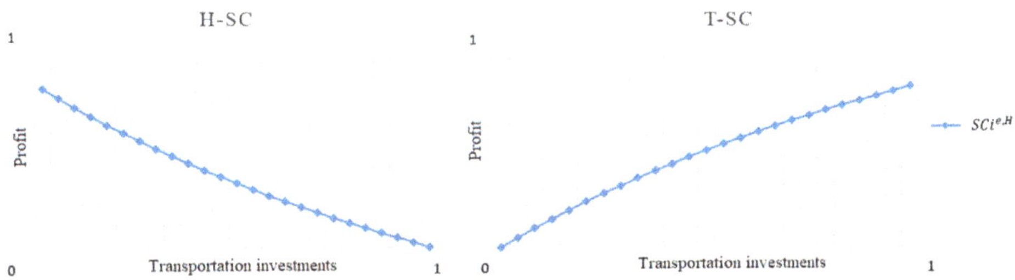

Figure 8. Analysis of profit variations (in billions of USD) resulting from changes in transportation investments (ranging between 0 and 1) for (**left**) H-SC and (**right**) T-SC.

5.1.4. Analysis of Profit Variations Resulting from Changes in Return Processing Time

Figure 9 shows that increasing the processing time ($ProcT^{t,\ H}$, $ProcT^{e,\ H}$, and$ProcT^{t,\ T}$) in H-SC's traditional and e-channels, as well as T-SC's traditional channel, results in a gradual decline in profit for both. This profit reduction indicates that consumers prefer faster processing, and extending the processing time may prompt consumers to rethink their preferences.

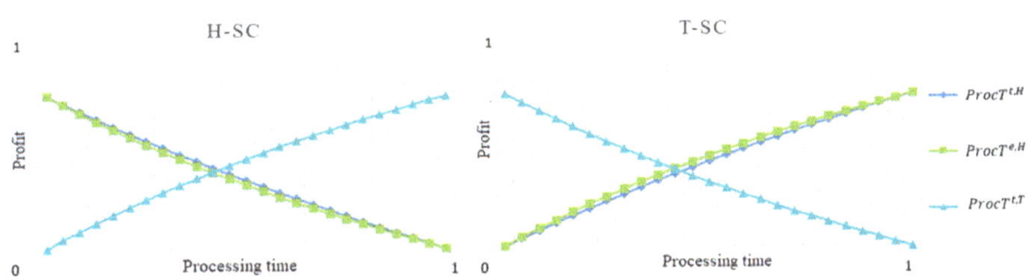

Figure 9. Analysis of profit variations (in billions of USD) resulting from changes in return processing time (ranging between 0 and 1) for (**left**) H-SC and (**right**) T-SC.

In the above analyses, it can be observed that profitability gains in one SC often result in corresponding losses in the competing SC, underscoring the competitive dynamics inherent in such systems. For instance, increased advertising investments can attract more consumers to one SC, thereby enhancing its profitability while diminishing that of its rival. However, this inverse relationship does not hold in the case of incentive strategies. Higher incentives offered by one SC may compel the rival to match or exceed these values, thereby increasing costs for both parties and ultimately reducing the overall profitability across both SCs.

Although the direction of profit variations has been investigated, the magnitude of these variations remains unclear. In the following section, we will address this by conducting Pareto and regression analyses.

5.2. Sensitivity Analysis of Policies: Insights from Pareto and Regression Approaches

This section uses a Pareto chart and regression analysis to evaluate model sensitivity, identifying the most effective policies and their direct impact (coefficients) and relative importance (weights) on profitability. The Pareto chart analysis (Figure 10) reveals that the foremost influential policy for the H-SC is the return quality level Q, which positively affects profit. Additionally, vital parameters such as $AiC^{t,H}$ and $AiC^{e,H}$ positively influence profit while $IncPr^{t,H}$, $IncPr^{e,H}$, $ProcT^{t,H}$, $ProcT^{e,H}$, and L_e negatively impact the H-SC's profitability, indicating areas for cost control. $Aic^{t,T}$ and $IncPr^{t,T}$ policies imposed by its rival, the T-SC, negatively impact the H-SC's profit, showing the need for a strategic response to these competitive pressures. For the T-SC, $Aic^{t,T}$ and Q exhibit a significant positive impact on profit, underscoring the importance of effective advertising and return quality management, whereas $IncPr^{t,T}$and$ProcT^{t,T}$ form negatively impact profitability, suggesting optimization opportunities. Both charts indicate that the initial policies have the most significant impact on profitability, with a diminishing trend in profit thereafter. This analysis underscores the importance of prioritizing key policies to optimize profit while mitigating negative influences, ensuring effective strategic management for both SCs.

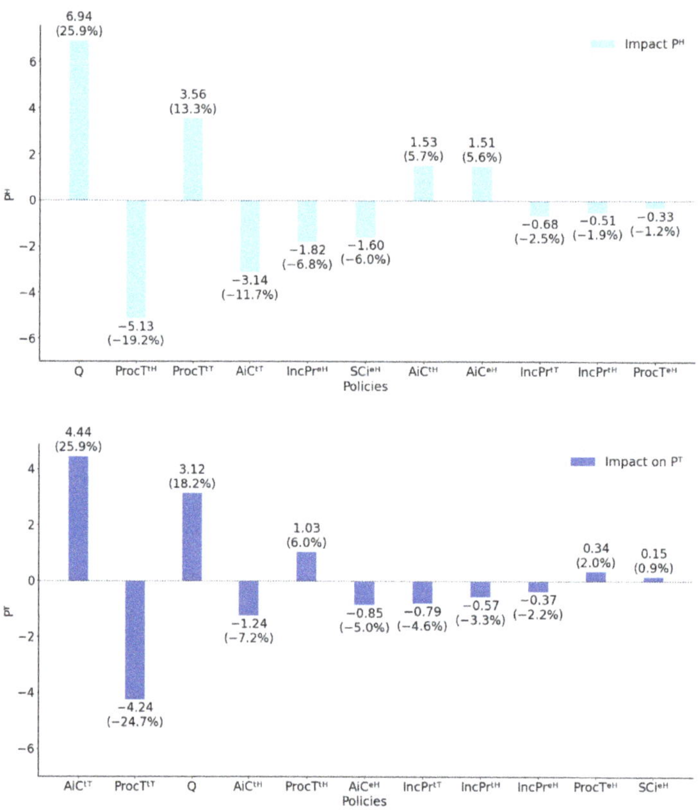

Figure 10. Impact of policies on profit: a Pareto perspective.

As illustrated in Figure 10, the return quality standards emerge as the most impactful policy (highest weight of 25.9%). SC managers should prioritize this policy while ensuring streamlined processes to maintain efficiency and prevent consumer disengagement. Conversely, due to the associated costs and limited operational flexibility, the T-SC structure gains comparatively less profitability from improving quality standards. Therefore, the T-SC should prioritize its most impactful policy, advertising (25.9% weight), to drive profitability. The T-SC can boost profitability by advertising its program's advantages, such as relaxed quality standards and faster processing times, which attract specific consumer segments.

The processing time emerges as the second critical policy affecting profitability in both the H-SC and T-SC (see Figure 10). For traditional channels, extended processing times substantially reduce profits (in H-SC: weight 19.2%; in T-SC: weight 24.7%), emphasizing the importance of time for the consumers who chose traditional channels for their quick process. However, in e-channels, where consumers prioritize convenience over speed, reducing processing time is less crucial (weight 1.22%).

As shown in Figure 10, transportation investments for the H-SC, ranked as the third most influential policy (weight 6%), initially reduce profitability up to a certain threshold, after which profitability stabilizes. However, in the T-SC, quality improvement is the third priority (weight 18.2%), which is the main reason for the contradiction between the higher return rates and lower profitability in traditional channels. By balancing stricter quality standards with cost-effective, consumer-friendly policies, they can enhance the profitability caused by qualified returns, without incurring excessive financial burdens.

In addition, based on Figure 10, advertising has been counted as the fourth influential policy (weights of 5.7% and 5.6%, respectively, for traditional and e-channels). Therefore, in this SC, digital campaigns should focus on personalized offers for e-channel users, while traditional promotions highlight less restrictive quality standards and faster processing, besides promoting environmental benefits.

Lastly, regarding the impact of incentive pricing as the fifth impactful policy (see Figure 10), the analysis reveals that while incentives boost return rates, they can reduce the overall profitability by increasing SC costs. This effect is particularly pronounced for the H-SC profitability (weights of 1.9% and 6.8%, respectively, for the traditional and e-channels), emphasizing the need for careful calibration.

These results highlight how H-SCs and T-SCs can strategically allocate resources and strategically balance these initiatives to boost profitability while effectively responding to competitive pressures.

6. Discussion

This study's findings highlight the dynamic interplay between competing SCs with different structures, where gains in one often result in trade-offs for the other. These results emphasize the critical importance of selecting optimal structures, policies, and competitive roles to balance environmental and economic objectives. The subsequent discussion provides some suggestions for effective SC management and strategic decision-making in competitive contexts.

When analyzing return rates across chains and channels, the H-SC consistently shows higher return rates than the T-SC. However, traditional channels within both chains exhibit better return rates overall, despite e-channels offering higher incentives. This contrast is driven by stricter quality checks and longer processing times in e-channels, which deter some consumers, while traditional channels benefit from faster processing and less restrictive quality standards. Consequently, the H-SC achieves a superior overall return rate by integrating the traditional channel's faster processing and more flexible quality standards with the e-channel's higher incentives and enhanced convenience.

Furthermore, when analyzing profits, the H-SC generates significantly higher profits than the T-SC, indicating the superiority of the H-SC dual-channel structures and offered policies. Therefore, for those prioritizing both economic gains (profit) and environmental benefits (return rate), the H-SC structure is the optimal choice. However, for budget-constrained SCs or scenarios where environmental concerns are the primary focus, the T-SC structure serves as a viable alternative. The findings are not restricted to the electronics sector and can be applied to other industries, such as fashion, where similar dynamics exist. For example, fashion SCs could implement H-SC strategies by combining physical and online channels to streamline returns and remanufacturing processes, as demonstrated by the Mac "Back to MAC Program" [73].

By analyzing the competitive scenarios, we observed that the H-SC achieves its highest return rate and profit under Nash–Stackelberg T-SC leadership, highlighting a positive relationship between the return rate and profit in this chain. In contrast, the T-SC achieves its highest return rate under Nash Equilibrium, coinciding with its lowest profit. This inverse relationship between return rate and profit suggests that increased operational costs and inefficiencies associated with managing higher returns negatively impact the profitability of this SC. In a competitive environment, the H-SC should adopt a follower role to effectively balance environmental and economic goals. Conversely, if the T-SC prioritizes budget savings and economic gains, it should take the leadership role. However, if the T-SC focuses solely on environmental benefits, competing under the Nash scenario would be more advantageous.

To better understand the individual and combined impacts of each policy on profitability and to optimize strategies for both chains, we proceed with a sensitivity analysis.

1. One of the most significant findings is the impact of incentive pricing, which demonstrates a delicate balance between increasing return rates and maintaining cost efficiency. Deviations from optimal incentive pricing reduce profitability in both SCs, with a more pronounced effect on the H-SC, particularly on its e-channel. Traditional channels demonstrate greater stability, maintaining profitability across a range of incentive values before declining more so in the T-SC and less in the H-SC. This stability allows traditional channels to adjust incentives strategically, attracting higher returns or addressing competitive pressures. These findings highlight the need for adaptive, data-driven strategies to manage incentive values, ensuring long-term profitability while avoiding excessive costs.
2. The return quality level, another key driver of profitability, significantly affects both SCs. Investments in advanced inspection technologies to meet quality criteria can reduce remanufacturing costs, enhance the value of remanufactured products, and address inspection errors, particularly in traditional channels, where manual intervention often leads to inefficiencies. Furthermore, streamlining more restricted quality standards and inspection protocols can achieve these goals without facing excessive financial burdens, especially for T-SCs, where operational flexibility is limited.
3. Advertising investments represent a policy with significant profitability implications for both SCs. These investments enhance consumer engagement and drive higher return rates, but their impact is maximized when integrated with complementary policies such as incentive pricing, quality standards, and processing times. Digital campaigns for e-channels should highlight personalized offers and consumer-friendly programs, while traditional channels can focus on promoting faster processing times and less restrictive quality standards. Additionally, both SCs could appeal to some sustainability-conscious consumers by emphasizing the impacts of their environmental activities, such as recycling and remanufacturing.
4. The processing time is another impactful policy, particularly in traditional channels, where consumers prioritize processing time efficiency. To sustain profitability, traditional channels should enhance processing times using automated inspection systems, leveraging technologies like AI or optical scanning. These systems not only accelerate

the process but also minimize manual intervention and reduce errors. For e-channels, where convenience and higher incentives outweigh speed, reducing the processing time is less critical. Instead, focusing on enhancing the customer experience through flexible return policies and real-time tracking systems can optimize consumer satisfaction and operational efficiency.

5. Transportation investments, unique to H-SCs, reveal an unexpected negative impact on profitability until a certain threshold, after which profitability stabilizes. This suggests that the e-channel reaches an efficiency plateau where further investments no longer result in significant financial losses, which means beyond a certain return volume, the revenue offsets incremental transportation costs, achieving an operational equilibrium where economies of scale mitigate increased expenses, ensuring consistent profitability.

Expanding the analysis to include other industries, such as automotive SCs [74], reveals that implementing stricter quality standards supported by technologies like AI and automated systems can significantly reduce the remanufacturing costs and enhance the operational efficiency. Moreover, targeted advertising investments, particularly in traditional channels, can boost consumer engagement in industries where brand trust and processing speed play a critical role in influencing purchasing and return decisions.

This study addresses a critical gap by explicitly analyzing the interplay between the economic and environmental impacts of policy variations on the SC performance. However, it has several limitations. First, it does not incorporate consumer behaviors, which are critical in shaping the SC's profitability and structure. Second, the dataset, while robust, is limited to Apple's practices over a short timeframe and a small selection of product models, reducing its generalizability. Third, the analysis is confined to Nash and Nash–Stackelberg scenarios, without exploring other game-theoretic frameworks, such as cooperative games, dynamic games, or repeated games, which could provide additional insights into SC interactions [75]. Lastly, other SC structures, such as those involving third-party collaborations or single online channels, have not been considered for comparison and evaluation.

Future research could address these gaps by integrating consumer behavior, exploring a broader range of game-theoretic scenarios and alternative SC structures, and applying the framework across diverse industries and market contexts to enhance its applicability and relevance.

7. Conclusions

Smart city initiatives, facilitated by the CE concept, emphasize sustainable development through resource reuse. This study explores the competitive dynamics between two SCs (H-SC and T-SC) in the context of remanufacturing and recycling, contributing to the goals of sustainable smart cities through enhanced resource efficiency and sustainability practices. Through the game-theoretic competitive approaches, including Nash equilibrium and Nash–Stackelberg leadership, we examine optimal pricing decisions, return rates (environmental objectives), and profitability (economic objectives) as key performance metrics for the SCs in this study. Specifically, this research aims to assess the economic and environmental impacts of policy variations on SC performance by leveraging decision-making frameworks and comparative analysis. This research fills a gap by developing a model that jointly assesses recycling and remanufacturing under five essential policies: incentive price, return quality level, advertisement investments, transportation investments, and return processing time.

In the context of SCs, the findings emphasize the superiority of the H-SC's dual channel in terms of return rates and profitability. The H-SC consistently demonstrates better performance than the T-SC, underscoring the effectiveness of its integrated structure and well-aligned policies.

Within both SCs, the e-channel, despite offering higher incentives, demonstrates lower return rates compared to those of traditional channels. This outcome is attributed to its rigorous quality inspections and extended processing times, which may discourage consumers. These stricter policies offset the e-channel's consumer-friendly features, such as higher incentives, free shipping, and convenient access. Conversely, traditional channels, despite their fewer consumer-friendly policies, achieve higher return rates, as consumers prioritize less restrictive quality inspections and shorter processing times. As a result, the higher overall return rate of the H-SC is attributed to its dual-channel structure, which leverages the combined performance of traditional and e-channels to surpass the single-channel strategy employed by the T-SC.

Consequently, For SCs prioritizing both economic (profit) and environmental (return rate) benefits, the H-SC is the most effective option. On the other hand, the T-SC structure serves as a suitable alternative for budget-constrained SCs or those primarily focused on environmental objectives.

An analysis of the competitive scenarios reveals a positive correlation between return rates and profit in the H-SC, with both reaching their highest levels under the Nash–Stackelberg T-SC leadership model. In contrast, the T-SC exhibits an inverse correlation, with the highest return rate occurring under the Nash scenario and the highest profit achieved under the Nash–Stackelberg T-SC leadership model. These correlations highlight the superior management of H-SCs, achieved by effectively balancing gains and operational costs, providing a clear advantage over the T-SC in the context of smart city goals, such as sustainable resource utilization and operational efficiency. Therefore, in a competitive setting, the H-SC is best suited to adopting a follower role to achieve an effective balance between environmental and economic objectives. On the other hand, if the T-SC prioritizes economic gains, it should take the leadership role, and if the T-SC focuses solely on environmental benefits, competing under the Nash scenario would be more advantageous.

The sensitivity analysis underscores the critical role of key policies in shaping profitability and optimizing SC strategies. The Pareto analysis shows that the H-SC's profit significantly depends on the return quality level and return processing time of its traditional channel, while the T-SC's profit hinges on advertisement investments and the return processing time. For H-SCs, stricter quality standards, efficient inspection protocols, and faster processing through automated systems (e.g., AI or optical scanning) are crucial for effective quality and time management. For the T-SCs prioritizing the advertising of their traditional channel is essential, highlighting its faster processing and less restrictive quality standards to enhance consumer engagement. Additionally, the next priority should be adopting time management strategies similar to those employed by H-SCs to improve operational efficiency.

Beyond SC-specific improvements, these findings hold broader societal significance by supporting sustainable smart cities and advancing CE principles. By reducing waste, optimizing resource utilization, and extending product life cycles through remanufacturing and recycling, these strategies contribute to lowering environmental footprints and fostering long-term sustainability. The implications extend to industries beyond electronics, such as fashion and automotives, offering a pathway for global industries to adopt sustainable practices while addressing competitive pressures in the context of a CE.

Author Contributions: Conceptualization, A.A., T.W. and F.C.; methodology, A.A., T.W. and F.C.; software, A.A.; validation, A.A.; formal analysis, A.A.; investigation, A.A., T.W. and F.C.; resources, A.A.; writing—original draft preparation, A.A., T.W. and F.C.; writing—review and editing, A.A., T.W.

and F.C.; visualization, A.A., T.W. and F.C.; supervision, T.W.; project administration, T.W.; funding acquisition, T.W. All authors have read and agreed to the published version of the manuscript.

Funding: This research was funded by the Japan Science and Technology Agency's Scientific Research Program, Grant-in-aid for Scientific Research (Basic Research B), Grant Number: 22H01717.

Data Availability Statement: Dataset available on request from the authors.

Conflicts of Interest: The authors declare no conflict of interest.

Appendix A

Appendix A.1. Nash Equilibrium

Under the Nash equilibrium scenario, the profit functions of T-SCs and H-SCs are parallelly determined. The decision process is displayed below.

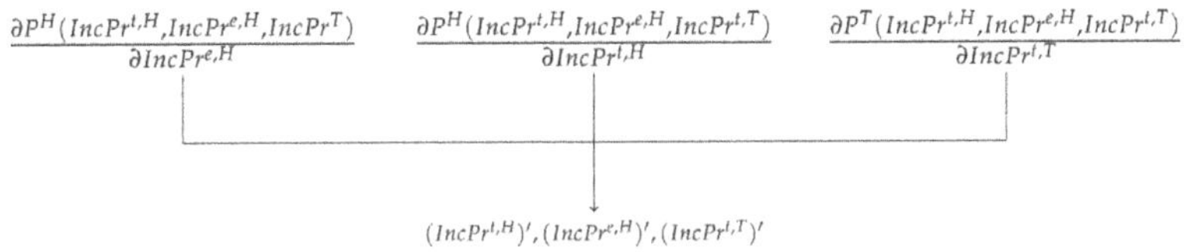

Figure A1. Illustration of decision-making structure under the Nash scenario.

The optimal values of $IncPr^{t,H}$ and $IncPr^{e,H}$ are determined based on the H-SC's profit function (P^H), and, at the same time, the optimal value of $IncPr^{t,T}$ is determined based on the T-SC's profit function (P^T) independently.

Proof of Proposition 1. To establish the concavity of P^H with respect to $IncPr^{t,H}$, we examine its second-order derivative:

$$\frac{\partial^2 P^H}{\partial\left(IncPr^{t,H}\right)^2} = N \times Q \times \left(-2 \times PercA^{t,H} \times \alpha_1^{t,H} \times ProbE - 2 \times PercA^{t,H} \times \alpha_1^{t,H} \times (1 - ProbE)\right),$$

which simplifies to $-2 \times N \times Q \times PercA^{t,\,H} \times \alpha_1^{t,\,H}$. Given that N, Q, $PercA^{t,\,H}$, and $\alpha_1^{t,\,H}$ are positive quantities within this problem context (representing quantities, rates, or coefficients that are inherently non-negative), the entire expression is negative. This result holds regardless of the value of $ProbE$ (provided $0 < ProbE < 1$). Hence, the expression is definitively negative, proving the concavity of P^H with respect to $IncPr^{t,\,H}$.

Furthermore, the second-order derivative of P^H with respect to $IncPr^{e,\,H}$ is

$$\frac{\partial^2 P^H}{\partial\left(IncPr^{e,H}\right)^2} = -2 \times N \times Q \times PercA^{e,\,H} \times \alpha_1^{e,\,H}$$

This expression is negative, given that N, Q, $PercA^{e,\,H}$, and $\alpha_1^{e,\,H}$ are positive quantities in this context (representing quantities, rates, or inherently non-negative coefficients). Thus, the negativity of the second-order derivative further confirms the concavity of P^H with respect to $IncPr^{e,\,H}$. □

Proof of Proposition 2. The second-order derivative of P^T with respect to $IncPr^{t,\,T}$ is given by

$$\frac{\partial^2 P^T}{\partial\left(IncPr^{t,\,T}\right)^2} = N \times Q \times \left(-2 \times PercA^{t,\,T} \times \alpha_1^{t,\,T} \times ProbE - 2 \times PercA^{t} \times \alpha_1^{t,\,T} \times (1 - ProbE)\right).$$

This expression simplifies in the same way as the second-order derivative with respect to $IncPr^{t,\,H}$ in P^H, resulting in $-2 \times N \times Q \times PercA^{t,\,T} \times \alpha_1^{t,\,T}$. Since N, Q, $PercA^{t,\,T}$, and $\alpha_1^{t,\,T}$ are positive quantities within this problem context, the entire expression is negative. This result holds regardless of the value of $ProbE$ (provided $0 < ProbE < 1$). Hence, the expression is negative, proving the concavity of P^T with respect to $IncPr^{t,\,T}$.

Given the concavity of profit functions with respect to their corresponding incentives, the optimal incentive values can be determined by equating the first derivative of the profit functions with respect to the respective incentives to zero. □

Appendix A.2. Nash–Stackelberg H-SC Leadership Scenario

In Nash–Stackelberg H-SC leadership, initially, the issue is approached from the T-SC's perspective, after which the H-SC makes optimal decisions (see Figure A2).

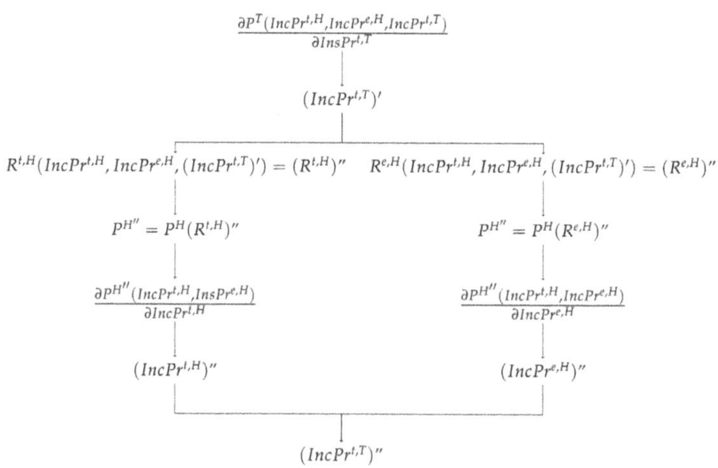

Figure A2. Illustration of decision-making structure under Nash–Stackelberg H-SC leadership scenario.

The value of $IncPr^{t,T}$ is determined by the T-SC's traditional channel as the optimal response to $IncPr^{t,H}$ and $IncPr^{e,H}$ derived simultaneously under the Nash game scenario. As demonstrated in Equation (17), the optimal reward of the T-SC is determined. For H-SCs, by substituting Equation (17) into Equation (1), the new values of $R^{t,H}$ and $R^{e,H}$ (return rates), labeled as $(R^{t,H})''$ and $(R^{e,H})''$, respectively, have been obtained. Integrating $(R^{t,H})''$ and $(R^{e,H})''$ into Equation (13), we derive an updated profit function, denoted as $(P^H)''$.

Proof of Proposition 3. The second-order derivative of P^H with respect to $IncPr^{t,\,H}$ is given by

$$\begin{aligned}\frac{\partial^2 (P^H)''}{\partial (IncPr^{t,H})^2} &= N \times Q \times \left(-2 \times PercA^{t,H} \times ProbE \times \left(\alpha_1^{t,H} - \frac{\alpha_2^{t,H} \times \alpha_2^{t,T}}{2 \times \alpha_1^{t,T}}\right)\right) \\ &\quad -\left(2 \times PercA^{t,H} \times (1 - ProbE) \times \left(\alpha_1^{t,H} - \frac{\alpha_2^{t,H} \times \alpha_2^{t,T}}{2 \times \alpha_1^{t,T}}\right)\right) \\ &= N \times Q \times \left(-2 \times PercA^{t,H} \times \left(\alpha_1^{t,H} - \frac{\alpha_2^{t,H} \times \alpha_2^{t,T}}{2 \times \alpha_1^{t,T}}\right)\right)\end{aligned}$$

Given the inherent positivity of the parameters involved, this outcome, due to the subtraction operation and the negative sign, is inherently negative, indicating the concavity of $(P^H)''$ with respect to $IncPr^{t,\,H}$.

In addition, the second-order derivative of $(P^{H})''$ with respect to $IncPr^{e,\,H}$ is

$$\frac{\partial^2 (P^H)''}{\partial \left(IncPr^{e,H}\right)^2} = -2 \times N \times Q \times PercA^{e,\,H} \times \left(\alpha_1^{e,\,H} - \frac{\alpha_2^{e,\,H} \times \alpha_2^{t,\,T}}{2 \times \alpha_1^{t,\,T}}\right).$$

The simplified expression matches the previous case, and under the given conditions, it is negative, affirming the concavity of $(P^{H})''$ with respect to $IncPr^{e,\,H}$.

After confirming the concavity of $(P^H)''$ over $IncPr^{t,H}$ and $IncPr^{e,H}$, we can determine a unique optimal value for $IncPr^{t,H}$ and $IncPr^{e,H}$, denoted as $(IncPr^{t,H})''$ and $(IncPr^{e,H})''$, by setting the first-order derivative of $(P^H)''$ with respect to $IncPr^{t,H}$ and $IncPr^{e,H}$ equal to zero. Subsequently, after substituting $(IncPr^{t,H})''$ and $(IncPr^{e,H})''$ into the formula for $\left(IncPr^{t,T}\right)'$, the optimal value for $\left(IncPr^{t,T}\right)'$ can be calculated. □

Appendix A.3. Nash–Stackelberg T-SC Leadership Scenario

In Nash-Stackelberg T-SC leadership, initially, the issue is approached from the H-SC's perspective, after which the T-SC makes optimal decisions (See Figure A3).

Despite the change in leadership, the analytical process, which involves solving from the follower's perspective before the leader finalizes their strategy, remains consistent, as illustrated in Figure A3. Consequently, after substituting Equations (15) and (16) into Equation (1) and deriving $(R^{t,T})'''$ as an updated return rate function and integrating it into Equation (14), a new formula for the profit of the T-SC, denoted as $(P^T)'''$, hasbeenestablished.

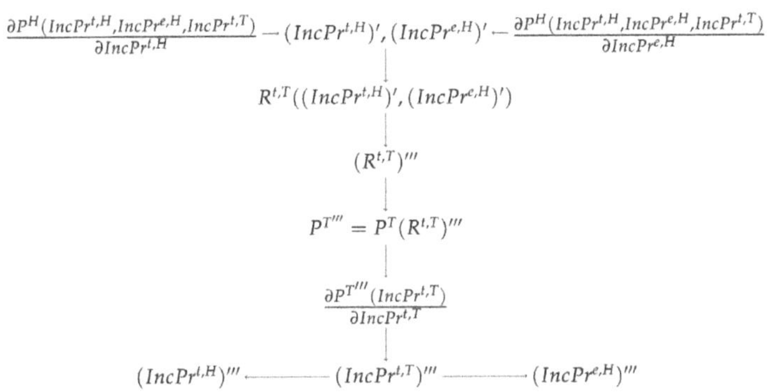

Figure A3. Illustration of decision-making structure under Nash–Stackelberg T-SC leadership scenario.

Proof of Proposition 4. The second-order derivative of $(P^{T})'''$ with respect to $IncPr^{t,\,T}$ is given by

$$\frac{\partial^2 (P^T)'''}{\partial \left(IncPr^{t,T}\right)^2} = N \times Q \times \left(-2 \times PercA^{t,\,T} \times ProbE \times \left(\alpha_1^{t,\,T} - \frac{\alpha_2^{t,\,H} \times \alpha_2^{t,\,T}}{2 \times \alpha_1^{t,\,H}} - \frac{\alpha_2^{e,\,H} \times \alpha_2^{t,\,T}}{2 \times \alpha_1^{e,\,H}}\right)\right)$$
$$-\left(2 \times PercA^{t,\,T} \times (1 - ProbE) \times \left(\alpha_1^{t,\,T} - \frac{\alpha_2^{t,\,H} \times \alpha_2^{t,\,T}}{2 \times \alpha_1^{t,\,H}} - \frac{\alpha_2^{e,\,H} \times \alpha_2^{t,\,T}}{2 \times \alpha_1^{e,\,H}}\right)\right)$$
$$= \frac{N \times Q \times PercA^{t,\,T} \times \left(-2 \times \alpha_1^{e,\,H} \times \alpha_1^{t,\,H} \times \alpha_1^{t,\,T} + \alpha_1^{e,\,H} \times \alpha_2^{t,\,H} \times \alpha_2^{t,\,T} + \alpha_1^{t,\,H} \times \alpha_2^{e,\,H} \times \alpha_2^{t,\,T}\right)}{\alpha_1^{e,\,H} \times \alpha_1^{t,\,H}}$$

Given that all the parameters are positive and considering $0 < ProbE < 1$, the sign of this expression depends on the numerator's terms. Specifically, the numerator includes both positive and negative terms. However, the presence of a negative sign in front of the largest term, $\left(2 \times \alpha_1^{e,\,H} \times \alpha_1^{t,\,H} \times \alpha_1^{t,\,T}\right)$, suggests that the expression is designed to be negative. This is unless the sum of the two smaller terms, $\left(\alpha_1^{e,\,H} \times \alpha_2^{t,\,H} \times \alpha_2^{t,\,T}\right)$ and $\left(\alpha_1^{t,\,H} \times \alpha_2^{e,\,H} \times \alpha_2^{t,\,T}\right)$, is greater than twice the product $\left(\alpha_1^{e,\,H} \times \alpha_1^{t,\,H} \times \alpha_1^{t,\,T}\right)$, which is unlikely under normal circumstances, given the context provided. Therefore, the expression is negative under the given conditions, where the coefficients are intended to signify the impact on their respective channels' return rates more than those of competitors. Therefore, the second-order derivative is negative, affirming the concavity of $(P^H)''$ is concave with respect to $IncPr^{e,\,H}$ under these conditions.

After confirming the concavity of $(P^T)'''$ over $IncPr^{t,T}$, we compute the optimal value for $IncPr^{t,T}$, known as $(IncPr^{t,T})'''$, by setting the first derivative of $(P^T)'''$ with respect to $IncPr^{t,T}$ to zero. Following this, by substituting $(IncPr^{t,T})'''$ into $\left(IncPr^{t,H}\right)'$ and $\left(IncPr^{e,H}\right)'$, we can obtain their optimal values, $\left(IncPr^{t,H}\right)'''$ and $\left(IncPr^{e,H}\right)'''$. □

References

1. Aceleanu, M.I.; Şerban, A.C.; Suciu, M.-C.; Biţoiu, T.I. The Management of Municipal Waste through Circular Economy in the Context of Smart Cities Development. *IEEE Access* **2019**, *7*, 133602–133614. [CrossRef]
2. Giffinger, R.; Fertner, C.; Kramar, H.; Kalasek, R.; Pichler-Milanovic, N.; Meijers, E. *Smart Cities—Ranking of European Medium-Sized Cities*; Vienna University of Technology: Vienna, Austria, 2007.
3. Fan, P.; Ouyang, Z.; Nguyen, D.D.; Nguyen, T.T.H.; Park, H.; Chen, J. Urbanization, economic development, environmental and social changes in transitional economies: Vietnam after Doimoi. *Landsc. Urban Plan.* **2019**, *187*, 145–155. [CrossRef]
4. MacArthur, F.E. *Towards the Circular Economy Vol. 1: An Economic and Business Rationale for an Accelerated Transition*; Ellen MacArthur Foundation: New York, NY, USA, 2013.
5. Genovese, A.; Acquaye, A.A.; Figueroa, A.; Koh, S.L. Sustainable supply chain management and the transition towards a circular economy: Evidence and some applications. *Omega* **2017**, *66*, 344–357. [CrossRef]
6. del Mar Alonso-Almeida, M.; Rodríguez-Antón, J.M.; Bagur-Femenías, L.; Perramon, J. Sustainable development and circular economy: The role of institutional promotion on circular consumption and market competitiveness from a multistakeholder engagement approach. *Bus. Strategy Environ.* **2020**, *29*, 2803–2814. [CrossRef]
7. Kirchherr, J.; Reike, D.; Hekkert, M. Conceptualizing the circular economy: An analysis of 114 definitions. *Resour. Conserv. Recycl.* **2017**, *127*, 221–232. [CrossRef]
8. Rouhani, S.; Wardley, L.J.; Amin, S.H. A Comprehensive Survey into Reverse Logistics and Closed-Loop Supply Chain Aspects to Provide Analyses and Insights for Implementation. *J. Clean. Prod.* **2025**, *490*, 144743. [CrossRef]
9. Khan, S.A.; Laalaoui, W.; Hokal, F.; Tareq, M.; Ahmad, L. Connecting reverse logistics with circular economy in the context of Industry 4.0. *Kybernetes* **2023**, *52*, 6279–6320. [CrossRef]
10. Trends, R.L. *Going Backwards: Reverse Logistics Trends and Practices*; Reverse Logistics Executive Council: Reno, NV, USA, 1999.
11. Saha, S.; Sarmah, S.P.; Moon, I. Dual Channel Closed-Loop Supply Chain Coordination with a Reward-Driven Remanufacturing Policy. *Int. J. Prod. Res.* **2016**, *54*, 1503–1517. Available online: https://www.tandfonline.com/doi/full/10.1080/00207543.2015.1090031 (accessed on 25 June 2024). [CrossRef]
12. Zheng, B.; Huang, S.; Jin, L. The bright side of online recycling: Perspectives of customer's channel preference and competition. *Electron. Commer. Res. Appl.* **2021**, *50*, 101102. [CrossRef]
13. Giutini, R.; Gaudette, K. Remanufacturing: The next great opportunity for boosting US productivity. *Bus. Horiz.* **2003**, *46*, 41–48. [CrossRef]
14. Yang, L.; Wang, G.; Ke, C. Remanufacturing and promotion in dual-channel supply chains under cap-and-trade regulation. *J. Clean. Prod.* **2018**, *204*, 939–957. [CrossRef]
15. Cao, J.; Chen, X.; Zhang, X.; Gao, Y.; Zhang, X.; Kumar, S. Overview of remanufacturing industry in China: Government policies, enterprise, and public awareness. *J. Clean. Prod.* **2020**, *242*, 118450. [CrossRef]
16. Dell Technologies Inc. *Dell Repair, Reuse, Recycle: The Circular Economy in Action*; Dell Technologies Inc.: Round Rock, TX, USA, 2022.
17. Alkahtani, M.; Ziout, A.; Salah, B.; Alatefi, M.; Elgawad, A.E.E.A.; Badwelan, A.; Syarif, U. An insight into reverse logistics with a focus on collection systems. *Sustainability* **2021**, *13*, 548. [CrossRef]

18. Zhao, S.; You, Z.; Zhu, Q. Quality choice for product recovery considering a trade-in program and third-party remanufacturing competition. *Int. J. Prod. Econ.* **2021**, *240*, 108239. [CrossRef]
19. Matsui, K. Dual-recycling channel reverse supply chain design of recycling platforms under acquisition price competition. *Int. J. Prod. Econ.* **2023**, *259*, 108769. [CrossRef]
20. Apple Inc. Apple Trade in Apple. Available online: https://www.apple.com/shop/trade-in (accessed on 5 February 2025).
21. Shi, Y.; Nie, J.; Qu, T.; Chu, L.K.; Sculli, D. Choosing reverse channels under collection responsibility sharing in a closed-loop supply chain with re-manufacturing. *J. Intell. Manuf.* **2015**, *26*, 387–402. [CrossRef]
22. Batarfi, R.; Jaber, M.Y.; Aljazzar, S.M. A profit maximization for a reverse logistics dual-channel supply chain with a return policy. *Comput. Ind. Eng.* **2017**, *106*, 58–82. [CrossRef]
23. Ghoreishi, N.; Jakiela, M.J.; Nekouzadeh, A. A cost model for optimizing the take back phase of used product recovery. *J. Remanuf.* **2011**, *1*, 1. [CrossRef]
24. Hong, X.; Xu, L.; Du, P.; Wang, W. Joint advertising, pricing and collection decisions in a closed-loop supply chain. *Int. J. Prod. Econ.* **2015**, *167*, 12–22. [CrossRef]
25. Taleizadeh, A.A.; Moshtagh, M.S.; Moon, I. Pricing, product quality, and collection optimization in a decentralized closed-loop supply chain with different channel structures: Game theoretical approach. *J. Clean. Prod.* **2018**, *189*, 406–431. [CrossRef]
26. Jolai, F.; Hashemi, P.; Heydari, J.; Bakhshi, A.; Keramati, A. Optimizing a reverse logistics system by considering quality of returned products. *Adv. Ind. Eng.* **2020**, *54*, 165–184.
27. Park, K.; Kim, J.; Ko, Y.D.; Song, B.D. Redesign of reverse logistics network with managerial decisions on the minimum quality level and remanufacturing policy. *J. Oper. Res. Soc.* **2021**, *72*, 1564–1577. [CrossRef]
28. Jena, S.K.; Sarmah, S.P.; Sarin, S.C. Joint-advertising for collection of returned products in a closed-loop supply chain under uncertain environment. *Comput. Ind. Eng.* **2017**, *113*, 305–322. [CrossRef]
29. Mishra, S.; Singh, S.P. Designing dynamic reverse logistics network for post-sale service. *Ann. Oper. Res.* **2022**, *310*, 89–118. [CrossRef] [PubMed]
30. Feng, L.; Govindan, K.; Li, C. Strategic planning: Design and coordination for dual-recycling channel reverse supply chain considering consumer behavior. *Eur. J. Oper. Res.* **2017**, *260*, 601–612. [CrossRef]
31. Gan, S.S.; Pujawan, I.N.; Widodo, B. Pricing decision for new and remanufactured product in a closed-loop supply chain with separate sales-channel. *Int. J. Prod. Econ.* **2017**, *190*, 120–132. [CrossRef]
32. Hong, X.; Wang, Z.; Wang, D.; Zhang, H. Decision models of closed-loop supply chain with remanufacturing under hybrid dual-channel collection. *Int. J. Adv. Manuf. Technol.* **2013**, *68*, 1851–1865. [CrossRef]
33. Wang, Z.; He, C.X. Decision-making of dual-channel reverse supply chain for end-of-life vehicles considering consumer preferences. *Environ. Dev. Sustain.* **2024**, 1–25. [CrossRef]
34. Dincă, G.; Milan, A.A.; Andronic, M.L.; Pasztori, A.M.; Dincă, D. Does Circular Economy Contribute to Smart Cities' Sustainable Development? *Int. J. Environ. Res. Public Health* **2022**, *19*, 7627. [CrossRef] [PubMed]
35. Luis, E.C.; Celma, D. Circular economy. A review and bibliometric analysis. *Sustainability* **2020**, *12*, 6381. [CrossRef]
36. El Saadany, A.M.; Jaber, M.Y. A production/remanufacturing inventory model with price and quality dependent return rate. *Comput. Ind. Eng.* **2010**, *58*, 352–362. [CrossRef]
37. Cai, G.G. Channel selection and coordination in dual-channel supply chains. *J. Retail.* **2010**, *86*, 22–36. [CrossRef]
38. Wu, C.-H. Price and service competition between new and remanufactured products in a two-echelon supply chain. *Int. J. Prod. Econ.* **2012**, *140*, 496–507. [CrossRef]
39. Gu, Q.; Gao, T. Management of two competitive closed-loop supply chains. *Int. J. Sustain. Eng.* **2012**, *5*, 325–337. [CrossRef]
40. Qiang, Q.; Ke, K.; Anderson, T.; Dong, J. The Closed-loop Supply Chain Network with Competition, Distribution Channel Investment, and Uncertainties. *Omega* **2013**, *41*, 186–194. Available online: https://digitalcommons.cwu.edu/cobfac/192/ (accessed on 5 February 2025). [CrossRef]
41. Govindan, K.; Popiuc, M.N. Reverse supply chain coordination by revenue sharing contract: A case for the personal computers industry. *Eur. J. Oper. Res.* **2014**, *233*, 326–336. [CrossRef]
42. Jena, S.K.; Sarmah, S. Price competition and co-operation in a duopoly closed-loop supply chain. *Int. J. Prod. Econ.* **2014**, *156*, 346–360. [CrossRef]
43. Fallah, H.; Eskandari, H.; Pishvaee, M.S. Competitive closed-loop supply chain network design under uncertainty. *J. Manuf. Syst.* **2015**, *37*, 649–661. [CrossRef]
44. Guo, J.; Ya, G. Optimal strategies for manufacturing/remanufacturing system with the consideration of recycled products. *Comput. Ind. Eng.* **2015**, *89*, 226–234. [CrossRef]
45. Xie, J.; Li, Z.J.; Yao, Y.; Liang, L. Dynamic acquisition pricing policy under uncertain remanufactured-product demand. *Ind. Manag. Data Syst.* **2015**, *115*, 521–540. [CrossRef]

46. Gao, J.; Han, H.; Hou, L.; Wang, H. Pricing and effort decisions in a closed-loop supply chain under different channel power structures. *J. Clean. Prod.* **2016**, *112*, 2043–2057. [CrossRef]
47. Jena, S.K.; Sarmah, S. Price and service co-opetition under uncertain demand and condition of used items in a remanufacturing system. *Int. J. Prod. Econ.* **2016**, *173*, 1–21. [CrossRef]
48. Xie, J.; Liang, L.; Liu, L.; Ieromonachou, P. Coordination contracts of dual-channel with cooperation advertising in closed-loop supply chains. *Int. J. Prod. Econ.* **2017**, *183*, 528–538. [CrossRef]
49. Liu, L.; Wang, Z.; Xu, L.; Hong, X.; Govindan, K. Collection effort and reverse channel choices in a closed-loop supply chain. *J. Clean. Prod.* **2017**, *144*, 492–500. [CrossRef]
50. Taleizadeh, A.A.; Sadeghi, R. Pricing strategies in the competitive reverse supply chains with traditional and e-channels: A game theoretic approach. *Int. J. Prod. Econ.* **2019**, *215*, 48–60. [CrossRef]
51. Ranjbar, Y.; Sahebi, H.; Ashayeri, J.; Teymouri, A. A competitive dual recycling channel in a three-level closed loop supply chain under different power structures: Pricing and collecting decisions. *J. Clean. Prod.* **2020**, *272*, 122623. [CrossRef]
52. Jin, L.; Zheng, B.; Huang, S. Pricing and coordination in a reverse supply chain with online and offline recycling channels: A power perspective. *J. Clean. Prod.* **2021**, *298*, 126786. [CrossRef]
53. Yanting, H.; Liang, Y. Exploring the strategies of online and offline recycling channels in closed-loop supply chain under government subsidy. *Environ. Sci. Pollut. Res.* **2022**, *29*, 21591–21602.
54. Wang, B.; Wang, N. Decision models for a dual-recycling channel reverse supply chain with consumer strategic behavior. *Sustainability* **2022**, *14*, 10870. [CrossRef]
55. Matsui, K. Optimal timing of acquisition price announcement for used products in a dual-recycling channel reverse supply chain. *Eur. J. Oper. Res.* **2022**, *300*, 615–632. [CrossRef]
56. Modak, N.M.; Sinha, S.; Ghosh, D.K. A review on remanufacturing, reuse, and recycling in supply chain—Exploring the evolution of information technology over two decades. *Int. J. Inf. Manag. Data Insights* **2023**, *3*, 100160. [CrossRef]
57. Kurata, H.; Yao, D.-Q.; Liu, J.J. Pricing policies under direct vs. indirect channel competition and national vs. store brand competition. *Eur. J. Oper. Res.* **2007**, *180*, 262–281. [CrossRef]
58. Giri, B.C.; Chakraborty, A.; Maiti, T. Pricing and return product collection decisions in a closed-loop supply chain with dual-channel in both forward and reverse logistics. *J. Manuf. Syst.* **2017**, *42*, 104–123. [CrossRef]
59. Taleizadeh, A.A.; Moshtagh, M.S.; Moon, I. Optimal decisions of price, quality, effort level and return policy in a three-level closed-loop supply chain based on different game theory approaches. *Eur. J. Ind. Eng.* **2017**, *11*, 486–525. [CrossRef]
60. Ziad, E.; Yang, Z.; Lu, Y.; Ju, F. Knowledge Constrained Deep Clustering for Melt Pool Anomaly Detection in Laser Powder Bed Fusion. In Proceedings of the 2024 IEEE 20th International Conference on Automation Science and Engineering (CASE), Puglia, Italy, 28 August–1 September 2024; pp. 670–675. [CrossRef]
61. De Giovanni, P.; Zaccour, G. A two-period game of a closed-loop supply chain. *Eur. J. Oper. Res.* **2014**, *232*, 22–40. [CrossRef]
62. Dridi, D.; Ben Youssef, S. *A Game Theoretic Framework for Competing/Cooperating Retailers Under Price and Advertising Dependent Demand*; University Library of Munich, Germany: Munich, Germany, 2015.
63. Mimouni, F.; Abouabdellah, A.; Mharzi, H. Study of the reverse logistics' break-even in a direct supply chain. *Int. Rev. Model. Simul.* **2015**, *8*, 277–283. [CrossRef]
64. Dhakal, M.; Smith, M.H.; Newbery, R. Secondary market: A significant aspect in reverse logistics and sustainability. *EmphThe Int. J. Sustain. Econ. Soc. Cult. Context* **2017**, *12*, 25. [CrossRef]
65. Giovanni, P.; Zaccour, G. A selective survey of game-theoretic models of closed-loop supply chains. *4OR* **2019**, *17*, 1–44. [CrossRef]
66. Wei, J.; Zhao, J. Pricing and remanufacturing decisions in two competing supply chains. *Int. J. Prod. Res.* **2015**, *53*, 258–278. [CrossRef]
67. Yang, Y.; Xu, X. A differential game model for closed-loop supply chain participants under carbon emission permits. *Comput. Ind. Eng.* **2019**, *135*, 1077–1090. [CrossRef]
68. Chu, X.; Xu, S.X.; Cai, F.; Chen, J.; Qin, Q. An efficient auction mechanism for regional logistics synchronization. *J. Intell. Manuf.* **2019**, *30*, 2715–2731. [CrossRef]
69. Chakraborty, T.; Chauhan, S.S.; Vidyarthi, N. Coordination and competition in a common retailer channel: Wholesale price versus revenue-sharing mechanisms. *Int. J. Prod. Econ.* **2015**, *166*, 103–118. [CrossRef]
70. Mukhopadhyay, S.K.; Yao, D.-Q.; Yue, X. Information sharing of value-adding retailer in a mixed channel hi-tech supply chain. *J. Bus. Res.* **2008**, *61*, 950–958. [CrossRef]
71. Xu, G.; Dan, B.; Zhang, X.; Liu, C. Coordinating a dual-channel supply chain with risk-averse under a two-way revenue sharing contract. *Int. J. Prod. Econ.* **2014**, *147*, 171–179. [CrossRef]
72. Saltelli, A.; Annoni, P. How to avoid a perfunctory sensitivity analysis. *Environ. Model. Softw.* **2010**, *25*, 1508–1517. [CrossRef]
73. MAC Cosmetics. Back to MAC. Available online: https://www.maccosmetics.com/back-to-mac (accessed on 5 February 2025).

74. Kumar, V.; Sutherland, J.W. Sustainability of the automotive recycling infrastructure: Review of current research and identification of future challenges. *Int. J. Sustain. Manuf.* **2008**, *1*, 145–167. [CrossRef]
75. Zarreh, M.; Yaghoubi, S.; Bahrami, H. Pricing of Drinking Water under Dynamic Supply and Demand based on Government Role: A Game-Theoretic Approach. *Water Resour Manag.* **2024**, *38*, 2101–2133. [CrossRef]

MDPI AG
Grosspeteranlage 5
4052 Basel
Switzerland
Tel.: +41 61 683 77 34

Smart Cities Editorial Office
E-mail: smartcities@mdpi.com
www.mdpi.com/journal/smartcities

MDPI

www.ingramcontent.com/pod-product-compliance
Lightning Source LLC
Chambersburg PA
CBHW041941260926
51850CB00018B/379
9783725856619